Reflection Seismology:
The Continental Crust

Geodynamics Series

Geodynamics Series

1 Dynamics of Plate Interiors
  A. W. Bally, P. L. Bender, T. R. McGetchin, and R. I. Walcott (Editors)

2 Paleoreconstruction of the Continents
  M. W. McElhinny and D. A. Valencio (Editors)

3 Zagros, Hindu Kush, Himalaya, Geodynamic Evolution
  H. K. Gupta and F. M. Delany (Editors)

4 Anelasticity in the Earth
  F. D. Stacey, M. S. Patterson, and A. Nicholas (Editors)

5 Evolution of the Earth
  R. J. O'Connell and W. S. Fyfe (Editors)

6 Dynamics of Passive Margins
  R. A. Scrutton (Editor)

7 Alpine-Mediterranean Geodynamics
  H. Berckhemer and K. Hsü (Editors)

8 Continental and Oceanic Rifts
  G. Pálmason, P. Mohr, K. Burke, R. W. Girdler, R. J. Bridwell, and G. E. Sigvaldason (Editors)

9 Geodynamics of the Eastern Pacific Region, Caribbean and Scotia Arcs
  Ramón Cabré, S. J. (Editor)

10 Profiles of Orogenic Belts
   N. Rast and F. M. Delany (Editors)

11 Geodynamics of the Western Pacific-Indonesian Region
   Thomas W. C. Hilde and Seiya Uyeda (Editors)

12 Plate Reconstruction From Paleozoic Paleomagnetism
   R. Van der Voo, C. R. Scotese, and N. Bonhommet (Editors)

13 Reflection Seismology: A Global Perspective
   Muawia Barazangi and Larry Brown (Editors)

Reflection Seismology: The Continental Crust

Edited by Muawia Barazangi
Larry Brown

Geodynamics Series
Volume 14

American Geophysical Union
Washington, D.C.
1986

Publication No. 0112 of the International Lithosphere Program

Published under the aegis of AGU Geophysical Monograph Board: Patrick Muffler, Chairman; Wolfgang Berger, Donald Forsyth, and Janet Luhmann, members.

Reflection Seismology: The Continental Crust

Library of Congress Cataloging in Publication Data

Main entry under title:

Reflection seismology.

(Geodynamics series ; v. 14) (Publication / International Lithosphere Program ; no. 112)
1. Earth—Crust—Congresses. 2. Continents—Congresses. 3. Seismic reflection method—Congresses. I. Barazangi, Muawia. II. Brown, Larry, 1951– . III. American Geophysical Union. IV. Series. V. Series: Publication (International Lithosphere Program) ; no. 112.

QE511.R36 1986   551.1'3'028   85-26684
ISBN: 0-87590-514-5
ISSN: 0277-6669

Copyright 1986 by the American Geophysical Union, 2000 Florida Avenue, NW, Washington, DC 20009

Figures, tables and short excerpts may be reprinted in scientific books and journals if the source is properly cited.

Authorization to photocopy items for internal or personal use, or the internal or personal use of specific clients, is granted by the American Geophysical Union for libraries and other users registered with the Copyright Clearance Center (CCC) Transactional Reporting Service, provided that the base fee of $1.00 per copy, plus $0.10 per page is paid directly to CCC, 21 Congress Street, Salem, MA 01970. 0277-6669/86/$01. + .10.
This consent does not extend to other kinds of copying, such as copying for creating new collective works or for resale. The reproduction of multiple copies and the use of full articles or the use of extracts, including figures and tables, for commercial purposes requires permission from AGU.

Printed in the United States of America

CONTENTS

Preface *Muawia Barazangi and Larry Brown* xi

List of Reviewers xii

CRUSTAL STRUCTURE AND EVOLUTION

Implications of Deep Crustal Evolution for Seismic Reflection Interpretation *David M. Fountain* 1

Interpretation of Seismic Reflection Data in Complexly Deformed Terranes: A Geologist's Perspective *Robert D. Hatcher, Jr.* 9

Continental Evolution by Lithospheric Shingling *Frederick A. Cook* 13

Crustal Reflections and Crustal Structure *Scott B. Smithson, Roy A. Johnson, and Charles A. Hurich* 21

Fluids in Deep Continental Crust *W. S. Fyfe* 33

Tectonic Escape in the Evolution of the Continental Crust *Kevin Burke and Celal Sengör* 41

Modern Analogs for Some Midcrustal Reflections Observed Beneath Collisional Mountain Belts *Robert J. Lillie and Mohammed Yousuf* 55

Reflections From the Subcrustal Lithosphere *Karl Fuchs* 67

Deep Crustal Signatures in India and Contiguous Regions From Satellite and Ground Geophysical Data *M. N. Qureshy and R. K. Midha* 77

PRECAMBRIAN CRUSTAL STRUCTURE: THE ORIGINAL SIGNATURE

Seismic Reflection Profiles of Precambrian Crust: A Qualitative Assessment *Allan K. Gibbs* 95

Composition, Structure and Evolution of the Early Precambrian Lower Continental Crust: Constraints From Geological Observations and Age Relationships *Alfred Kröner* 107

Precambrian Crustal Structure of the Northern Baltic Shield From the Fennolora Profile: Evidence for Upper Crustal Anisotropic Laminations *Kenneth H. Olsen and Carl-Erik Lund* 121

Evidence for an Inactive Rift in the Precambrian From a Wide-Angle Reflection Survey Across the Ottawa-Bonnechere Graben *Robert Mereu, Dapeng Wang, and Oliver Kuhn* 127

A Possible Exposed Conrad Discontinuity in the Kapuskasing Uplift, Ontario *John A. Percival* 135

Seismic Crustal Structure Northwest of Thunder Bay, Ontario *Roger A. Young, Jeffrey Wright, and G. F. West* 143

PALEOZOIC CRUSTAL STRUCTURE: THE EVOLUTION OF A MATURE CRUST

A Seismic Cross Section of the New England Appalachians: The Orogen Exposed   *Robert A. Phinney*   157

Moho Reflections From the Long Island Platform, Eastern United States   *D. R. Hutchinson, J. A. Grow, K. D. Klitgord, and R. S. Detrick*   173

The Quebec-Western Maine Seismic Reflection Profile: Setting and First Year Results   *D. B. Stewart, J. D. Unger, J. D. Phillips, R. Goldsmith, W. H. Poole, C. P. Spencer, A. G. Green, M. C. Loiselle, and P. St-Julien*   189

Structural Interpretation of Multichannel Seismic Reflection Profiles Crossing the Southeastern United States and the Adjacent Continental Margin—Decollements, Faults, Triassic (?) Basins and Moho Reflections   *John C. Behrendt*   201

Crustal Thickness, Velocity Structure, and the Isostatic Response Function in the Southern Appalachians   *Leland T. Long and Jeih-San Liow*   215

Nature of the Lower Continental Crust: Evidence From BIRPS Work on the Caledonides   *Jeremy Hall*   223

The Hercynian Evolution of the South West British Continental Margin   *G. A. Day*   233

WESTERN NORTH AMERICA CORDILLERA AND OTHER REGIONS: THE EVOLVING AND REACTIVATED CRUST

The Deep Crust in Convergent and Divergent Terranes: Laramide Uplifts and Basin-Range Rifts   *George A. Thompson and Janice L. Hill*   243

Phanerozoic Tectonics of the Basin and Range—Colorado Plateau Transition from COCORP Data and Geologic Data: A Review   *Richard W. Allmendinger, Harlow Farmer, Ernest Hauser, James Sharp, Douglas Von Tish, Jack Oliver, and Sidney Kaufman*   257

Seismic Profiling of the Lower Crust: Dixie Valley, Nevada   *David A. Okaya*   269

Reflection Profiles From the Snake Range Metamorphic Core Complex: A Window Into the Mid-Crust   *Jill McCarthy*   281

Shallow Structure of the Southern Albuquerque Basin (Rio Grande Rift), New Mexico, From COCORP Seismic Reflection Data   *Zhengwen Wu*   293

Geometries of Deep Crustal Faults: Evidence From the COCORP Mojave Survey   *M. J. Cheadle, B. L. Czuchra, C. J. Ando, T. Byrne, L. D. Brown, J. E. Oliver, and S. Kaufman*   305

Structure of the Lithosphere in a Young Subduction Zone: Results From Reflection and Refraction Studies   *Ron M. Clowes, George D. Spence, Robert M. Ellis, and David A. Waldron*   313

The Victoria Land Basin: Part of an Extended Crustal Complex Between East and West Antarctica   *Yeadong Kim, L. D. McGinnis, and R. H. Bowen*   323

Whole-Lithosphere Normal Simple Shear: An Interpretation of Deep-Reflection Profiles in Great Britain   *Brian Wernicke*   331

The companion to this volume, Geodynamics Series Volume 13, *Reflection Seismology: A Global Perspective* (Muawia Barazangi and Larry Brown, Editors), contains the following:

A GLOBAL SURVEY

A Global Perspective on Seismic Reflection Profiling of the Continental Crust   *Jack Oliver*

Deep Reflections from the Caledonides and Variscides West of Britain and Comparison with the Himalayas   *Drummond H. Matthews and Michael J. Cheadle*

Deep Seismic Profiling of the Crust in Northern France: The ECORS Project   *C. Bois, M. Cazes, B. Damotte, A. Galdéano, A. Hirn, A. Mascle, P. Matte, J. F. Raoult, and G. Torreilles*

Nature and Development of the Crust According to Deep Reflection Data From the German Variscides   *Rolf Meissner and Thomas Wever*

Detailed Crustal Structure From a Seismic Reflection Survey in Northern Switzerland   *P. Finckh, W. Frei, B. Fuller, R. Johnson, St. Mueller, S. Smithson, and Chr. Sprecher*

Characteristics of the Reflecting Layers in the Earth's Crust and Upper Mantle in Hungary   *Károly Posgay, István Albu, Géza Ráner, and Géza Varga*

A Review of Continental Reflection Profiling in Australia   *F. J. Moss and S. P. Mathur*

Recent Reflection Seismic Developments in the Witwatersrand Basin   *R. J. Durrheim*

Recent Seismic Reflection Studies in Canada   *A. G. Green, M. J. Berry, C. P. Spencer, E. R. Kanasewich, S. Chiu, R. M. Clowes, C. J. Yorath, D. B. Stewart, J. D. Unger, and W. H. Poole*

Seismic Reflection Studies by the U.S. Geological Survey   *Robert M Hamilton*

The First Decade of COCORP: 1974–1984   *Larry Brown, Muawia Barazangi, Sidney Kaufman, and Jack Oliver*

Crustal Structure Studies in New Zealand   *T. A. Stern, F. J. Davey, and E. G. C. Smith*

Tectonic Framework of Narmada-Son Lineament—A Continental Rift System in Central India From Deep Seismic Soundings   *K. L. Kaila*

A Geophysical Investigation of Deep Structure in China   *Xuecheng Yuan, Shi Wang, Li Li, and Jieshou Zhu*

Deep Crustal Knowledge in Italy   *C. Morelli*

Long-Range Seismic Refraction Profiles in Europe   *St. Mueller and J. Ansorge*

PROCESSING AND MODELING

Crustal Studies in Central California Using an 800-Channel Seismic Reflection Recording System   *Mark D. Zoback and Carl M. Wentworth*

Interpretive Processing of Crustal Seismic Reflection Data: Examples From Laramie Range COCORP Data   *Roy A. Johnson and Scott B. Smithson*

Aspects of COCORP Deep Seismic Profiling   *Larry D. Brown*

Enhanced Imaging of the COCORP Seismic Line, Wind River Mountains   *J. Sharry, R. T. Langan, D. B. Jovanovich, G. M. Jones, N. R. Hill, and T. M. Guidish*

An Expanding Spread Experiment During COCORP's Field Operation in Utah   *Char-Shine Liu, Tianfei Zhu, Harlow Farmer, and Larry Brown*

Crustal Reflection and Refraction Velocities: A Comparison   *Z. Hajnal*

The Continental Mohorovičić Discontinuity: Results From Near-Vertical and Wide-Angle Seismic Reflection Studies   *L. W. Braile and C. S. Chiang*

Reassessing Seismic Refraction on the Edwards Plateau   *H. J. Dorman, T. H. Crawford, J. W. Stelzig, and P. J. Tarantolo*

Deep Seismic Reflection Profiling the Continental Crust at Sea   *M. R. Warner*

Modeling Lower Crust Reflections Observed on BIRPS Profiles   *D. J. Blundell and B. Raynaud*

Comparison of Deep Reflection and Refraction Structures in the North Sea   *P. J. Barton*

Interpreting the Deep Structure of Rifts With Synthetic Seismic Sections   *Carolyn Peddy, Larry D. Brown, and Simon L. Klemperer*

PREFACE

That deep seismic reflection profiling has become a "necessary tool" to explore the deep basement of the continental crust is now well established. The question is: Will deep reflection profiling become a "standard tool" that is routinely used on all continents, as is surface geological mapping? This question can be answered in the affirmative. A clear indication of this trend are the results of the "International Symposium on Deep Structure of the Continental Crust: Results from Reflection Seismology", which was held June 26-28, 1984 on the Cornell University campus and which forms the basis for these two Geodynamics Series volumes. This was the first such international meeting, and plans are already under way to hold such a meeting every two years in a different country.

Four major observations emerged from the conference. First, there is nearly universal awareness and acceptance of deep seismic reflection profiling as a necessary tool for exploring the deep basement of continents. The rapid and widespread increase in its use in the early 1980's by numerous countries is a clear manifestation of such awareness and acceptance. Second, a new kind of earth scientist has emerged as the traditional fields of geology and geophysics are of necessity unified to achieve the best interpretation of deep reflection data. Third, results presented during the symposium demonstrated how much we still have to learn about the deep basement of continents. However, the results from deep reflection profiling to date already constitute a cornerstone in our understanding of the evolution of continents. Fourth, because of the global nature of the deep reflection profiling programs it is already possible to decipher similar seismic "signatures" for different parts of the continental crust with similar geological environments. With the rapid growth of available deep reflection data, it is not unreasonable to expect that within our lifetime a clear understanding of the structure and evolution of the continental lithosphere may be achieved.

Because the majority of the speakers at the Cornell symposium agreed to submit papers for publication, two volumes were necessary. Although each volume stands on its own, the two together constitute an integral review of the field. Most of the papers contribute new data or interpretations; others have elements of review and/or overview that are, in our judgment, essential to the completeness of the two volumes.

Over 270 geoscientists from 17 countries attended the conference, with delegates from ten of them presenting summaries of results of deep reflection programs in their own countries. Several of these summaries represent the first reports from major new national programs. The conference was organized by the Institute for the Study of the Continents (INSTOC) of Cornell University and was sponsored and financially supported by the Cornell Program for Study of the Continents (COPSTOC), an industrial associates group of INSTOC. The conference was also co-sponsored by the American Geophysical Union, Geological Society of America, Society of Exploration Geophysicists, International Lithosphere Program, and the International Association of Seismology and Physics of the Earth's Interior. The Steering Committee of the conference included Muawia Barazangi (Coordinator, Cornell University), Albert Bally (Rice University), Robert Hamilton (U.S. Geological Survey), Leonard Johnson (U.S. National Science Foundation), Robert Phinney (Princeton University), Donald Turcotte (Cornell University).

The success of the conference and the subsequent task of organizing and editing these two volumes depended on numerous individuals, but primary among them were Theresa Alt and Judy Healey. These two volumes could not have been produced on schedule without their dedication and extra care. We also especially thank the reviewers of the papers who "volunteered" their services; a list of their names and institutions is published in both volumes.

We sincerely hope that these two volumes benefit geoscientists and that they will be only the first of many such volumes to be published in the years to come.

Muawia Barazangi and Larry Brown, Editors
Institute for the Study of the Continents
 (INSTOC)
Cornell University
Ithaca, New York 14853

LIST OF REVIEWERS

The following is a list of scientists who generously gave their time to review papers for this volume.

A. Al-Shanti, Abdulaziz U., Jeddah, Saudi Arabia
R. Allmendinger, Cornell Univ., Ithaca, New York
C. Ando, Shell Development Co., Houston, Texas
A. Bally, Rice University, Houston, Texas
M. Berry, Dept. of Energy, Mines, Res., Ottawa
C. Burchfiel, MIT, Cambridge, Massachusetts
C. Chapin, Inst.Mining, Tech.,Socorro, New Mexico
N. Christensen, Purdue, West Lafayette, Indiana
K. Condie, Inst.Mining, Tech.,Socorro, New Mexico
P. Coney, Univ. of Arizona, Tucson, Arizona
F. Cook, Univ. of Calgary, Calgary, Canada
R. Crosson, Univ. of Wash., Seattle, Washington
G. Davis, U. of Southern California, Los Angeles
J. Dorman, Exxon Prod. Res. Co., Houston, Texas
A. Ford, USGS, Menlo Park, California
D. Fountain, Univ. of Wyoming, Laramie, Wyoming
K. Fuchs, Univ. Karlsruhe, Karlsruhe, W. Germany
W. Fyfe, U. of Western Ontario, London, Ontario
A. Gibbs, Cornell University, Ithaca, New York
J. Grow, USGS, Denver, Colorado
W. Hamilton, USGS, Denver, Colorado
R. Hatcher, Jr., U. South Carolina, Columbia
T. Hauge, Exxon Production Res., Houston, Texas
E. Hauser, Cornell University, Ithaca, New York
D. Hayes, Lamont-Doherty Obs., Palisades, N.Y.
D. Hutchinson, USGS, Woods Hole, Massachusetts
R. Johnson, Univ. of Wyoming, Laramie, Wyoming

T. Jones, Union Oil Co., Calif., Brea, California
R. Kay, Cornell University, Ithaca, New York
S. Kay, Cornell University, Ithaca, New York
K. Klitgord, USGS, Woods Hole, Massachusetts
A. Kröner, Gutenberg-Univ. Mainz, W. Germany
P. Maguire, Univ. of Leicester, Leicester, U.K.
D. Matthews, Bullard Lab, Cambridge, U.K.
A. Maxwell, University of Texas, Austin, Texas
R. Mereu, U. of Western Ontario, London, Ontario
H. Mooney, Univ. of Minnesota, Minneapolis
W. Mooney, USGS, Menlo Park, California
J. Mutter, Lamont-Doherty Obs., Palisades N.Y.
D. Nelson, Cornell University, Ithaca, New York
K. Olsen, Los Alamos Nat. Lab., Los Alamos, N.M.
J. Percival, Geol. Survey of Canada, Ottawa, Ont.
R. Phinney, Princeton Univ. Princeton, New Jersey
C. Potter, Cornell University, Ithaca, New York
R. Price, Geol. Survey of Canada, Ottawa, Ontario
C. Prodehl, Geophys. Inst., Karlsruhe, W. Germany
L. Russell, ARCO Exploration Co., Dallas, Texas
A. Ryall, Univ. of Nevada, Reno, Nevada
R. Shackleton, Open Univ., Milton Keynes, U.K.
J. Sharry, Sun Expl. and Prod. Co., Dallas, Texas
P. Sims, USGS, Denver, Colorado
R. Smith, Univ. of Utah, Salt Lake City, Utah
D. Snyder, Cornell University, Ithaca, New York
P. Talwani, Univ. of South Carolina, Columbia
P. Tapponnier, Inst. de Physique du Globe, Paris
W. Travers, Cornell University, Ithaca, New York
B. Wernicke, Harvard U., Cambridge, Massachusetts
M.L. Zoback, USGS, Menlo Park, California

IMPLICATIONS OF DEEP CRUSTAL EVOLUTION FOR SEISMIC REFLECTION INTERPRETATION

David M. Fountain

Program for Crustal Studies, Department of Geology and Geophysics
University of Wyoming, Laramie, Wyoming 82071

Abstract. Inference of the nature and evolution of the lower continental crust from seismic reflection profiling requires calibration of reflection events in terms of lithology and structure. One approach to provide this calibration is to construct synthetic reflection seismograms of known crustal cross-sections. Synthetic records coupled with geologic data for the well-known Ivrea and Strona-Ceneri zones of northern Italy indicate that highly reflective lower crust can result from complex lithologic layering generated by underplating of mafic and ultramafic magmas, high grade metamorphism and partial melting of metasedimentary rocks, and deformation. Upper crustal levels appear transparent although structural and lithologic complexity is prevalent. The similarity of the synthetic records to the COCORP Kansas line suggests this approach has great potential in assisting in interpretation of lower crustal reflection data.

Introduction

Recent success in detection of seismic reflections from deep levels of the continents provides an opportunity to unravel processes of crustal evolution by geophysical methods. This goal can be achieved if we find methods of correctly interpreting individual seismic events in terms of structure and lithology. Absence of appropriate deep drill holes for direct calibration forces employment of indirect methods. One fruitful approach is to use exposed terrains, which at one time resided in the deep crust, as models of structures and lithologies that might be imaged by continental reflection profiling. Among the most revealing of these terrains are those regarded as relatively intact cross-sections of the continental crust [Fountain and Salisbury, 1981]. These terrains not only provide clues about the nature of deep crustal lithologies and their geometry, but also yield information concerning the processes responsible for crustal evolution. Furthermore, seismic properties of rocks from these terrains can be measured and used to generate synthetic reflection seismograms. The resultant records are useful because individual events can be related to specific features and therefore can ultimately be tied to pertinent aspects of the evolution of the crust.

Perhaps the best studied and most suitable case to use in this approach is the terrain consisting of the Ivrea zone (IZ) and Strona-Ceneri zone (SCZ) of northern Italy [Menhert, 1975; Fountain and Salisbury, 1981]. Not only is rock property data available for this section, but the particulars of its crustal evolutionary sequence are reasonably well understood. This paper briefly reviews that evolution, develops a seismic velocity model of the resultant geometry, and finally presents a synthetic reflection seismogram which provides valuable lessons about the interpretation of seismic reflections from the deep continental crust.

General Geology and Crustal Evolution of the Ivrea and Strona-Ceneri Zones

General Geology

Located in the South Alps of Northern Italy (Figure 1), the terrain encompassed by the IZ and SCZ has long been recognized as a near-complete cross-section of the continental crust [Berckhemer, 1969] and has been studied in some depth by petrologists, geochemists and geophysicists. A comprehensive review of research on the two zones is presented in Zingg [1983]. The IZ lies to the southeast of the Insubric line (Figure 1) and is comprised of a variety of upper amphibolite and granulite facies rocks including mafic, ultramafic, carbonate and pelitic compositions [Menhert, 1975; Zingg, 1980; 1983]. Several distinct groups of mafic rocks can be identified on the basis of petrologic and geochemical characteristics. The largest group consists of the "Mafic Formation" which dominates the southwest and northeast portions of the zone. The sequence is commonly layered, exhibits relict igneous textures and is associated with the ultramafic rocks [Rivalenti et al., 1975]. Garnet granu-

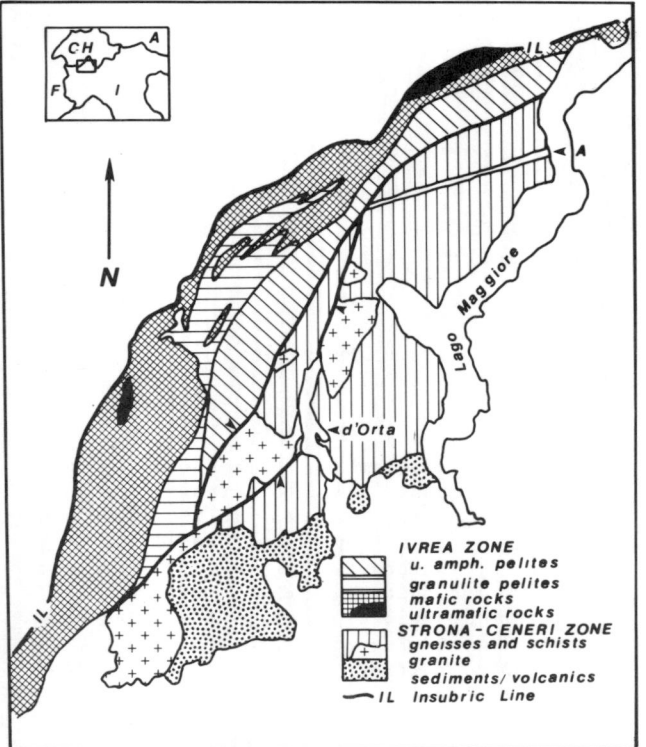

Fig. 1. General geologic map of the IZ and SCZ. Small arrowheads point to Pogallo line and associated faults. The amphibolite unit in the SCZ is labelled "A". Map modified from Hunziker and Zingg (1980). Countries in location map are Italy (I), France (F), Austria (A) and Switzerland (CH).

lites in the area may be part of this sequence. Amphibolites associated with paragneisses comprise a second category of mafic rocks and are common in the central part of the zone. Geochemical analysis of these amphibolites suggests that they were once oceanic basalts which were tectonically intercalated with paragneisses [Sills and Tarney, 1984]. Pelitic metasedimentary rocks which dominate the paragneiss assemblage vary in composition but are similar to graywackes and associated sedimentary lithologies [Menhart, 1975; Sills and Tarney, 1984]. Paragneisses in the granulite facies are depleted apparently as a result of a partial melting episode during peak metamorphic conditions [Schmid, 1978/79]. The age of high-grade metamorphism is still debated. Rb-Sr whole rock data [Hunziker and Zingg, 1980] place metamorphism at 478 Ma, whereas U-Pb and Sm-Nd data [Koppel, 1974; Polvé, 1983] point to a 275-300 Ma metamorphism. Other workers prefer the younger metamorphic date for geological reasons [Pin and Vielzeuf, 1983; Rivalenti et al., 1984]. In general, structural trends are approximately parallel to lithologic layering and several phases of deformation are observed in the zone. Importantly, the granulite facies rocks exhibit isoclinal folds which developed before the peak metamorphic event [Schmid, 1967; Lensch, 1968]. An early phase of mylonitization was associated with this deformation [Zingg, 1983].

Southeast of the IZ, the Pogallo line and associated fault zones separate the IZ from the lower grade SCZ. Mylonites in the fault zones are apparently younger than the previously mentioned deformation and developed under relatively low grade (greenschist) metamorphic conditions [Zingg, 1983]. Although generally regarded as late Paleozoic structures, recent analyses [Zingg, 1983; Hodges and Fountain, 1984] postulate an early Mesozoic age for the fault zones. Hodges and Fountain [1984] proposed that the zones developed during early Mesozoic rifting as low-angle normal faults which were subsequently rotated into their present near-vertical position.

The SCZ is in angular discordance to the Pogallo line and represents the upper levels of the crustal section. This zone is dominated by middle to lower amphibolite facies schists and orthogneisses but also includes less abundant lithologies such as amphibolites and rare ultramafic rocks. Relict high pressure assemblages have been reported from the SCZ [Zingg, 1983]. Hunziker and Zingg [1980] place the peak metamorphic event at 473 Ma. Post-metamorphic Permian granites intrude the SCZ and compositionally equivalent Permian volcanics are found to the east [Hunziker, 1974]. The zone is structurally characterized by large folds with steeply dipping fold axes [Reinhard, 1964].

Crustal evolution

What geological events were instrumental in developing lithologic and structural features which might cause reflections in the deep crust? Available data allow construction of a tentative crustal evolution scenario for the crustal section represented by the IZ and SCZ. As suggested above, controversy continues concerning timing of key events and there is no consensus regarding the nature of tectonic environments through time. Regardless of timing, certain individual geologic events had significant geophysical consequences because lithology and structural geometry were determined during these episodes. In the following interpretive discussion of the evolution of the IZ and SCZ I will point out which events were instrumental in developing lithologic and structural characteristics of potential interest to reflection seismologists.

The earliest recognizable event in the IZ and SCZ is the deposition and deformation of the pelitic sequence of sedimentary rocks (Figure 2a). U-Pb zircon data [Koppel, 1974] indicates that the sediments were derived from older continental basement and Rb-Sr data constrain deposi-

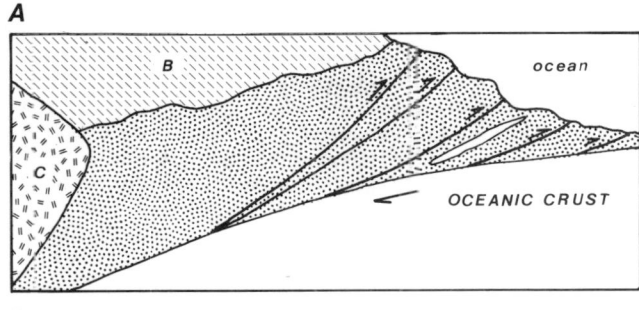
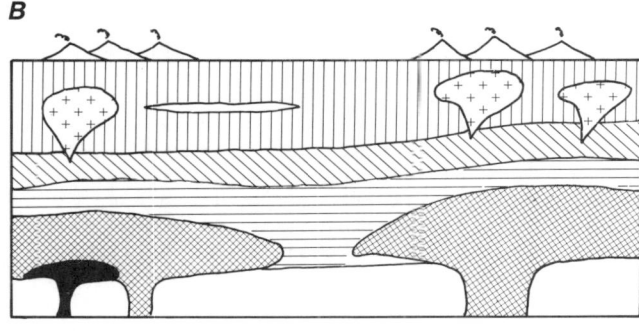
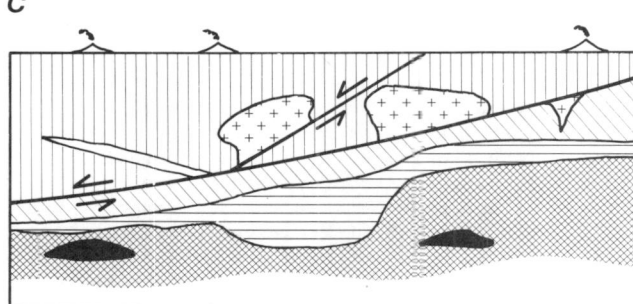

Fig. 2. Speculative crustal evolution sequence for IZ and SCZ. (a) late Precambrian-early Paleozoic sediment deposition and deformation in an accretionary wedge (stippled pattern) and forearc basin (F). C refers to old crust or an arc complex. (b) Permian underplating, high-grade metamorphism, plutonism and volcanism. (c) early Mesozoic extension and volcanism. Patterns in (b) and (c) are the same as in Figure 1.

tion to the late Precambrian-early Paleozoic [Hunziker and Zingg, 1980]. Tectonic intercalation of these rocks with amphibolites of oceanic basalt parentage coupled with the sedimentary rock composition suggests that the sediments were deposited and deformed in an accretionary wedge [Sills and Tarney, 1984]. Deformational style and presence of relict high pressure assemblages [Zingg, 1983] are consistent with this interpretation. Compositional layering and sub-horizontal structures developed in this early episode of crustal evolution.

Granulite facies metamorphism, partial melting within high-grade meta-pelites, intrusion of the mafic-ultramafic series, intrusion of granites and eruption of silicic volcanics followed (Figure 2b). Because of their Rb-Sr whole rock date, Hunziker and Zingg [1980] argue for a 478 Ma metamorphic date. However, Sm-Nd, cm-scale Rb-Sr and U-Pb zircon data [Polvé, 1983; Graeser and Hunziker, 1968; Koppel, 1974] coupled with geologic arguments [Pin and Vielzeuf, 1983; Rivalenti et al., 1984] suggest peak metamorphic conditions lasted until or occurred at about 275-300 Ma. If the latter interpretation is accepted, the following sequence can be developed. Peak metamorphic conditions were reached in the IZ as a consequence of high temperatures associated with intrusion of the mafic-ultramafic igneous complex. The complex intruded during a significant episode of Variscan (Permian) crustal underplating as postulated by Pin and Vielzeuf [1983] and Herzberg et al. [1983]. Consequent partial melting of metasedimentary rocks generated granitic melts which intruded shallow crustal levels and erupted at silicic volcanic centers. Low initial $^{87}Sr/^{86}Sr$ ratios of the granites and volcanics would seemingly preclude this interpretation [Hunziker and Zingg, 1980] but I speculate that contamination of granite melts by mafic magmas might produce the same unradiogenic isotopic signature.

Thus, Permian intrusion of mafic magmas, lower crustal partial melting, silicic volcanism and plutonism and high-grade metamorphism significantly modified the crustal column. From a geophysical point of view, these events generated a highly stratified crust in that a thick section

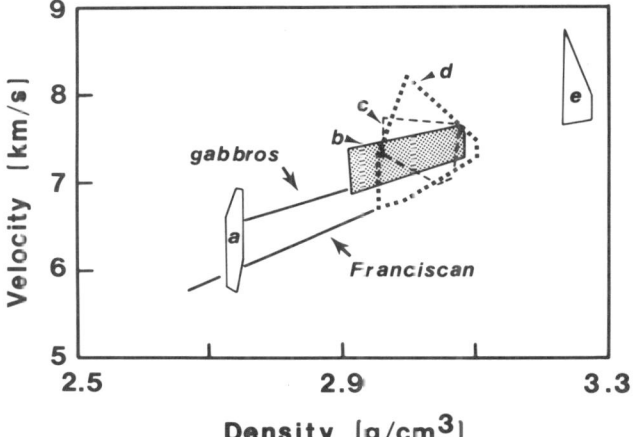

Fig. 3. Velocity versus density for various lithologies from IZ and SCZ from 0.6 to 1.0 MPa based on data from Fountain (1976). (a) SCZ schists and gneisses and IZ upper amphibolite facies metapelites; (b) IZ "Mafic Formation" rocks: (c) IZ amphibolites; (d) IZ granulite facies metapelites; and (e) ultramafic rocks. Also shown are linear trends for Franciscan metagraywackes (Stewart and Peselnick, 1977) and ophiolite metagabbros (Christensen, 1978) at 0.6 MPa.

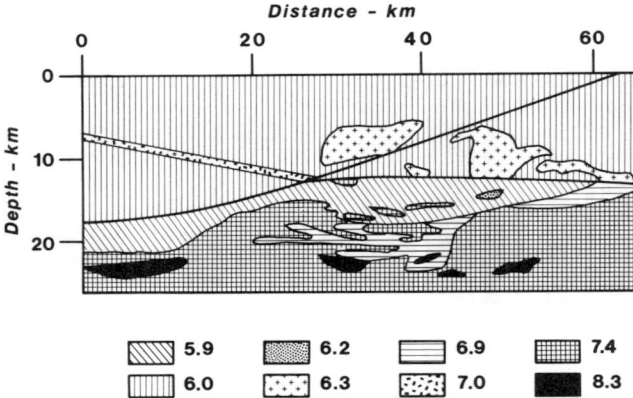

Fig. 4. Seismic velocity model of the IZ-SCZ based on surface geology and data from Fountain (1976). Velocities indicated in legend below the model are in km/s.

of high velocity mafic and ultramafic rocks (Figure 3) was introduced into the lower crust, compositional layering was enhanced by intrusive activity and chemical depletion of metasedimentary rocks resulted in increased density and velocity (Figure 3) for these lithologies [Fountain, 1976]. To a large extent the geophysical nature of the lower crust was well established by this time.

This complex crustal section was further modified in the early Mesozoic in response to normal faulting associated with early opening of western Tethys and development of the continental margin in northern Italy [Trumpy, 1975]. Hodges and Fountain [1984] speculated that this event developed low-angle normal faults in the IZ-SCZ (Figure 2c). These mylonitic fault zones, the Pogallo line and associated faults, thinned the crustal section with respect to its pre-Mesozoic thickness and placed relatively shallow crustal levels of the SCZ on the deeper IZ rocks. Crustal thinning enhanced the vertical crustal zonation established in the Permian. Eventually the complex was tectonically rotated from its horizontal position in the crust to its present near-vertical orientation perhaps during phases of Alpine deformation [Zingg, 1983; Hodges and Fountain, 1984].

Seismic Reflection Characteristics of the Ivrea and Strona-Ceneri Zones

The final map plan of the IZ and SCZ (Figure 1) is a consequence of the cumulative events summarized above and is generally regarded as a cross-section of the crust. To construct a velocity model of this section (Figure 4), I rotated the map plan so that it appears as a vertical section and assigned seismic velocities and densities to individual units based on data presented by Fountain [1976]. Although the mylonite zones could be reflective [Fountain et al., 1984; Jones and Nur, 1984], the effect was not included in the velocity model because of lack of appropriate measurements of the mylonites and difficulty in assessing the detailed structural geometry of the zones. This model was used to generate a synthetic reflection seismogram by convolving a 25-Hz Ricker wavelength with the appropriate relection coefficient series [e.g., Dennison, 1960] through use of AIMS 2-D modelling software. This produces a normal-incidence seismogram with no attenuation due to anelastic, geometric, reflectivity or transmitivity effects. Amplitudes for all traces were scaled to the maximum value and thus represent relative, not true, amplitudes. The traces in the final profile (Figure 5a) exhibit no background noise.

There are several important features evident in the synthetic profile. First, there is a preponderance of relatively short, discontinuous, near-horizontal reflections from the deeper levels of the crust (two-way travel-times greater than 5 seconds). Inspection of the ray trace diagram shows that most of these events are from contacts characterized by relatively high velocity contrasts such as boundaries between mafic and pelitic rocks and mafic and ultramafic rocks. The discontinuous and hyperbolic nature of these events reflects the laterally discontinuous and irregular character of the contacts imaged. Because the model does not simulate array geometry and does not include attenuation effects, the sloping tails of the hyperbolic reflections appear stronger than they may in actual reflection records. The amplitude of the tails would diminish if array geometry and attenuation effects were included and the horizontal events would become more prominent. The horizontal nature of the events results from the dominantly horizontal structures in the model. In general, these events tend to occur at certain horizons and could potentially be interpreted as events from single layers or structures. The model presented in Figure 5a only depicts the large-scale structures in the IZ. Examination of detailed mapping in the zone [Schmid, 1967; Bertolani, 1968] reveals fine-scale interlayering of the various rock types. Hale and Thompson [1982] generated synthetic reflection records for this fine-scale laminated zone based on the geology exposed in Val d'Ossola [Schmid, 1967] and found abundant complex events would be generated. If these events were included in Figure 5a, the lower crust would appear much more reflective than depicted. The model in Figure 5a and the results of Hale and Thompson [1982] suggest that structurally enhanced lithologic layering coupled with the abundance of mafic and ultramafic units in a lower crustal section can produce the reflective character of the lower crust commonly observed in reflection records [Meissner, 1973; Meissner et al., 1983; Brewer et al., 1983].

Secondly, the upper crust of the model appears transparent with respect to seismic reflection

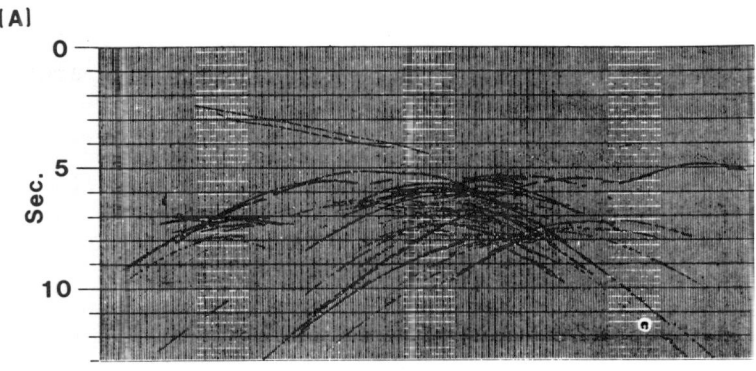

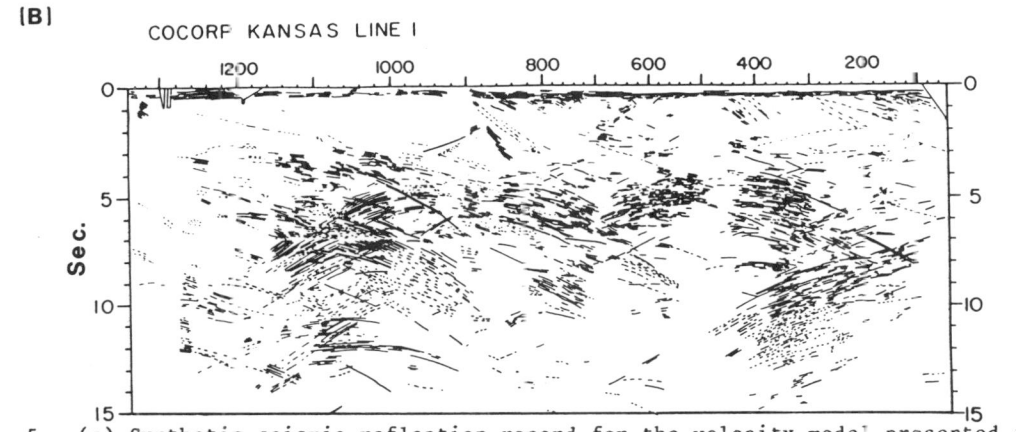

Fig. 5. (a) Synthetic seismic reflection record for the velocity model presented in Figure 4. (b) Line drawing of Kansas COCORP line from Brown et al. (1983). Both diagrams are shown with approximately same vertical and horizontal scales.

events. This is because the velocity variation in this section (SCZ) is not large, although there is certainly lithologic variation and structural complexity. In Figure 5a the granites are not imaged because of their low acoustic impedance contrast with surrounding schists and gneisses. Importantly, the absence of reflections from this zone does not point to the existence of a homogeneous layer. The zone is actually lithologically heterogeneous and structurally complex, but there are, in general, low acoustic impedances between the constituent lithologies, in distinct contrast to the seismic nature of the lower crust. Reflections from mylonite zones, if added to the model, would stand out in this level of the crust.

The applicability of this model to interpretation of crustal reflection records is demonstrated in Figure 5b where the synthetic profile is displayed at the same scale as the recent results from the COCORP Kansas line [Brown et al., 1983]. There are a number of common features in these two profiles, notably the numerous diffractions and the short, discontinuous events from a reflective lower crust. I do not offer a reinterpretation of the Kansas line, but simply suggest that the similarity of the model and actual field data may point to common threads of crustal evolution between the IZ-SCZ and the crust of Kansas.

Conclusion

Understanding the nature and evolution of the continental crust through use of seismic reflection data depends upon proper interpretation of reflection events in terms of lithology and structure. In this paper, I demonstrated how crustal cross-sections might provide calibration in that their geophysical characteristics can be connected to aspects of crustal evolution. Utilizing the IZ and SCZ of northern Italy as an example, we see that highly reflective lower crust can be caused by a combination of underplating of mafic and ultramafic magmas, high-grade metamorphism and resultant partial melting of lower crustal metasedimentary rocks and structurally induced compositional layering. In contrast, the upper crust appears transparent, not because of its lack of lithologic heterogeneity and structural complexity, but because constituent rock types generally have relatively low acoustic impedance contrasts compared to lower crustal lithologies. The IZ-SCZ is but one of several cross-sections identified to date [Fountain and Salisbury, 1981; Percival and Card, 1983] and differs significantly from the other examples. This suggests considerable variation of the geophysical nature of the crust and points out the need to examine these other cross-sections in the manner presented here.

Acknowledgments. I would like to thank C. Hurich for computer work on the seismic model and S. Roberts and C. Frost for helpful discussions. GeoQuest International Inc. donated the AIMS software package. Computing work was done on the Department of Geology and Geophysics VAX 11/780 computer. This work was supported by NSF Grants ISP-8011449 and EAR-8300659.

References

Berckhemer, H., Direct evidence for the composition of the lower crust and Moho, Tectonophysics, 8, 97-105, 1969.

Bertolani, M., La petrogafia della Valle Strona, Schweiz. Min. Petr. Mitt., 48, 696-732, 1968.

Brewer, J.A., D.H. Matthews, M.R. Warner, J. Hall, D.K. Smythe, and R.J. Whittington, BIRPS deep seismic reflection studies of the British Caledonides, Nature, 305, 206-210, 1983.

Brown, L., L. Serpa, T. Setzer, J. Oliver, S. Kaufman, R. Lillie, D. Steiner, and D.W. Steeples, Intracrustal complexity in the United States midcontinent: Preliminary results from COCORP surveys in northeastern Kansas, Geology, 11, 25-30, 1983.

Christensen, N.I., Ophiolites, seismic velocities and oceanic crustal structure, Tectonophysics, 47, 131-157, 1978.

Dennison, A.T., An introduction to synthetic seismogram techniques, Geophys. Prosp., 8, 231-241, 1960.

Fountain, D.M., The Ivrea-Verbano and Strona-Ceneri zones, northern Italy: A cross-section of the continental crust-New evidence from seismic velocities of rock samples, Tectonophysics, 33, 145-165, 1976.

Fountain, D.M., and M.H. Salisbury, Exposed cross sections through the continental crust: implications for crustal structure, petrology and evolution, Earth. Planet. Sci. Lett., 56, 263-277, 1981.

Fountain, D.M., C.A. Hurich, and S.B. Smithson, Seismic reflectivity of mylonite zones in the crust, Geology, 12, 195-198, 1984.

Graeser, S., and J.C. Hunziker, Rb-Sr- und Pb-Isotopenbestimmungen an Gesteinen und Mineralien der Ivrea-Zone, Schweiz. Min. Petr. Mitt., 48, 189-204, 1968.

Hale, L.D., and G.A. Thompson, The seismic reflection character of the continental Mohorovicic Discontinuity, J. Geophys. Res., 87, 4625-4635, 1982.

Herzberg, C.T., W.S. Fyfe, and M.J. Carr, Density constraints on the formation of the continental Moho and crust, Contrib. Mineral. Petrol., 84, 1-5, 1983.

Hodges, K.V., and D.M. Fountain, Pogallo line, South Alps, northern Italy: An intermediate crustal level, low-angle normal fault?, Geology, 12, 151-155, 1984.

Hunziker, J.C., Rb-Sr and K-Ar age determination and the alpine tectonic history of the Western Alps, Mem. Ist. Geol. Mineral. Univ. Padova, 31, 1-54, 1974.

Hunziker, J.C., and A. Zingg, Lower Paleozoic amphibolite to granulite facies metamorphism in the Ivrea zone (southern Alps, northern Italy), Schweiz. Min. Petr. Mitt., 60, 181-213, 1980.

Jones, T.D., and A. Nur, The nature of seismic reflections from deep crustal fault zones, J. Geophys. Res., 89, 3153-3171, 1984.

Koppel, V., Isotopic U-Pb ages of monazites and zircons from the crust-mantle transition and adjacent units of the Ivrea and Ceneri zones (southern Alps, Italy), Contr. Mineral. Petrol., 43, 55-70, 1974.

Lensch, G., Die Ultramafitite der Zone von Ivrea und ihre geologische Interpretation, Schweiz. Min. Petr. Mitt., 48, 91-102, 1968.

Mehnert, K.R., The Ivrea Zone, a model of the deep crust, N. Jb. Mineral. Abh., 125, 156-199, 1975.

Meissner, R., The Moho as a seismic transition zone, Geophys. Surv., 1, 195-216, 1973.

Meissner, R., E. Luschen, and E.R. Fluh, Studies of the continental crust by near-vertical reflection methods: A review, Phys. Earth Planet. Int., 31, 363-376, 1983.

Percival, J.A., and K.D. Card, Archean crust as revealed in the Kapuskasing uplift, Superior province, Canada, Geology, 11, 323-326, 1983.

Pin, C., and D. Vielzeuf, Granulites and related rocks in Variscan median Europe: A dualistic interpretation, Tectonophysics, 93, 47-74, 1983.

Polvé, M., Les isotopes du Nd et Sr dans les lherzolites orogeniques: Contribution a la determination de la structure et de la dynamique du manteau superieur, PhD. Diss., Univ. of Paris, 1983.

Reinhard, M., Uber das Grundgebirge des Sottoceneri im Sud-Tessin und die darin auftretenden Ganggesteine, Beitr. Geol. Karte Schweiz, N.F. 117, 89pp., 1964.

Rivalenti, G., G. Garuti, and A. Rossi, The origin of the Ivrea-Verbano basic formation (Western Italian Alps) - Whole rock geochemistry, Boll. Soc. Geol. Ital., 94, 1149-1186, 1975.

Rivalenti, G., A. Rossi, F. Siena and S. Sinigoi, The layered series of the Ivrea-Verbano igneous complex, western Alps, Italy, Tschermaks Min. Pet. Mitt., 33, 77-99, 1984.

Schmid, R., Zur Petrographie und Struktur der Zone Ivrea-Verbano zwischen Valle d'Ossola und Val Grande (Prov. Novara, Italien), Schweiz. Min. Petr. Mitt., 47, 935-1117, 1967.

Schmid, R., Are the metapelites of the Ivrea-Verbano Zone restites?, Mem. Ist. Geol. Mineral. Univ. Padova, 33, 67-69, 1978/79.

Sills, J.D., and J. Tarney, Petrogenesis and tectonic significance of amphibolites interlayered with metasedimentary gneisses in the Ivrea Zone, southern Alps, Northwest Italy, Tectonophysics, 107, 187-206, 1984.

Stewart, R., and L. Peselnick, Velocity of compressional waves in dry Franciscan rocks to 8 kbar and 300°C, J. Geophys. Res., 82, 2027-2039, 1977.

Trumpy, R., Penninic-Austroalpine boundary in the Swiss Alps: A presumed former continental boundary and its problems, Am. J. Sci., 275A, 209-238, 1975.

Zingg, A., Regional metamorphism in the Ivrea Zone (Southern Alps, N-Italy): Field and microscopic investigations, Schweiz. Min. Petr. Mitt., 60, 153-179, 1980.

Zingg, A., Regional metamorphism in the Ivrea Zone (Southern Alps, N-Italy) - A review, Schweiz. Min. Petr. Mitt., 63, 361-392, 1983.

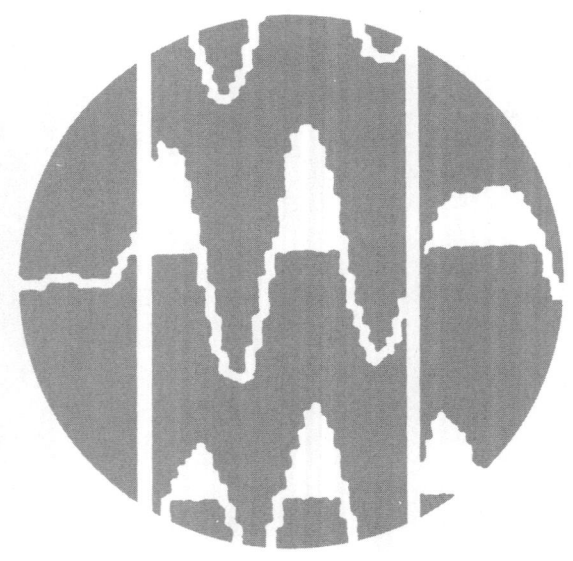

INTERPRETATION OF SEISMIC REFLECTION DATA IN COMPLEXLY DEFORMED TERRANES: A GEOLOGIST'S PERSPECTIVE

Robert D. Hatcher, Jr.

Department of Geology, University of South Carolina Columbia, S.C. 29208

Abstract. Horizontal seismic reflectors have been used recently to speculate on the existence and continuity of major tectonic features such as thrust faults in crystalline rocks, or as boundaries between crystalline basement and cover sedimentary rocks. A large body of published seismic reflection profiles, in areas where structure can be verified using drilling and/or downplunge projection, supports this interpretation. However, it has been recently shown with the Arizona A-1 hole and elsewhere that prominent, continuous horizontal reflectors in crystalline rocks do not necessarily prove a major break is present, indicating a great deal of work remains before the nature of seismic reflection events in the deeper crust is understood.

The seismic reflection method commonly detects gently dipping layers having contrasting acoustic properties, although it may be theoretically possible to detect steeply dipping layers. Correct interpretation of recurrent crustal reflection patterns could provide enormous insight into crustal structure and evolution. However, geologic interpretations from the same sets of reflectors are numerous. Single layered subhorizontal reflectors may be thrusts, unconformities, mylonite/cataclasite zones, stratigraphic contacts, or facies boundaries. Inclined reflector packages may be as imbricate thrusts or normal faults, duplexes, root zones or ramps. Curved reflectors may indicate the tops of plutons, broad open folds or refolded folds. Transparent zones are the most difficult to interpret and may indicate zones of structural complexity, structurally simple areas or rocks that contain no acoustic contrast.

The interiors of mountain chains contain a complex assemblage of rocks of low to high metamorphic grade which have markedly different mechanical and probably acoustic properties. Reflection coefficients occur at interfaces between rocks of different densities, and anisotropies or compositions without presence of tectonic discontinuities. Sub-horizontal tectonic discontinuities, such as thrust faults, may provide additional acoustic contrasts, provided there is a difference in properties of the rocks on either side of the discontinuity. Later thrusts in orogenic belts are less deformed and, if they juxtapose rocks of contrasting acoustic properties, should provide excellent reflectors. However, unless independently verifiable, continuous reflectors in orogenic terranes may not prove to be tectonic or even lithologic boundaries.

Models of crustal structure should, of necessity, include geologic, seismic reflection, and data from other geophysical techniques, such as potential fields. These should increase understanding of the nature of the deeper continental crust.

Introduction

Geologists have for many years been aware of the great diversity both in geometry, and in the character of rocks in the interiors of mountain chains. Mountain chains become incorporated into continental cratons and their roots are either later exposed in shields, or remain covered beneath younger sedimentary rocks. These complexly deformed regions constitute one of the greatest challenges for both geophysicists and geologists, for they likely hold the keys to understanding crustal structure and evolution of the continents. Until recently, geophysical models of crustal structure have been relatively unsophisticated, owing to the nature and the precision of the techniques used to probe the deeper crust. Application of the VIBROSEIS technique of seismic reflection to problems of the deeper crust, principally by the Consortium for Continental Reflection Profiling (COCORP), has resulted in a great deal of new data for both geologists and geophysicists to interpret. The purpose here is to discuss some of these data and outline some of the difficulties and pitfalls, from a geologist's point of view, that may affect the interpretation of seismic reflection data in structurally complex regions.

The most obvious problem in interpretation of seismic reflection data in the continental crust is the design of the technique and its existing maximum efficiency in detecting horizontal or near horizontal layering. This has lead to both erroneous and correct interpretations of structurally complex regions where corroboration by drilling, or with a combination of drilling and projection of surface geologic data is possible.

Seismic Reflections

Seismic reflection occurs where acoustic properties change and energy is reflected at a convenient angle to be detected by a surface receiver array. However, obvious geologic boundaries frequently lack acoustic contrast and thus do not always produce reflections. At the same time, steeply-dipping reflectors can rarely be preserved in processed seismic sections. The standard textbooks on exploration geophysics and seismic reflection techniques (e.g., Telford et al., 1976; Dobrin, 1976) outline the kinds of artifacts and pitfalls which may appear in seismic records. They range from coupling problems, artifacts of acoustic properties which may develop because of the nature of the energy, the design of the geophone array and/or the frequency of energy being utilized (diffractions, multiples, reflected refractions, ringing, ground roll, etc.). Geologic possibilities for crustal reflection include major lithologic contacts, unconformities and faults. In addition, lithologic contrasts within rock bodies and not at contacts, may produce reflections.

Crustal Structure and Seismic Reflector Geometry

The earth's continental crust consists of a heterogeneous and diverse assemblage of rock bodies and structures of varying sizes and extents. Decades of geologic mapping in the exposed continental shields and orogenic belts have demonstrated the extraordinary geometric and lithologic complexity of these zones. Of all the geophysical techniques presently available, seismic reflection provides the best data needed to unravel structure of the deeper crust. Yet, it too is limited by the pitfalls mentioned above and by the design of most seismic reflection experiments.

The number of possibilities for resolution of crustal structure using

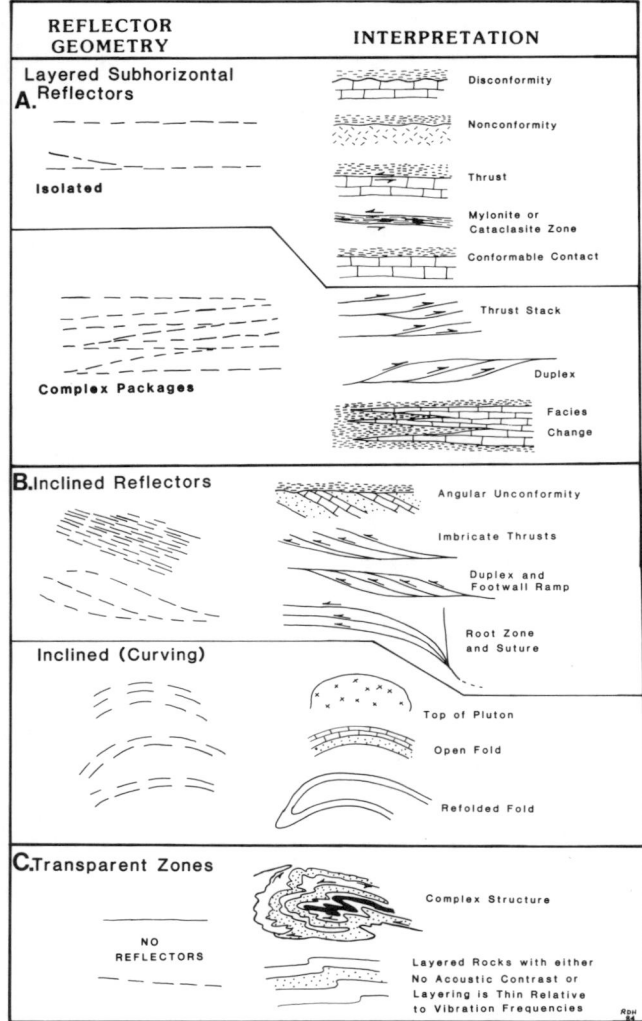

Fig. 1. Commonly recurring crustal reflector geometries and several possible geologic interpretations. Note that no scale is specified and that particular interpretations would be favored or eliminated by indicating scale, as would be the case in reflection profiles.

seismic reflection is large. Yet particular deep crustal reflector patterns recur in reflection profiles which may provide some insight into crustal structure, if they can be correctly interpreted. These are: 1) layered subhorizontal reflectors; 2) complexes of inclined, curved or dipping reflectors; and 3) transparent nonreflecting zones (Figure 1). Each of these types will be discussed below in an attempt to relate them to geologic features in the middle to lower crust.

Layered Subhorizontal Reflectors

Layered reflectors consist of one or more reproducible events which are continuous over an appreciable distance in the profile. Interest in the interpretation of layered reflections (Figure 1A) in the deeper crust has existed for a number of years, even before sophisticated processing techniques and digital seismic records became available. For example, Widess and Taylor (1959) attempted to interpret seismic reflection data from the Precambrian rocks in Oklahoma. They were able to calibrate their seismic interpretation with drilling which indicated the layered reflections were produced by alternating layers of silicic and gabbroic igneous rocks. These rocks would be expected to produce strong reflections because of their marked differences in density. The COCORP reflection profiles across the Michigan Basin (Brown et al., 1982) contain an excellent group of layered reflectors in the middle to upper crust. These probably represent sedimentary rocks deposited in the Keweenawan rift basin.

The remarkably continuous single reflector beneath the southern Appalachian Blue Ridge and Inner Piedmont (Cook et al., 1979) may confirm the existence of a great crystalline thrust sheet, originally suggested by Hatcher (1971, 1972) and Hatcher and Zietz (1978, 1980) from surface geologic studies and potential field data, respectively. This remarkably clear record in the middle to upper crust has given rise to some controversy, in a region to the southeast where the reflectors become inclined and less continuous. Hatcher and Zietz (1980), Hatcher (1981) and Iverson and Smithson (1982) concluded that the zone of inclined reflectors is the root zone for the master detachment, whereas Harris and Bayer (1979), Cook and Oliver (1981), and Cook et al. (1983) interpreted the reflection profiles as indicating the detachment is continuous far to the east beneath the Coastal Plain and possibly the continental shelf. This controversy remains unresolved, and even the existence of the detachment beneath the Blue Ridge and western Piedmont remains unresolved until the reflection profile and the other interpretations are calibrated by drilling.

An even more spectacular reflection profile which resolves a large amount of middle to upper crustal layered structure is the COCORP Basin and Range Utah line (Allmendinger et al., 1983) Here, several subhorizontal reflectors were interpreted as detachment faults and the structure from the western edge of the Colorado Plateau to the Nevada-Utah border can be interpreted.

The spectacular results of the COCORP southern Appalachian and Utah lines are clouded somewhat by Phillips-Anschutz experience using seismic reflection data to locate a well for petroleum exploration in crystalline rocks in southeastern Arizona (Reif and Robinson, 1981; Robinson, 1982). Here, seismic reflection and magnetotelluric data revealed several horizontal reflectors and a magnetotelluric anomaly at different depths. The reflecting horizon and the magnetotelluric anomaly were interpreted as a thrust which carried crystalline rocks over unmetamorphosed and, possibly, petroliferous sediments. Drilling to about 6 km revealed only granitic rocks of different textures and slightly different compositions. A sonic log of the hole and construction of a synthetic seismogram indicated that the drill had penetrated the reflector horizons. One of these reflectors is still interpreted as a thrust fault, but one which transported more massive granitic rocks over strongly foliated granitic gneiss.

Horizontal to subhorizontal reflectors may be produced in layered crystalline rocks for a variety of reasons. Obviously, all the processes and features by which these reflectors are produced are not well understood and, obviously, not all have a geologic origin. The abundant recurrence of laminated reflector zones in the lower crust and upper mantle at or below the Moho may be further indicators of different processes occurring at these depths. It is likely that most of these formed within or below the ductile-brittle transition zone and may represent zones of ductile flow.

Complexes of Inclined (Dipping) and Curved Reflectors

Complex zones of overlapping inclined reflectors (Fig. 1B) occur in the middle to upper crust in several of the COCORP profiles. The Wind River profile contains a single zone of inclined reflectors which, when interpreted together with gravity data, indicate that the Wind River thrust may cut the entire crust along a remarkably planar 35 to 40 degree northeast-dipping fault (Smithson et. al., 1979; Brewer et al., 1980). Additional interpretation of stacked and over lapping curved reflectors by Smithson et al. (1980) indicates that portions of large refolded isoclinal folds may be identified in the reflection profiles.

Very complex overlapping inclined layered reflectors occur in the COCORP Kansas lines (Brown et al., 1983). Similar reflectors are also present beneath the western parts of the COCORP New England traverse (Ando et al., 1983) and the interior (central) portions of the southern Appalachians lines (Cook et al., 1983).

Interpretation of complexes of inclined or curving reflectors is difficult at best (Fig. 1B). They probably indicate that the crust in these areas is different from that in areas dominated by subhorizontal layered reflectors. Some of these inclined reflections also are probably artifacts, such as diffractions, as indicated by Brown et al.(1983). However, potential field data (Zietz, 1982); Lyons and O'Hara, 1982) indicate the crust in these regions contains a greater number of plutons along with a greater number of magnetic lineaments, which may be interpreted as either dike complexes or as tectonic boundaries.

Transparent (Homogeneous) Zones

Probably the most difficult regions to interpret are the seismically transparent zones (Figure 1C), which represent acoustic homogeneity. Zones of this type are present in many of the COCORP and other profiles, and each may represent a geologically different situation

The COCORP southern Appalachians line (Cook et al., 1983) contains a transparent zone from the base of the surface noise to depths of 5 to 10 km. The surface structure of this region is very complex, consisting of large early refolded isoclinal recumbent folds and early folded thrusts cut by later, more planar faults (Hatcher, 1972, 1981), some of which (e.g., the Brevard) appear to be expressed in the seismic reflection profile. It may be that the structural complexity is so great here that the rocks are rendered acoustically homogeneous, despite the fact that rocks, e.g., amphibolites, schists, granites and metasandstones, that should possess abrupt acoustic contrasts, are present in these transparent zones. Another factor may be the scale of repetition by folding may be small enough so that the acoustic energy is either dispersed or passes through unimpeded at the frequencies employed for this study. Transparent zones in the BIRPS MOIST line (Smythe et al., 1982) occur in a lithologically and structurally similar region in the offshore projection of the Scottish Highlands.

Discussion

The rocks of the interior portions of mountain chains and shields contain rocks with contrasting mechanical and chemical properties and should yield strong contrasts in acoustic properties, provided these mechanical or chemical differences occur on a scale which is detectable by the seismic reflection technique. Ideally, geologic materials, such as rocks of layered character in the cores of orogenic belts, including strongly layered gneisses, interlayered gneisses and amphibolites or interlayered igneous rocks of different character, should yield good reflectors, assuming that the structure is not unduly complex. There is generally a chronological order of development of structures in most orogenic belts in which structural complexity is increased by overprinting several generations of early ductile structures by later brittle structures. The superposition of structures within an orogen is further complicated by intrusions of various types of igneous bodies and by the overprinting of metamorphic/thermal events onto these structures. Gently inclined to subhorizontal late brittle faults may provide the best structures to yield reflectors and, consequently, be the easiest to interpret in zones of complex deformation. The COCORP Utah line (Allmendinger et al., 1983) provides a good example of interpretable complex structure.

If it could be known that seismically transparent zones (Figure 1C) in crystalline crust are frequently caused by complexly folded zones, it would be possible to delineate their boundaries but not their internal structure. However, all complexly folded zones are not obviously related to surface structure. Their lack of continuity may be some indicator of structural complexity.

Models of crustal structure should probably employ projection of as much surface geology as possible to help interpret seismic reflection data, along with potential field and other types of geophysical data. Contrasts in crustal type may be readily identified using all these data in a heterogeneous region in Kansas, whereas further south in southern Oklahoma and Texas, the crust appears more homogeneous to laminated. Drilling should also be employed where possible as an additional test of any hypothsis of crustal structure.

Conclusions

1. Numerous conditions that produce seismic reflections occur in the continental crust. Lithologic and tectonic contacts, unconformities and lesser acoustic inhomogeneities would become reflectors if they are geometrically and acoustically suitable. These must be distinguished from various artifacts if they are to be correctly interpreted.

2. Crustal processes ranging from orogenic activity to ductile flow in the lower crust and upper mantle serve to produce a large variety of structural features. Crustal seismic reflector patterns of three types recur: continuous subhorizontal isolated or multiply layered reflectors, inclined (dipping) and curved reflectors, and transparent zones

3. Layered reflectors may represent lithologic packages, broad deformation zones (laminae), or fault zones (isolated)

4. Inclined or curved reflector packages may represent fold and intrusive complexes.

5. Some transparent zones may represent tightly refolded zones where deformation has been both ductile and intense enough to render the zone acoustically homogeneous.

6. Models of crustal structure are best developed by utilization of all types of geologic and geophysical data (seismic, potential field, conductivity, etc.). Where possible, these models should be tested with drilling.

Acknowledgements. Research support was provided by National Science Foundation Grants 76-15564, 79-11802, 81-0852 and 82-06949. Conversations and discussions with many seismologists and other geophysicists during the last decade have provided me with the proverbial little knowledge in geophysics. Comments by W.B. Travers and an anonymous reviewer resulted in considerable improvement in the manuscript. However, I remain responsible for all errors of fact or interpretation.

References

Ando, C. J., F.A. Cook, J.E. Oliver, L.D Brown, and S. Kaufman, Crustal geometry of the Appalachian orogen from seismic reflection studies, in *Contributions to the Tectonics and Geophysics of Mountain Chains*, Edited by R.D. Hatcher, H. Williams, and I. Zietz, *Geol. Soc. America Memoir 158*, 1983.

Allmendinger, R.W., J.W. Sharp, D. Von Tish. L. Serpa, L. Brown, S. Kaufman, and J. Oliver, Cenozoic and Mesozoic structure of the eastern Basin and Range Province, Utah, from COCORP seismic reflection data, *Geology, 11*, 532-536, 1983.

Brewer, J.A., S.B. Smithson, J.E. Oliver, S. Kaufman, and L.D. Brown, The Laramide orogeny: Evidence from COCORP deep crustal seismic profile in the Wind River Mountains, Wyoming, *Tectonophysics, 62*, 165-189, 1980.

Brown, L., L. Serpa, T. Setzer, J. Oliver, S. Kaufman, R. Lillie, and D. Steiner, Intracrustal complexity in the United States mid-continent: Preliminary results from COCORP surveys in northeastern Kansas, *Geology, 11*, 25-30, 1983.

Brown, L., L. Jensen, J. Oliver, S. Kaufman, and D. Steiner, Rift structure beneath the Michigan Basin from COCORP profiling, *Geology, 10*, 645-649, 1982.

Cook, F.A., and J.E. Oliver, The early Paleozoic continental edge in the Appalachian orogen, *Am. Jour. Sci., 281*, 993-1008, 1981.

Cook, F.A., D.S. Albaugh, L.D. Brown, R.D. Hatcher, Sidney Kaufman, and J.E. Oliver, Thin-skinned tectonics in the crystalline southern Appalachians; COCORP seismic-reflection profiling of the Blue Ridge and Piedmont, *Geology, 7*, 563-567, 1979.

Cook, F.A., L.D. Brown, Sidney Kaufman, and J.E. Oliver, The COCORP seismic reflection traverse across the southern Appalachians, *Amer.*

Assoc. Petroleum Geologists Studies in Geology, 1A, 61p., 1983.

Dobrin, M.B., *Introduction to Geophysical Prospecting,* McGraw-Hill Publishing Company, New York, 630p., 1976.

Harris, L.D., and K.C. Bayer, Sequential development of the Appalachian orogen above a master decollement - a hypothesis, *Geology, 7,* 568-572, 1979.

Hatcher, R.D., Jr., Structural Petrologic and Stratigraphicevidence favoring a thrust solution to the Brevard problem: *Am. Jour. Sci. 270,* 177-202, 1971.

Hatcher, R.D., Jr., Developmental model for the southern Appalachians, *Geol. Soc. America Bull., 83,* 2735-2760, 1972.

Hatcher, R.D., Jr., Thrusts and nappes in the North American Appalachian Orogen, in *Thrust and Nappe Tectonics,* edited by K. R. McClay and N. J. Price, *Geol. Soc. London Special Pub. 9,* 491-499, 1981.

Hatcher, R.D., Jr., and Isidore Zietz, Thin crystalline thrust sheets in the southern Appalachian Inner Piedmont and Blue Ridge: Interpretation based upon regional aeromagnetic data, *Geol. Soc. America Abs. with Programs, 10,* 417, 1978.

Hatcher, R.D., Jr., and I. Zietz, Tectonic implications of regional aeromagnetic and gravity data from the southern Appalachians, in *The Caledonides in the U.S.A.,* edited by D.R. Wones, *Virginia Polytechnic Institute Memoir No. 2,* 235-244, 1980.

Iverson, W.P., and S.B. Smithson, Master decollement root zone beneath the southern Appalachians and crustal balance, *Geology, 10,* 241-245, 1982.

Lyons, P.L., and N.W. O'Hara, Gravity anomaly map of the United States, Society of Exploration Geophysicists: scale 1/2, 500,000, 1982.

Reif, D.M., and J.P. Robinson, Geophysical, geochemical, and petrographic data and regional correlation from the Arizona State A-1 well, Pinal County, Arizona, *Arizona Geol. Soc. Digest, 13,* 99-109, 1981.

Robinson, J.P., Petroleum exploration in southeastern Arizona: Anatomy of an overthrust play, *Rocky Mtn. Assoc. Geologists,* 665-674, 1982.

Smithson, S.B., J.A. Brewer, S. Kaufman, J.E. Oliver, and C.A. Hurich, Structure of the Laramide Wind River uplift, Wyoming, from COCORP deep reflection data and from gravity data, *Jour. Geophys. Res., 84,* 5955-5972, 1979.

Smithson, S.B., J.A. Brewer, S. Kaufman, J.E. Oliver, and R.L. Zawislak, Complex Archean lower crustal structure revealed by COCORP crustal reflection profiling in the Wind River Range, Wyoming, *Earth and Planetary Sci. Letters, 46,* 295-305, 1980.

Smythe, D.K., A. Dobinson, R. McQuillin, J.A. Brewer, D.H. Matthews, D.J. Blundell, and B. Kelk, Deep Structure of the Scottish Caledonides revealed by the MOIST reflection profile, *Nature, 299,* 338-340, 1982.

Telford, W.M., L. P. Geldart, R. E. Sheriff, and D.A. Keys, *Applied Geophysics,* Cambridge University Press, New York, 860 p., 1976.

Widess, M.B. and G. L. Taylor, Seismic reflections from layering within the Pre-Cambrian basement complex, Oklahoma, *Geophysics, 24,* 417-425, 1959.

Zietz, I., Composite magnetic anomaly map of the United States, Part A: Conterminous United States: U.S. Geological Survey Map GP-954-A, Scale 1/2,500,000, 1982.

CONTINENTAL EVOLUTION BY LITHOSPHERIC SHINGLING

Frederick A. Cook

Department of Geology and Geophysics, University of Calgary,
Calgary, Alberta, Canada T2N 1N4

Abstract. Crustal geometries observed on seismic reflection data indicate that the boundaries of terranes accreted to the margins of continental cratons often have a low dip angle (30° or less). Where terranes have encroached upon a subducted passive margin, the boundaries are typically listric into upper or mid-crustal detachments and rarely, if ever, penetrate the entire crust. Outboard of the subducted margin, terrane boundaries may penetrate the lower crust and upper mantle. Examples of the former type include the west margin of the Bronson Hill arc in New England, the Rheno-Hercynian zone in France and the Piedmont Carolina slate belt in the southern Appalachians. Examples of the latter type may include the Brunswick terrane in the southern Appalachians and the Gander terrane in the northern Appalachians. Such observations suggest a model in which the areal extent of continental cratons is increased in collisional orogens by a process of lithospheric "shingling".

Introduction

Major advances have recently taken place in our understanding of how collisional orogens evolve. Foremost among these is that many orogens are comprised of a collage of geologically distinct terranes which have been accreted to the margins of cratons [Coney et al., 1980; Williams and Hatcher, 1982]. Data bearing on the deep structure of these accreted terranes have, in some areas, been successful in delineating the nature and orientation of some of the boundaries of the terranes and in mapping the structure of major faults which are the products of the accretionary process. A common characteristic is the development of an extensive thrust belt verging toward the craton. The belt usually includes a thin-skinned foreland composed dominantly of shortened miogeoclinal strata, and a thin-skin interior zone which consists of metamorphosed and shortened eugeoclinal strata. Outboard of the compressed eugeocline are the accreted terranes, with boundaries and internal faults which are usually low angle (or listric) and which generally verge toward the craton. In many cases it thus appears the continental crust has been enlarged laterally by a process of accretion along low-angle boundaries, and thus exhibits a 'shingled' appearance.

Crustal structure data, particularly from controlled source seismology, are now being extensively used to map detailed features of the continental lithosphere. Seismological data have been obtained in several mountain belts which are believed to have developed through the accretion of many distinct terranes. Although it is premature for sweeping generalizations, some patterns are beginning to emerge which may be applicable to similar orogens worldwide. Of particular significance, and the focus of this paper, is the observation that many "accreted terranes" are both bounded by, and internally deformed by, craton-verging, low-angle detachment faults [Cook et al., 1979; 1981: Allegré et al., 1984; Monger et al., in press]. Examples to be discussed include the Himalayas, the northern Appalachians, and the southern Appalachians. Other ancient orogens, such as the Canadian Cordillera, and more modern systems, such as Australia-Timor, may also have developed along similar lines.

The purpose of this paper is to elucidate a model for continental evolution which has been implicit in some previous work. The model is developed with a heavy emphasis on results from the Appalachian-Caledonide orogen, for, at the present time there are more crustal scale seismic reflection data available for this orogen than for any other. Recently acquired data from some other orogens [e.g. Allegré et al., 1984; Bois et al., 1984] show strikingly similar geometries and suggest such a model may be more widely applicable than the Appalachians. To begin, we discuss some recent findings in the Himalayas which have a bearing on this problem.

Lithospheric Geometry - Himalayas

In the Himalaya, many workers have believed that the northward underthrusting of the Indian lithosphere beneath the Asian lithosphere was largely responsible for the development of the orogen. However, during the approach of India to Asia, separate blocks [e.g. the Lhasa, Quantang

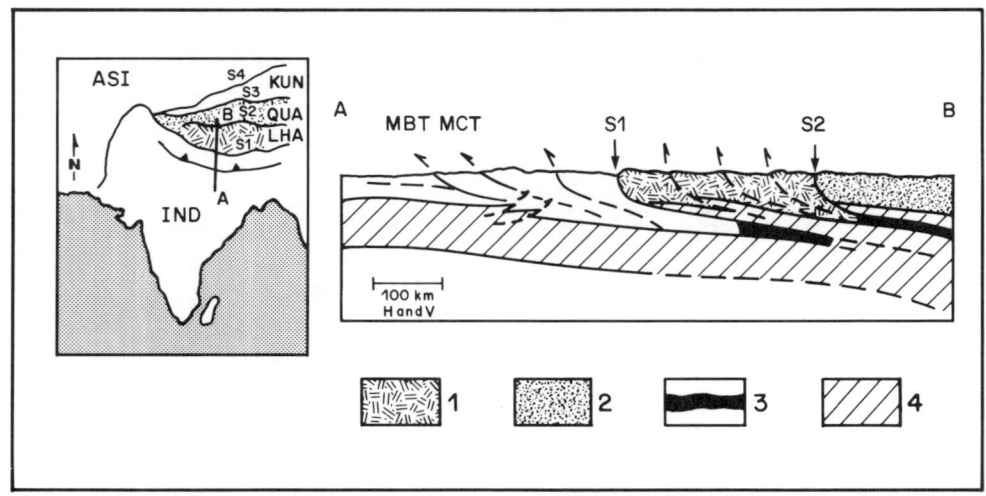

Fig. 1. Location map and cross section of the Himalayas [modified from Allegré et al., 1984]. The major tectonic units include India (IND), the Lhasa Block (LHA), the Quantang block (QUA), and the Kunlun block (KUN) and Asia (ASI). The cross section shows the predominance of south verging thrusts, such as the Main Boundary thrust (MBT) and the Main Central Thrust (MCT), and the listric nature of sutures (S1, S2). The patterns are as follows: 1 = accreted terrane 1; 2 = accreted terrane 2; 3 = oceanic crust; 4 = lithosphere.

and Kunlun blocks - Allegré et al., 1984] were caught between the two continental lithospheres. During the interval 140 m.y. - 40 m.y., these terranes were accreted to the Asian lithospheric plate.

The subsurface geometries of these terranes and their boundaries with the Indian and Asian lithospheres had been largely unconstrained until seismic data were acquired [Hirn et al., 1984]. Interpretations of these data in conjunction with surface geological information strongly imply that the upper crusts of the Indian plate, the Lhasa block, and the Quantang block are shortened along south verging thrust faults which are generally low angle at depth (Fig. 1). Examples include the Main Boundary Thrust (MBT) and Main Central Thrust (MCT) which are both surface faults positioned above low angle detachments.

Gravity data and seismic refraction data have long been interpreted as indicating the crust beneath the high Himalaya has an abnormally large thickness of 50-70 km [Powell and Conaghan, 1973]. Broadside refraction data further indicate that the Moho exhibits a complex topography beneath India and Tibet. Indeed there may even be the superposition of Mohos by thrusting [Fig. 1 - Allegré et al., 1984]. This structure is most easily interpreted as resulting from crustal repetition by thrust slivering. Furthermore, thrust slivers in the lower crust may verge in the opposite direction to the south-verging faults near the surface [Hirn et al., 1984].

The sutures which bound the blocks (terranes) also have a listric appearance and verge southward (Fig. 1). Thus, the evolutionary scenario which emerges is one in which the Indian continental lithosphere (and perhaps some old oceanic lithosphere) was subducted northward beneath the Lhasa block. The Lhasa lithosphere (and perhaps some old oceanic lithosphere) was previously subducted northward beneath the Quantang block. The increased crustal thickness beneath the high Himalaya apparently occurred by intracrustal and upper mantle thrusting, rather than simply by crustal doubling. Hence the upper crust in the area is 'shingled' along detachments which verge in the opposite direction to the subduction direction.

Lithospheric Geometry - Appalachians

In the southern Appalachians, crustal structure studies and geologic information have produced a similar picture to that described for the Himalayas. As shown in Figure 2 (section A-B), seismic reflection data indicate that Paleozoic accreted terranes are sutured to North America along boundaries which generally have a low dip angle and verge westward toward the old craton.

The detailed structural geometry across this portion of the Appalachian orogen includes a craton-verging thin-skin foreland thrust belt (Valley and Ridge) and an extensive allochthon of metamorphosed eugeoclinal rocks to the southeast (Blue Ridge and Inner Piedmont). This allochthon is also thin-skin (up to 15 km thick) and is thrust west above essentially undeformed North American basement [Cook et al., 1979]. Southeast of the Inner Piedmont (IP) is the accreted

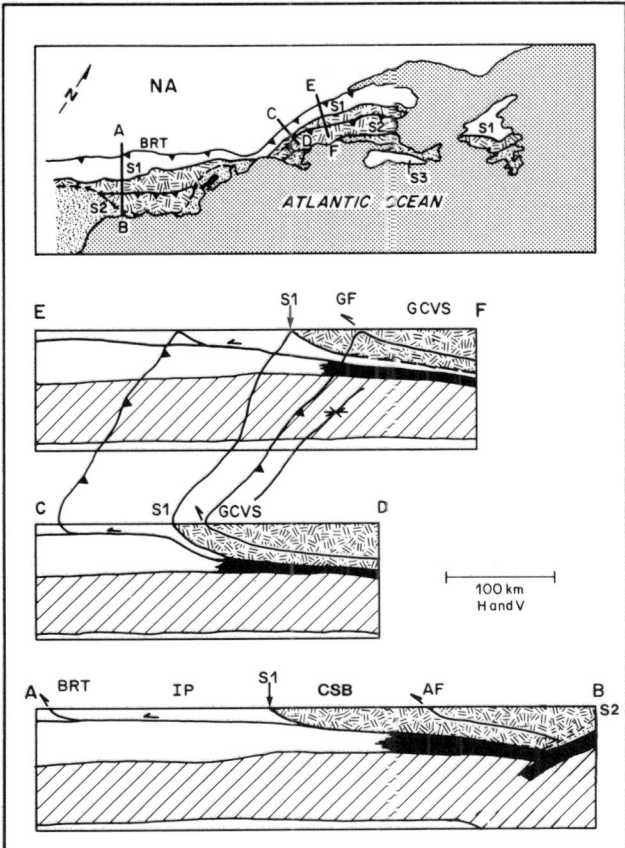

Fig. 2. Location map and cross sections of the Appalachians. The map is modified from Williams and Hatcher [1982]. Cross section for the southern Appalachians (A-B) modified from ref. 3 and 4. The abbreviations are: BR = Blue Ridge, IP = Inner Piedmont, S1 = Suture between arc and continent, CSB = Carolina slate belt arc, AF = Augusta fault. A suture (S2) may be east of the Augusta fault as shown. The reversal of structures at depth is shown on the east side of the section. The northern Appalachian cross sections are from St. Julien et al., [1983] (profile C-D), Green and Berry [1984] and Stewart [1984] (profile E-F). Interconnecting lines are shown between the northern Appalachian cross sections in an effort to illustrate similarities of structural style. Abbreviations used are GCVS - Gaspé-Connecticut Valley Synclinorium, GF - Guadaloupe fault.

Carolina slate belt (CSB) arc terrane which is bounded on the west by a suture which verges westward and flattens at mid-crustal depths [Cook et al., 1979; 1981]. Southeast of the arc terrane, the Augusta fault (AF) is an east dipping fault (Fig. 3; apparent dip of 8° - 10°) which can be traced on reflection data for a distance of nearly 80 km from its surface position [Cook et al., 1981; Petersen et al., 1984].

The development of the southern Appalachians probably began with the partial eastward subduction of Precambrian crust (and overlying miogeocline) during the Taconic (Ordovician) orogeny [Rankin, 1975; Hatcher, 1978]. Detachments formed at that time were reactivated in the late Paleozoic (Alleghanian - Hercynian), at the same time as major intra-crustal thrusts (such as the Augusta fault) were formed. It is not known at this time if the Augusta fault is a suture, but it is clearly a major crustal thrust [Petersen et al., 1984].

The northern Appalachians of New England and Quebec exhibit a similar, albeit narrower, crustal structure. Two crustal seismic reflection transects have been acquired which image the pertinent structures (Fig. 2). Profile C-D is along the COCORP New England Traverse [Ando et al., 1983]; the west half of profile E-F is along Quebec line 2001 [St. Julien et al., 1983] and the east half of E-F is interpreted from a recent southeastward extension of line 2001 across Maine [Green and Berry, 1984; Stewart, 1984]. As in the southern Appalachians, Precambrian basement can be followed for a substantial distance eastward beneath the allochthonous orogen; hence, the major surface thrusts above the basement must be either listric or truncated at depth.

Furthermore, boundaries of the accreted terranes, such as the Ordovician Bronson Hill arc are underlain by moderately east-dipping (10°-40°) reflections which in most places flatten in the middle of the crust (Fig. 2). In some cases low angle thrusts which may be in the interior of accreted terranes (e.g. - the Guadaloupe fault) can be traced at depth for great distances (up to 80 km) eastward [St. Julien et al., 1983; Green and Berry, 1984; Stewart, 1984]. Hence, in addition to suturing along low angle (listric) boundaries, the data indicate that some crustal-scale thrust faults causing significant crustal shortening are also low angle and verge toward the craton.

In terms of the Paleozoic accretionary history, the northern Appalachians appear to have been initiated by eastward subduction of North American basement beneath the Ordovician arc system. Following the suturing of the arc to North America, succeeding orogenies, such as the Acadian (about 350 m.y.) may also have been due to accretion, but clearly resulted in the development of craton verging thrust faults such as the Guadaloupe fault (GF). This portion of the Appalachians thus also evolved and was enlarged by lateral accretion along low-angle boundaries.

A Simplified Model

The observations presented, along with similar structures observed elsewhere, have important geodynamic implications. The repetition of crust and lithosphere by craton-verging intra-crustal

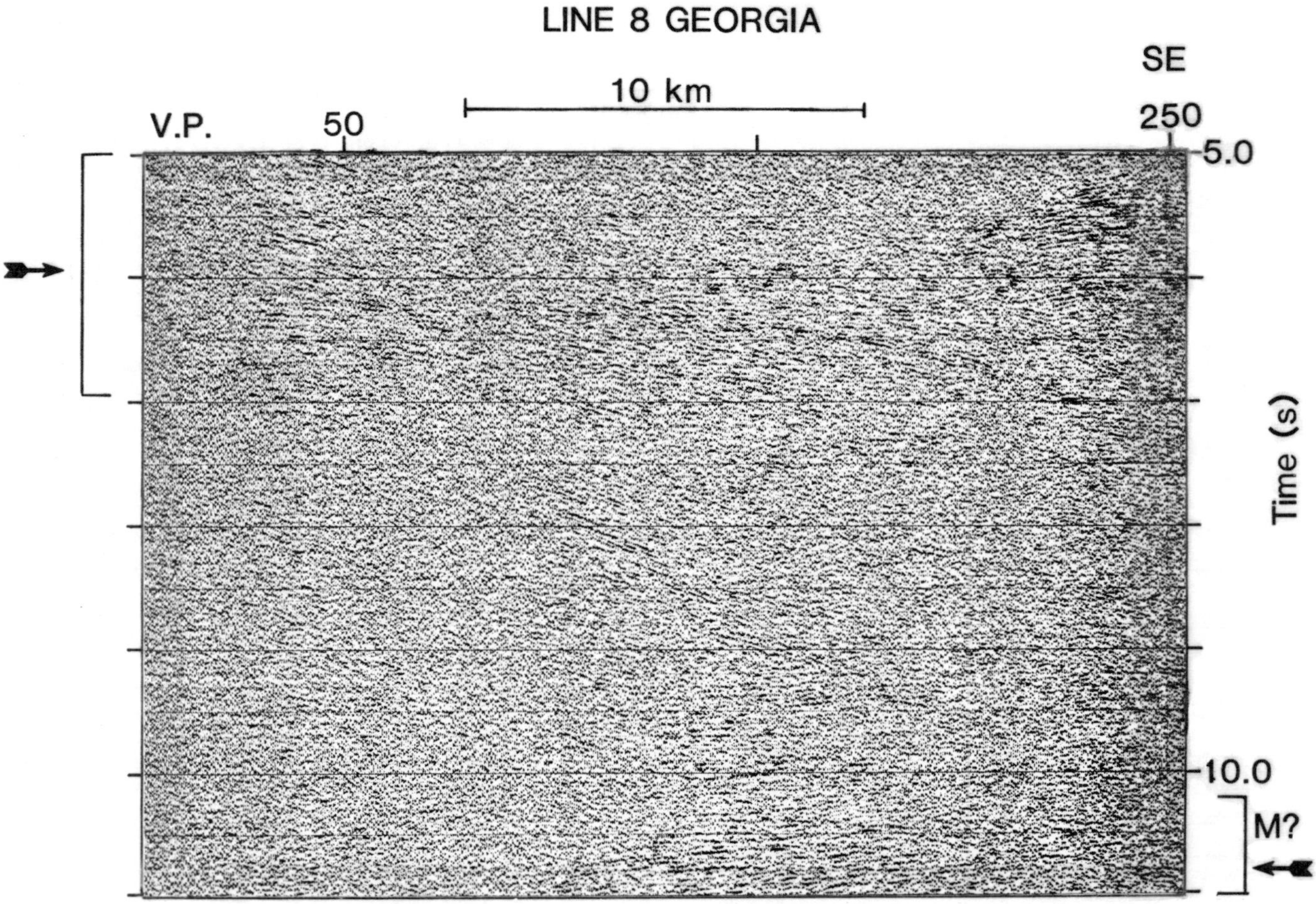

Fig. 3(a). Section of COCORP seismic reflection data from the Coastal Plain of northeast Georgia. On the west side (arrow) the boundaries have a dominantly east dip and are likely related to the Augusta fault. On the east side (near V.P. 150) the boundaries have a dominantly west dip. The reversal may indicate a reversal of subduction orientation as discussed in the text.

thrusting and low angle sutures implies that, in some instances, a continent may have its areal extent enlarged by 'shingling' as exotic terranes are juxtaposed with the subduction margin. Mechanisms similar to this have been proposed previously, but require 'flaking' of thin sheets emplaced over a margin as subduction takes place beneath the margin [Oxburgh, 1972; Ben-Avraham et al., 1982]. The mechanism proposed here is not intended as an alternative to 'flaking', but rather is likely a partner to it. Figure 4 illustrates the concept.

Subduction of lithosphere beneath a terrane results in the convergence of a continental craton with the arc. As the continent is partially subducted, ocean basin sediments (accretionary wedge) are compressed and emplaced above the craton along with the shortened cratonal cover (miogeocline). Figure 4 also shows that small blocks, such as seamounts or microcontinents, may be sandwiched between the arc and continent and decapitated as the continent attempts to subduct [Ben-Avraham et al., 1982]. The buoyancy constraints of the continental lithosphere cause subduction to cease with the suturing (S1) of the arc complex. Subduction is transferred to a new zone (SZ2) beneath a second lithospheric block or terrane which in turn is eventually sutured along S2. This suturing event may result in reactivation of previously formed thrusts (e.g., T1), and may also produce new crustal thrusts (e.g., T2). In addition, this collisional event may produce some crustal thickening beneath the craton along deep crustal and upper mantle thrusts (e.g., T3) as suggested for the Himalayas and Scottish Caledonides

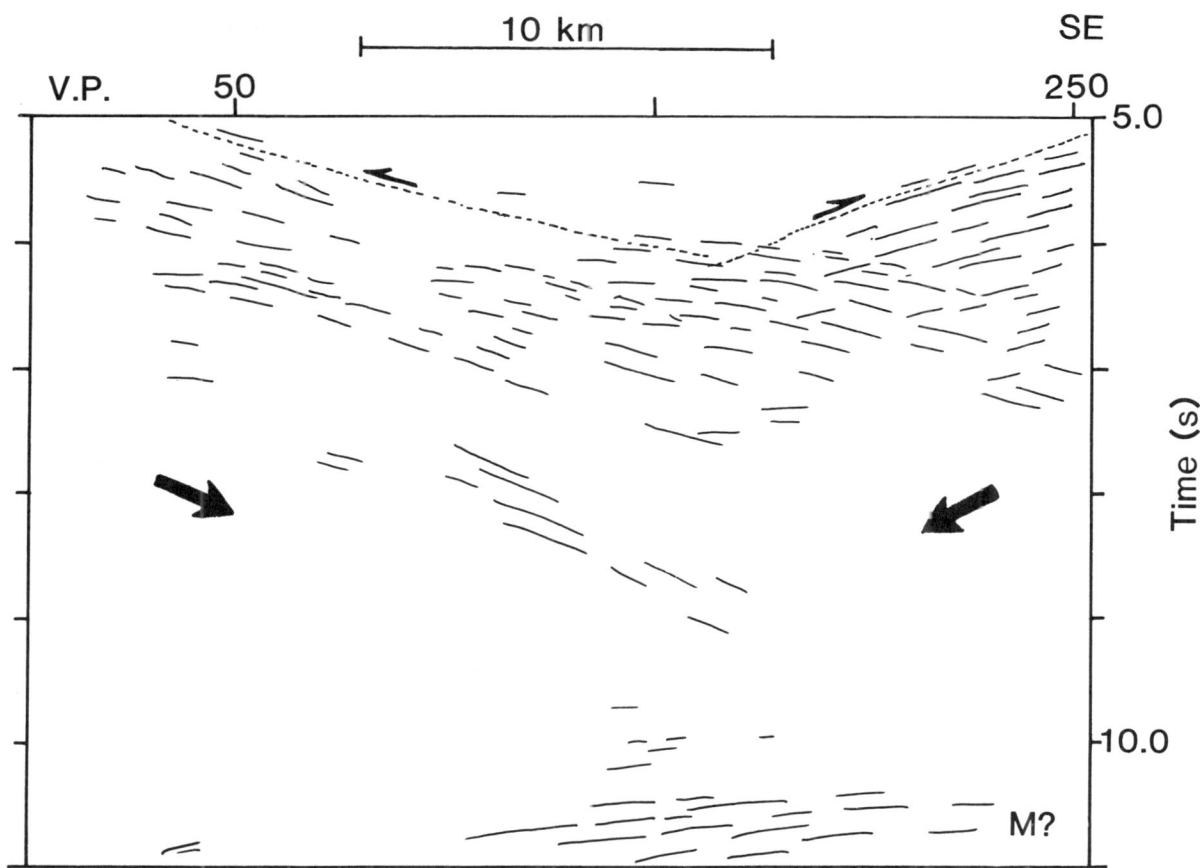

Fig. 3(b). Line drawing interpretation of (a) showing possible detachment surfaces (dotted lines). Large arrows indicate change in underthrusting direction. Presumably, the initial direction was from northwest to southeast.

[Matthews and Hirn, 1984]. The process continues with the development of a third subduction zone (SZ3).

As illustrated, the sequence could be appropriate for the Appalachians with the initial development resulting from the eastward subduction of North American crust beneath the early accreted terranes. In the Himalaya, on the other hand, the subduction of Indian lithosphere was the final event in the sequence; hence, it is clear that the sequence in Figure 4 could be reversed (with S3 forming first, followed by S2 and S1).

Reversal of the process in midstream (for example with SZ3 dipping toward the craton) may produce a crustal structure in which the major boundaries exhibit a reversal of orientation. Figure 3 (near V.P. 150 on COCORP Line 8) illustrates such a reversal in the southern Appalachians beneath the Coastal Plain. Phinney [1982] and Phinney et al. [1984] have also shown a reversal of dip in crustal boundaries on seismic data from Long Island sound in the New England Appalachians. To the west of these reversals, the geologic (and crustal) boundaries have a dominantly east dip; to the east of the reversals the boundaries have a dominantly west dip. Such crustal-scale reversals may represent zones at which lithospheric plate boundaries changed their orientation and dipped toward the craton (and presumably away from a craton or arc in the opposite direction).

The model presented here provides an explanation for the apparent disparity between high-angle (often near-vertical) structures observed at some surface boundaries and the extensive low-angle boundaries observed on many seismic data. Examples include the Green Mountain/Bron-

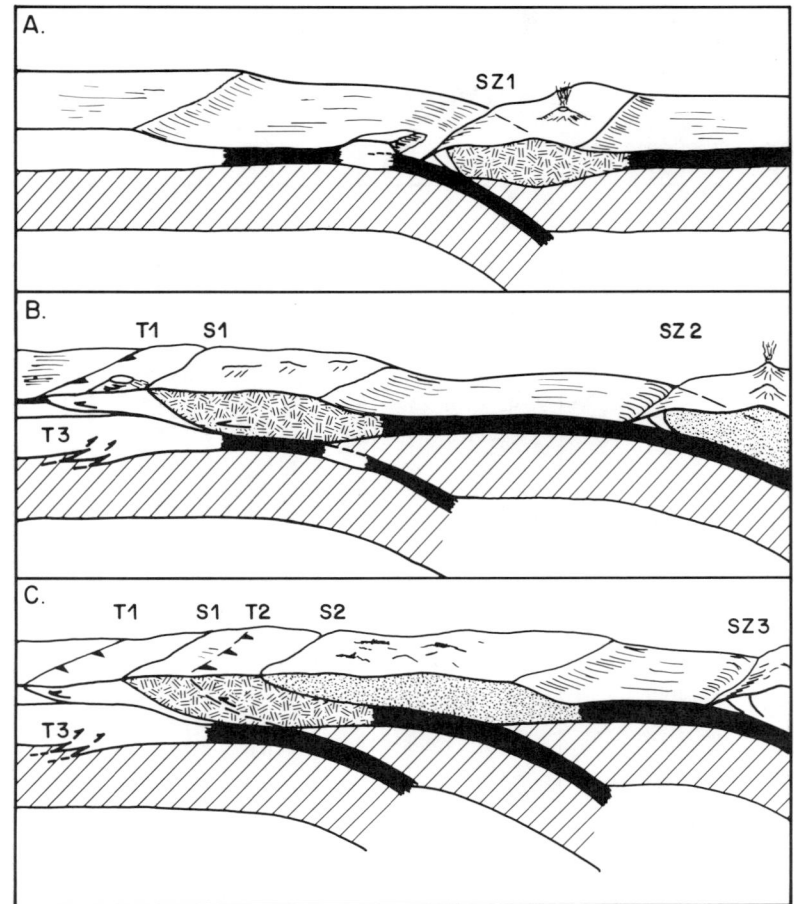

Fig. 4. Suggested model for continental shingling. In A, the subduction of ocean lithosphere (SZ1) beneath an arc causes some features such as seamounts, to be decapitated. Partial subduction of continental lithosphere (B) results in the emplacement of extensive allochthons above the continental margin, and the development of a low angle suture (S1). This event may also produce some thrusting in the lower crust (T3). Cessation of subduction results in the creation of a new subduction zone (SZ2) outboard of the previous one. Juxtaposition of the enlarged continent with SZ2 results in the suturing along S2 verging toward the craton possibly reactivation of pre-existing thrusts (T1), and formation of new thrusts (T2). A new subduction zone (SZ3) then develops outboard of the enlarged continent. The reader should note that the sequence could be reversed, as discussed in the text.

son Hill arc transition in New England and the Inner Piedmont/Carolina slate belt transition in the southern Appalachians. The continuity and low angle geometries of seismic reflectors beneath these important boundaries imply that the steep surface features either flatten with depth into shallow dipping structures or are truncated at depth by these structures.

Such information thus suggests that the crust of collisional orogens may, in some cases, respond to compression in a manner similar to other layered media. Hence, concepts developed in the extensive study of foreland thrust belts may be applicable to crustal-scale deformation in compressive orogens. Such ideas include décollements, duplexing, and ramps and flats [and high angle structures often associated with ramps -- Ando et al., 1983]. The inclined reflectors observed near such transitions as the Green Mountain/Bronson Hill arc (Fig. 2) are likely lower plate features which were subducted beneath the accreted complexes. They are significant in so far as they act as ramp structures which have the effect of localizing stresses and producing high angle upper plate structures.

Furthermore, large amounts of new continental crust may not be formed with this mechanism. Much previously formed crust is eroded from the old continents and then reworked in the arc complexes. The terranes are then shingled onto the continent such that the apparently young near surface rocks may overlie older, subducted rocks. In this way, the subduction process can be related to, and is largely responsible for, the lateral growth of continental crust.

Acknowledgments. This research is partially supported by a grant from the Natural Sciences and Engineering Research Council of Canada. The manuscript was typed by Tari Forrest.

References

Allegré, C. et al., Structure and Evolution of the Himalaya-Tibet orogenic belt, Nature, 307, 17-22, 1984.

Ando, C., F. Cook, J. Oliver, L. Brown and S. Kaufman, Crustal geometry of the Appalachian orogen from seismic reflection studies, Contributions to the Tectonics and Geophysics of Mountain Chains, edited by R. Hatcher Jr., H. Williams, and I. Zietz, Geol. Soc. Amer. Memoir 158, 83-102, 1983.

Bois, C., M. Cazes, B. Damotte, A. Galdeano, A. Hirn, A. Mascle, P. Matte, J. Raoult, and G. Torreilles, Deep seismic profiling of the crust in France: The ECORS project (abstract), International Symposium on Deep Structure of the Continental Crust: Results from Reflection Seismology, Cornell Univ., 9-10, 1984.

Ben-Avraham, Z., A. Nur, and D. Jones, The emplacement of ophiolites by collision, J. Geophys. Res., 87, 3861-3868, 1982.

Coney, P., D. Jones, and J. Monger, Cordilleran suspect terranes, Nature, 288, 329-333, 1980.

Cook, F., D. Albaugh, L. Brown, S. Kaufman, J. Oliver, and R. Hatcher Jr., Thin-skinned tectonics in the crystalline southern Appalachians: COCORP seismic reflection profiling of the Blue Ridge and Piedmont, Geology, 7, 563-567, 1979.

Cook, F., L. Brown, S. Kaufman, J. Oliver, and T. Petersen, COCORP seismic profiling of the Appalachian orogen beneath the Coastal Plain of Georgia, Geol. Soc. Amer. Bulletin, 92, 738-748, 1981.

Green, A., and M. Berry, The third dimension of geology from seismic reflection studies in Canada (abstract), International Symposium on Deep Structure of the Continental Crust: Results from Reflection Seismology, Cornell Univ., 9-10, 1984.

Hatcher, R., Jr., Tectonics of the western Piedmont and Blue Ridge, southern Appalachians, Amer. J. Sci., 278, 276-304, 1978.

Hirn, A., J. Lepine, G. Jobert, M. Sapin, G. Wittlonger, X. Xin, G. Yuan, W. Jing, T. Wen, X. Bai, M. Pandey and J. Tater, Crustal structure and variability of the Himalayan border of Tibet, Nature, 307, 23-25, 1984.

Matthews, D. and A. Hirn, Crustal thickening in Himalayas and Caledonides, Nature, 308, 497-498, 1984.

Monger, J., R. Clowes, R. Price, P. Simony, R. Riddihough, and G. Woodsworth, Continent-Ocean Transect B2: Juan de Fuca Plate to Alberta, Geol. Soc. Amer. Continent-Ocean Transects, in press.

Oxburgh, E., Flake Tectonics and continental collision, Nature, 239, 202-204, 1972.

Petersen, T., L. Brown, F. Cook, S. Kaufman, and J. Oliver, Structure of the Riddleville basin from COCORP seismic data and implications for reactivation tectonics, J. Geology, 92, 261-271, 1984.

Phinney, R., Structure of the Appalachian orogen on the Long Island platform (abstract), Trans. AGU, 63, 1112, 1982.

Phinney R., K. Chowdhury, and J. Leven, Accretional architecture of the continental crust (abstract), International Symposium on Deep Structure of the Continental Crust: Results from Reflection Seismology, Cornell Univ., 71, 1984.

Powell, C. and P. Conaghan, Plate tectonics and the Himalayas, Earth and Planet. Sci. Lett., 20, 1-12, 1973.

Rankin, D., The continental margin of eastern North America in the southern Appalachians: the opening and closing of the proto-Atlantic ocean, Amer. J. Sci., 275-A, 298-336, 1975.

St. Julien, P., A. Slivitsky, and T. Feininger, A deep structural profile across the Appalachians in southern Quebec, Contributions to the Tectonics and Geophysics of Mountain Chains, edited by R. Hatcher Jr., H. Williams and I. Zeitz, Geol. Soc. Amer. Memoir 158, 103-113, 1983.

Stewart, D., The Quebec-western Maine profile: first year results (abstract), International Sympoisum on Deep Structure of the Continental Crust: Results from Reflection Seismology, Cornell Univ., 84, 1984.

Williams, H. and R. Hatcher Jr., Suspect terranes and accretional history of the Appalachian orogen, Geology, 10, 530-536, 1982.

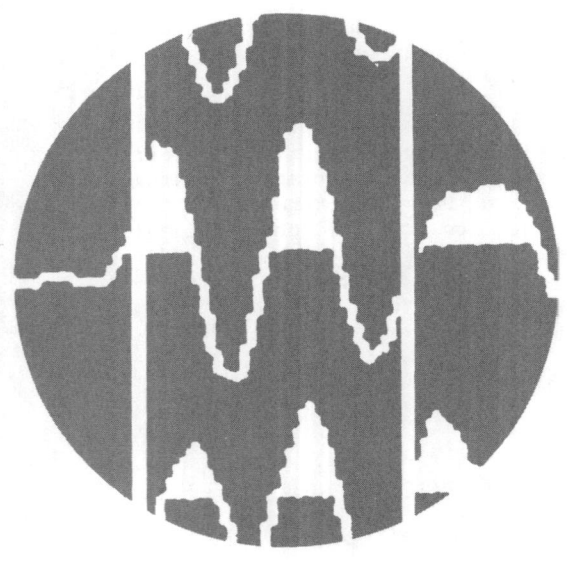

CRUSTAL REFLECTIONS AND CRUSTAL STRUCTURE

Scott B. Smithson, Roy A. Johnson, and Charles A. Hurich,

Department of Geology and Geophysics, Program for Crustal Studies, University of Wyoming, P.O. Box 3006, Laramie, Wyoming 82071

Abstract. The nature of crustal reflectors is still a major question, especially since typical structures in the crystalline crust are so complex that they would generate short, possibly arcuate events. Evidence starting with the Wind River thrust suggests that crustal fault zones (mylonites) may be good reflectors, and recent profiling over mylonites and accompanying detachment faults in the Kettle Dome and the Ruby Mountains, metamorphic core complexes, shows consistent subhorizontal reflections that can be correlated with the lithology of the mylonite zones. Reflections are so abundant within these sillimanite-grade mylonite zones as to resemble a sedimentary terrain. These specially designed reflection experiments conclusively demonstrate for the first time that mylonites are good reflectors. On the other hand, the flanking thrust fault on the Laramie Range only generates a minimal reflection in the relatively shallow zone where brittle deformation predominates. These results show that mylonite zones may be the best reflectors in the crust because of their layered, planar geometry that commonly includes low dip, and crustal scale deformation may thus be mapped seismically. This means that such major features as thrusting, crustal doubling and crustal extension may be recognized through crustal reflection profiling. Crustal-scale duplexes may exist and give the appearance of dipping sedimentary reflections generated from their numerous mylonite zones. The best crustal reflections in the U.S. seem to be related to crustal extension in the western U.S.

Introduction

The question of the cause of crustal reflections is a major one that has remained elusive. Possibly starting with the Wind River thrust (Smithson, et al., 1978), it has become common to attribute crustal reflections to fault zones (Allmendinger et al., 1982; Lynn et al., 1983) without knowing why or whether this is generally correct. The situation becomes more special because in industry seismic reflection interpretations, faults are usually picked on the basis of offset reflectors, diffractions, degraded data quality, and misties but not on fault plane reflections themselves (Bally, 1983). On the other hand, typical structures found in the crust (Berthelsen, 1960) give complex reflection patterns (Smithson et al., 1980, Fig. 5, p. 302). With usual recording and processing techniques, the only part of the reflection wave field that is visible above noise might be short segments where interfaces are near horizontal or the near-horizontal part of arcuate events. This could be an explanation for short subhorizontal reflections that are typical of the reflection response of much of the crust. Another explanation is overprocessing so that noise appears coherent (Howard and Danbom, 1983). But we are left with the paradox that most of the crystalline crust with the exception of some Proterozoic basins (Oliver et al., 1976; Brewer et al., 1981) is so complex that it should not produce strong and continuous reflections. Nevertheless, a number of crustal reflection studies starting with the Wind River thrust (Smithson et al., 1978; 1979), the Bay of Biscay (Montadert et al., 1979), the Southern Appalachians (Cook et al., 1979), the Sevier Desert detachment (Allmendinger et al., 1983), the Outer Isles thrust (Brewer and Smythe, 1984), and to a lesser extent the poorly understood and contradictory data from the Phillips-Anschutz well in Arizona (Reif and Robinson, 1981), show good continuous reflections through much of the crust or, in the case of the Outer Isles thrust, through the entire crust. The reflections are typically strong and multicyclic even in field records (Fig. 1). It is principally the correlation of dipping reflections with surface geology that allowed the interpretation that crustal fault zones were reflective.

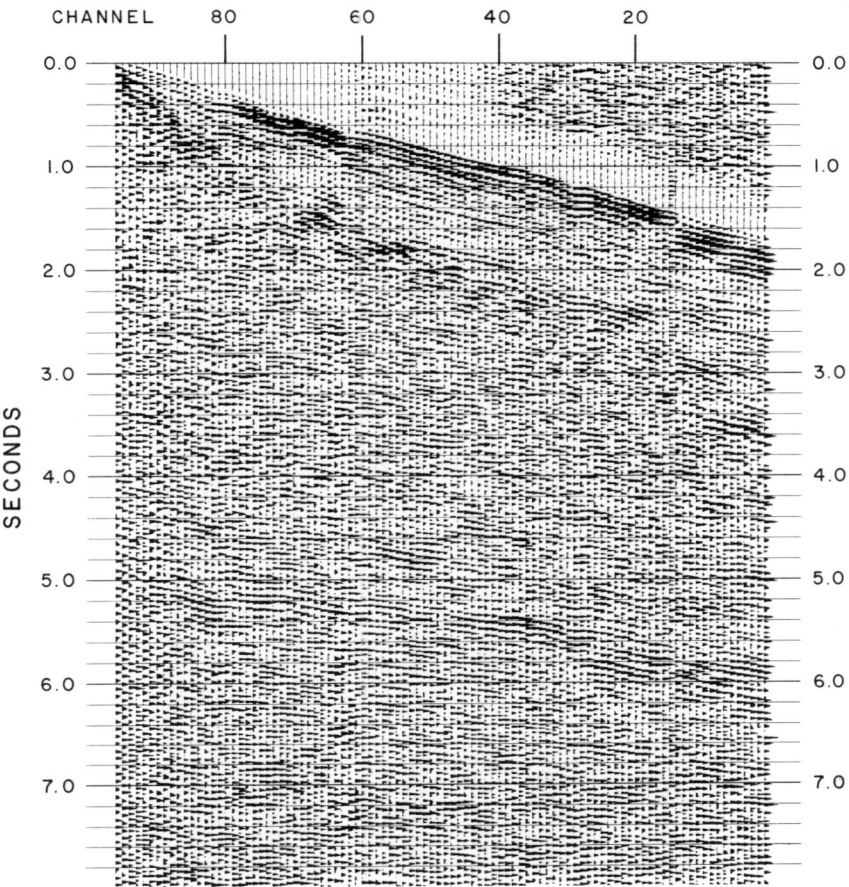

Fig. 1. Correlated common-shot gather from COCORP Wind River Line 1A, Sta. 170. Fault-zone reflections from 5 to 6 s are multicyclic indicating layering. Note the high amplitude and character of the reflection even in single-fold data. Similar reflections are found in common-source gathers from other areas containing mylonites.

Mylonite Zone Reflectivity

These results have focused attention on mylonites, the product of shearing in the crystalline crust, as the probable source of fault-zone reflections. Mylonites are rocks that are distinctly layered on all scales and that have a strong fabric or preferred orientation of minerals because of shearing (Simpson, 1982, Fig. 10B, p. 513). They range in thickness from centimeters to several kilometers. Mylonites should be highly reflective for the following reasons: 1) compositional layering, 2) fabric, 3) planar and continuous geometry, 4) chemical alteration. Of the above four reasons, compositional layering and planar geometry are probably by far the most important, although the importance of the fabric is not yet fully understood (Jones and Nur, 1982). Among compositional variations in layering, variations in fabric (anisotropy) undoubtedly also exist and contribute to reflectivity. The importance of layering in generating reflections was pointed out by Fuchs long ago (1969). Velzeboer, (1981) showed that layering resulting in average reflection coefficients as low as 0.03 can generate usable reflections in Gulf Coast sediments, and Hurich et al., (in press) suggested that layering can increase reflection amplitude by a factor of 2 to 3. If a mylonite zone forms in rocks of variable composition, then the different rocks are sheared and drawn into the mylonite zone, and a series of discontinuous layers of different composition will comprise the mylonite. In the general case, the mylonite will have layers of varying composition, and a fabric is

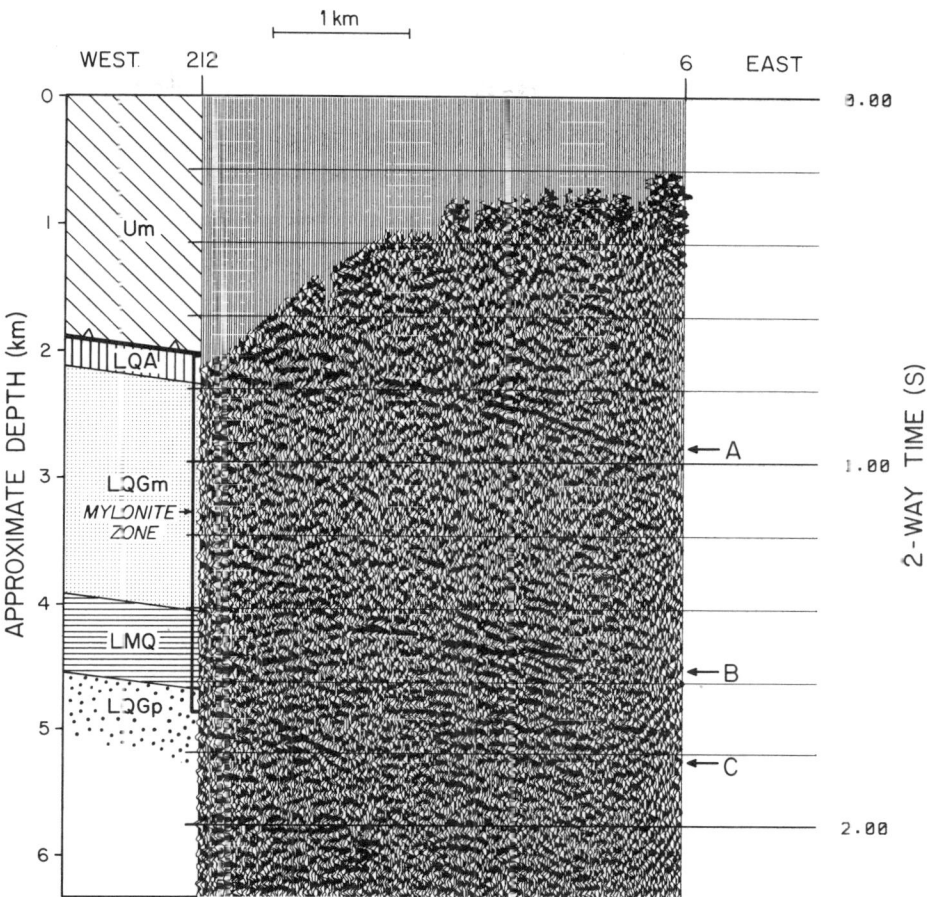

Fig. 2. 48-fold CDP seismic section from the east flank of the Kettle dome metamorphic core complex, Washington. A: Reflections from cataclastic Kettle River detachment fault and top of mylonite zone; A-B: Transparent zone correlative with homogeneous mylonitic gneiss; B-C: Reflections from interlayered mylonitic marble, quartzite, paragneiss and schist. The more reflective parts of the mylonite zone can be correlated with compositional layering. The number of reflections is unusual for crystalline crust. Um: Low-grade phyllite, marble and greywacke in the upper plate of the Kettle River detachment fault; LQA: Mylonitic amphibolite and quartzite; LQGm: Mylonite quartzo-feldsphthic gneiss; LMQ: Mylonitic marble, quartzite, paragneiss and schist; LQGp: Porphyroclastic gneiss with mylonitic overprint near the top. Rocks of the mylonite zone contain upper amphibolite facies assemblages.

generated within the layers by ductile deformation. Laboratory studies of rock velocity (Birch, 1960; Christensen, 1965; Christensen and Fountain, 1975; Kern and Fakhimi, 1975; Meissner and Fakhimi, 1977) all show a strong dependence of velocity on composition in igneous and metamorphic rocks, but these studies also demonstrate that the metamorphic rocks definitely show velocity anisotropy as did the recent study of Jones and Nur (1982). Velocity anisotropy is particularly related to concentration and preferred orientation of biotite or other phyllosilicates, minerals which exhibit extreme velocity anisotropy (Alexandrov and Rhyzhova, 1961). Wong et al. (1982) relied on the effect of layering to demonstrate mylonite reflection response in a synthetic seismogram that showed the same character as the reflections from the Wind River thrust; Jones and Nur (1982) proposed first very special conditions such as fluid pressure before finally recognizing the effect of layering (Jones and Nur, 1984). Thus, while the relative importance of velocity anisotropy in causing mylonite reflectivity is still uncertain, abundant evidence exists that thick (about 100 m) compositionally layered mylonites are good reflectors.

Depending on the signal-to-noise ratio and velocity contrasts, a mylonite zone with a

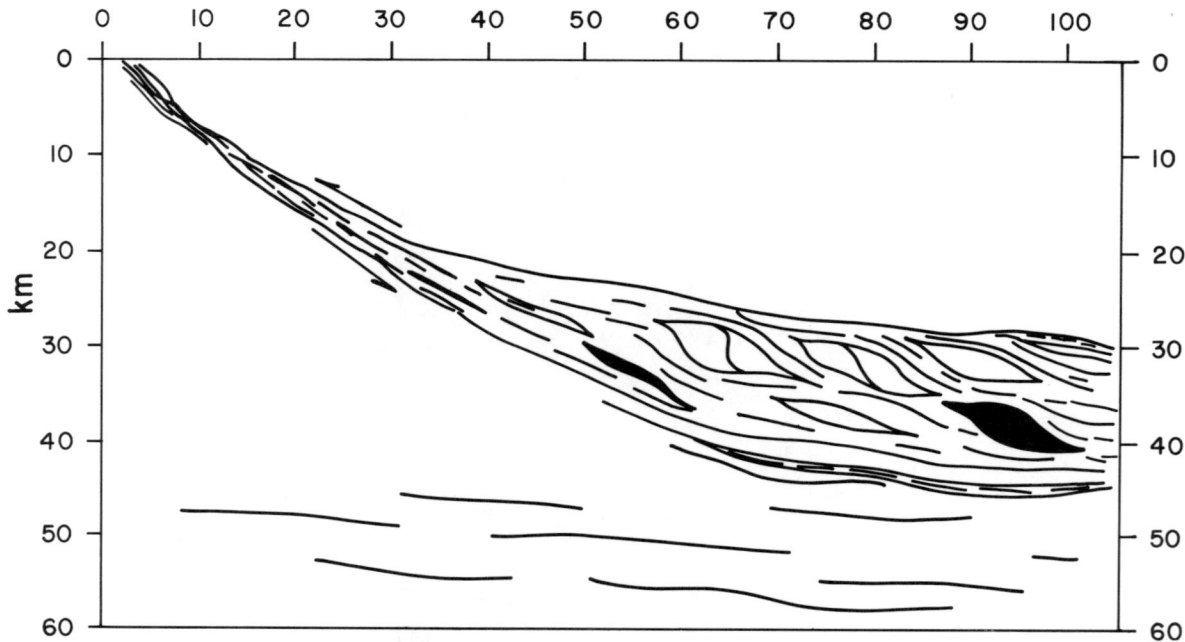

Fig. 3. Hypothetical diagram showing shallow brittle fault passing into restricted mylonite zone in middle crust, passing into diffuse mylonite zone in deep crust. Adapted from drawing in the ETH-Zürich Geological Museum.

minimum thickness of 30-50 m might generate detectable reflections. While our studies concentrate on layered mylonite zones, readers should remember that any compositionally layered sequence of gneisses and/or schists of similar dimensions would produce similar reflections. Mylonite zones, however, are more likely to be planar or subplanar over greater distances and thus produce the most continuous reflections. The proof of the hypothesis lies in reflection profiling from a mylonite outcrop and tracing any possible reflection into depth. This test has recently been accomplished by the University of Wyoming seismograph crew in two different areas, the Kettle dome, Washington and the Ruby Mountains, Nevada. Both of these areas consist of metamorphic gneiss complexes, thick (about 1 km or more) mylonite zones, and overlying detachment faults (Cheney, 1980; Snoke, 1980). The mylonite zone in the Kettle dome is about 3 km thick and consists of such variable rock types as quartzo-feldspathic gneiss, quartzite, marble and amphibolite. The 48-fold CDP seismic section over this mylonite, shows an extraordinary set of reflections and an encouraging correspondance with the lithologies in the mylonite zone (Fig. 2). The first continuous reflection comes from the detachment fault or just beneath it. Reflections within the mylonite zone, itself, include a relatively transparent zone corresponding to the quartzo-feldspathic gneiss and abundant reflections corresponding to the part with interlayered rock types (Fig. 2) based on up-dip correlation with the exposed mylonite. The seismic section, in fact, shows such numerous reflections as to resemble a seismic section from sedimentary rocks. These reflections attributed to subhorizontal layering through the upper 15 km (5 s) of the crust may be caused by a deeper mylonite zone or the inverted limb of a crystalline nappe similar to the Shuswap terrain further north in Canada (Brown and Reed, 1983). In any case, the seismic section is extremely unusual for a high-grade crystalline terrain in its abundance of reflections. Similar results showing even more reflections have just been obtained from the mylonite zone in the Ruby Mountains in Nevada (Hurich et al., 1984).

The COCORP seismic section from the Laramie Range (Line 3) provides other important conclusions about mylonite reflectivity because it represents the other end of the spectrum for fault zone reflectivity (Johnson and Smithson, in press). In the Laramie Range, the simple Laramide arch that exposes a Precambrian core is overturned and thrust to the east; the deformation took place under a sedimentary overburden of about 4 km, and dip slip on the major thrust fault is about 1200 m. A reprocessed version of Line 3 shows a short, fairly weak dipping reflection that can be projected to the surface trace of the major

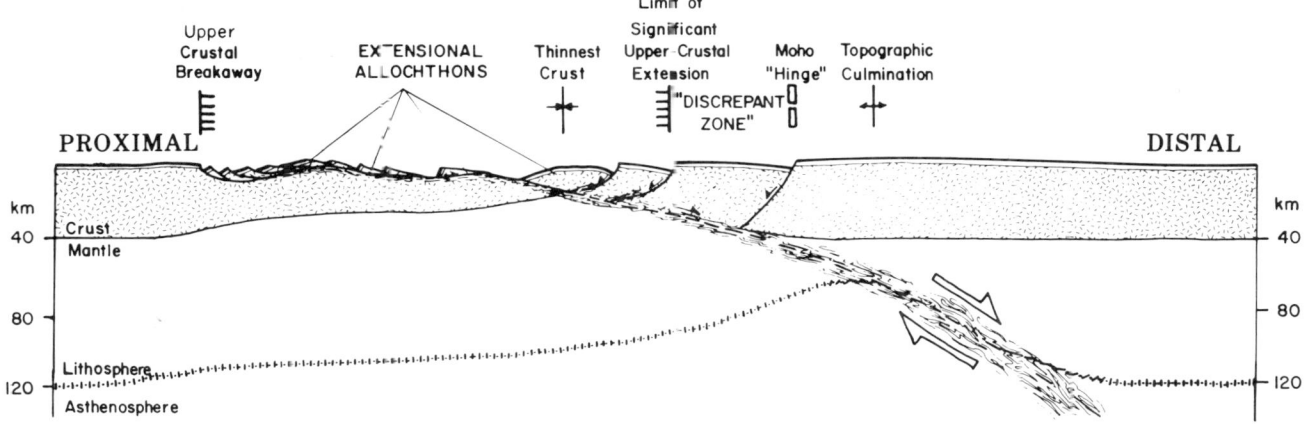

Fig. 4. Illustration of how extensional listric faulting might extend into asthenosphere allowing extreme extension of the lithosphere. An extensional fault with such extreme movement might generate a thick mylonite that is a good reflector, allowing testing of the hypothesis by crustal reflection profiling. (After Wernicke, 1985).

thrust fault (Johnson and Smithson, this volume, Fig. 5) so that this single-cycle reflection is attributed to the thrust fault near the surface where Precambrian anorthosite is overthrust onto Paleozoic sedimentary rocks. The important feature of this data set is that we do not see a continuation of the thrust fault reflection into the crystalline crust. Because slip of 1200 m is probably enough to form a mylonite zone about 100 m wide (Ramsay and Allison, 1979; Simpson, 1982), we attribute the lack of a thrust reflection to the shallow level of exposure view, the thrust fault cuts across Paleozoic sedimentary rocks into the Precambrian where it consists of a crushed and altered zone about 10-100 m wide (Johnson and Smithson, in press). However, single-fold reflection data collected by the University of Wyoming suggests that the fault may be reflective at a depth of about 8 km (Iltis, 1983). Thus the Wind River thrust and Laramie Range thrust form extreme examples; the Wind River thrust showing about 20 km of slip and formed under 12 km of cover is a good reflector; the Laramie Range thrust showing 1200 m of slip and formed under 4 km of cover is not, at least not in its shallower levels.

A model of crustal scale faulting is thus proposed (Fig. 3) that shows brittle fracturing at shallow depth, a discrete mylonitized zone at intermediate depth, and more homogeneous ductile deformation at great depth. A zone of cataclasis would grade from fracturing into mylonite. Homogeneous ductile deformation implies widespread recrystallization (metamorphism) and accompanying resetting of at least some radiological clocks.

We must conclude then that mylonite zones containing compositional layering are excellent reflectors within the crystalline crust; we do not, however, yet know the efficacy of fabric alone in generating usable reflections. We will thus discuss the tectonic implications of mylonite reflectivity.

Tectonic Consequences

A number of hypotheses concerning crustal scale deformations have been proposed related to crustal thickening (Hsu, 1979; Giese and Pavlenkova, 1976; Trümpy, 1980; Cook et al., 1979; Newton and Perkins, 1982) and crustal thinning (Keen and Hyndman, 1979; Wernicke, 1985; Allmendinger et al., 1983; Brewer and Smythe, 1984) related to plate motions. All of these hypotheses imply long-distance movement of crustal- or subcrustal-sized blocks of continental crust and possibly underlying uppermost mantle.

Crustal attenuation may take place along a continental margin (Keen and Hyndman, 1979; Brewer and Smythe, 1984) or on a continental interior (Wernicke, 1985; Allmendinger et al., 1983) during an extensional episode. In either case, extensions and concomitant thinning on the order of 100 to 200% may be achieved (Wernicke, 1985). Listric normal faulting would seem to be the mechanism by which this extension is accomplished in at least some cases (Brewer and Smythe, 1984; Allmendinger et al., 1983). Crust on the passive continental margin may be thinned to about 15 km (Keen and Hyndman, 1979, Fig. 8, p. 724). Strain of this amount should be concentrated in zones that develop mylonites which may be detectable in reflection profiles as illustrated by the studies of Montadert et al., (1979) and Brewer and Smythe (1984). Thus the exact mechanism of crustal thinning

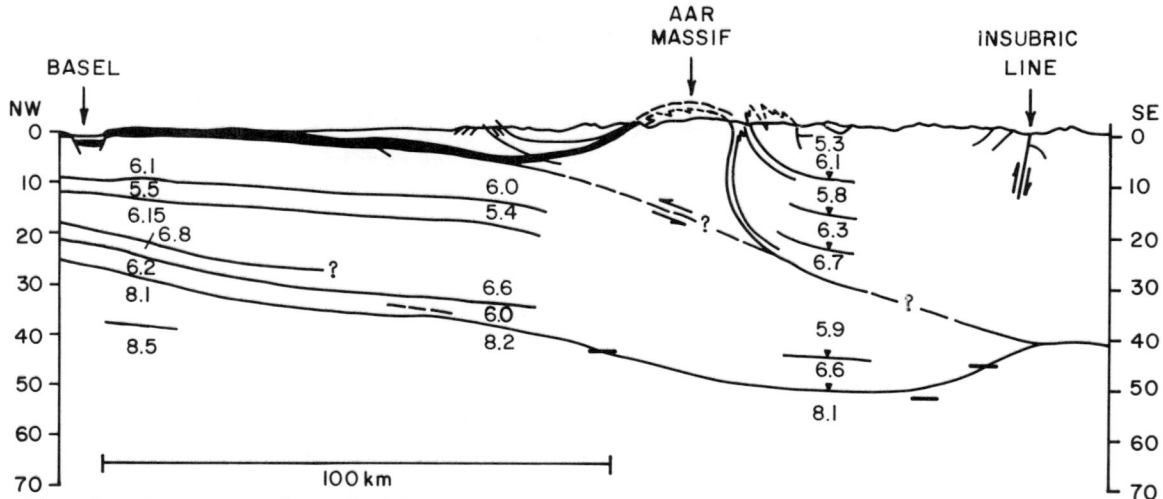

Fig. 5. Interpretation of thickened continental crust under the Swiss Alps. Crust is thickened by crustal doubling as one crustal mass is thrust underneath another block of continental crust. Transport of one crustal block past another should generate a moderately dipping complex mylonite zone that is reflective, allowing testing of the hypothesis by crustal reflection profiling (After Hsu, 1979).

may be quantitized through interpretations of crustal reflection profiles. Wernicke (1985) has proposed a mechanism of crustal denudation and lithospheric thinning along listric extensional faults that penetrate into the top of the asthenosphere (Fig. 4). To test this hypothesis, these listric faults might be mapped seismically through much of the crust as is suggested by the bright multicyclic reflector that continues to a depth of at least 18 km in the crust in the Sevier Desert COCORP section (Allmendinger et al., 1983). The reason for its disappearance is not clear; i.e., whether it passes into a broad zone of homogeneous deformation or whether it disappears because of recording and/or processing problems. A fascinating possibility is that shear zones might also be followed through at least part of the upper mantle (Brewer and Smythe, 1984; Warner, 1985).

The question of crustal thickening is accentuated by the possibility that, for continent-continent collision, the edges of the colliding continental block are thin passive margins. Hsu (1979) has proposed a generalized scheme for explaining the 50-km thick Alpine crust by underthrusting of a northern continental block beneath a southern continental block (Fig. 5). This interpretation pictures two continental blocks of approximately normal (30 km) crustal thickness separated by a gently (20-30°) dipping suture zone. Such a suture would be marked by a mylonite zone on a grand scale, and in fact, the mylonite might well be complex and anastomozing around kilometer-sized rhombs. The reflection pattern for anastomozing mylonite zones could, however, be very complex and discontinuous. As in the case of strong crustal extension, well developed mylonite zones should be present and act as crustal reflectors. One interpretation (Trümpy, 1980) indicates that the Pennine zone together with the internal massifs in the Alps consists of a stack or crustal-scale duplex of attenuated crustal slabs that may have once been separate microcontinents in a Pennine Sea. This interpretation is promoted by the ophiolites and associated ocean-floor sediments (Bündnerschiefer) mapped as envelopes around and intercalations between slices of crystalline nappes. If the crystalline nappes of the Alps are formed of separate thinned crustal slabs, their boundaries might well be marked by seismically reflective mylonite zones containing greater or lesser amounts of Bündnerschiefer. Similarly if attenuated continental crust is thickened to normal or to orogenic thicknesses, mylonite zones should have formed in between the crustal slabs as they were stacked (Fig. 6). In addition, suture zones should typically be expected to be gently dipping rather than vertical, and docking of exotic terrains to form continental accretion should take place along moderately dipping sutures if the terrains are stacked by underthrusting. Clearly all these major questions in continental deformation and growth may lead to formation of seismically detectable mylonite zones. Thus these zones would form major targets for reflection seismology as a means to distinguish between various tectonic hypotheses.

Island arcs constitute the generally

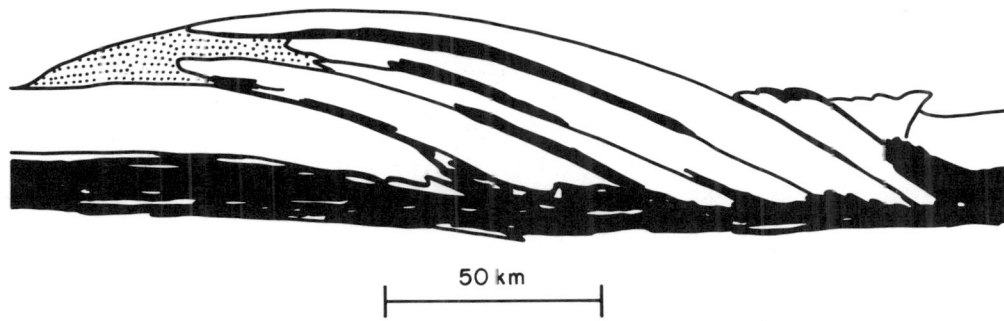

Fig. 6. Hypothetical sketch of crustal duplex formed by stacking of attenuated slabs of continental crust along a former continental margin. Could represent the Pennine zone of the Alps. Mylonites could mark the boundary zones between the stacked continental slabs and allow resolution of this structure in seismic profiles (Adapted from Trümpy, 1980).

accepted means of crustal growth, and they are closely associated with Benioff zones, which are the sites of strong simple shear. Numerous thrust faults are found in the sediments in the forearc basin, continental rise, and trench (Hamilton, 1977; Bally, 1983) and similar thrusts are expected in the more consolidated rocks of the arc. These thrusts are roughly (allowing for oblique subduction) parallel to the Benioff zone or are conjugate to the major thrusts. At depth in the igneous and metamorphic rocks of the arc, mylonite zones might form and be common. While the role of plate motions and even vertical versus horizontal tectonics in the Precambrian is under some dispute, some Precambrian crustal boundaries are marked by mylonite zones which may be kilometers thick. Several examples of this are the Nagssugtoqidian deformation in West Greenland (Escher et al., 1976) and the Cheyenne belt in SE Wyoming (Johnson et al., 1984). The Nagssugtoqidian zone is a shallowly dipping complex mylonite zone tens of kilometers thick and is a late-Archean early Proterozoic deformation zone superposed on older Archean rock by thrusting. It marks the border between old Archean crust and Proterozoic crust and thus marks a zone of crustal remobilization and accretion. The Cheyenne belt is a steeply dipping mylonite 1-7 km thick. This zone marks the border between the Archean Wyoming province and Proterozoic rocks to the south believed to have been accreted as island arcs (DePaolo, 1981; Karlstrom and Houston, 1984). Both of these thick mylonites are associated with crustal accretion and possibly with island arcs, and they would most likely form good reflectors.

Another possible example of Archean mylonite formation from thrusting comes from the Archean of Minnesota. Here the Great Lakes tectonic zone marks the boundary between the ancient Minnesota gneiss terrain to the south and a sequence of younger Archean greenstone belts that may represent vestiges of Precambrian island arcs. COCORP crustal reflection data shows a reflection dipping moderately northward underneath the greenstone terrain at the approximate position of the boundary (Fig. 7). Gibbs et al. (1984) have interpreted this reflection as a thrust fault, and Pierson (1984) found evidence for recumbent folding in the seismic data and used magnetic interpretation to demonstrate that older Archean crust has different magnetic properties over a thickness of about 20 km. Although crustal extension cannot be ruled out, the results suggest the zone is a major crustal boundary marked by a mylonite formed by compressive movements to generate the recumbent folding and mylonite. This could therefore represent late Archean crustal accretion by overthrusting and fusion of island arcs to the ancient Archean nucleus.

Mylonite reflectivity has important implications for interpretation of a thick band of dipping reflections underneath the outer Piedmont in the COCORP Southern Appalachian data set (Fig. 8). This band of reflections is the most correlatable feature in seismic sections in the Appalachians and was initially interpreted as a thick wedge of sediments overthrust by 500 km of continental crust during continent-continent collision during the Alleghenian (Cook et al., 1979; Cook, 1984). Iverson and Smithson (1983) later proposed in an alternate interpretation that this zone consists of crystalline rock that represented a root zone along the extension of the Kings Mountain belt, a zone of high strain (mylonite) in gneisses. A better term, however, would be to call this feature a crustal duplex. We suggest that the dipping reflections in the COCORP line are generated from mylonites and their intervening packets of less strained gneissic rocks so that a large-scale duplex is formed in the upper half

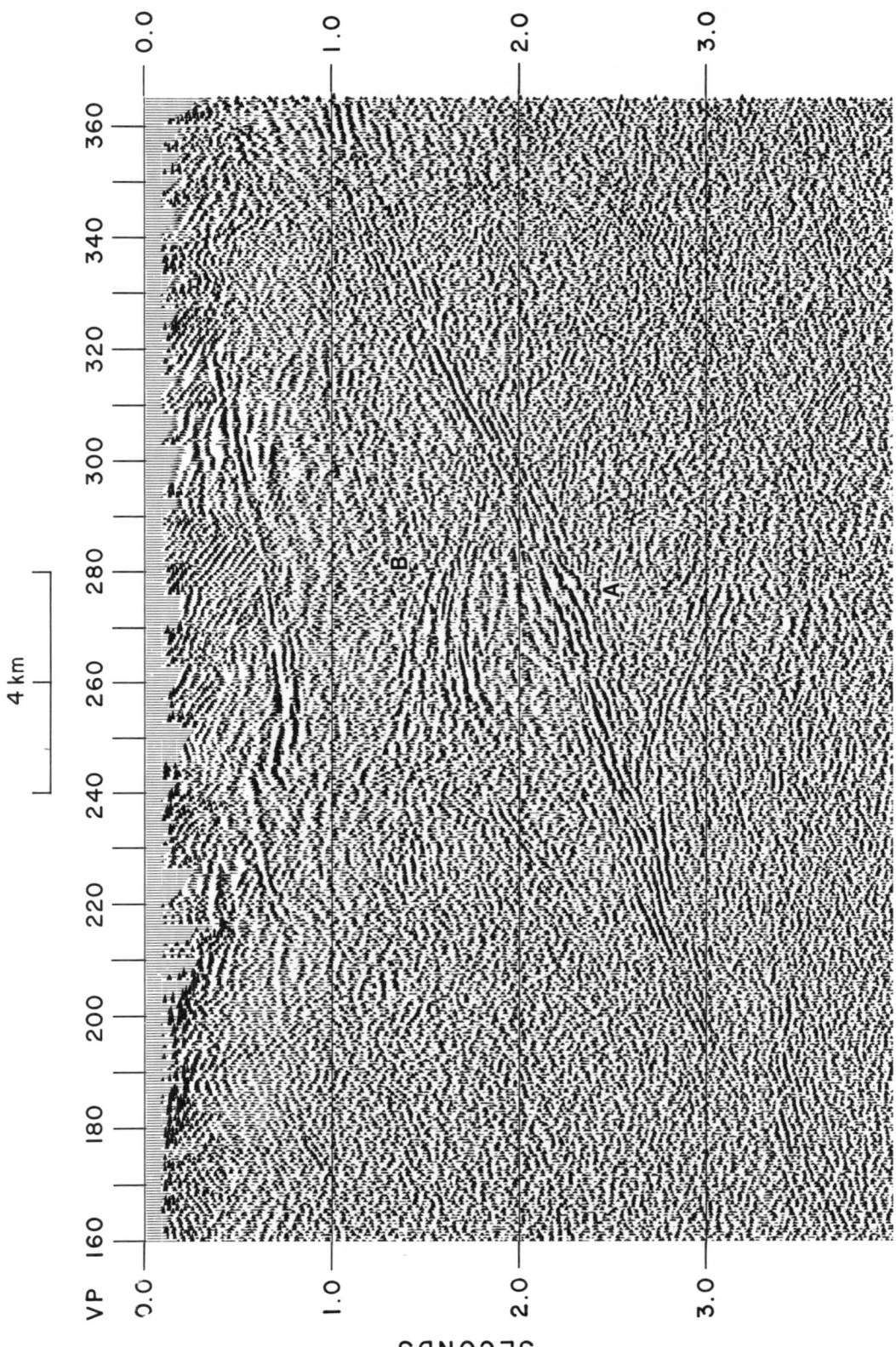

Figure 7. Reprocessed seismic section of COCORP Line 3 in Minnesota. Strong dipping reflection may come from a mylonite zone that marks the crustal boundary between an overlying late Archean greenstone belt terrain and the underlying ancient Minnesota gneiss terrain. (After Pierson, 1984).

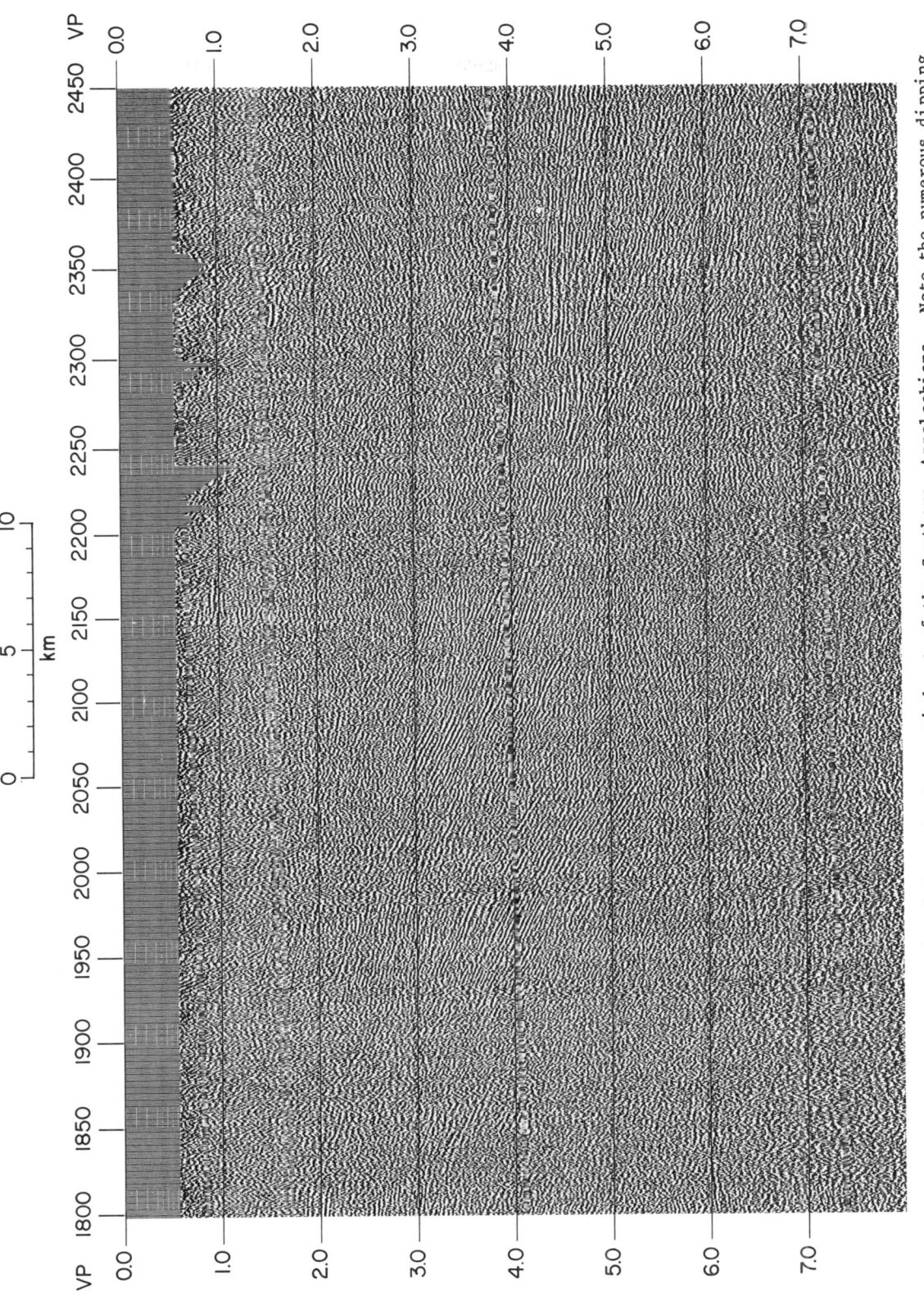

Figure 8. Reprocessed COCORP Line 1 from the Inner Piedmont of the Southern Appalachians. Note the numerous dipping events that have been called a wedge of sediments on the continental margin in other interpretations (Cook et al., 1979). (After Iverson and Smithson, 1983).

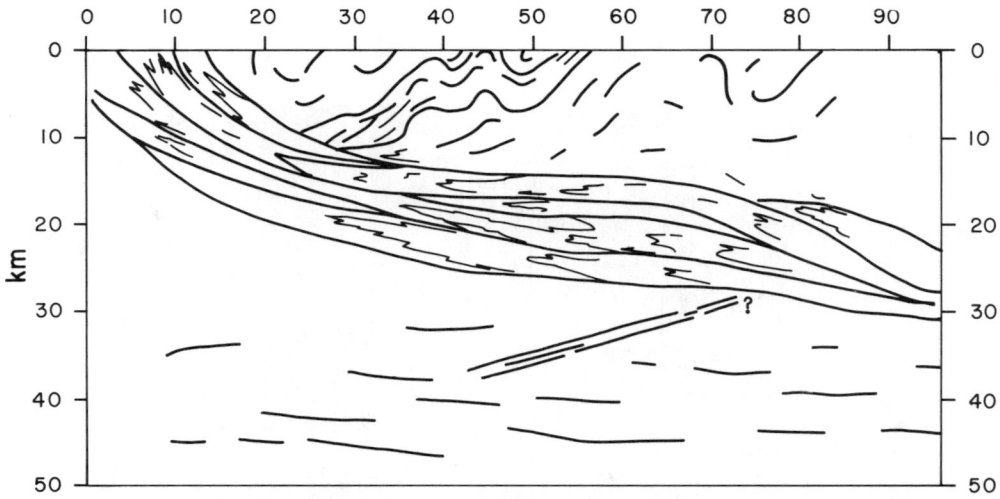

Fig. 9. Interpretation of dipping reflections as a crustal duplex zone in which mylonites separate tectonized slabs of gneiss.

of the crust (Fig. 9) i.e., deeper erosion would reveal a complex stack of variably strained crystalline thrust sheets, a crustal duplex. The results of seismic profiling over mylonites (Kettle dome and Ruby Mountains) demonstrate such strong and numerous reflections that the need for sedimentary reflectors in this area is obviated. While not as sensational, the above hypothesis is also more consistent with other structural and tectonic constraints. The deep crust below the zone of reflections (crustal duplex) may have been deformed by more homogeneous ductile strain. Butler and Coward (1984) have recently postulated a similar structure beneath the Moine thrust in Scotland.

A pattern of highly reflective mylonite zones and unusually reflective crust is beginning to emerge in metamorphic core complexes, areas where strong extension has been the most recent deformation. Such good subhorizontal reflections do not seem to be as common in areas affected solely by compression. It thus seems plausible that the reflective crust and even Moho in the western U.S. may be produced by extension that generated subhorizontal ductile deformation in the deeper parts of the crust.

Conclusions

Mylonite zones marking faults in the crystalline crust may be reflective where they are thick enough, and they are probably the best crustal reflectors because of their planar, layered, continuous geometry. In fact, they may be such good reflectors as to resemble a sedimentary section (Fig. 3) and might even be confused with sedimentary successions from their reflection pattern. Fault zones at shallow levels within the crystalline crust may develop as fractures and may, therefore, not be reflective. Mylonites provide us with a means to trace crustal deformation by applying the techniques of reflection seismology to outline crustal blocks and consequently gives us a means to test tectonic hypotheses. As a word of warning, however, if crustal slabs are commonly thinned by extension along mylonite zones and then thickened by stacking along mylonite zones, we might expect fairly continuous crustal reflections to be more common than they are. The reason for lack of continuous reflectors in most segments of continental crust is not clear at present. Recording problems in data acquisition undoubtedly contribute to this observation but the extent of recording effects is not presently known. Precambrian crustal boundaries are marked by mylonites that are reflective and that indicate accretion in the Archean associated with thrusting and by comparison may be interpreted based on accretion of island arcs. Mylonites and other good subhorizontal reflectors in the deeper crust of the western U.S. may be generated by ductile flow related to extension.

Acknowledgments. Financial support was received from U.S. National Science Foundation grants EAR-8306542 and EAR-8300659. M.C. Humphreys, Allen Tanner, and Robert Tweed carried out the seismic recording. Barbara Cox receives thanks for help with many aspects of the projects. Processing was carried out on the DISCO VAX 11/780 computer system in the Program for Crustal Studies.

References

Alexandrov, K.S. and T.V. Rhyzhova, Elastic properties of rock forming minerals 2. Layered silicates: Bull. Acad. Sci., USSR, Geophys. Ser., 9, 1165-1168, 1961.

Allmendinger, R.W., J.A. Brewer, L.D. Brown, S. Kaufman, J.E. Oliver and R.S. Houston, COCORP profiling across the Rocky Mountain front in southern Wyoming, Part 2: Precambrian basement structure and its influence on Laramide deformation: Geol. Soc. Amer. Bull., 93, 1253-1263, 1982.

Allmendinger, R.W., J.W. Sharp, D. Von Tish, L. Serpa, L.D. Brown, S. Kaufman, J.E. Oliver, and R.B. Smith, Cenozoic and Mesozoic structures of the eastern Basin and Range province, Utah, from COCORP seismic-reflection data: Geology, 11, 532-536, 1983.

Bally, A.W., Seismic expression of structural styles: Amer. Assoc. Petroleum Geologists Studies in Geology Series, No. 15, 3, 1983.

Berthelsen, A., Structural studies in the Precambrian of western Greenland: Med. Gronland, 123, 222 p., 1960.

Birch. F., The velocity of compressional waves in rocks to 10 kilobars, 1: Jour. Geophys. Res., 65, 1083-1102, 1960.

Brewer, J.A., L.D. Brown, D. Steiner, J.E. Oliver, S. Kaufman, and R.E. Denison, Proterozoic basin in the southern Midcontinent of the United States revealed by COCORP deep seismic reflection profiling: Geology, 9, 569-575, 1981.

Brewer, J.A. and D.K. Smythe, MOIST and the continuity of crustal reflector geometry along the Caledonian-Appalachian orogen: Jour. Geological Society: 141, 105-120, 1984.

Brown, R.L. and P.B. Read, Shuswap terrane of British Columbia: A Mesozoic "core complex,": Geology, 11, 164-168, 1983.

Butler, R.W.H., and M.P. Coward, Geological constraints, structural evolution and deep geology of the northwest Scottish Caledonides: Tectonics, 3, 347-366, 1984.

Cheny, E.S., Kettle dome and related structures of northeastern Washington: in M.D. Crittenden, Jr., P.J. Coney, and G.H. Davis, eds., Cordilleran metamorphic core complexes: Geol. Soc. Amer. Memoir 153, 463-484, 1980.

Christensen, N.I., Compressional wave velocities in metamorphic rocks at pressures to 10 kilobars: Jour. Geophys. Res., 70, 6147-6164, 1965.

Christensen, N.I. and D.M. Fountain, Constitution of the lower continental crust based on experimental studies of seismic velocities in granulite: Geol. Soc. Amer. Bull., 86, 227-236, 1975.

Cook, F.A., Geophysical anomalies along strike of the Southern Appalachian Piedmont: Tectonics, 3, 45-62, 1984.

Cook, F.A., D.S. Albaugh, L.D. Brown, S. Kaufman, J.E. Oliver and R.D. Hatcher, Thin-skinned tectonics in the crystalline southern Appalachians: COCORP seismic reflection profiling of the Blue Ridge and Piedmont: Geology, 7, 563-567, 1979.

De Paolo, D.J., Neodynium isotopes in the Colorado Front Range and crust-mantle evolution in the Proterozoic: Nature, 291, 193-196, 1981.

Escher, A., K. Sorensen, and H.P. Zeck, Nagssugtoqidian mobile belt in West Greenland in A. Escher and W.S. Watt, eds., Geology of Greenland: Grøn. Geol. Undersøg., 76-95, 1976.

Fountain, D.M., C.A. Hurich, and S.B. Smithson, Seismic reflectivity of mylonite zones in the crust: Geology, 12, 195-198, 1984.

Fuchs, K., On the properties of deep crustal reflections: J. Geophys., 35, 133-149, 1969.

Gibbs, A.K., B. Payne, T. Letzer, L.D. Brown, J.E. Oliver and S. Kaufman, Seismic reflection study of the Precambrian crust of central Minnesota: Geol. Soc. Amer. Bull., 95, 280-294, 1984.

Giese, P. and N.I. Pavlenkova, Some particularities of crustal structure in young orogenic systems: Geol. Rundsch., 65, 1109-1129, 1976.

Hamilton, W., Subduction in the Indonesian region: in M. Talwani and W.C. Pitman III, eds., Island arcs, deep sea trenches and back-arc basins: Amer. Geophys. Union, Maurice Ewing Series, 1, 15-32, 1977.

Howard, M.S. and S.H. Danbom, Random noise example in A.W. Bally, ed., Seismic expression of structural styles: Amer. Assoc. Petroleum Geologists Studies in Geology Series, No. 15, 1, 1983.

Hsu, K.J., Thin-skinned plate tectonics during neo-Alpine orogenesis: Am. J. Sci., 279, 353-366, 1979.

Hurich, C.A., M.C. Humphreys, A.W. Snoke, R.A. Johnson, D.M. Fountain, and S.B. Smithson, Crustal reflection profiling in the metamorphic core complex of the Ruby Range, Nevada (Abs.): EOS, Trans. Amer. Geophys. Union, 985, 1984.

Hurich, C.A., S.B. Smithson, D.M. Fountain, and M.C. Humphreys, Mylonite Reflectivity: Evidence from Kettle Dome, Washington: Geology, in press.

Iltis, S.T., Processing and interpretation of seismic reflection data from the Precambrian of the central Laramie Range, Albany County, Wyoming: Unpublished M.S. Thesis, Univ. of Wyoming, Laramie, 94 p., 1983.

Iverson, W.P. and S.B. Smithson, Reprocessing and reinterpretation of COCORP southern Appalachian profiles: Earth Planet. Sci. Lett., 62, 75-90, 1983.

Johnson, R.A., K.E. Karlstrom, S.E. Smithson, and R.S. Houston, Gravity Profiles across the Cheyenne Belt, a Precambrain Crustal Suture in

Southeastern Wyoming: Jour. of Geodynamics, 1, 445-471, 1984.

Johnson, R.A. and S.B. Smithson, Thrust faulting in the Laramie Mountains, Wyoming from reanalysis of COCORP data: Geology, in press.

Johnson, R.A. and S.B. Smithson, Interpretive Processing of Crustal Seismic Reflection Data: Examples from Laramie Range COCORP Data: Int'l Symposium on Deep Structure of the continental Crust: Amer. Geophys. Union, Geodynamics Series, in press.

Jones, T.D. and A. Nur, Seismic velocity and anisotropy in mylonites and the reflectivity of deep crustal faults: Geology, 10, 260-263, 1982.

Jones, T.D. and A. Nur, Nature of seismic reflections from deep crustal fault zones: Jour. Geophys. Res., 89, 3153-3171, 1984.

Karlstrom, K.E. and R.S. Houston, The Cheyenne belt: analysis of a Proterozoic suture in southern Wyoming: Precambrian Research, 25, 415-446, 1984.

Keen, C.E. and R.D. Hyndman, Geophysical review of the continental margins of eastern and western Canada: Canad. J. Earth Sci., 16, 712-747, 1979.

Kern, H. and M. Fakhimi, Effect of fabric anisotropy on compressional wave propagation in various metamorphic rocks for the range 20-700°C at 2 kbars: Tectonophysics, 28, 227-244, 1975.

Lynn, H.B., S. Quam, and G.A. Thompson, Depth migration and interpretation of the COCORP Wind River, Wyoming, seismic reflection data: Geology, 11, 462-469, 1983.

Meissner, R., and M. Fakhimi, Seismic anisotropy as measured under high-pressure, high-temperature conditions: Geophys. J., Roy. Astron. Soc., 49, 133-143, 1977.

Montadert, L., D.G. Roberts, O. DeCharpol and P. Guemoc, Rifting and subsidence of the Northern Continental Margin of the Bay of Biscay: Initial Rep. Deep Sea Drill. Prol., 48, 1025-1060, 1979.

Newton, R.C., and D. Perkins, Ancient granulite terrains, Eight KBAR metamorphism: EOS Trans. Amer. Geophys. Union, 62, 420, 1981.

Oliver, J.E., M. Dobrin, S. Kaufman, R. Meyer, and R. Phinney, Continuous seismic reflection profiling of the deep basement, Hardeman County, Texas: Geol. Soc. Amer. Bull., 87, 1537-1546, 1976.

Pierson, W.R., A geophysical study of the contact between the greenstone-granite terrain and the gneiss terrain in central Minnesota: Unpublished M.S. Thesis, Univ. of Wyoming, Laramie, 84 p., 1984.

Ramsay, J.G. and I. Allison, Structural analysis of shear zones in an alpinized Hercynian granite: Schweiz. mineral. petrogr. Mitt., 59, 251-279, 1979.

Reif, D.M. and J.P. Robinson, Geophysical, geochemical, and petrographic data and regional correlation from the Arizona State A-1 well, Pinal County, Arizona: Arizona Geologial Society Digest, 13, 99-109, 1981.

Simpson, C., The structure of the northern lobe of the Maggia nappe, Ticino, Switzerland: Eclog. Geol. Helvetiae, 75, 495-516, 1982.

Smithson, S.B., J.A. Brewer, S. Kaufman, J.E. Oliver, and C.A. Hurich, Nature of the Wind River thrust, Wyoming, from COCORP deep reflection data and from gravity data: Geology, 6, 648-652, 1978.

Smithson, S.B., J.A. Brewer, S. Kaufman, J.E. Oliver, and C.A. Hurich, Structure of the Laramide Wind River uplift, Wyoming from COCORP deep reflection data and gravity data: Jour. Geophys. Res., 84, 5955-5972, 1979.

Smithson, S.B., J.A. Brewer, S. Kaufman, J.E. Oliver, and R.L. Zawislak, Complex Archean lower crustal structure revealed by COCORP crustal reflection profiling in the Wind River Range, Wyoming: Earth Planet. Sci. Lett., 46, 295-305, 1980.

Snoke, A.W., Transition from infrastructure to suprastructure in the northern Ruby Mountains, Nevada: in M.D. Crittenden, Jr., P.J. Coney, and G.H. Davis, eds., Cordilleran metamorphic core complexes: Geol. Soc. Amer. Memoir, 153, 287-334, 1980.

Trümpy, R., Geology of Switzerland, Basel, Wepf and Co., 104 p., 1980.

Velzeboer, C.J., The theoretical seismic reflection response of sedimentary sequences: Geophysics, 46, 843-853, 1981.

Warner, M.R., Seismic reflections from below the crust (abs.): Terra Cognita, 5, 185, 1985.

Wernicke, B., Uniform-sense normal simple shear of the continental lithosphere: Canad. J. Earth Sci., 22, 108-125, 1985.

Wong, Y.K., S.B. Smithson, and R.L. Zawislak, The role of seismic modeling in deep crustal reflection interpretation. Part I: Univeristy of Wyoming Contributions to Geology, 20, 91-109, 1982.

FLUIDS IN DEEP CONTINENTAL CRUST

W. S. Fyfe

Geology Department, University of Western Ontario,
London, Canada, N6A 5B7

Abstract. The production of high grade metamorphic rocks of the amphibolite-granulite facies in continental crust requires an increase in temperature associated with magma underplating or tectonic thickening. Underplating of continental crust by dense mantle magmas may lead to assimilation of wet, heavy crustal components and their degassing (H_2O-CO_2-S). Deep subduction of ocean crust spilites, serpentinites and sediments rich in H_2O-CO_2-S, may also cause fluid release into the basal crust. Major continental collisions and minor overthrust events associated with transform faults can lead to massive degassing. The fluid mobilization in the Himalayan event may involve a fluid mass similar to the ice caps. In addition, such a large scale event may cause mobilization of significant quantities of CO_2 and salt depending on the lithology of the thickened crust. Whenever large volumes of fluids move there is ore potential. In general, the environmental consequences of such tectonic events are little understood.

Introduction

Over the past decade or so we have become intensely aware that the earth is a water cooled planet and that water and other volatiles are recycled to considerable depths in the mantle. Studies of rock mechanics has also shown that the properties of earth materials are extremely dependent on volatiles and that the development of many major structures, such as thrusts, faults, folds, involves fluids at high pressures (Fyfe et al., 1978). The gradual quantification of plate tectonic phenomena has further demonstrated the scale of fluid processes during the major parts of this convective process. Thus water cooling at ridges cycles the ocean mass through the near ridge systems every few million years (Wolery and Sleep, 1976) and plays a major role in heat transport: continuous slow convection possibly occurs almost to the time of subduction (Anderson et al., 1979) and during subduction large masses of water and volatiles such as CO_2, S, are returned to the mantle at significant rates (Fyfe, 1983). The scale of these processes in a slowly cooling planet must cast doubt on any purely steady state model of crust-hydrosphere distribution. The total mass of surface fluids that interact with rocks annually, on continents and in the sea floor is very large. Surface rocks are hydrated and plate tectonics show us that such surface rocks are often recycled to depth. As most hydrated near-surface phases degass at depth, fluids must be generated in most deep tectonic processes and as these high temperature fluids move, they carry elements in solution in highly variable quantities depending on the exact solute chemistry. Just as our planet is a dynamic system, its geochemistry is also dynamic in relation to tectonic processes.

This symposium is focussed on the deep continental crust and in these notes I wish to focus attention on fluids in the deeper continental regimes.

The Normal State

Common wisdom tells us that average, light quartz-feldspar dominated, continental crust is about 30 km thick and from the heat flow in stable regions would have a temperature in the range 400-500°C near the Moho. Given such numbers, rocks near the continental Moho would be under conditions of the high greenschist low amphibolite facies of metamorphism. Such rocks would contain a wide range of hydrated and carbonated minerals (epidote, chlorite, micas, amphiboles and carbonates) with water contents in the range 2-4% by weight.

But when we examine many crustal regimes we find almost anhydrous granulites, amphibolites and rocks which show signs of partial melting. Such rocks often have a mineral phase assemblage indicative of temperatures in the range 600-1000°C. How do such rocks form? As was stressed by Fyfe et al. (1958) they cannot represent any 'normal" regime. These rocks must have formed under conditions where heat flow has been augmented or crustal thickness has been increased. Given the heat production of continental crust it is clear that ultimately the thickness is limited by melting. Recent seismic studies of deep regimes

Fig. 1. A dyke rock, intruding greenstone, Archaean, near Timmins, Ontario. The large plagioclase crystals (5 cm) occur in a fine grained basaltic matrix. This rock shows how plagioclase may accumulate in a deeply ponded intrusion which, in this case, has become unstable. It is not difficult to see how an anorthosite could form by floating of plagioclase.

(Oliver, 1982) increasingly show the complexity of the continental Moho and this complexity must be related to processes which create the thermal perturbations. The recent observation of coesite in metamorphic rocks by Chopin (1984) and Smith (1984) shows that continental materials can attain depths of 90 km or more and forces us to contemplate processes which drag light siliceous precursors to mantle depths.

An equally intriguing problem is that of the retrograde metamorphism and rehydration of granulitic rocks so often seen pervasively or in limited domains of high grade rocks exposed at the surface. What is the source of the fluids which are involved in such retrograde processes? There is a tendency to call on primordial sources (as for ^{3}He) but given recycling, is this necessary?

Recycling-Ingassing

Geochemists have long been concerned with the outgassing history of our planet, an essentially unidirectional differentiation. But as understanding of the subduction process increases it is clear that such a simple concept is invalid. Recent studies (see Hilde and Uyeda, 1983) have clearly shown that when old ocean floor is subducted, initially light sediments may be trapped in the descending slab and carried to as yet unknown depths. The subducted oceanic lithosphere, following its chemical modification at ridges and along the conveyer belt, is enriched in volatiles (CO_2, H_2O, S) and in heat producing elements (K, U) and carries an as yet unquantified amount of serpentinized peridotite. The present water subduction rate (1.5×10^{15} g a^{-1}, Fyfe, 1983) would recycle the ocean mass in a billion years or so and without doubt can easily account for the volatiles in andesites and ridge basalts and possibly in exotic kimberlites, etc. In fact, the quantity is so large as to suggest that hydrosphere volumes may decrease slowly with time in a cooling, stirred planet like earth (cf. Mars). Common thermodynamic wisdom suggests that heating bodies degass while cooling bodies ingass.

Much is to be learned about the details of degassing of subducted slabs as ocean crust spilite and hydrated peridctite passes to blueschist assemblages and eventually to eclogites but high resolution seismic studies as are now in progress beneath Vancouver Island (Clowes et al., 1983) should throw light on such phenomena. But it is certain that as various portions of the slab release volatiles along their metamorphic paths, complex mechanical decoupling must occur (Price and Audley-Charles, 1983).

Magma Underplating

Most who study volcanism above subducting slabs would agree that degassing plays a critical role in promoting melting in or above the slab. While there is debate about the details of material transfer in the process, the fact that volatiles are important is little in doubt (see Thorpe, 1982).

Recent data on the compressibility of melts (Herzberg et al., 1983) shows that at Moho pressures of about 10 kbar or so, most mafic mantle melts are more dense than many of the common constituents of continental crust (diorite, granodiorite, granite). Thus, if there is decoupling of continental crust from the mantle, such dense liquids should underplate rather than intrude or extrude continental materials. The decoupling mechanism can be created by the process itself for where there is high heat flow associated with rising mantle magmas, partial melting of continental crust and degassing near the Moho will lead to a low viscosity, low strength boundary.

Large scale continental underplating by dense mantle magmas ($\rho > 2.8$ g cm^{-3}) with temperatures near 1200°C, must lead to a giant prograde metamorphic event which will be associated with crustal weakening and melting. The recent recognition of magma mixing above subduction zones must come as no surprize. Thus, in such situations of underplating, rocks of the granulite facies may form from continental crust of normal thickness. One would also anticipate that the underplate primitive basaltic magmas might differentiate quite perfectly by floating of plagioclase and sinking of olivine-pyroxene, to form the layered anorthosite bodies commonly found in some high grade terranes (Fig. 1).

Given underplating by dense mantle magmas and decoupling near the continental Moho by meta-

Fig. 2. Graphite filling fractures in a granulite from Sri Lanka (1 cm scale).

morphic degassing and partial melting of the overplate, interesting "continental subduction" processes may result. Any large continental components with density 2.9 g cm^{-3} or larger must tend to sink through, or be assimilated by, the underplate magmas. Thus mafic rocks ($\rho > 3.0$ g cm^{-3}) or even rocks like dolomites or massive anhydrite or dense sulphide-oxide ore bodies, may be removed. A massive lower crust density filter may thus operate. It is frequently noted that high grade terrains are rather monotonous chemically. Such processes must lead to a complex sub Moho chemistry (see Fuchs, this volume).

Imagine that a typical volume of Archaean granite-greenstone terrain is subjected to a massive basaltic underplate event. The mafic-peridotitic domains would tend to sink while the more silicic components would float and tend to melt. The sinking and partially assimilated greenstones would degass at the very high temperature of the underplate magmas and produce high temperatures gases of the H_2O-CO_2-S system. At such temperatures, species such as CO-COS would be significant. When these exceed the saturation pressures for the underplate melts there would thus be produced a flow of gas which would move into the hot overlying continental crust recycling the volatiles fixed in originally near surface materials. Do such processes explain features like the graphite veins so frequently seen in granulites (Fig. 2) via reactions like

$$2CO \rightarrow C + CO_2$$
$$COS + FeO(sil) \rightarrow CO_2 + FeS?$$

And species such as CO, COS in fluids, would be excellent complexers and transport agents of a wide range of metals. Do some of the gold deposits associated with magmatic events contain a spike of highly metal-enriched fluids due to such phenomena? We know little about such very high temperature fluids but recent work on "cluster molecules" in reduced, CO systems (Muetterties, 1982) suggests that such species could be significant.

Left Over Slabs

The subduction process may stop or a subduction zone may move. What happens to the upper portion of a subducted slab, after underthrusting beneath a continent in a case in which the spilite-eclogite transition has not occurred and low density material is still present (e.g. micas and chlorites in sediments, serpentine, talc in ultramafics, etc.)? Should descent cease, the slab will slowly come to thermal equilibrium and the entire hydrated portion must dehydrate partially or totally depending on the angle of subduction and depth. Many views of the upper mantle perhaps indicate that such left-over slab fragments are common (e.g. Davies, 1981). Imagine a slab as shown in Fig. 3. Given that it is emplaced with temperatures less than 10°/km^{-1}, a blueschist regime, then at 50 km depth, temperatures will be

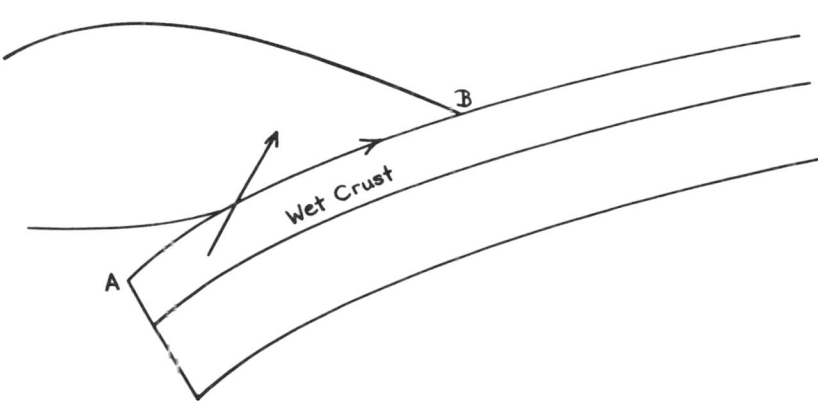

Fig. 3. Cartoon of degassing of a left-over fragment of subducted crust with a surface of wet sediments and spilitic rocks.

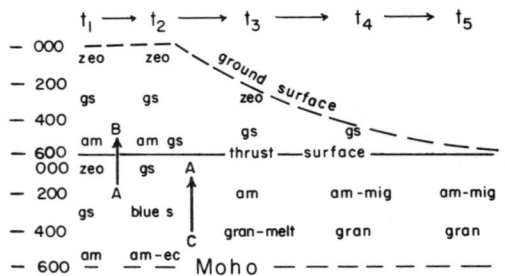

Fig. 4. The types of metamorphic sections which may develop following a major continental overthrust. Along paths like A-B, rising fluids will retrograde the overthrust and will not form quartz veins due to the inverted gradient. Along C-A, normal veining may occur.

near 500°C. As it heats up the deeper portions will undergo dehydration to amphibolite or eclogite facies. The mass of fluid evolved, mainly H_2O at shallow depths but H_2O-CO_2 at extreme depths, may move up the thrust plane or penetrate the overplate by hydraulic fracture. The quantities are impressive and could be of the order of 10^{16} g per km strike length (Fyfe and Kerrich, 1984). Such fluids would move silica in quantities amounting to almost 1% of the solvent mass, produce carbonates in colder rock and even form ore deposits. Is this the type of process which produced the Mother Lode gold deposits of California with their spectacular silica deposition and carbonate alteration?

Continental Thrusts

Modern seismic studies of deep continental structures are constantly revealing the presence of major low angle thrust structures which frequently penetrate the Moho (see this volume). But perhaps the most spectacular case at present under intense investigation involves the Himalayan collision (Barazangi and Ni, 1982; Allegre et al., 1984; Hirn et al., 1984). While the structure of this vast area of thickened crust is slowly being resolved, the geochemical consequences of the process have not been adequately considered.

According to present views on the Himalaya-Tibet region, a crustal area of about 4.5×10^6 km^2 has a thickness in the range 60-80 km (Allegre et al., 1984). Whatever the detailed mechanism of stacking of thrust sheets, all deep portions of underthrust crust must undergo prograde metamorphism. In a general way, the question we must ask is what will happen if we take an area of continental crust of normal thickness and thermal gradient (basal T 450°C) and load it with 30 km of continental overburden and raise its average temperature to 700°C? Or to put the question another way, what would happen if we were to take about 60% of the mass of continental Australia, heat it by an additional 300°C and compress it by an additional 10 kb? Given a rapid collision, after stacking, local areas will have a thermal and metamorphic structure as shown in Fig. 4. Prograde metamorphism will raise the lower 30 km to amphibolite-granulite facies with some degree of partial melting near the base. The young granites of the Himalayas with their extremely high $^{87}Sr/^{86}Sr$ initial ratios testify to such melting and granite production in continental crust with minimal input from mantle sources (Hamet and Allegre, 1976; Ferrara et al., 1983). The pre-metamorphic crust would have a water content in hydrated minerals close to 5% while the post metamorphic amphibolitic-granulitic deep crust would have a water content closer to 1%. Assuming a 4% water loss from 13.5×10^7 km^3 of rock, the H_2O mass released is of the order of 1.6×10^{22} g, a mass similar to the modern ice caps (2.5×10^{22} g) and about 1% of the ocean mass.

An equally interesting gas is carbon dioxide. For CO_2, the mass involved depends critically on the stratigraphic nature of the crust that is forced to deep levels and in particular the contribution from CO_2-bearing ophiolites and calcareous sediments. Reactions such as calcite + quartz $\longrightarrow$ wollastonite + CO_2 will release massive quantities of CO_2. To place this in context imagine that a layer of limestone 1 km thick was involved in the thrusting process. The degassing of such a layer with an area of 4.5×10^6 km^2 could liberate about 0.5×10^{20} g of CO_2. The present atmosphere contains 2.3×10^{18} g of CO_2, while the oceans contain 1.1×10^{20} g mostly as bicarbonate. It is estimated that the total CO_2 in carbonate sediments is around 6.7×10^{20} g (Wehmiller, 1972). While the figure of 1 km is probably too large, the numbers do show that a significant perturbation of global CO_2 could result from a Himalayan event, with the impact depending on the processing rate or the rate of thermal equilibration.

Not all volatiles released in the thrust process will reach the surface. Depending on the mechanism of fluid movement, in shear zones or hydraulic fractures, some will cause massive retrograde metamorphism in the overplate and one would anticipate large zones of carbonate alteration. Such zones must occur in the Himalayan system and will show the pathways of fluid release. But it is unlikely that there will be total absorption of such fluids which would require pervasive penetration into the total volume of overthrust rocks.

An intriguing question involves the possibility of salt release caused by such a collision. Any inspection of the distribution of evaporites in the continental crust points to their widespread occurrence (see Feldman and Cruft, 1972). Evaporite formations are commonly in the range 1000-4000 ft thick. It is certain that if a major salt basin was involved in a collision, most of the salt would reach the surface in solution though the cation chemistry might be modified during transport. To place this problem in context, the

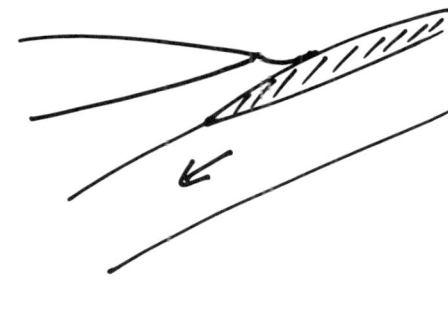
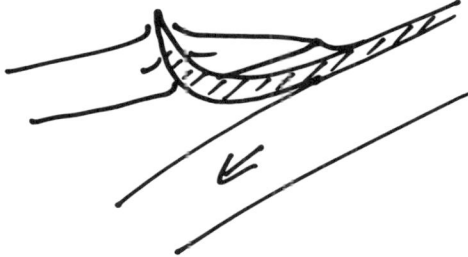

Fig. 5. The type of mechanism which could lead to coesite formation. Rather light and wet sediments are dragged to depth. As degassing starts, the wet section is floated off and may rise via hydrofracture mechanisms similar to the mechanism of Price and Audley-Charles (1983). The process may be similar to the formation of sandstone dykes.

oceans contain about 4×10^{22} g of salt. The useful salt reserves of the U.S.A. alone are 6×10^{19} g (Kostick, 1980). Depending on stratigraphy, a Himalayan event could cause a significant change in ocean salinity and the process need not be slow for the remobilization could be almost as fast as the collision due to the very high solubility of salt.

One thing is certain, a mountain building event of Himalaya scale must increase global erosion. I am intrigued by the pattern of changes of $^{87}Sr/^{86}Sr$ in the oceans over time (Burke et al., 1982). The data show a sharp increase in oceanic ^{87}Sr since the Jurassic with a remarkable build up from the Tertiary to the present. Does this pattern reflect mountain building and the increased output of evolved continental strontium as might be supplied by erosion of the high ^{87}Sr granites? And if the strontium isotope ratios change, is it possible that there will be changes in other cationic and isotopic ratios in the oceanic input system?

A final effect of a major thrust event on which I would like to comment involves the mechanical strength of different parts of the thrust system as a function of time. At the time of thrusting, and before major metamorphic events are initiated, most of the crust of Fig. 4 will be relatively free of a fluid phase and will be strong. But once degassing and melting commences at deep levels, the rocks will become porous and weak with fluid pressures similar to lithostatic pressure. Thus deep sections will pass through stages, strong→weak→strong. Above the thrust plane where retrograde processes occur, rocks will tend to remain relatively strong except for sites involving the development of shear zones and hydraulic fractures, or in zones where rising plutons degass their aureole rocks.

Ultra high Pressures

One of the most spectacular recent findings in metamorphic rocks of the Alps (Chopin, 1984) and the Caledonides (Smith, 1984) has been the evidence for the preservation of coesite in blueschist-eclogite facies rocks. This implies metamorphic pressures near 30 kbar and depths of burial of 90-100 km. Such depths are close to the maximum crustal thicknesses reported in the Himalayas (Hirn et al., 1984). As some of these rocks are quartzites, there can be no doubt about their primary origin as surface rocks. Such metamorphic rocks clearly show how rapidly surface rocks may be transported to depth and equally rapidly returned to the surface (Fig. 5). We know little about such mechanisms. It seems possible that careful study of reaction kinetics in such systems (see Brown et al., 1962) can provide information on the reaction paths but given fast plate motions (10 cm a^{-1}) rather short times may be involved both during descent and uplift or penetration to high levels, a few million years.

Thrusts and Ore Formation

The dewatering events that must be associated with major thrust events occurs on so large a scale that major local metasomatism and at times ore deposition must occur. As mentioned above, the water release in the Himalaya event must be of the order of 10^{22} g. Given that an element is carried at the 1 ppm level, a million ton ore deposit would require 10^{18} g of fluid (the annual flow of the Yangtse river). There must be ore potential in the Himalayas, particularly for metals such as gold where metamorphic processes appear to be the most efficient for concentration (Fyfe and Kerrich, 1984b). We have previously reported major uranium mineralization associated with a thrust event in Brazil (Lobato et al., 1983). Mineralization associated with thrusting is little understood at this time (see Fyfe and Kerrich, 1984).

Transcurrent Faults and Local Overthrusting

Recently Allis (1981) has proposed an interesting model for the great Alpine fault of the New Zealand Alps. In this model very thick crust is produced by local lateral zones of overthrusting in the order of 100 km wide associated with a dominantly transform motion. The metamorphic phenomena which must occur during such events will be similar in type to the Himalaya event but will be more localized and will depend critically on the transform velocity. If the velocity is fast and overthrusting localized, only the first stages of overloading (e.g. eclogite formation) will be observed. But if the motions are slow or periodic, even melting may occur. I recently had the privilege of being shown some of the fault and thrust structures in Portugal by Professor Antonio Ribeiro of Lisbon University. In particular I was intrigued by the VilaRica fault system, a transform which shows wide zones of silicification and even localized production of granitic rocks. This fault has been active for some 300 Ma. Possibly it represents a much deeper level of exposure of the New Zealand process. It is interesting to note that many of the ancient mines of Saudi Arabia (Cu-Au) were sited on the great fault systems of the Proterozoic shield. Periodic overthrusting on a moving transform may explain the abrupt longitudinal transitions from normal to high pressure metamorphics (including units like aragonite marbles) observed in the Franciscan of California. To preserve minerals like aragonite, extremely rapid unloading is required (Brown et al., 1962).

Concluding Statement

In these notes I have attempted to draw attention to the mobilization of large fluid volumes which are associated with subduction and collision events. They may have profound significance for slow environmental change and the formation of ore deposits. The connections between geochemical transport processes and tectonics are as yet poorly quantified.

References

Allegre, C.J. et al, Structure and evolution of the Himalaya-Tibet orogenic belt. Nature, 307, 17-22, 1984.

Allis, R.G., Continental underthrusting beneath the Southern Alps of New Zealand. Geology, 9, 303-307, 1981.

Anderson, R.N., Hobart, M.A. and Langseth, M.C., Geothermal convection through oceanic crust sediments in the Indian Ocean. Science, 204, 828-832, 1979.

Barazangi, M. and Ni, J., Velocities and propogation of Pn and Sn beneath the Himalayan arc and Tibetan plateau: Possible evidence of underthrusting of Indian continental lithosphere beneath Tibet. Geology, 10, 179-185, 1982.

Brown, W.H., Fyfe, W.S. and Turner, F.J., Aragonite in California glaucophane schists, and the kinetics of the aragonite-calcite transformation. J. Petrol., 3, 566-582, 1962.

Burke, W.H. et al., Variation of sea water $^{87}Sr/^{86}Sr$ throughout Phanerozoic time. Geology, 10, 516-519, 1982.

Chopin, C., Coesite and pure pyrope in high-grade blueschists of the Western Alps: a first record and some consequences. Contrib. Mineral. Petrol., 86, 107-118, 1984.

Clowes, R.M. et al., Seismic reflections from subducting lithosphere. Nature, 303, 668-670, 1983.

Davies, G.F., Earth's neodymium budget and structure of evolution of the mantle. Nature, 290, 208-213, 1981.

Feldman, S. and Cruft, E.F., Evaporite processes, in The encyclopedia of geochemistry and environmental sciences, R.W. Fairbridge, ed., Van Nostrand Reinhold Co., New York, 351-361, 1972.

Ferrara, G. et al., Rb/Sr geochronology of granites and gneisses from the Mount Everest region, Nepal Himalaya. Geol. Rundschau, 72, 119-136, 1983.

Fyfe, W.S., Subduction and the geochemical cycle. Tectonophysics, 99, 271-278, 1983.

Fyfe, W.S. and Kerrich, R., Fluids and thrusting. Chem. Geol.(in press) 49, 1985.

Fyfe, W.S. and Kerrich, R., Gold: Natural concentration processes, in Gold 82: The geology, geochemistry and genesis of gold deposits, R.P. Foster, ed. A.A. Balkema, Rotterdam, 99-128, 1984b.

Fyfe, W.S., Price, N.J. and Thompson, A.R., Fluids in the Earth's Crust. Elsevier Scientific Publishing Co., Amsterdam, 383 pp, 1978.

Fyfe, W.S., Turner, F.J. and Verhoogen, J., Metamorphic reactions and metamorphic facies. Geol. Soc. Am. Memoir 73, 259 pp, 1958.

Hamet, J. and Allegre, C.J., Rb-Sr systematics in granite from central Nepal (Manaslu): significance of the Oligocene age and high $^{87}Sr/^{86}Sr$ in Himalayan orogeny. Geology, 4, 470-472, 1976.

Herzberg, C.T., Fyfe, W.S. and Carr, M.J., Density constraints on the formation of the continental Moho and crust. Contrib. Mineral. Petrol., 84, 1-5, 1983.

Hilde, T.W.C. and Uyeda, S., Convergence and subduction. Tectonophysics, special issue, 99, 85-400, 1983.

Hirn, A. et al., Crustal structure and variability of the Himalayan border of Tibet. Nature, 307, 23-25, 1984.

Kostick, D.S., Salt, in Mineral Facts and Problems. U.S. Dept. of the Interior, Bureau of Mines Bull. 671, 769-780, 1980.

Lobato, L.M. et al., Uranium enrichment in Archaean crustal basement associated with overthrusting. Nature, 303, 235-237, 1983.

Muetterties, E.L., Metal clusters. Chem. Eng. News, 60, No. 35, 28-41, 1982.

Oliver, J., Changes at the crust-mantle boundary. Nature, 299, 398-399, 1982.

Price, N.J. and Audley-Charles, M.G., Plate rupture by hydraulic fracture resulting in overthrusting. Nature, 306, 572-575, 1983.

Smith, D.C., Coesite in clinopyroxene in the Caledonides and its implications for geodynamics. Nature, 310, 641-644, 1984.

Thorpe, R.S., Andesites: orogenic and related rocks. John Wiley and Sons, 724 pp, 1982.

Wehmiller, J., Carbon cycle, in The encyclopedia of geochemistry and environmental sciences, R.W. Fairbridge, ed., Van Nostrand Reinhold Co., New York, 124-128, 1972.

Wolery, T.J. and Sleep, N.H., Hydrothermal circulation of geochemical flux at mid-ocean ridges. J. Geol., 84, 249-275, 1976.

TECTONIC ESCAPE IN THE EVOLUTION OF THE CONTINENTAL CRUST

Kevin Burke[1] and Celal Sengör[2]

Lunar and Planetary Institute, 3303 NASA Road One, Houston, Texas

Abstract. The continental crust originated by processes similar to those operating today and continents consist of material most of which originated long ago in arc-systems that have later been modified, especially at Andean margins and in continental collisions where crustal thickening is common. Collision-related strike-slip motion is a general process in continental evolution. Because buoyant continental (or arc) material generally moves during collision toward a nearby oceanic margin where less buoyant lithosphere crops out, we call the process of major strike-slip dominated motion toward a 'free-face' "Tectonic Escape."

Tectonic escape is and has been an element in continental evolution throughout recorded earth-history. It promotes: (1) rifting and the formation of rift-basins with thinning of thickened crust; (2) pervasive strike-slip faulting late in orogenic history which breaks up mountain belts across strike and may juxtapose unrelated sectors in cross-section; (3) localized compressional mountains and related foreland-trough basins.

Introduction

The processes by which continents are assembled and modified are beginning to be well understood. A major development has been the acquisition of deep reflection seismic profiles which have done much to illuminate the deep structure of the continents. We pose the question: "Are there processes influencing deep continental structure in ways that may be specially significant for the reflection seismologist?" and we respond that there exists a range of phenomena, associated mainly with continental collision and dominated by strike-slip faulting, which record the operation of such a process for which we suggest the name "Tectonic Escape." We will attempt to show that Tectonic Escape is now and has been through much of earth history an important process in continental evolution and we suggest that direct evidence of its operation may be particularly difficult to discern in deep seismic reflection records.

We first review the evolution of the continents in order to establish a framework within which Tectonic Escape may be considered to operate.

Origin and Modification of the Continental Crust

It has long been widely held that continents originate by the sweeping together of island arcs (e.g., Dewey and Windley, 1981, p. 201). The main reasons for this are: (1) that continents and island arcs are dominated by rocks of comparable compositional and density ranges (SiO_2 55–70%; 2.6–2.9 gm/cc), (2) that evidence of the operation of this process is abundant in the geological record (e.g., in Arabia, Gass, 1981), and (3) that the process can be seen in action today in Indonesia (Hamilton, 1979). Evidence from detailed geology has been reviewed elsewhere (e.g., Burke et al., 1976). Understandably, misgivings have been voiced about the simple idea of assembling continents by colliding arcs. These misgivings are mainly based on observations that: (1) some arcs do not closely resemble average continental material (e.g., Karig and Kay, 1981) and (2) much continental material is very old (> 2.5 Ga) and the world can be assumed to have operated in a different way in the remote past (e.g., Hargraves, 1976).

The observation that some arcs will not make typical continent does not invalidate the inference that continents are made by the assembly of arcs. Morgan (1984) has, for example, shown how the thermal structure of continental crust indicates substantial compositional diversity, and considerable differences from average continental composition are to be expected in material representative of individual arcs accreted to continents. Moreover, processes likely to modify crust after its addition to the continent, such as the formation of Andean arcs and Tibetan-style collision may tend to drive the bulk composition of the upper crust accessible to observation closer to that of typical continent.

The other idea, that the world operated in a different way in the very remote past, is obviously right, but it is increasingly being recognized that the record of the rocks as far back as 3.8 Ga ago, although it reveals some unidirectional changes such as biological evolution, is consistent with the idea "that many processes have apparently not changed at all" (Thompson et al., 1984, p. 404). For this reason, we will here employ the hypothesis that as far back as the record goes, continents have been assembled by processes fundamentally comparable to those acting today although rates and thermal regimes were different. Specifically, the idea that greenstone belt and granodiorite terranes, which are typical of very old continents, represent collided arcs and remnants of ocean floor is one of long standing (Burke and Dewey, 1972; Burke et al., 1976; de Wit, 1982; Talbot, 1973)

There are within the oceans today shallow-water objects distinct from active arcs, extinct arcs and micro-continents. It has been suggested that some of these might be of a kind capable of contributing to the continental crust (e.g., Nur and Ben-Avraham, 1982). One population of such objects is that formed at intra-oceanic hot-spots, e.g., Hawaii and Iceland. Although small objects a few cubic km to a few tens of cubic km in volume representative of oceanic islands have long been reported from mountain belts of all ages (e.g., Berklund, 1972, Karson and Dewey, 1978, Cooke and Moorhouse, 1970) there are no recognized Iceland-or Hawaii-like objects with volumes in excess of a thousand cubic km in the mountain belts of the worlds. Since these mountain belts are the products of the closing of oceans (Wilson, 1968) we infer that it is the fate of such hotspot-generated material to be largely removed from the surface of the earth by either subduction or

---

[1] Also at Department of Geosciences, University of Houston, Houston, Texas.

[2] Also at Istanbul Technical University, Istanbul, Turkey.

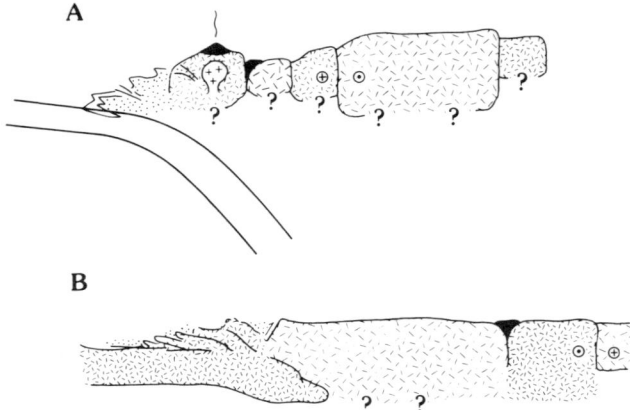

Figure 1. a. The assembly of arcs and microcontinents of varied thickness and composition will not make for a uniform Moho, indeed it is not clear what the Moho is below an island arc. Strike-slip motion will also tend to emphasize differences in crustal thickness. b. Continental collision and thickening also produces a crust of varied thickness.

obduction (cf. Vogt et al., 1977). The Siletz terrane (Ben-Avraham et al., 1981) may have been formed by hot-spot volcanism but its location at the edge of North America makes it uncertain whether it would survive in an intracontinental mountain belt. The alternative idea that older oceans did not contain hotspot volcanics seems less likely. A similar argument applies to the oceanic plateaus (e.g., Caribbean, Manihiki, Ontong Java, Shatsky Rise, Hess Rise) which extend over areas of millions of km^2. Evidence both from places where slivers of some of these objects have been caught up in accretionary prisms (Burke et al., 1984; Hughes and Turner, 1977) and from their relationship to old spreading centers (Winterer, 1976) indicates that they do not represent continental fragments (as has been suggested) but very thick accumulations of basalt interbedded with pelagic sediments. The mountain belts of the world contain slivers of material of this type but no large volumes, and we therefore conclude that, like oceanic hotspot products, oceanic plateaus do not contribute significant volumes of material to the continents. The Bahama Bank is the largest of a final class of shallow water objects consisting of great thicknesses (>5km) of carbonate sediment over oceanic crust (Dietz, 1973). Again, objects of this type are not preserved in the mountain belts of the world except as exotic blocks and thrust slivers and we infer that they are ultimately either subducted or obducted.

We conclude that the materials which contribute and have contributed substantial volumes of material to the continents are parts of island arc systems, including sediments of accretionary wedges which in some areas, for example, the Songpan-Ganzi (Sengör, 1984), the Makran (McCall and Kidd, 1983) and the Barbados wedge (Lawrence et al., in press) may be of enormous volume and contribute significantly to continental growth.

Once a continent is assembled by arc collision, the vicissitudes to which it can be subjected are limited. We distinguish six main processes (short of destruction) in which continents are modified and summarize them thus:

(1) The construction of an Andean arc at the continental margin. This involves a variety of phenomena, including: (a) addition of magmatic material from the mantle; (b) partial melting of older continental crust and related vertical fractionation; (c) thickening of crust by up to a factor of 2 (to 70 km) and consequent elevation; (d) compressional tectonism (especially on the borders of the arc) with formation of both a forearc and a foreland trough; (e) extensional tectonism (especially at the crest of the arc).

(2) Continental collision, as now in Tibet and the Himalayas, produces effects similar to those of Andean collision, but over a much wider area and on a larger scale. Pioneering efforts by Chinese geologists and their French colleagues are beginning to reveal some of the ways in which continental crust is modified at collision (Hirn et al., 1984; Xuan et al., this volume).

Tremendous thrusting of the kind first described from the Alps, Scandinavia and from the North of Scotland about 100 years ago and widely recognized in Deep Reflection Profiles (this volume, passim) has accommodated the extensive doubling of continental thickness that characterizes Andean and collisional mountain belts. Partial melting of the thickened crust at collision produces compositional differentiation with shallow "minimal-melt" granites and deep anhydrous residues (Dewey and Burke, 1973; Fyfe, this volume). Thickened continents, which typically stand 5 km above sea level are unstable, being subject to rapid erosion as well as to gravitational collapse, two processes that restore continental thicknesses to normal values (30–40 km) in a few tens of millions of years after the cessation of convergence.

A typical mature continent assembled from arcs and microcontinents, parts of which have been subjected to Andean or Tibetan style modification 10^8 or more years ago, is likely to be a low-lying (up to 2 km relief) object with a variety of crustal thicknesses and deep structural styles reflecting the local incidence of arc accretion and Andean and Himalayan-Tibetan collision. Australia represents such an object.

(3) Flooding of the continents by the waters of the ocean produces a sediment thickness typically of a few hundred meters. This flooding is simultaneous on the continents as has been demonstrated for the Russian, North American and West African platforms (Sloss, 1972) and reflects sea level changes dominated by the age of the ocean floor.

(4) The mature continent records bombardment from outer space, but since the beginning of the rock record (at 3.8 Ga), the flux has been low (see papers in Silver and Schultz, 1982) and the most complete record is on the oldest continents with Vredefort in South Africa and Sudbury on the Canadian Shield.

(5) Sporadic non-plate margin (or hot-spot) volcanism modifies the continental crust, but these effects as seen for example in Africa and the Canadian Shield generally produce only local modification.

(6) Rifting (the formation of large scale elongate depressions related to rupture of the lithosphere in extension) is one of the dominant ways in which once assembled continents are modified. Rifts form in diverse tectonic environments (Sengör and Burke, 1978), for example: Andean arcs may split along the crestal volcanoes to form marginal basins; continents like Africa now and about 200 Ma ago may come to rest over the convective circulation of the mantle and rupture (Burke and Wilson, 1972) and rifts may also form in association with continental collision

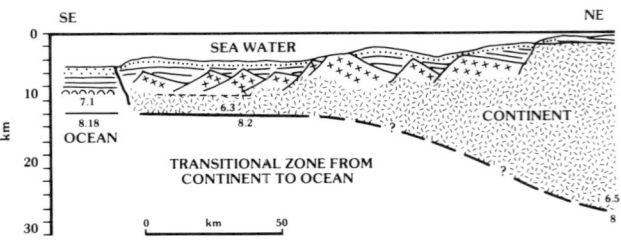

Figure 2. This classic section from the Bay of Biscay suggests a smooth Moho produced by flow of the lower crust during rifting (modified from Charpal et al., 1978). Rifting appears to contrast with most other processes in smoothing the Moho.

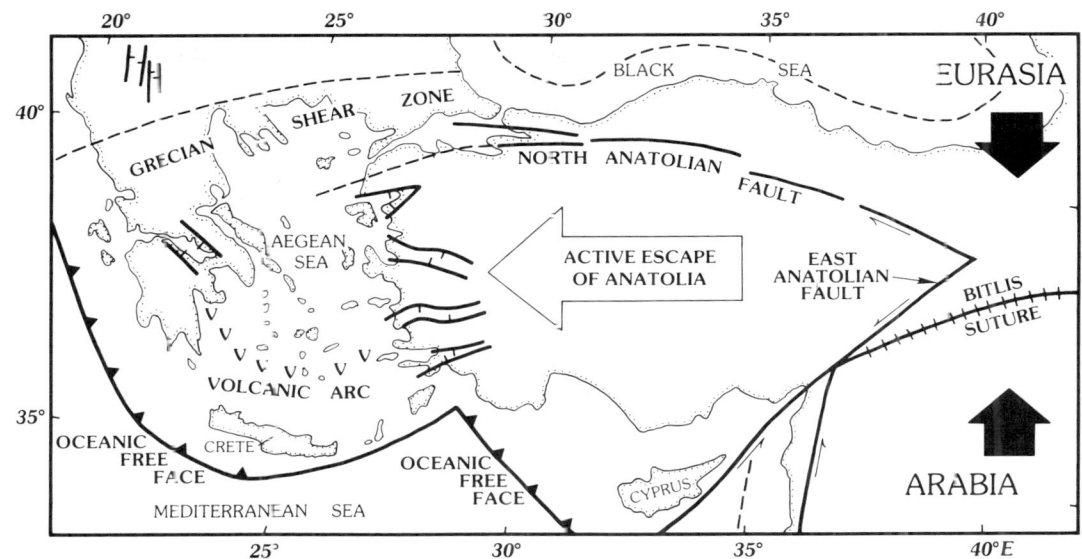

Figure 3. The type example of the process here called Tectonic Escape. Anatolia has been moving west toward the Aegean since the Bitlis collision at about 10 Ma. Only the bounding faults and some grabens are depicted.

(as in the Rhine and Baikal; Molnar and Tapponier, 1975; Sengör et al., 1978).

Continental assembly from island arcs, Andean arc formation, continental collision and rifting are the dominant processes of continental evolution which leave their mark in the deep structure of the continents. We illustrate in Figure 1 a tendency (based on reported profiles) for all these processes except rifting to generate crust of varied thickness. By contrast, the process of rifting is frequently associated with the production of a smooth Mohorovicic discontinuity in some places approaching the horizontal. This smoothness, which may also be associated with continuous, deep, smooth reflections within the continental crust and uppermost mantle, we suggest results from flow of deep crustal rocks under rifting conditions.

During rifting, hot asthenospheric mantle is close enough to the lower continental crust to elevate the temperature into the range in which flow rather than fracture dominates the response to stress of the lower crustal rocks (see for example, Bott, 1979). We suggest that flow under these conditions is a major mechanism by which smooth extensive, sharply defined, Mohorovicic discontinuities such as have been recognized on many deep seismic profiles (Hutchinson, 1984) are formed from more irregular crust-mantle boundaries. The reason for making this suggestion is the observation that smooth well-defined Mohorovicic discontinuities are common in areas that have suffered rifting (papers presented at this meeting) but not elsewhere. In Figure 2 we illustrate a familiar cross-section from the northern shores of the Bay of Biscay and suggest that the smooth rise of the Moho from a depth of ~27 km under Brittany to ~12 km, 150 km to the south-southwest at an average slope of 1 in 10 represents a typical example of a crustal structure modified by the flow of deep crustal rocks during rifting.

Tectonic Escape

In this paper we emphasize one process to which continents assembled and modified by the variety of processes described above are particularly susceptible. This process is lateral motion toward an oceanic (readily subductable) area during arc or continental collision. The lateral motion is dominantly attained by movement on strike-slip faults and is, we suggest, likely to be very hard to recognize in deep reflection data. We briefly describe a number of examples of Tectonic Escape from widespread geographic areas and of a range of ages in order to characterize the process.

Turkey

McKenzie (1972) elaborating an idea of Ketin (1948) suggested that following collision between Arabia and Asia about 10 Ma ago (Figure 3), Anatolia, because of its buoyancy, began to move westward into the Aegean along the bounding Northern and Eastern Anatolian faults. Deep reflection data are not available for Turkey, but the complex pattern of rifting in western Anatolia (Sengör and Gorur, in press) reflects and typifies the style of breakup of continental and arc-assembled Anatolia as it has moved into the larger general area of the Aegean Sea. Widespread late Neogene basaltic volcanism, a sporadic distribution of high ground and the occurrence of both strike-slip and normal faults are typical products of Tectonic Escape that have been recognized in Western Anatolia. The strongest evidence of the close association of the escape with collision on the Bitlis suture comes from (1) the geometric and kinematic pattern of the faulting and (2) the timing of onset of rifting (10 Ma) being contemporary with the collision. In examining older examples it will become clear that these two types of criterion are generally those that define the escape process.

An important observation is the relative topographic elevations of the area from and towards which escape occurred. During the earlier late Miocene (Serravallian-Tortonian boundary, 11.5 Ma) eastern Anatolia was the site of quiet, reef carbonate deposition (the Adilcevaz Limestone and its equivalents) indicating that this region lay just below sea-level. By contrast, a tremendous crustal thickness inherited from the Eocene to early Miocene thrusting in western Turkey and in the Aegean region characterized the areas towards which escape took place. The localized lacustrine and fluviatile sediments, closely associated with silicic volcanics (mainly rhyolites and ignimbrites), characterized the high ground whereas marine sediments were laid down around its periphery (Luttig and Steffens, 1974).

The disintegration of this high region by E-W striking normal fault

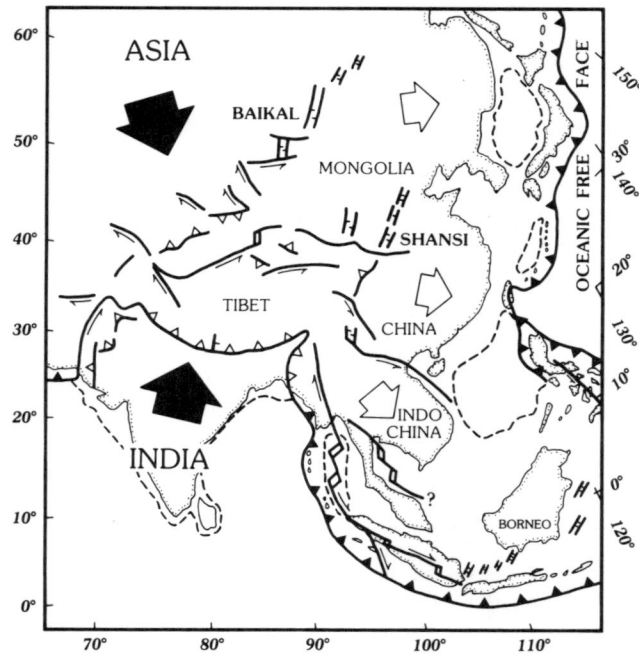

Figure 4. Tectonic Escape in South-East Asia. Modified and simplified from Tapponnier et al., 1982. Note that evidence of timing indicates an association with the Himalayan collision for rifts as far apart as Baikal and the Sumatran coal basins (at 113°E, 3°S).

systems began during the Tortonian, contemporaneous within the power of resolution of the Mediterranean marine stratigraphy of the Miocene, with the initiation of uplift of the east Anatolian plateau (youngest marine sediments in the E. Anatolian plateau are of Serravallian age: Gelati, 1975) and with the onset of movement along the North and the East Anatolian faults (Sengör, 1979). Thus, it seems difficult to ascribe this escape to the gravitational stresses generated by the high east Anatolian plateau as the model of England and McKenzie (1982, 1983) and England (1982) would require. Such stresses, if at all influential, could not have been the main motor behind the escape.

China and Southeastern Asia

Although recognition by Molnar and Tapponnier in 1975 of operation of the process here called Tectonic Escape in association with the active Himalayan collision post-dated McKenzie's recognition of the process in Turkey by three years, studies of Himalayan related escape phenomena have come to dominate the field because of areal extent and vast scale (Figure 4).

An early suggestion (Molnar and Tapponnier, 1975) that the effects of collision could be interpreted either as analogous to those of an instantaneous plastic response to rigid indentation or to a time dependent response with motion toward free faces was followed by papers leaning heavily toward the first alternative (e.g., Molnar and Tapponnier, 1979), but simulation of motion of China to the east in an experiment with modelling clay (Tapponnier et al., 1983) produced a fault pattern very similar to that recognized on the basis of earthquake mechanisms and LANDSAT imagery (Molnar and Tapponnier, 1975) and, for this reason, the time dependent interpretation (called "tectonic extrusion" by the French workers) has come to be preferred. England and McKenzie (1982, 1983) interpreted all deformation in the escape zone as a response to topographic elevation induced by collision and continuing convergence. This is in marked contrast to slip-line theory which is not only instantaneous, but is also two-dimensional and can take no account of topography.

As in the case of Turkey, application of the Tectonic Escape interpretation to Asia depends mainly on recognizing both a pattern of faulting and appropriate timing for the initiation of faulting. The timing of the Himalayan collision cannot be characterized as precisely as that along the shorter Bitlis suture, but Himalayan collision took place over about 10 Ma between 45 and 35 Ma (Molnar and Deng, 1984; Tapponnier et al., in press). Effects of escape are recorded in sediments deposited in extensional fault troughs in areas as remote as Lake Baikal (now 2000 km from the collision site) during the early Oligocene (Florensov, 1969), the Pattani trough in the Gulf of Thailand (Sclater et al., 1983) and the Sumatra Coal Basins (Koesoemadinata, 1978; Tapponnier et al., in press) beginning in the Latest Eocene (Figure 4).

Elaboration of the details of Tectonic Escape in relation to the

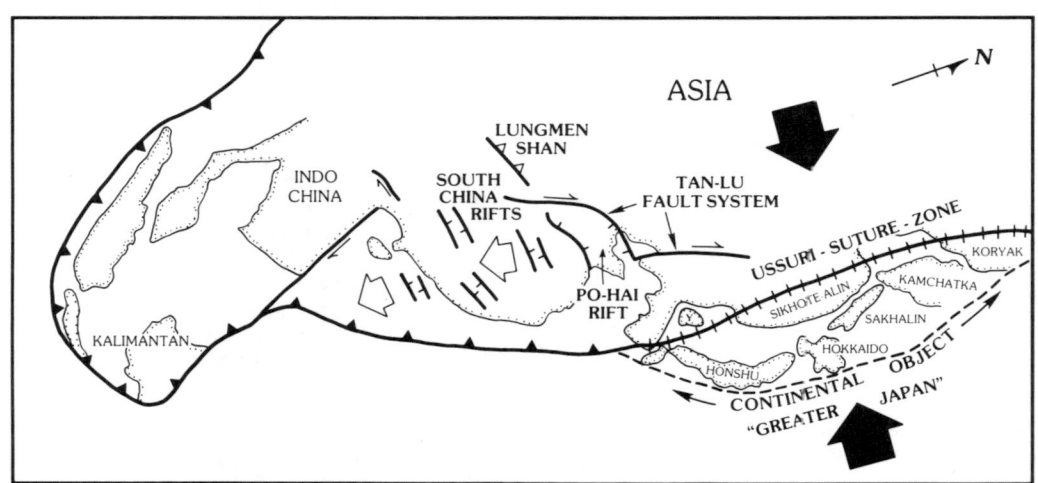

Figure 5. An earlier episode of Tectonic Escape in the Cenozoic of Eastern Asia is related to the Po Hai and South China rifts as well as the Tan Lu strike-slip fault system. It is here associated with the collision between "Greater Japan" and Asia.

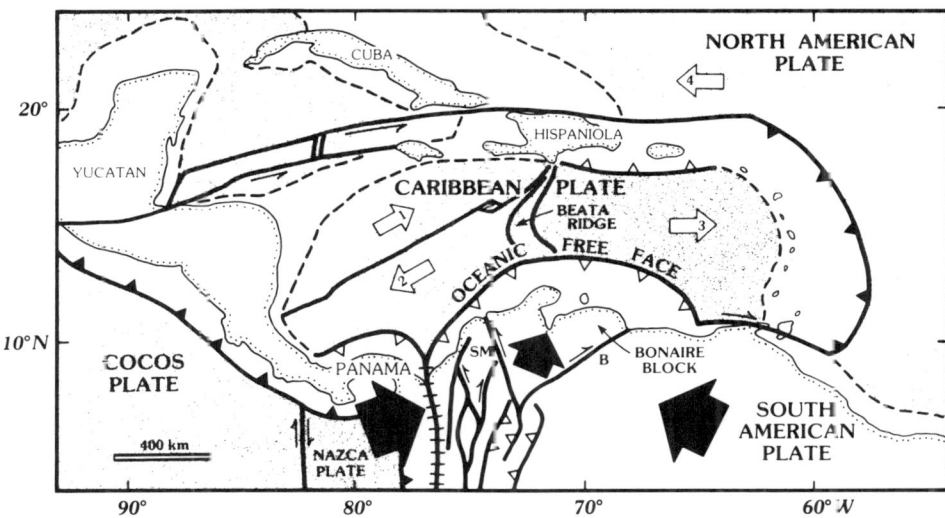

Figure 6. Collision between the Panama-island-arc and Colombia about 5 Ma ago was associated with initiation of an episode of Tectonic Escape within the Colombian Andes and in the Bonaire block bounded by the Santa Marta (SM) and Bocono (B) faults. Internal deformation of the Caribbean includes: uplift of the Beata ridge, eastward escape of the Venezuelan Basin (arrow 3) south-westward escape of the Colombian basin (arrow 2) and north-eastward escape of the Nicaraguan Ridge (arrow 1). All this internal deformation is related to the collision.

Himalayan collision has only just begun and, because much of the record lies in non-marine strata in China, must await establishment of detailed stratigraphic successions and correlations. Studies of large and major earthquakes of the last 80 years by Molnar and Deng (1984) have shown that "The overall strain field is consistent with a large part of India's penetration into Eurasia being absorbed by the extrusion of material out of India's way." These authors suggest that Southeast Asia is moving southeastward with respect to the rest of Asia at about 20 mm/yr (within a factor of 2). Deng et al., (1984) have illustrated by field studies in Ningsia-Hui the characteristic interrelations of strike-slip and thrust-faulting in a part of the escape zone and Deng's (Deng and Fan, 1980) study of the Shansi graben provides evidence from bore-holes indicating that these major extensional elements in the Tectonic Escape of Southeast Asia did not exist until the beginning of the Quaternary (about 2 ma ago). Deng's evidence for a youthful age of the Shansi graben contrasts with a suggestion that they represent a relatively old element in south-east Asian escape (Tapponnier et al., 1983) and illustrates the kind of complex time dependent behavior that is to be expected as blocks (scholle) jostle one another in the plate boundary zone.

Reconstruction of the geological situation in China and neighboring southeast Asia immediately prior to the Himalayan collision (Figure 5) shows features leading to the recognition of a different fault pattern, but one that is also suggestive of Tectonic Escape. The key to interpretation of this pattern is that the Po Hai basin is a giant pull-apart on the Tan Lu fault system. The sense of motion on the Tan Lu fault system has been much debated, but active motion, which relates to the Himalayan collision is right-lateral (Molnar and Deng, 1984) and we suggest (from the shape of Po Hai subbasins) that early Cenozoic motion was in the same sense. The age of formation of Po Hai rifts is difficult to pin down because they are occupied by nonmarine strata, but the rift fill overlies rocks as young as Cretaceous and its greatest age is probably Paleocene (Li Desheng, 1984). Rifting thus appears to have begun in the Late Cretaceous prior to the Himalayan collision.

South of latitude 35°N the Tan Lu system is not well defined (Figure 5) and its place is taken by a more diffuse belt of deformation with numerous half-graben (Geological Map of China, 1976) whose fill ranges in age from Late Cretaceous to Oligocene. Stratigraphic successions are mainly nonmarine in the rifts and an Eocene marine transgression recognized in the Funing Formation about 1 km above the base of the Subei basin succession (at 30°E 31°N) provides the earliest marine control on the age of rifting (Huang Vianzhi, personal communication, 1983). The geological map of China at 1:4 M shows that the strike of the Paleogene rifts of South China changes from about E-W at 35°N to about NE-SW close to the coast (Figure 5). These rifts are also known offshore where they subsided substantially some 30 Ma after their initiation during formation of the South China Sea, an event which took place between 31 and 17 Ma ago (Taylor and Hayes, 1980) apparently under the influence of the Himalayan collision (Tapponnier et al., in press).

We interpret the Tan Lu fault system, the Po Hai rifts and the South China rifts as all recording a Paleogene episode of Tectonic Escape with the free face off the south coast of China where, significantly, Andean arc igneous activity shut off at the end of the Cretaceous (Li Desheng, 1984). The site of collision lay to the north where a substantial object (1500 km long by 500 km wide) collided with Asia along the Wusuli (Ussuri) River in Late Cretaceous time (Figure 6). We identify this suture on the basis of the contrasting stratigraphies and intrusive history on either side (Kosygin and Parfenov, 1975). The boundary between the Kyongsang Cretaceous arc province and the rest of Korea (Sillitoe, 1977) we regard as its southern extension.

In summary we characterize the Cenozoic geology of China as dominated by two episodes of Tectonic Escape, the older a response to collision between Greater Japan and Asia along the Ussuri (Wusuli) River and the younger a response to the more familiar collision between India and Asia along the Indus-Yarlung Tsangpo suture zone. Ideas somewhat similar to those expressed here about the internal deformation of the Chinese continental lithosphere were developed by J. S. Lee (1939), but in the absence of the kind of knowledge that now exists about the timing of tectonic events, he interpreted what we would regard as sequential though similar events as a continuous style of behavior through Phanerozoic time.

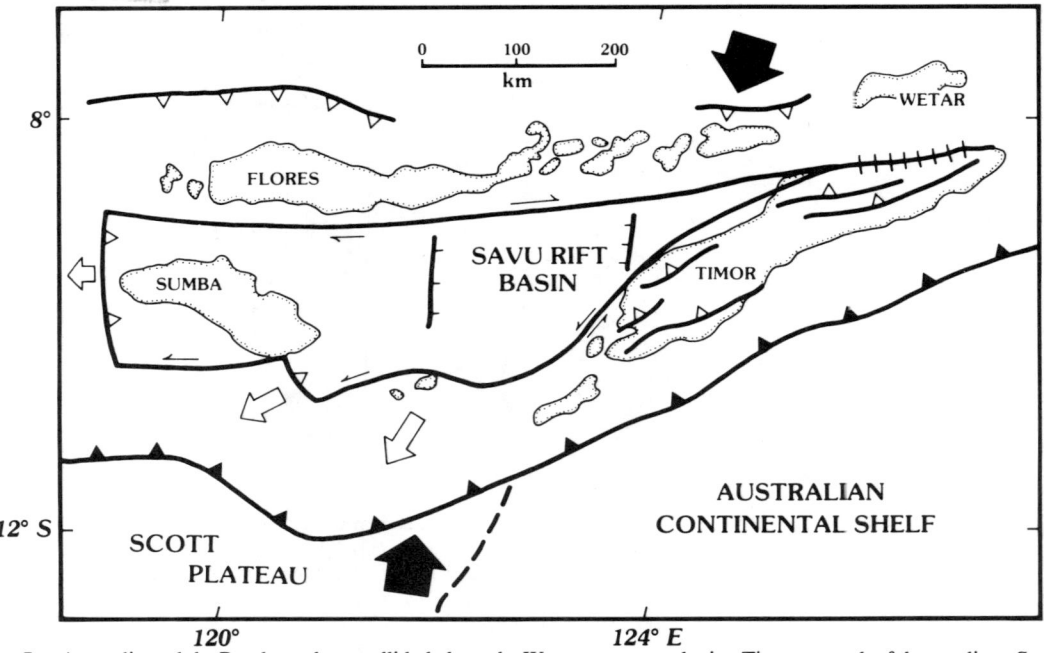

Figure 7. Australia and the Banda arc have collided along the Wetar suture producing Timor, a stack of thrust-slices. Sumba is escaping toward the west and south-west and the Savu rift has formed behind it.

Colombia-Panama

Our discussion of Tectonic Escape concentrates on the geometry of fault patterns and the timing of events. We do not attempt to address the nature and distribution of the forces involved except to observe (1) that structures within areas of Tectonic Escape can readily be treated as accommodation features related to "jostling" of blocks and (2) that forces besides those usually thought of as driving plates (such as slab pull and ridge push) need to be considered. The great variation in the areas over which Tectonic Escape takes place that is clear from the three examples described so far and becomes even clearer in the case of the arc-continent collision presently taking place in Colombia and neighboring parts of the Caribbean (Wadge and Burke, 1983; Mann and Burke, in press) perhaps helps to illuminate the varied scale of the forces involved.

About 5 Ma ago the end of an island arc at that time forming southern Central America began to impinge on the then west coast of Colombia (Wadge and Burke, 1983, Figure 1). Continuing convergence has been accommodated by the bending of the arc to the south, forming a suture zone which marks the boundary between older South America and the newly accreted arc, roughly along the 77th meridian (Figure 6).

Since the collision began, Panama on the western side of the suture has been deformed into its familiar "S" shape while a major episode of Tectonic Escape toward a free face in the Caribbean has begun on the eastern side. The effects of this Tectonic Escape can be seen in three main areas, each with a different structural style. These areas are the Northern Andes of Colombia, the Bonaire block and the Caribbean ocean floor itself.

The Colombian Andes accomodate escape largely by strike-slip motion on faults trending approximately N-S parallel to the fabric of the older Cenozoic and Mesozoic areas. The Bonaire block is bounded on the east and west by the Bocono and Santa Marta faults (Figure 6), each with an offset of about 100 km and on the north by a broad zone of compression which lies largely offshore and in which it is possible to discern evidence of a change (within the last 10 Ma) from right-lateral motion of the Caribbean with respect to South America to northward thrusting of South America over the Caribbean (Mann and Burke, 1984).

The Caribbean ocean floor is about 1 km shallower than its age would lead one to expect and this has long been interpreted (Burke et al., 1978) as indicating that it represents an oceanic plateau. Crustal thickness in most of the Caribbean is 15 to 20 km and perhaps because of the large proportion of basalt in that thickness the Caribbean lithosphere is responding to the escape motion of the Bonaire block by extensive internal deformation. In this respect the Caribbean is behaving much more like continental material such as China than like typical ocean floor. In typical ocean floor, the lithosphere is predominantly made of olivine and rigid plates with sharp boundaries are the rule.

The response of the Caribbean to the collision is depicted in Figure 6: the Beata ridge is being uplifted, the Venezuela basin is deforming with a fault pattern that indicates motion toward the east (Burke et al., 1978, Figure 3 and arrow 3 in Figure 6). The Colombian basin is moving south-west with respect to the Hess Escarpment (arrow 2 in Figure 7), a feature separating it from the Nicaraguan Rise, which is moving to the north-east (arrow 1 in Figure 6).

Timor and Sumba

The collision between Australia and the Banda arc in Timor (Hamilton, 1979; Audley-Charles, 1983) is the best known of a small set of current examples of an active collision between an island arc and a continent, but ancient examples show that this has been a common process throughout earth history (Burke et al., 1984).

Burke and Dewey (1974), in considering the consequences of the headland and embayment structure of Atlantic type margins for arc-continent collisions predicted that strike-slip motion mainly in the accretionary prism of the arc would be a common condition early during collision where a headland had impinged on an arc, but

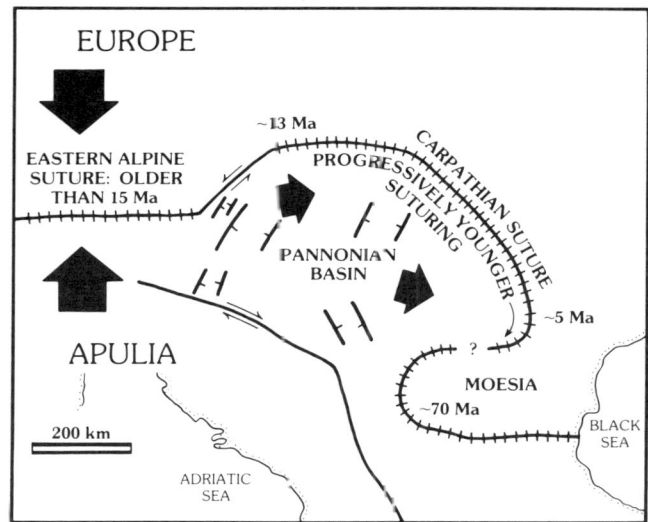

Figure 8. Rocks in the Pannonian basin have moved eastward escaping the effects of Alpine suturing as the Carpathian suture has progressively closed. Rifts in the basin formally indicated.

embayments still faced a trench. This geometry exists at Timor and Figure 7 represents the preliminary results of a study based on interpreting teleseismic and reflection seismic data and on land geology which indicates that the island of Sumba is escaping westward from the Timor collision and that the Savu basin represents a late Neogene rift formed in the escape process.

Pannonian Basin

Tectonic Escape, although it has not been given that name, is a familiar process in the Alpine system and numerous examples illustrating movement toward free-faces have been recognized (see for example, Tapponnier, 1977, for a comprehensive discussion). We here confine our treatment to one well-studied and impressive example. The Carpathian mountains (Figure 8) record progressively younger Neogene suturing best documented by the shutting off of andesitic volcanism and by the timing of thrusting (Royden et al., 1983, Burchfiel et al., 1982). Inside the curve of the Carpathian mountains motion of the lithosphere eastward, toward the free face provided by the closing ocean has been accomodated on strike-slip faults and by the formation of the numerous rifts of the Pannonian basin. As in previous examples the two basic reasons for identifying an episode of Tectonic Escape are (1) the recognition of a fault pattern suggestive of motion toward the free face and (2) the timing of the onset of the escape episode which, in this case, was about 15 Ma ago, subsequent to the East Alpine collision (Dietrich, 1976).

Sung Liao Basin

The recognition of episodes of Tectonic Escape becomes more difficult as the tectonic environment in which the episode took place becomes older because less of the elements that define the episode are preserved. The free face is less likely to be preserved and no pre-Cenozoic free faces persist as ocean floor. For this reason the recognition of Mesozoic and earlier episodes of Tectonic Escape becomes difficult. The main reason for suggesting that Tectonic Escape is recorded in these older cases is that it provides a hypothesis whose validity can be tested with existing data or by the acquisition of new data (including perhaps reflection seismic data) in the area under study. The examples of Tectonic Escape outlined in the following sections are all from areas where more detailed mapping of fault patterns and acquisition of better evidence on the timing of events should go far toward resolving whether or not Tectonic Escape has happened.

The Jurassic Sung Liao basin in China (Figure 9) is occupied by a complex of rifts containing up to 6 km of nonmarine Late Jurassic and Cretaceous sediments overlying older Jurassic volcanic rocks (Chin, 1980). Reconstruction of the tectonics of China and neighboring areas at this time, shortly after North and South China had been sutured along the Qinling-Dabieshan, is far from certain and we have chosen to depict in Figure 9 a version similar to that of Klimetz (1983) in which

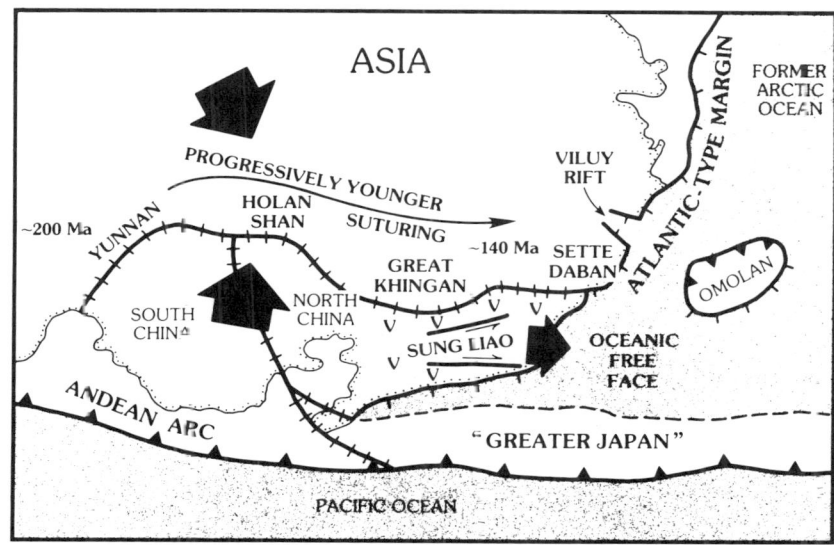

Figure 9. The Sung Liao basin depicted as forming by Tectonic Escape toward the then Arctic Ocean during medial Jurassic time before north-eastern Siberia was assembled.

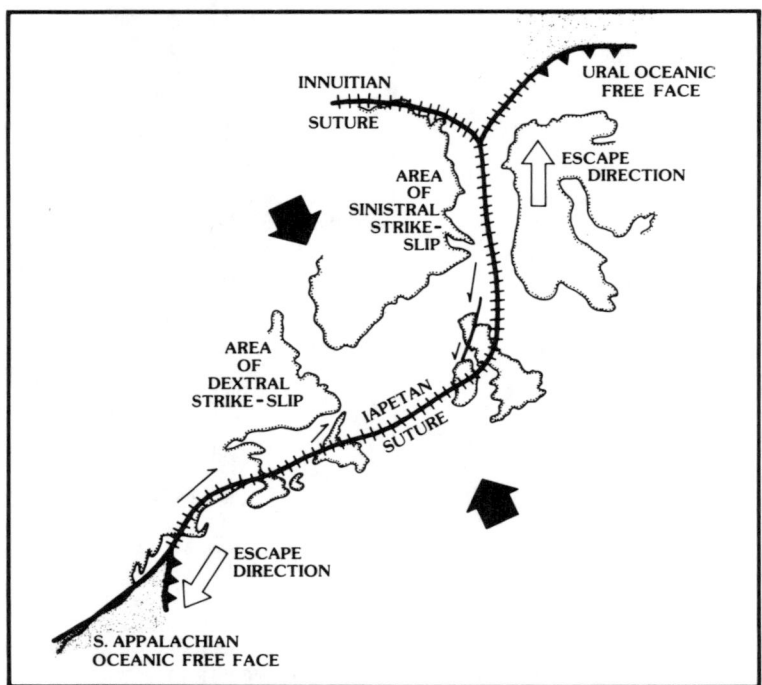

Figure 10. Iapetan suturing during the Late Silurian and Devonian progressed from Scandinavia, through the British Isles to New England. Tectonic Escape toward free faces on the site of the Urals in one direction and the Southern Appalachians in the other may have defined areas of sinistral and dextral strike-slip motion, just as similar areas occur on either side of the Pamir node in Asia today.

the extensive medial Jurassic volcanic and granitic igneous activity are linked to continental thickening following a possible collision in the Great Khingan mountains. Both Klimetz (1983) and Sengör and Hsu (in press) attribute the origin of the Sung Liao basin to east-west strike-slip motion on the old Tumen suture zone, but because at this time the assembly of northeastern Siberia had not yet taken place and because Greater Japan had not yet reached Asia (Figure 9), the postulated collision could have been accompanied by an episode of escape toward the present site of the Sea of Okhotsk. We suggest that the Sung Liao basin formed during such an episode of escape. Our hypothesis requires that the bounding faults of the Sung-Liao basin and many of its internal faults are strike-slip faults. We are not aware that earlier students of Sung Liao geology have considered this possibility (see for example, Fan, 1980; and Klimetz, 1983), but our examination of small-scale maps and LANDSAT images indicates that it is a possibility. Structures established in the oil exploration of the Sung-Liao basin may indicate whether strike-slip motion has been important in its evolution just as those mapped in the younger Po Hai basin further south (Li Desheng, 1980, 1984) are clearly indicative of right lateral motion on the Tan-Lu system.

West Siberian Basin Rifts

The West Siberian basin consists of a broad depression filled with Jurassic and Cretaceous sediments overlying a group of north-south trending Triassic rifts which are only locally exposed and which have been mapped by airborne magnetometry (Shablinskaya, 1977). The key to recognition of an episode of Triassic Tectonic Escape in the formation of these Triassic rifts is that the collision of Siberia and Europe in the Urals is diachronous beginning about 265 Ma ago and being progressively younger in Novaya Zemlya to the north and in the Taymyr peninsula to the northeast where intrusions associated with suturing may be as young as 230 Ma (Pogrebitski, 1971).

At the time of the Permian collision in the Urals the margin of Siberia was occupied by an Andean arc and the coast of Europe by an Atlantic margin. Post collisional thickening of Tibetan style is indicated by the occurrence of sporadic granitic intrusions of Permian age for up to 300 km into Siberia (Geol. Map of Asia 1:5M, Peking, 1976) and by the occurrence of highly potassic volcanic rocks on the eastern slopes of the Urals (Ivanov, 1979). We suggest that post-collisional convergence continuing into the Triassic was accommodated by northward (present coordinates) escape toward Novaya Zemlya and Taymyr and that the Triassic rifts of western Siberia with their thick volcanic and sedimentary sequences record this motion.

Rifting along N-S normal faults continued throughout the Triassic and earliest Jurassic extending farther south into the area of the Porte of Turgay where enormous outpourings of basalt were associated with normal faulting (Koltchanoff, 1964). This later Triassic and early Jurassic rifting was no longer associated with Tectonic Escape related to the Uralide collision, but was an impactogenal development at high angles to the Cimmeride collisions to the south in northern Iran and Afghanistan (Sengör, 1984). This impactogenal event exploited the escape-related rifts and led to renewed faulting on the site of the West Siberian Basin area. Only in the middle Jurassic did fault activity cease and postrifting subsidence of the entire basin complex begin.

Caledonides and Appalachians

In their analysis of the Himalayan collision Molnar and Tapponier (1975) showed that while Asia east of the Pamirs is moving toward the

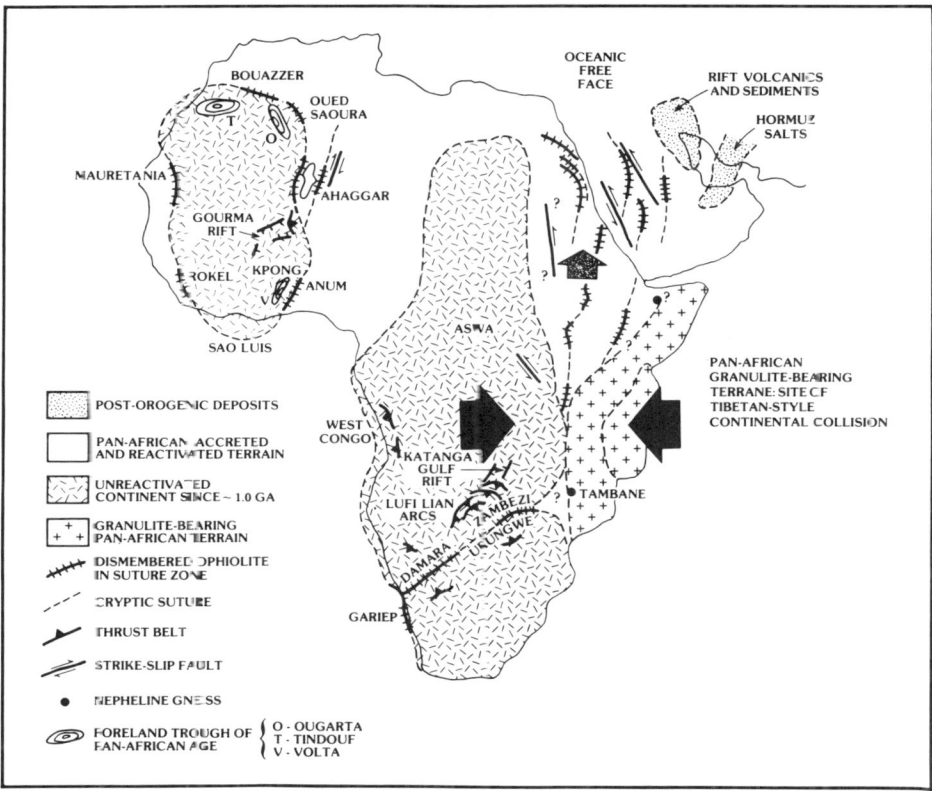

Figure 11. Pan African granulites (indicative of Tibetan style thickening) characterize Kenya and the Mozambique belt. The area farther north in Sudan, Egypt and Saudi Arabia was characterized by arc collisions and late strike-slip faulting indicating Tectonic Escape. Another episode of Tectonic Escape of similar age is recorded in the Ahaggar.

Pacific free-face in response to collision, a smaller area west of the Pamirs is moving westward toward the free-face represented by the Caspian Sea. Strike-slip faults accommodating the motion east of the Pamir are dominantly sinistral and those west of the Pamir are dominantly dextral.

A possibly analogous situation with strike-slip motion in opposite directions is recorded in the mid-Paleozoic Caledonides and Appalachians (Figure 10). Suturing in the Caledonides of the British Isles was complete by about 410 Ma ago (Dewey, 1982, Figure 36) having gradually given way to sinistral strike-slip motion (Watson, 1984, Figure 9). Along strike in the northern Appalachians suturing was progressively later south-westward (present coordinates) and was complete by about 380 Ma giving way to dextral strike-slip by late Devonian times (about 370 Ma). This progressive suturing from Scandinavia to New England carried the Caledonide-Appalachian mountain belt through a 50° change of strike (Figure 10; see also Dewey, 1982, Figure 36) on the site of the British Isles. We interpret this change of strike as marking a nodal collisional area (cf. Dewey and Kidd, 1974) with motion on one side toward an oceanic free-face on the site of the Urals accommodated by sinistral strike-slip faults (Watson, 1984) and motion on the other side toward an oceanic free-face on the site of the Southern Appalachians accommodated by dextral strike-slip faults (Bradley, 1982). We point out that the former area is characterized more by igneous activity indicative of continental thickening (Watson, 1984) and the latter by extension and the development of major pull-apart basins (Bradley, 1982).

Pan-African

The Pan-African (Kennedy, 1964) has long been recognized as recording an episode of Tibetan-style continental thickening mainly because of its vast areal extent and because of the amount of reworking of older continental material that it involves (e.g., Burke and Dewey, 1972). However, it has become clear during the last decade that large areas of Pan-African terrane in Egypt and Saudi Arabia consist of is and arcs that were sutured to the continent during the interval 900 Ma to 600 Ma and, in contrast to much of the rest of Africa, do not involve substantial reactivation of older continental crust (Gass, 1977; Fleck et al., 1980). We have sketched the relationship of this newly added arc material to the rest of Africa in Figure 11 in which we have also shown: (1) that major Pan-African granulites (indicating an episode of continental thickening to about 70 km) lie south of the Arabian-Egyptian accreted zone in the Mozambique belt of Kenya and more southerly East Africa; (2) that although there are numerous Pan-African suture zones in Egypt, Sudan and Saudi Arabia only one (in the Cherengani Hills) has been recognized in Kenya farther south; (3) that the Najd fault system of Arabia is a strike-slip system dominated by NW-striking faults but with a complementary N-S to NNE-SSW fault set; (4) that continental collision indicating the obliteration of an ocean took place during Ordovician time in southwestern Turkey which was then located in the present eastern Mediterranean (Sengör et al., in press). This collision took place some 100 to 150 Ma later than the main Pan-African collisions of northeastern Afro-Arabia and is located

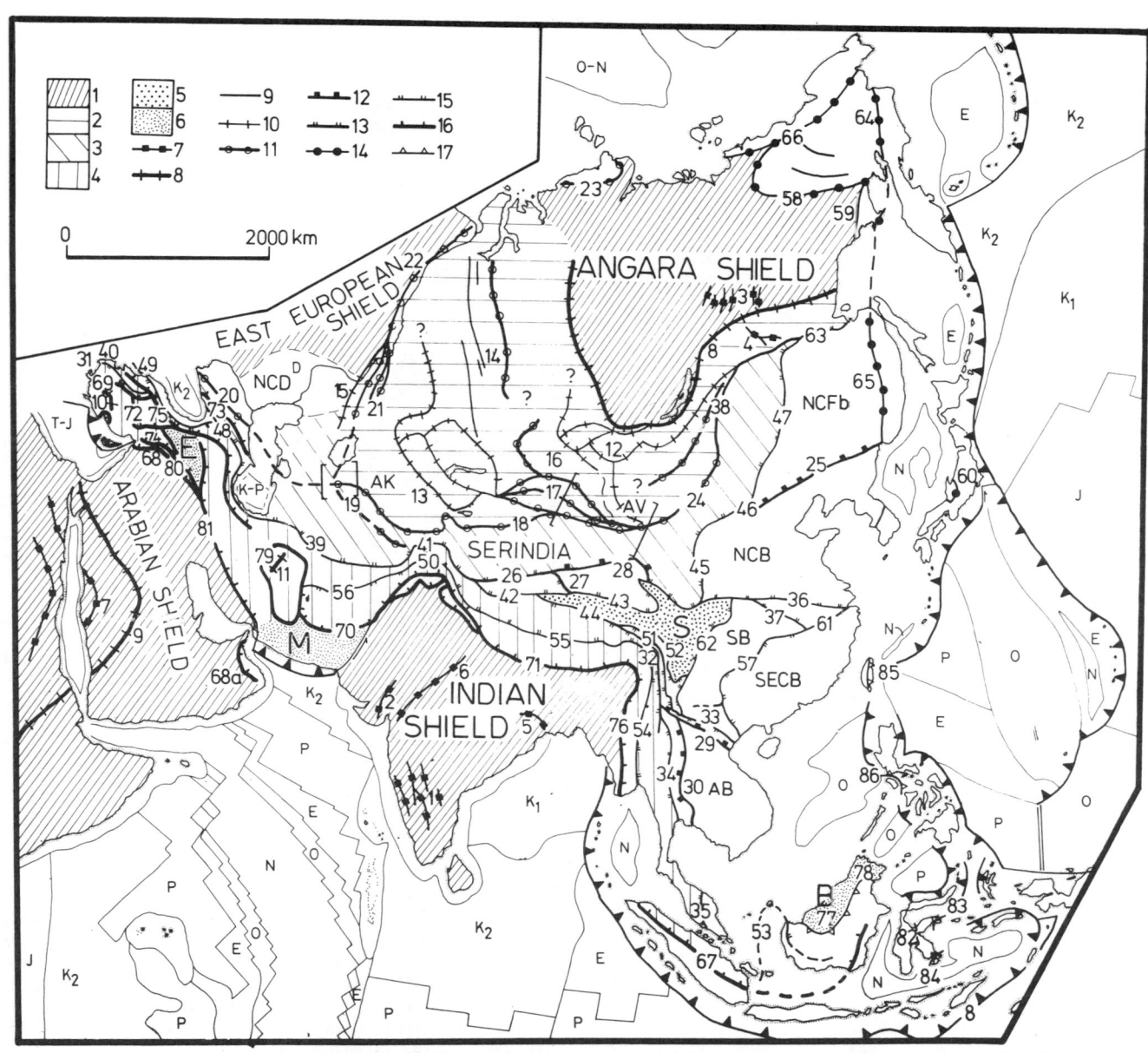

Figure 12. Eurasia is illustrated as a continent that has been assembled during the Phanerozoic. The oldest parts of Eurasia representing Archean and older Proterozoic continental and island arc fragments are shown with an oblique line ornament. Suture zones, marking places where oceans have closed and arcs and continental fragments have been added to the larger continents, are numbered and ornamented according to age. The Songpan-Ganzi (S) and Makran (M) both of which are vast areas of dominantly accretionary wedge sediments are also identified.

"in front" of the pieces driven along the Najd faults (Figure 11). The ocean that vanished as a result of the Ordovician closure in Turkey was perhaps the "free-face" towards which escape occurred. This "belated closure" was thus a part of the Pan African system (Sengör et al., in press) and not an ill-defined "Caledonian" event, as has been suggested. The rifts associated with this escape produced a strong bimodal volcanism and formed the basins in which the Hormuz salts were deposited. As Sengör et al., (in press) pointed out, the depositional basins containing the Hormuz salt related to the Pan-African collisional events, and thus the salt was an "ocean-closing salt" and not, as hitherto hypothesized, an "ocean-opening salt."

Interpreting these observations in terms of collision, we would regard the granulite area as an analogue of Tibet today and Sudan, Egypt and Saudi Arabia as a terrane of assembled arcs escaping to the north (present coordinates). This episode of escape was largely accommodated on the Najd faults and was accompanied by extension in the northeasterly trending rifts within which the Hammamat sediments and Dokhan volcanics of the eastern desert of Egypt were deposited (Stern et al., 1984).

A second episode of Tectonic Escape of Pan-African age has been recognized by Caby and his colleagues (Caby et al., 1981) who have mapped huge strike-slip faults in the Ahaggar and compared them to those of China today (Caby et al., 1981, Figure 16.8).

Older Examples of Tectonic Escape

Older examples of Tectonic Escape are increasingly difficult to recognize simply because of the limited preservation of evidence. Perhaps the best case for an episode of escape has been made by King and Zietz (1978) who compared the New York-Alabama lineament (marking a 1200-km-long truncation of gravity and magnetic trends) to the Altyn Tagh strike-slip fault. This Albany-Alabama lineament appears to have been established in Grenville times during an episode of Tibetan style collision (Dewey and Burke, 1973) and, like the Altyn Tagh fault, could well represent a major fracture of the continent accommodating escape.

Major strike-slip faults of older Proterozoic and Archean age are well known (e.g., Borradaile, 1984) and judging by their map patterns some may record episodes of Tectonic Escape. The difficulty lies in correlating them with near contemporary collisional events and identifying the sites of free-faces which existed at the appropriate time.

Summary and Conclusions

It is unlikely that any of the continental material preserved on earth today was produced by processes significantly different from those that operate now and which have operated in the assembly of continents, such as Asia during the Phanerozoic (Figure 12). Most of these processes produce crustal material of quite variable composition and variable depths to the Mohorovicic discontinuity, but rifting is a widespread process that smooths the Mohorovicic discontinuity by deep crustal flow and may be dominant in producing extensive areas of continental crust of uniform thickness.

Tectonic Escape is a process that is and has been widespread in its incidence and may have been very important in the evolution of the continents. Recognition of the extent to which it has operated may be very difficult, especially in older continents. For this reason, attempting to restore the structure of a mountain belt by such methods as that of the drawing of balanced cross-sections may be unrealistic (e.g., Williams and Fischer, 1984). It may also be very hard to recognize the incidence of Tectonic Escape on deep seismic profiles.

To prevent this paper becoming unduly long, we have not attempted to describe all examples of Tectonic Escape, only to provide a set of illustrative examples and, for the same reason, we have not attempted to discuss the forces involved in Tectonic Escape. It is appropriate to point out here that the mechanism considered responsible for escape by McKenzie and England (1983), who relate the escape process to collisional elevation, is quite different from either of the two introduced by Molnar and Tapponnier (1975), although collisional elevation does not appear to have provided the driving force in the present well-documented Turkish episode of Tectonic Escape.

An obvious path for future research in areas of ancient Tectonic Escape is in refining evidence for progressive suturing in one direction. Evidence of the required kind will have to come from detailed field studies and especially stratigraphic studies. It is perhaps appropriate to emphasize that in these cases application of the classical techniques of geological research will be providing an essential framework for the interpretation of sophisticated deep seismic reflection data.

Acknowledgments. Part of this work was undertaken at the Lunar and Planetary Institute, which is operated by the Universities Space Research Association under Contract No. NASW-3389 with the National Aeronautics and Space Administration. This paper is Lunar and Planetary Institute Contribution No. 565.

References

Audley-Charles, M. G., Comments on Analogous tectonic evolution of the Ordovician foredeeps, southern and central Appalachians. *Geology, 11*, 490–491, 1983.

Beijing, Geological Map of China, Scale 1:4M, 1976.

Ben-Avraham, Z., A. Nur, D. Jones, and A. Cox, Continental Accretion: From Oceanic Plateaus to Allochthonous Terranes, *Science, 213*, 47–54, 1981.

Berkland, J. O., Paleogene "frozen" subduction zone in the coast ranges of northern California, in *International Geological Congress, 24th session, Section 3, Tectonics*, edited by J. E. Gill, pp. 99–105, IGC, Ottawa, Canada, 1972.

Borradaile, G. J. and W. M. Schwerdtner, Horizontal shortening of upward-facing greenstone structures in the southern Superior Province, Canadian Shield, *Can. J. Earth Sci., 21*, 611–615, 1984.

Bott, M. H. P., Subsidence mechanisms at passive continental margins, in *Geological and Geophysical Investigations of Continental Margins*, edited by J. S. Watkins, L. Montadert and P. W. Dickerson, pp. 3–9, The American Association of Petroleum Geologists, Tulsa, Oklahoma, 1979.

Bradley, D. C., Subsidence in late Paleozoic basins in the northern Appalachians, *Tectonics, 1*, 107–123, 1982.

Burchfiel, B. C. and L. Royden, Carpathian Foreland fold and thrust belt and its relation to Pannonian and other basins, *AAPG Bulletin, 66* (9), 1179–1195, 1982.

Burke, K., C. Cooper, J. F. Dewey, P. Mann, and J. L. Pindell, Caribbean tectonics and relative plate motions. *Geol. Soc. Am. Memoir 162*, 31–63. 1984.

Burke, K. C. and J. F. Dewey, Orogeny in Africa, in *Proceedings of the Conference on African Geology*, edited by T. F. J. Dessauvagie and A. J. Whiteman, pp. 583–608. Univ. of Ibadan, Ibadan, Nigeria, 1972.

Burke, K. and J. Dewey, Hot spots and continental breakup: implications for collisional orogeny, *Geology, 2*, 57–60, 1974.

Burke, K., J. F. Dewey and W. S. F. Kidd, Dominance of horizontal movements, arc and microcontinental collisions during the later permobile regime, in *The Early History of the Earth*, edited by B. F. Windley, pp. 113–129, Wiley Interscience, London, 1976.

Burke, K, P. J. Fox and A. M. C. Sengör, Buoyant ocean floor and the evolution of the Caribbean, *Journal of Geophysical Research, 83*, 3949–3954, 1978.

Burke, K. and J. T. Wilson, Is the African Plate stationary? *Nature, 239*, 387–390, Sketch maps, 1972.

Caby, R., J. M. L. Bertrand and R. Black, Pan-African closure and continental collision in the Hoggar-Iforas segment, central Sahara, in *Precambrian Plate Tectonics*, edited by A. Kroner, pp. 407–434, Elsevier, Amsterdam, 1981.

Charpal, O. de, P. Guennoc, L. Montadert and D. C. Roberts, Rifting, crustal attenuation, and subsidence in the Bay of Biscay. *Nature, 275*, 706–711, 1978.

Chin, C., Non-marine setting of petroleum in the Songliao Basin, China, *J. Pet. Geol., 2*, 300–322, 1980.

Cooke, D. L. and W. W. Moorhouse, Timiskaming volcanism in the Kirkland Lake area, Ontario, Canada, *Canadian J. Earth Sci., 6*, 117–132, 1970.

Deng, Q. and F. Fan, Cenozoic and recent geotectonic characteristics of the North China fault block region, in *Formation and Development of the North China Fault Block Region*, Chinese Transl., 192–205, 1980.

Dewey, J. F. Plate tectonics and the evolution of the British Isles, *J. Geol. Soc. London, 139*, 371–412, 1982.

Dewey, J. F. and W. S. F. Kidd, Continental collisions in the Appalachian-Caledonian orogenic belt: variations related to complete and incomplete suturing, *Geology, 2*, 543–546, . sketch map, 1974.

Dewey, J. F. and B. F. Windley, Growth and differentiation of the continental crust, *Phil. Trans. R. Lond. A, 301*, 189–206, 1981.

de Wit, M. J., Gliding and overthrust nappe tectonics in the Barberton Greenstone Belt, *J. Struc. Geol., 4*, 117–136, 1982.

Dietrich, V., Evolution of the Eastern Alps; a plate tectonics working hypothesis, *Geology, 4*, 147–152, illus. (sects., sketch map), 1976.

Dietz, R. S., Morphologic fits of North America/Africa and Gondwana, A review, in *Implications of Continental Drift to the Earth Sciences 2*, edited by D. H. Tarling and S. K. Runcorn, pp. 865–872, Academic Press London, 1973.

England, P., Some numerical investigations of large-scale continental deformation, in *Mountain Building Processes*, edited by K. J. Hsu, pp. 129–139, Academic Press, London, 1982.

England, P. and D. McKenzie, A thin viscous sheet model for continental deformation, *Geophys. Jour. R. Astron. Soc.*, 70, 295–321, 1982.

England, P. and D. McKenzie, Correction to: a thin viscous sheet model for continental deformation, *Geophys. Jour. Roy. Astron. Soc., 73*, 523–532, 1983.

Fan, P. F., Geology of the Sungliao Basin, China, evolution of a failed rift system, *Geol. Soc. Am. Abstr. Programs, 11*, (7), 423, 1980.

Fleck, R. J., W. R. Greenwood, D. G. Hadley, R. E. Anderson and D. L. Schmidt, Rubidium-strontium geochronology and plate-tectonic evolution of the southern part of the Arabian Shield, *U.S. Geol. Surv. Prof. Pap., 1131*, 38 pp ., GPO, Washington, 1980.

Florensov, N. A., Rifts of the Baikal Mountain region, *Tectonophysics, 8*, 443–456, 1969.

Gass, I. G., The evolution of the Pan African crystalline basement in NE Africa and Arabia, *J. Geol. Soc. London, 134*, 129–138, 1977.

Gass, I. G., Pan-African (upper Proterozoic) plate tectonics of the African-Nubian Shield, in *Precambrian Plate Tectonics*, edited by A. Kroner, pp. 387–405, Elsevier, Amsterdam, 1981.

Gelati, R., Miocene marine sequence from Lake Van, Eastern Turkey: *Riv. ital. Palaeont. Stratigr., 81*, 477–490, 1975.

Hamilton, W., Tectonics of the Indonesian region, *U. S. Geol. Surv. Prof. Pap., 1078*, 345 pp, 1979.

Hirn, A., A. Nercessian, M. Sapin, G. Jobert, X. Z. Xin, G. E. Yuan, L. D. Yuan, and T. J. Wen, Lhasa block and bordering sutures-a continuation of a 500-km Moho traverse through Tibet, *Nature, 307*, 25–27, 1984.

Hughes, G. W. and C. C. Turner, Upraised Pacific floor, southern Malaita, Solomon Islands, *Geol. Soc. Am. Bull., 88*, 412–414, 1977.

Hutchinson, D. Structure and Tectonics of the Long Island Platform. Ph.D. Thesis, 289 pp., University of Rhode Island, 1984.

Ivanov, K. S., Ultrapotassic liparite-porphyries in the Urals (in Russian), *Doklady Akademii Nauk SSSR, 247*, 908–912, 1979.

Karig, D. E. and R. W. Kay, Fate of sediments on the descending plate at convergent margins, *Phil. Trans. R. Soc. Lond. A., 301*, 233–251, 1981.

Karson, J. A. and J. F. Dewey, Coastal complex, western Newfoundland: an early Ordovician oceanic fracture zone, *Geol. Soc. Am. Bull, 89*, 1037–1049, 1978.

Kennedy, W. Q., The structural differentiation of Africa in the Pan-African (+ 500 m. y.) tectonic episode, in *Leeds University Research Institute of African Geology 8th Annual Report on Scientific Results*, pp. 48–49, 1964.

Ketin, I., Uber die tektonisch-mechanischen folgerungen aus den grossen anatolischen erdbeben des letzten dezenniums, *Geol. Rundsch., 36*, 77–83, 1948.

King, E. R. and I. Zietz, The New York-Alabama lineament; geophysical evidence for a major crustal break in the basement beneath the Appalachian Basin, *Geology, 6*, 312–318, sketch maps, 1978.

Klimetz, M. P., Speculations on the Mesozoic plate tectonic evolution of eastern China, *Tectonics, 2*, 139–166, 1983.

Koesoemadinata, R. P., Sedimentary framework of tertiary coal basins of Indonesia, in *Third Regional Conference on Geology and Mineral Resources of Southeast Asia*, edited by P. Nutalaya, Bangkok, Thailand, pp. 621–639, 1978.

Koltchanoff, V., The Turgay Trough, in *Tectonics of Europe*; Explanatory note to the tectonic map of Europe, Scale 1:2,500,000, edited by A. A. Bagdanoff, M. V. Mouratov and N. S. Shatskiy, pp. 129–132, State Printing House, Moscow, 1964.

Kosygin, Uy. A. and L. M. Parfenov, Structural evolution of eastern Siberia and adjacent areas, *Am. Jour. Sci., 275–A*, 187–208, 1975.

Lawrence, S. R., L. H. Barker, A. Soulsby and P. Payne, A geological and geophysical investigation of Barbados and the application of an accretionary prism model, presented at 10th Caribbean Geological Conference, Cartagena, Colombia, August, 1983.

Lee, J. S., *The Geology of China*, 399 pp. Thomas Murby and Co., London, 1939.

Li Desheng, Geological structure and hydrocarbon occurrence of the Bohai Gulf oil and gas basin (China), in *Petroleum Geology in China*, edited by J. F. Mason, pp. 180–192, Pennwell Books, Tulsa, Oklahoma, 1980.

Li Desheng, Geologic evolution of petroliferous basins on continental shelf of China, *AAPG Bull., 68*, 993–1003, 1984.

Liittig, G. and P. Steffens, Explanatory notes for the palaeogeographic atlas of Turkey from the Oligocene to the Pleistocene, *Burdesanstalt fur geowiss enschafter und Rohstaffe*, 64 pp. and 7 maps, Hannover, 1976.

Mann, P. and K. Burke, Neotectonics of the Caribbean, *Rev. Geophys. Space Phys., 22*, 309–362, 1984.

McCall, G. J. H. and R. G. W. Kidd, The Makran, Southeastern Iran: the anatomy of a convergent plate margin active from cretaceous to present, in *Trench-Forearc Geology: Sedimentation and Tectonics on Modern and Ancient Active Plate Margins*, edited by J. K. Leggett, pp. 387–397, Geol. Soc. Lond., Blackwell, Oxford, 1982.

Molnar, P. and Q. Deng, Faulting associated with large earthquakes and the average rate of deformation in central and eastern Asia, *J. Geophys. Res., 89*, 6203–6227, 1984.

Molnar, P. and P. Tapponnier, Cenozoic tectonics of Asia: effects of a continental collision, *Science, 189*, 419–425, 1975.

Morgan, P., The thermal structure and thermal evolution of the continental lithosphere, *Phys. Chem. of the Earth*, in press, 1984.

Nur, A. and Z. Ben-Avraham, Oceanic plateaus, the fragmentation of continents and mountain building, *J. Geophys. Res., 87*, 3644–3661, 1982.

Pogrebitskiy, Yu. E., Paleotectonic analysis of Taymyrian folded system, *Transactions of the Arctic Institute of Geology, 166*, 248 pp., Leningrad, 1971.

Royden, L., F. Horváth, and J. Rumpler, Evolution of the Pannonian Basin System, 1. Tectonics, *Tectonics, 2*, 63–90, 1983.

Sengör, A. M. C., The Cimmeride Orogenic system and the tectonics of Eurasia, *Geol. Soc. America Spec. Pap.*, 75 pp., 1984.

Sengör, A. M. C., K. Burke and J. F. Dewey, Rifts at high angles to orogenic belts: tests for their origin and the upper Rhine Graben as an example, *Am. Jour. Sci., 278*, 24-40, 1978.

Sengör, A. M. C. and K. J. Hsu, The Cimmerides of Eastern Asia: History of the eastern end of Palaeo-Tethysin, *Mem. Soc. Geol. France*, in press., 1985.

Sengör, A. M. C., M. Satir and R. Akkok, Timing of tectonic events in the Menderes massif, western Turkey implications for tectonics evolution and evidence for Pan-African basement in Turkey: *Tectonics 3*, 693-707, 1984.

Shablinskaya, N. V., Newly discovered very large structure of northwestern Siberia, the buried Yamal-Pul aulacogen, *Doklady Akademii Nauk SSSR, 227*, 71-74, 1977.

Sillitoe, R. H., Metallogeny of an Andean-type continental margin in Korea: implications for opening of the Japan Sea, in *Island Arcs, Dead Sea Trenches and Back-Arc Basins, Maurice Ewing Ser. 1*, edited by M. Talwani and W. C. Pitman, III, pp. 303-310, Am. Geophys. Union, Washington, 1977.

Silver, L. T. and P. H. Schultz, Geological implications of impacts of large asteroids and comets on the Earth. *GSA Special Paper 190*, 528 pp., GSA, Boulder, Colo., 1982.

Sloss, L. L., Synchrony of Phanerozoic sedimentary-tectonic events of the North American craton and the Russian platform, *Int. Geol. Cong.*, 24th Session, Sec. 4, 24-32, 1972.

Stern, R. J., D. Gottfried and C. E. Hedge, Late Precambrian rifting and crustal evolution in the northeastern desert of Egypt, *Geology, 12*, 168-172, 1984.

Talbot, C. J., A plate tectonic model for the Archean crust, *Phil. Trans. Roy. Soc. London, A273*, 413-427, 1973.

Tapponnier, P., Evolution tectonique du Systeme Alpin en Mediterranee: poinconnement et ecrasement rigide-plastique, *Bull. Soc. Geol. France, Ser. 7, 19*, 437-460, 1977.

Tapponnier, P., G. Peltzer and R. Armijo, On the mechanics of the collision between India and Asia, *Geol. Soc. London Journal*, (William Smith meeting on collision), in press, 1985.

Tapponnier, R., G. Peltzer, A. Y LeDain, Roland Armijo, and P. Cobbold, Propagating extrusion tectonics in Asia: new insights from simple experiments with plasticine, *Geology, 10*, 611-616, 1982.

Taylor, B. and D. E. Hayes, The tectonic evolution of the South China Basin, in *The Tectonic and Geologic Evolution of Southeast Asian Seas and Islands*, edited by D. E. Hayes, pp. 89-104, Geophysical Monograph No. 23, AGU, Washington, D.C., 1980.

Thompson, A. B., F. M. Richter, E. Ahrendt, M. J. Bickle, K. Burke, R. F. Dymek, W. Frisch, R. D. Gee, A. Kroner, R. K. O'Nioris, E. R. Oxburgh, and K. Weber, The long-term evolution of the crust and mantle in *Patterns of Change of Earth Evolution*, edited by Holland and Trendall, pp. 389-406, 1984.

Wadge, G. and K. Burke, Neogene Caribbean plate rotation and associated Central American tectonic evolution, *Tectonics, 2*, 633-643, 1983.

Watson, J., The ending of the Caledonian orogeny in Scotland, *J. Geol. Soc. London, 141*, pp. 193-214, 1984.

Williams, G. D. and M. W. Fischer, A balance section across the Pyrenean Orogenic Belt, *Tectonics, 3*, 773-780, 1984.

Wilson, J. T., Static or mobile earth: the current scientific revolution, *Proc. Am. Phil. Soc., 112*, 309-320, 1968.

Winterer, E. L., Marine geology and tectonics, anomalies in the tectonic evolution of the Pacific, in *The Geophysics of the Pacific Ocean Basin and Its Margin*, edited by G. H. Suttor, M. H. Manghnani, and R. Moberly, pp. 269-278, AGU, Washington, D.C., 1976.

MODERN ANALOGS FOR SOME MIDCRUSTAL REFLECTIONS OBSERVED BENEATH COLLISIONAL MOUNTAIN BELTS

Robert J. Lillie

Department of Geology, Oregon State University, Corvallis, Oregon, 97331

Mohammed Yousuf

Oil and Gas Development Corporation, Markaz F/7, Islamabad, Pakistan

Abstract. Seismic reflection data across collisional mountain belts often show prominent events at midcrustal levels. Through an example in the southern Appalachian mountains, three types of sequences are identified and compared to reflection data from modern compressional and extensional settings. Beneath foreland areas, dipping events which lie entirely beneath underthrusted shelf strata are interpreted as late-Precambrian rift graben fill, analogous to Mesozoic rift grabens observed on seismic profiles from the east coast of the United States. More typical features of the foreland, however, are small normal fault offsets of the underthrusted shelf strata. These normal faults commonly result in thrust ramping but apparently are not associated with continental rifting processes. Rather, the faults are similar to features contemporaneous with thrust faulting which have been observed on seismic profiles across areas of modern collisional deformation, such as the Himalayan foreland in Pakistan. The third type of reflection sequence consists of eastward dipping events which lie along the Appalachian gravity and magnetic gradients associated with the edge of precollisional continental basement. Although these dipping sequences may be a product of collisional deformation, it is possible that they represent original volcanic stratigraphy at the early Paleozoic continent/ocean boundary.

Introduction

Many recent seismic studies have revealed prominent reflections at midcrustal levels beneath collisional mountain belts. In some areas it is clear that these reflections lie immediately below the detachment zone that separates highly deformed rocks on the upper thrust plate from rocks of the lower plate. Examples occur on seismic profiles recorded across foreland areas of the Bavarian Alps in Germany [Bachmann et al., 1982], the southern Appalachians in Tennessee, Georgia and North Carolina [Cock et al., 1983; Harris et al., 1981] and the Quebec Appalachians [St. Julien et al., 1983; LaRoche, 1983]. In other areas it is not clear if the midcrustal reflections are the product of collisional deformation, or if they represent precollisional structure and/or stratigraphy preserved intact on the lower thrust plate. Seaward dipping sequences observed on seismic profiles across interior portions of the southern Appalachians [Cook et al., 1979, 1983; Iverson and Smithson, 1982], New England Appalachians [Ando et al., 1984] and Ouachitas [Nelson et al., 1982; Lillie et al., 1983] are in this latter category.

This study analyzes seismic events recognized at midcrustal levels beneath a portion of a collisional mountain belt, the southern Appalachians. Specifically, previous suggestions that some of the seismic events may be due to late Precambrian rift features preserved beneath thrust sheets are evaluated by comparing reflection profiles from the mountain belt with reflection profiles from modern rifting settings.

In interpreting reflection data from ancient compressional tectonic settings, it is also helpful to consider observations from areas of modern convergence. Figure 1 shows a portion of a marine multichannel line across the deep-sea trench associated with the Lesser Antilles island arc [Westbrook et al., 1982]. Above the major thrust detachment, strata within the accretionary wedge are so deformed that they are unresolvable on the section. Beneath the detachment, however, the seismic expression of oceanic basement (hyperbolic diffraction pattern) and overlying deep-marine strata (sub-horizontal reflections) remain unchanged because features on the lower thrust plate are essentially undeformed. Likewise, even though seismic resolution is generally poor within deformed rocks of major overthrusts, reflection events are often observed at midcrustal levels in collisional mountain belts. Analogy with Figure 1 might suggest that, in some cases,

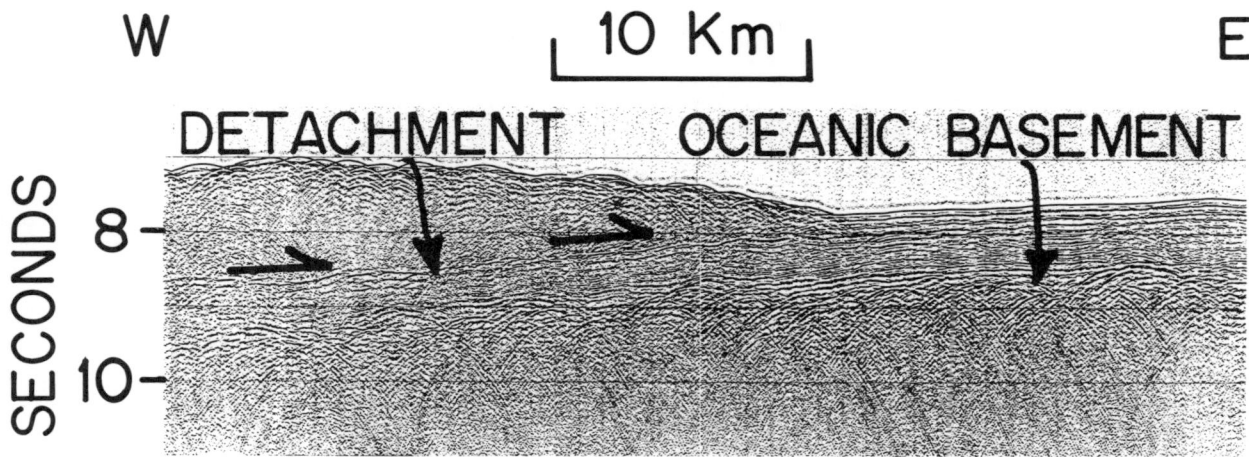

Fig. 1. Example of unmigrated seismic section across deep-sea trench. While complex structures on upper thrust plate are unresolved on section, seismic expressions of undeformed oceanic sediments and basement on lower plate remain unchanged during underthrusting. Durham University, Discovery 109, line 17 across Barbados ridge complex (Westbrook et al., 1982).

the midcrustal events are attributable to pre-existing structure and stratigraphy preserved intact on the lower thrust plate.

Example From Southern Appalachian Mountains

Seismic reflection studies of the southern Appalachian mountains suggested that crystalline, as well as sedimentary, rocks were thrust for great distances over the North American continent [Clark et al., 1978; Cook et al., 1979, 1981]. This brief report will not reproduce all of the results; rather, general interpretations of reflection events recognized at midcrustal levels are evaluated in the context of seismic profiles across modern tectonic settings. Publications of entire seismic profiles, as well as detailed discussions of southern Appalachian traverses, are found in Cook et al. [1983] and in Harris et al. [1981].

Of the many types of reflection sequences observed at midcrustal levels beneath the southern Appalachians, three characteristic types (Figures 2a, 2b, and 2c) can be compared with modern examples. The positions of the sequences relative to major tectonic elements of the southern Appalachians are illustrated schematically in Figure 3a. The first two types of sequences occur toward the foreland; that is, beneath the Valley and Ridge, Blue Ridge and Inner Piedmont areas (Figures 2a and 2b). Seismic resolution in these areas is sufficient to document that the sequences lie entirely below the lowermost thrust detachment, because the detachment can be traced eastward from foreland areas overlain by sedimentary strata in the Valley and Ridge [see Cook et al., 1979].

The third type of reflection sequence is wedge-shaped and lies along the prominent gravity and magnetic gradients associated with interior portions of the Appalachian/Ouachita orogenic belt (Figure 2c). Many workers associate the gravity and magnetic gradients with the present-day edge of North American continental basement which pre-dates Paleozoic collisional deformation. However, it is not clear if the edge represents the original continent/ocean transition [Cook and Oliver, 1981], or if the edge is an abrupt truncation of the proto-North American continent resulting from the collisional processes [Hatcher and Zietz, 1980]. Because the detachment zone is not clearly traceable to the wedge-shaped sequence, it is uncertain whether the reflection sequence represents structure formed during collision, or if the sequence results from precollisional structure and/or stratigraphy of the North American plate preserved intact beneath the detachment.

Features Beneath Foreland Areas

Two distinct types of midcrustal features beneath the Valley and Ridge, Blue Ridge and Inner Piedmont are considered in this study. The first type are dipping sequences that lie entirely below underthrusted shelf strata (Figure 2a), while small normal fault offsets of the underthrusted shelf strata constitute the second type (Figure 2b). Because it has been suggested that some subthrust structures beneath the foreland may be extensional features originally formed during late Precambrian continental rifting and ocean basin formation (e.g. Thomas, 1983), a schematic illustration of seismic reflection features observed across a passive continental margin is given for comparison (Figure 3b).

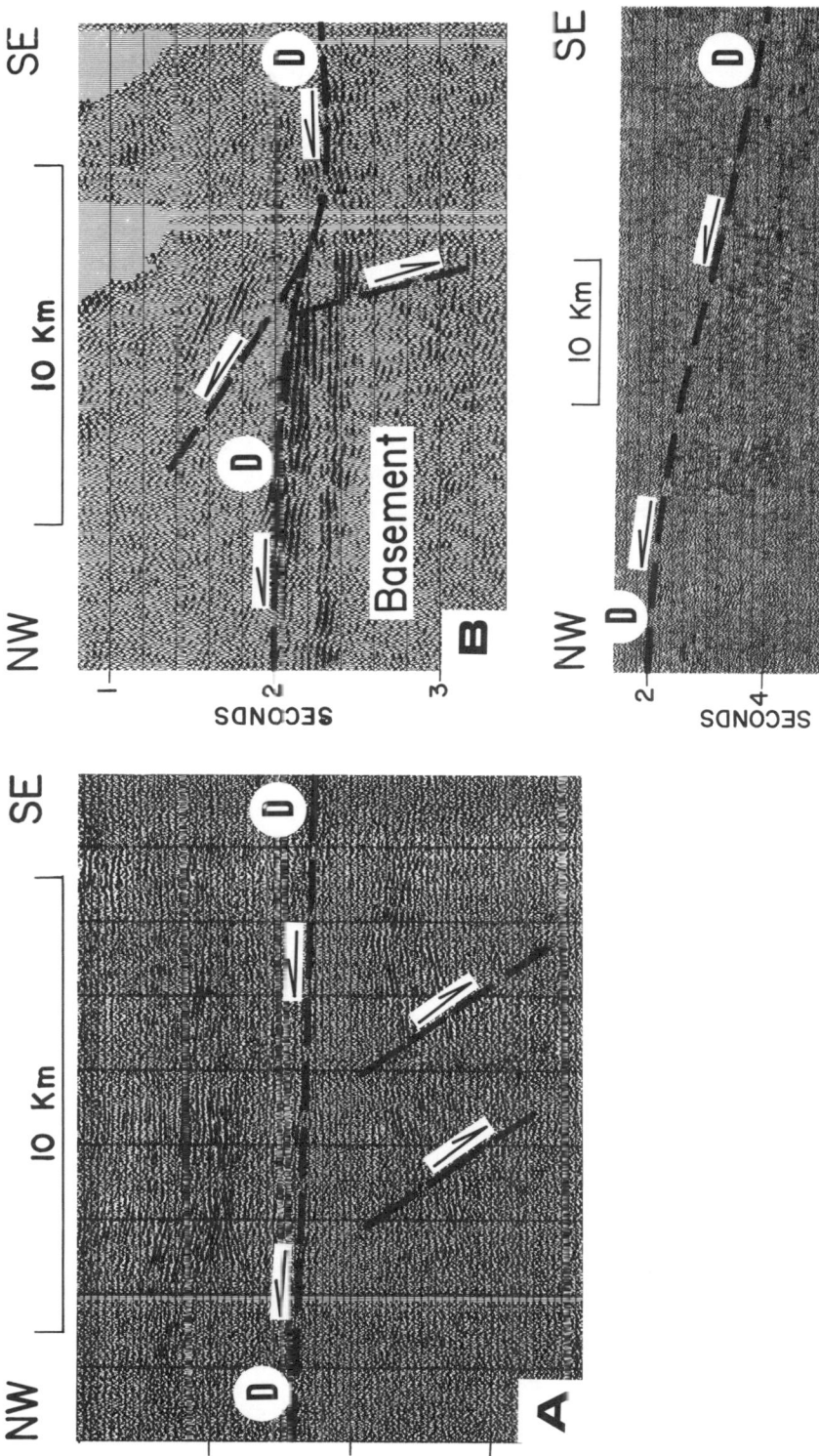

Fig. 2. Examples of midcrustal features recognized through seismic profiling of southern Appalachians. "D" represents detachment. (a) Dipping events entirely beneath underthrusted shelf sequence, interpreted as Precambrian syn-rift stata (Harris et al., 1981). Portion of USGS, North Carolina line 5 near northwest edge of Inner Piedmont. (b) Normal fault offset of underthrusted shelf strata. Relief on normal fault resulted in thrust ramping. Portion of COCORP Georgia line 1 across western Blue Ridge, reprocessed by University of Wyoming. (c) Seaward dipping sequence occurring along Appalachian gravity gradient. Portion of COCORP Georgia line 1 across Kings Mountain Belt, reprocessed by University of Wyoming (Iverson and Smithson, 1982).

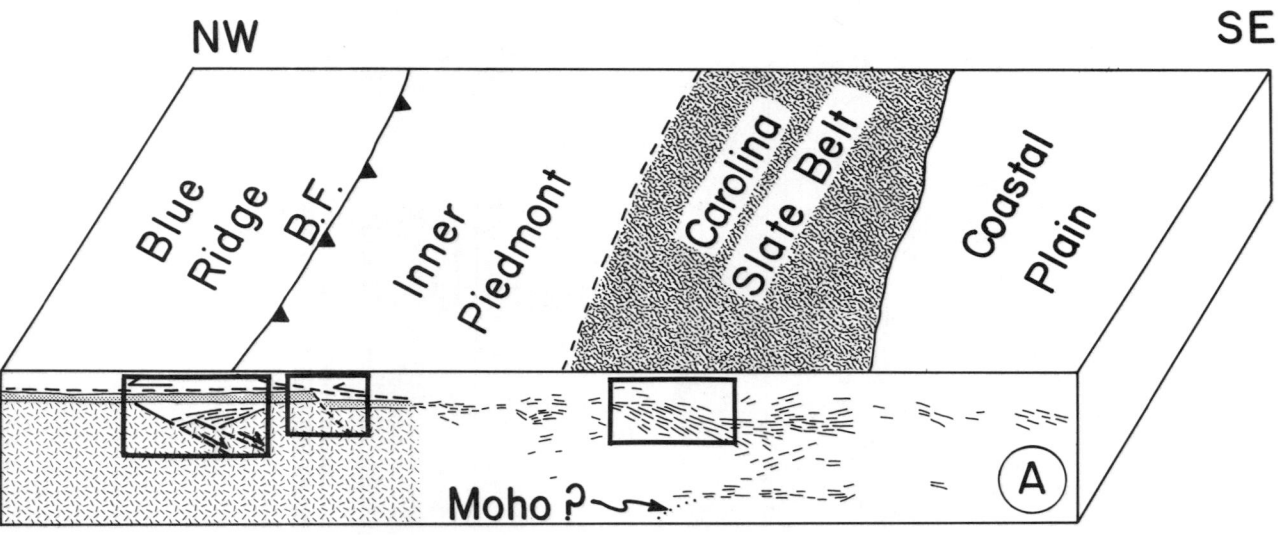

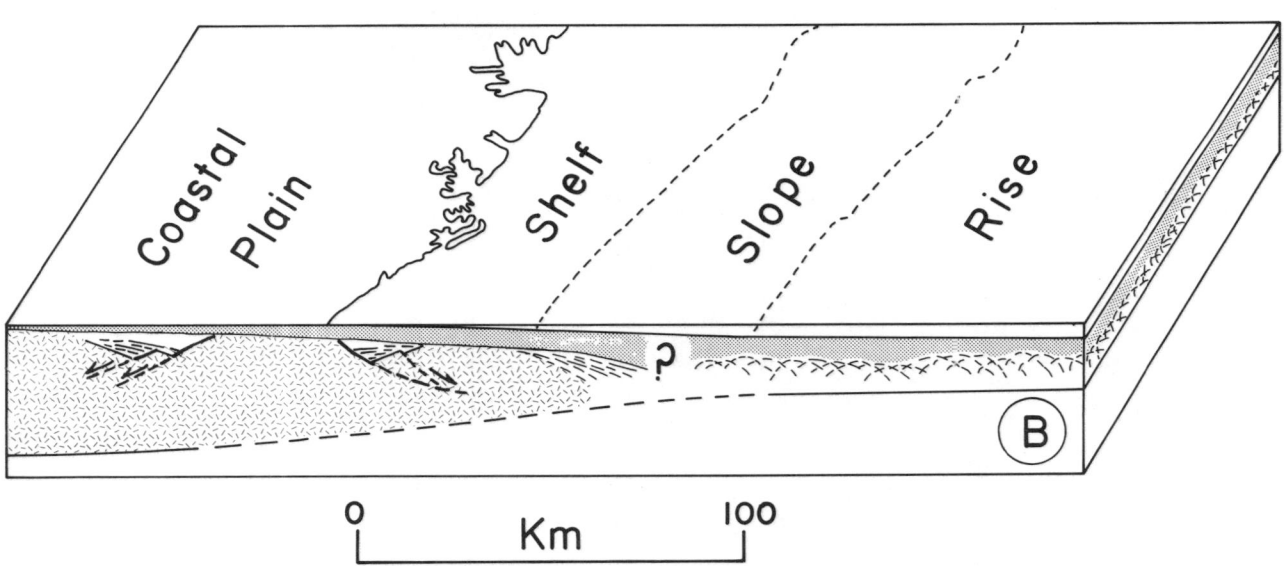

Fig. 3. (a) Conceptual diagram emphasizing midcrustal features recognized through reflection profiling of southern Appalachian mountains. Dashed pattern - Precambrian North American basement; stippled pattern - underthrusted early Paleozoic shelf strata. Seismic sections illustrating features highlighted by rectangles are shown in Figure 2. Generalized from Cook et al. (1981, 1983); Cook and Oliver (1981); Hatcher and Zietz (1980); and Harris et al. (1981). (b) Conceptual diagram of passive continental margin, emphasizing features recognized beneath post rift sediments through reflection profiling. Figures 4a and 4c show examples of seismic profiles across extended continental crust and continent/ocean boundary, respectively. Dashed pattern, continental crust; stippled pattern, post-rift sediments; hyperbolic diffractions, top of oceanic basement. Breakup unconformity is at base of post-rift sediments. Generalized from Grow et al. (1979); Bally (1981); Hinz (1981); Mutter et al. (1982).

Harris et al. [1981] have suggested that dipping seismic sequences, such as the one shown in Figure 2a, represent late Precambrian rifting strata preserved from the early Paleozoic continental margin. The results of this study are in agreement with that interpretation, in that rift grabens are generally identifiable in areas of continental rifting [Serpa et al., 1984] and passive margins [Bally, 1981] by moderately dipping reflection sequences. In the example from a modern passive margin shown in Figure 4a, the dipping sequences representing syn-rift strata lie entirely below flat-lying post-rift strata (i.e. within a half-graben below a flat breakup unconformity, Figure 3b). During later ocean basin closure, if the lowermost thrust detachment were to remain within the post-rift strata, then the dipping sequences would be preserved intact.

Basement normal faults associated with continental rifting on modern passive margins typically do not result in offset of post-rift strata (Figures 3b and 4a). Accordingly, it is reasonable to interpret normal fault offsets, such as that shown in Figure 2b, as features formed after continental rifting and deposition of strata along the subsiding continental margin. This interpretation is clear in the foreland of the Ouachita Mountains of Arkansas and Oklahoma where drilling has demonstrated that underthrusted shelf strata are offset but do not thicken across basement normal faults [Buchanan and Johnson, 1968]. Furthermore, syn-orogenic clastic strata (Carboniferous Lower Atoka Formation) thicken by as much as a factor of three across the Ouachita normal faults, demonstrating that the faults formed just prior to thrust emplacement.

An example of a similar normal fault offset from the Himalayan foreland in Pakistan (Figure 4b) serves as a modern analog for the normal fault offset beneath the Blue Ridge of the southern Appalachians shown in Figure 2b. The normal fault in Pakistan offsets basement and an Eocambrian evaporite sequence (Salt Range Formation). There is no indication of dipping sequences representing syn-rift graben fill, as in the rifting normal fault shown in Figure 4a. Rather, the evaporite sequence serves as the detachment zone associated with Quaternary thrusting [Seeber et al., 1981; Yeats et al., 1984], and the basement offset facilitates thrust ramping. Molnar et al. [1976] attribute the 1966 Ganga Basin earthquake to normal faulting associated with flexural loading of the Indian lithospheric plate. Nearby in Pakistan, the normal fault beneath the Salt Range is apparently of similar origin, and is therefore analogous to normal faults associated with the subduction of oceanic [Isaacs et al., 1968] and continental [Montecchi, 1976] lithosphere. Balanced cross sections currently in preparation will help to determine if the normal fault had any history of movement prior to Cenozoic collision, or if the movement is totally associated with lithospheric flexure during collision.

In the southern Appalachians most of the normal fault offsets which serve as thrust ramps are similar in seismic appearance to that depicted in Figures 2b and 4b. This is also true for features observed beneath the foreland of the Ouachita Mountains, the Quebec Appalachians [St. Julien et al., 1983; LaRoche, 1983] and the Bavarian Alps [Bachmann et al., 1982]. Dipping events beneath underthrusted shelf strata, which may represent buried rift grabens, are far less common. Besides the examples shown by Harris et al. [1981] beneath the Piedmont in North Carolina (e.g. Figure 2a), other possible examples of syn-rift strata on published reflection profiles include V.P. 350-500 of COCORP Tennessee line 1 across the western Blue Ridge [e.g. see Cook et al., 1983] as well as V.P. 629-689 of line 2003 and V.P. 1445-1509 of line 2001 on the SOQUIP data across the Quebec Appalachians [e.g. see St. Julien et al., 1983; LaRoche, 1983]. Therefore, although it is reasonable to suspect that continental rift features analogous to the Rome trough and Rough Creek graben underlie thrust sheets beneath the Appalachian foreland [Thomas, 1983], only a few possible examples can be found on reflection profiles at this time (e.g. Figure 2a). Basement offsets which facilitate thrust ramping [see Wiltschko, 1983] are quite common (e.g. Figure 2b), but there is no conclusive evidence that these features are related to earlier rifting normal faults.

Features Beneath Interior

In both the southern Appalachians [Cook et al., 1979, 1981] and Ouachitas [Nelson et al., 1982; Lillie et al., 1983, Lillie, 1984, 1985a, b], COCORP reflection profiles reveal seaward dipping sequences along prominent gravity and magnetic gradients ("seaward" refers to the direction of early Paleozoic ocean basins or marginal seas relative to proto-North America). Less prominent, seaward dipping sequences observed east of the Green mountains in the New England Appalachians may be similar in origin [Ando et al., 1984]. Thompson et al. [1983] show that the southern Appalachian sequence (Figure 2c) corresponds to a zone of high conductivity, perhaps indicating water trapped within the midcrust. As suggested by Cook et al. [1981], this sequence may be at least partly attributable to collisional processes. Two previous interpretations are: (1) that the sequence is due to sedimentary strata and basement imbricated against the edge of continental basement during west-directed thrusting [Cook et al., 1979], and (2) that intense basement deformation related to steep suturing during collision accounts for the reflections [Cook et al., 1981; Iverson and Smithson, 1982].

An alternative interpretation is that the sequence represents a precollisional feature preserved intact on the lower thrust plate. Through analogy with seismic profiles from modern passive margins, it is suggested that the

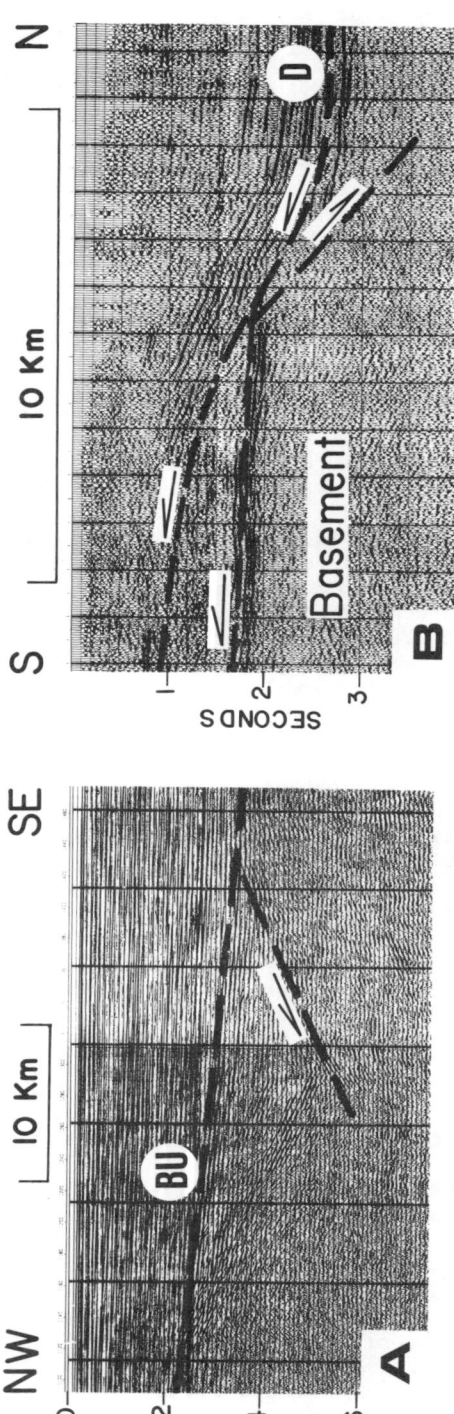
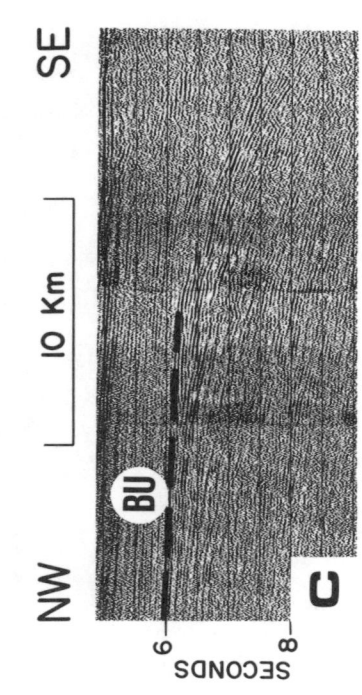

Fig. 4. Possible modern analogs for midcrustal structures beneath southern Appalachians. D, detachment; BU, breakup unconformity. (a) Seismic section showing tilted strata within Mesozoic half-graben off Atlantic coast of the United States. Portion of USGS line 5, after Schlee et al. (1976). Compare with Figure 2a. (b) Normal fault acting as thrust ramp in modern collisional setting, the Himalayan foreland in northern Pakistan. Portion of Oil and Gas Development Corporation seismic line across Salt Range, available to Oregon State University through joint project with Pakistan government. Compare with Figure 2b. (c) Seaward dipping reflection wedge near the continent/ocean boundary off the Atlantic coast of the United States. Portion of USGS line 25, after Grow (1980) and Grow et al. (1983). Compare with Figure 2c.

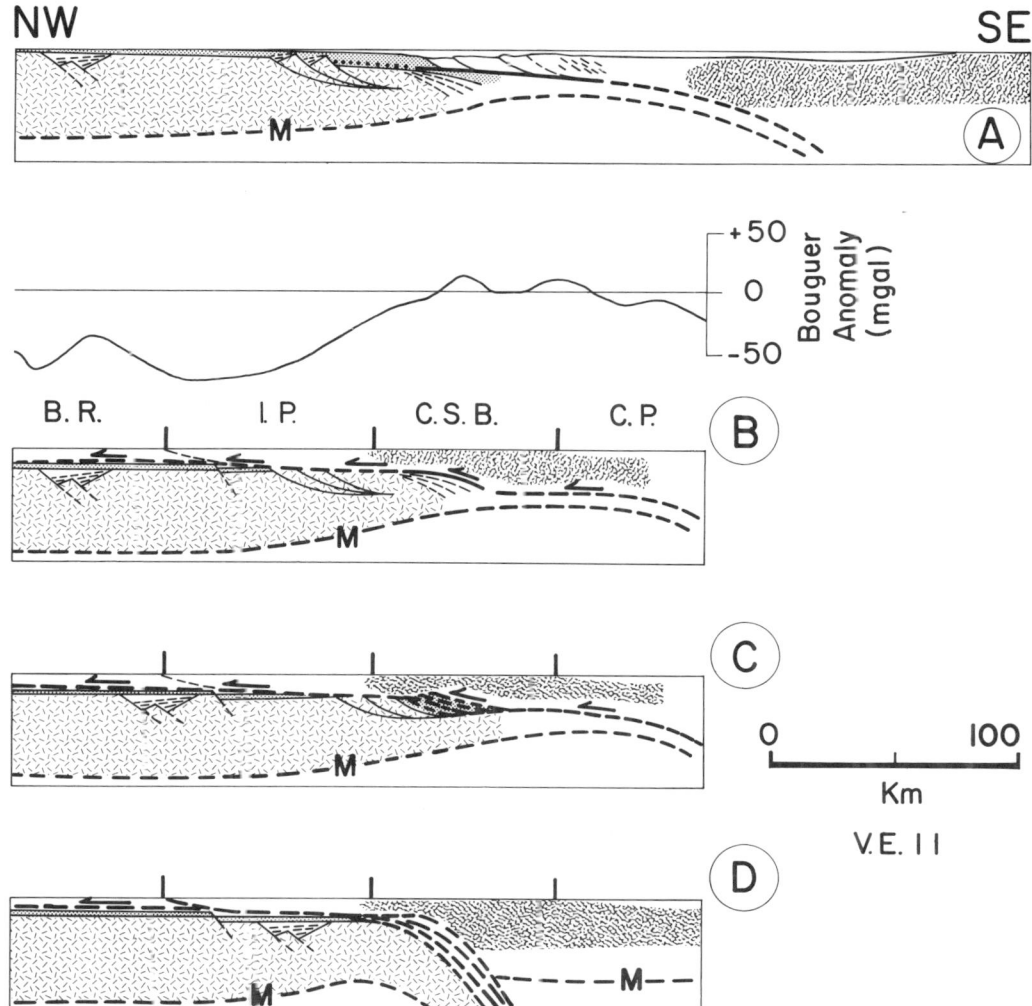

Fig. 5. Suggested interpretations for some midcrustal reflections observed beneath the southern Appalachians. (a) Situation just prior to arc emplacement during Ordovician Taconic orogeny. Figures 5b, 5c and 5d illustrate different degrees of underthrusting and structural deformation of the earlier continental margin. In each of the three alternatives the wedge of seaward dipping reflections (Figure 2c) beneath the Carolina Slate Belt arc (CSB) lies essentially at the edge of (late Precambrian) continental basement of the lower thrust plate. (b) Margin underthrusted with only slight off-scraping of continental basement. Wedge-shaped sequence interpreted as volcanic strata at early Paleozoic continent/ocean boundary; (c) Wedge-shaped sequence represents post-rift sediments and perhaps underlying basement imbricated against continental edge during underthrusting (Cook et al., 1979, 1981); (d) Wedge-shaped sequence represents truncated edge of earlier North American continent, perhaps due to steep suturing during collision (Hatcher and Zietz, 1980; Cook et al., 1981; Iverson and Smithson, 1982). Figure 5 is similar to that presented by Cook et al. (1981, 1983) and Cook (1983), but with 5b added as a further possibility. M, Moho; IP, Inner Piedmont; BR, Blue Ridge; CP, Coastal Plain. Observed Bouguer gravity anomaly after Woollard and Joesting, 1964.

sequence is original volcanic stratigraphy at the early Paleozoic continent/ocean boundary [Lillie, 1984, 1985b]. The modern example shown in Figure 4c occurs in a narrow zone separating continental from oceanic basement along the east coast of the United States [Grow, 1980; Grow et al., 1983]. Drilling of similar, seaward dipping sequences on the Voring plateau off Norway [Hinz, 1981;

Mutter et al., 1982] and the Rockall Plateau [D.S.D.P. leg 81, 1982; Roberts et al., 1984] suggest that the sequences consist of volcanic strata deposited at the time of continental breakup, thereby marking the modern continent/ocean boundary [see also Smyth, 1982].

Figure 5 is a series of plate tectonic cartoons representing the three suggested interpretations of the seaward dipping sequences beneath the southern Appalachians. The three models (Figures 5b, 5c, and 5d) are similar in the following aspects:

1. Syn-rift volcanics and crystalline basement now exposed in the western Blue Ridge were scraped off the underthrusting North American margin during the Ordovician Taconic orogeny [Rankin, 1975; Hatcher, 1978; Wehr and Glover, 1985]. After additional westward transport during the Carboniferous Alleghenian orogeny, these syn-rift structures now overlie similar, but undeformed, rift structures (e.g. Figure 2a; Thomas, 1983). New normal faults offsetting shelf strata and underlying basement were formed during the Paleozoic thrust emplacement.

2. In the early Paleozoic, basement of the Piedmont was separated from the North American margin by an ocean basin or marginal sea, as suggested by deformed rocks now exposed in the eastern Blue Ridge [Rankin, 1975; Hatcher, 1978].

Figure 5a illustrates a possible tectonic configuration just prior to arc/continent collision associated with the Taconic orogeny [e.g. see Hatcher, 1978]. In each of the three suggested interpretations (Figures 5b, 5c, and 5d), the seaward dipping sequence lies at the present-day edge of continental basement associated with the precollisional North American continent. The interpretations imply different amounts of deformation of the continental edge during collision, as well as different amounts of underthrusting of the early Paleozoic continent/ocean boundary. Each of the models incorporates a northwest to southeast shallowing of the Moho to account for the prominent gravity gradient [e.g., see Cook and Oliver, 1981; Hutchinson et al., 1983; Cook, 1984].

Figure 5b suggests minimum amounts of deformation and underthrusting, in that the continent/ocean boundary is preserved intact on the lower thrust plate. Figure 5c involves thrust imbrication of sedimentary strata and possible basement deformation along the seaward edge of continental basement [Cook et al., 1979, 1981]. The new interpretation (Figure 5b) is an important perturbation of the earlier model (Figure 5c); although both interpretations imply preservation of the broad zone of transition from continental to oceanic basement beneath a shallowly dipping suture zone, the new model suggests that a specific component of the transition (i.e. the actual continent/ocean boundary) is preserved intact and is represented seismically by the seaward dipping wedge. Underthrusting in each case was not enough to remove all of the original Moho shallowing associated with the earlier margin [e.g. see gravity models by Cook and Oliver, 1981, for the southern Appalachians and Lillie et al., 1983, for the Ouachitas, as well as crustal section by Price, 1981, for the Canadian Rockies].

Figure 5d differs greatly from the other two models, in that the seaward dipping sequence represents collisional structures developed along a steep suture zone [Cook et al., 1981; Iverson and Smithson, 1982]. This interpretation implies large amounts of structural deformation and underthrusting (possibly subduction) of the early Paleozoic continent/ocean boundary. While the lower crust southeast of the seaward dipping sequence in the other two models is suggested to have been part of the North American plate prior to collision, the entire crust in this position in the third model was accreted to North America during Paleozoic collisions. West to east Moho shallowing, in this case, can be attributed to emplacing exotic terranes with thinner crust than that of the original North American continental crust (e.g. see gravity models by Fountain and Salisbury, 1981, for ancient collisional mountain belts, and by Chamalaun et al., 1976, for the Banda Arc).

Note that, in a general sense, these models represent possibilities for the "subsurface load" suggested in the lithospheric flexure studies of Karner and Watts [1983]. In Figures 5b and 5c the load is shallow mantle material preserved from the earlier rifted margin, while the load in Figure 5d is shallow mantle material associated with the overriding plate.

Although interpretation of seaward dipping sequences as volcanic strata related to the early Paleozoic continent/ocean boundary are plausible for the southern Appalachian and Ouachita mountains [Lillie, 1984, 1985a, b], other interpretations may be more appropriate in other areas. In the New England Appalachians, Ando et al. [1984] suggest that seaward dipping events east of the Green Mountain massif project to the surface within a zone of thrust imbricated (late Precambrian) Grenville basement and younger metasediments. In the central Appalachians of Virginia, Wehr and Glover [1985] suggest that continental basement related to pre-collisional North America restores eastward to a position well beyond the on-strike continuation of the southern Appalachian sequences (i.e. Figure 2c). Brewer [1985] suggests that the seaward dipping sequences along the Appalachian/Ouachita trend represent different degrees of deformation of the former continental edge, similar, in a general sense, to the possibilities presented in Figure 5. Because of these ambiguities in interpretation, the three models (Figures 5b, 5c and 5d) should be considered working hypotheses for interpretation of mid-crustal reflection sequences observed beneath the interior of the Appalachians and other collisional mountain belts.

Hinz [1981] suggests that seaward dipping wedges representing ancient continent/ocean boundaries might be preserved somewhere in the geologic record. These sequences are thicker and differ in internal geometry from normal oceanic layer 2 [Mutter et al., 1982]. In addition to the suggestions presented here that some of these sequences might have been underthrust beneath the southern Appalachians and Ouachitas, we should also consider the possibility that some "ophiolites" within deformed overthrust rocks may contain the actual continent/ocean boundary rather than normal oceanic layer 2.

Conclusions

Seismic reflection data across deep-sea trenches have shown that, in many cases, the seismic expression of structures can be retained during underthrusting. By analogy, certain seismic sequences observed at midcrustal levels beneath collisional mountain belts may represent precollisional structures preserved intact on the lower thrust plate.

By comparing midcrustal structures observed across collisional mountain belts with structures observed in modern settings, criteria have been established to distinguish types of subthrust features based on seismic expression. Beneath foreland areas of the southern Appalachians, a few examples of moderately dipping sequences entirely beneath underthrusted shelf strata may be indicative of rift-grabens related to late Precambrian continental breakup. More common normal fault offsets of the shelf strata, which typically facilitate thrust ramping, are probably much later structures formed in conjunction with flexural loading during thrust emplacement. While normal faults associated with continental rifting are observed to have up to 7 km of offset, flexural normal faults are much smaller, with less than 2 km of offset.

Beneath interior portions of the southern Appalachians, seaward dipping sequences lie along prominent potential field gradients. These gradients have been suggested to mark the present-day extent of continental crust associated with the early Paleozoic North American continent. In addition to previous suggestions that the dipping sequences are at least partly due to Paleozoic collisional deformation, it is possible that they represent an earlier feature preserved intact on the lower thrust plate. By analogy with reflection profiles from modern passive margins, it is suggested that the sequences represent a wedge of volcanic strata at the early Paleozoic continent/ocean boundary.

Athough the analogies and interpretations presented here are somewhat speculative, they may by useful guides to the interpretation of reflection data from other collisional mountain belts. It is hoped that the comparisons with reflection data from modern settings will be helpful in the planning of future studies of the Appalachians and other areas through the continental deep drilling program.

Acknowledgments. We are grateful to the COCORP project, the U.S. Geological Survey, Durham University and the Oil and Gas Development Coorporation of Pakistan for the use of seismic profiles illustrated in this study. Work on the Himalayan foreland is a cooperative project involving Oregon State University and the Geological Survey of Pakistan, supported by NSF grants EAR-83-18194, INT-81-18403, and gifts fom CONOCO, Inc. and CHEVRON International. We thank Karen Lund, Bob Yeats and two anonymous reviewers for suggestions which improved the original manuscript.

References

Ando, C., B. Czuchra, S. Klemperer, L. Brown, M. Cheadle, F. Cook, J. Oliver, S. Kaufman, T. Walsh, J. Thompson, Jr., J. Lyons and J. Rosenfeld, Crustal profile of mountain belt: COCORP deep seismic reflection profiling in New England Appalachians and implications for architecture of convergent mountain chains, Amer. Assoc. Pet. Geol. Bull., 68, 819-837, 1984.

Bachmann, H., G. Dohr, and M. Muller, Exploration in a classic thrust belt and its foreland: Bavarian Alps, Germany, Amer. Assoc. Petrol. Geol. Bull., 66, 2529-2542, 1982.

Bally, A., Geology of passive continental margins: History, structure and sedimentologic record (with special emphasis on the Atlantic margin), Course Note Series, 19, pp. 1-48, American Association Petroleum Geologists, Tulsa, Okla., 1981.

Brewer, J., Deep structure of orogenic belts inferred from crustal reflection profiling, Tectonics, in press, 1985.

Buchanan, R., and F. Johnson, Bonanza gas field - A model for Arkoma Basin growth faulting, in Geology of the western Arkoma Basin and Ouachita Mountains, edited by L. Cline, pp. 75-85, Oklahoma City Geological Society, Oklahoma City, 1968.

Chamalaun, F., K. Lockwood, and A. White, The Bouguer gravity field and crustal structure of eastern Timor, Tectonophysics, 30, 241-259, 1976.

Clark, H., J. Costain, and L. Glover, III, Structure and seismic reflection studies on the Brevard ductile deformation zone near Rosman, North Carolina, Am. J. Sci., 278, 419-441, 1978.

Cook, F., Some consequences of palinspastic reconstruction in the southern Appalachians, Geology, 11, 86-89, 1983.

Cook, F., Geophysical anomalies along strike of the southern Appalachian Piedmont, Tectonics, 3, 45-61, 1984.

Cook, F., and J. Oliver, The late Precambrian - early Paleozoic continental edge in the Appalachian orogen, Amer. Jour. Sci., 281, 993-1008, 1981.

Cook, F., D. Albaugh, L. Brown, S. Kaufman, J. Oliver, and R. Hatcher, Jr., Thin-skinned tectonics in the crystalline southern Appalachians: COCORP seismic-reflection profiling of the Blue Ridge and Piedmont, Geology, 7, 563-567, 1979.

Cook, F., L. Brown, S. Kaufman, J. Oliver, and T. Peterson, COCORP seismic profiling of the Appalachian orogen beneath the Coastal Plain of Georgia, Geol. Soc. Amer. Bull., 92, 738-748, 1981.

Cook, F., L. Brown, S. Kaufman, and J. Oliver, The COCORP seismic reflection traverse across the southern Appalachians, 14, 60 pp., American Association of Petroleum Geologists, Tulsa, Okla., 1983.

DSDP Leg 81 Scientific Party, Leg 81 drills west margin, Rockall Plateau, Geotimes, 27, 21-23, 1982.

Fountain, D., and M. Salisbury, Exposed cross-sections through continental crust: Implications for crustal structure petrology and evolution, Earth Planet. Sci. Lett., 56, 263-277, 1981.

Grow, J., Deep structure and evolution of the Baltimore Canyon trough in the vicinity of the Cost No. B-3 well, U. S. Geol. Surv. Circ., 833, pp. 117-132, 1980.

Grow, J., R. Mattick, and J. Schlee, Multichannel seismic depth sections and interval velocities over outer continental shelf and upper continental slope between Cape Hatteras and Cape Cod, in Geological and Geophysical Investigations of Continental Margins, Mem. 29, edited by J. Watkins, L. Montadert, and P. Dickerson, pp. 65-83, American Association of Petroleum Geologists, Tulsa Okla., 1979.

Grow, J., D. Hutchinson, K. Klitgord, W. Dillon and J. Schlee, Representative multichannel seismic profiles over the U.S. Atlantic margin, in Seismic Expression of Structural Styles, Stud. Geol., 15 (2), edited by A. Bally, pp. 2.2.3.1-2.2.3.19, American Association Petroleum Geologists, Tulsa, Okla., 1983.

Harris, L., A. Harris, W. deWitt, Jr., and K. Bayer, Evaluation of southern eastern Overthrust Belt beneath Blue Ridge - Piedmont thrust, Am. Assoc. Pet. Geol. Bull., 65, 2497-2505, 1981.

Hatcher, R., Jr., Tectonics of the western Piedmont and Blue Ridge, southern Appalachians: Review and speculation, Am. J. Sci., 278, 276-304, 1978.

Hatcher, R., Jr., and I. Zietz, Tectonic implications of regional aeromagnetic and gravity data from the southern Appalachians, in International Geological Correlation Program -- Caledonide Orogen Project Symposium, Mem. 2, edited by D. Wones, pp. 235-244, Virginia Polytechnic Institute, Blacksburg, Virg., 1980.

Hinz, K., A hypothesis on terrestrial catastrophies. Wedges of very thick oceanward dipping layers beneath passive continental margins -- their origin and paleoenvironmental significance, Geol. Jahrb., Reihe E., 22, 3-28, 1981.

Hutchinson, D., J. Grow, and K. Klitgard, Crustal structure beneath the Southern Appalachians: Nonuniqueness of gravity modeling, Geology, 11, 611-615, 1983.

Isaacs, B., J. Oliver, and L. Sykes, Seismology and the new global tectonics, J. Geoph. Res., 73, 5855-5899, 1968.

Iverson, W., and S. Smithson, Master decollement root zone beneath the southern Appalachians and crustal balance, Geology, 10, 241-245, 1982.

Karner, G.D., and A.B. Watts, Gravity anomalies and flexure of the lithosphere at mountain ranges, J. Geoph. Res., 88, 10449-10477, 1983.

LaRoche, P., Appalachians of southern Quebec seen through seismic line no. 2001, in Seismic Expression of Structural Styles, Stud. Geol., edited by A. Bally, pp. 3.2.1.7 - 3.2.1.22, American Association Petroleum Geologists, Tulsa, Okla., 1983.

Lillie, R., Tectonic implications of subthrust structures revealed by seismic profiling of Appalachian-Ouachita orogenic belt, Tectonics, 3, 619-646, 1984.

Lillie, R., Tectonically buried continent/ocean boundary, Ouachita Mountains, Arkansas, Geology, 13, 18-21, 1985a.

Lillie R., Correction to "Tectonic implications of subthrust structures revealed by seismic profiling of Appalachian-Ouachita orogenic belt", Tectonics, 4, 263-265, 1985b.

Lillie, R., K. Nelson, B. deVoogd, J. Brewer, J. Oliver, L. Brown, S. Kaufman, and G. Viele, Crustal structure of Ouachita Mountains, Arkansas: A model based on integration of COCORP reflection profiles and regional geophysical data, Am. Assoc. Pet. Geol. Bull., 67, 907-931, 1983.

Molnar, P, W. Chen, T. Fitch, P. Tapponier, W. Warsi, and F. Wu, Structure and tectonics of the Himalaya: A brief summary of relevant geophysical observations, Colloques International du Centre National de la Recherch Scientific, 268, 269-294, 1976.

Montecchi, P., Some shallow tectonic consequences of subduction and their meaning to the hydrocarbon explorationist, in Circum-Pacific Energy and Mineral Resources, Mem. 25, edited by M. Halbouty, J. Maher, and H. Lian, pp. 189-202, American Association Petroleum Geologists, Tulsa, Okla., 1976.

Mutter, J., M. Talwani, and P. Stoffa, Origin of seaward dipping reflectors in oceanic crust off the Norwegian margin by "subaerial sea-floor spreading," Geology, 10, 353-357, 1982.

Nelson, K., R. Lillie, B. deVoogd, J. Brewer, J. Oliver, S. Kaufman, L. Brown and G. Viele, COCORP seismic reflection profiling in the Ouachita Mountains in western Arkansas: Geometry and geologic interpretation, Tectonics, 1, 413-430, 1982.

Price, R.A., The Cordilleran foreland thrust and fold belt in the southern Canadian Rocky Mountains, in Thrust and Nappe Tectonics, Spec. Pub. 9, edited by K. McClay and N. Price, pp. 427-448, Geological Society London, 1981.

Rankin, D., The continental margin of eastern North America in the southern Appalachians: The opening and closing of the proto-Atlantic ocean, Am. J. Sci., 275A, 258-336, 1975.

Roberts, D., J. Backman, A. Morton, J. Murray, and J. Keene, Evolution of volcanic rifted margins: Synthesis of leg 81 results on the west margin of Rockall Plateau, Init. Rep. Deep Sea Drill. Proj., 81, 883-911, 1984.

Schlee, J., J. Behrendt, J. Grow, J. Robb, R. Mattick, P. Taylor, and B. Lawson, Regional geologic framework off northeastern United States, Am. Assoc. Pet. Geol. Bull., 60, 926-951, 1976.

Seeber, L., J. Armbruster, and R. Quitmeyer, Seismicity and continental subduction in the Himalayan arc, in Zagros-Hindu Kush-Himalayan Geodynamics Evolution, Geodynamics Ser. 1, edited by H. Gupta and F. Delany, pp. 215-242, American Geophysical Union, Washington D.C., 1981.

Serpa, L., T. Setzer, H. Farmer, L. Brown, J. Oliver, S. Kaufman, J. Sharp, and D. Steeples, Structure of the southern Keweenawan Rift from COCORP surveys across the Midcontinent Geophysical Anomaly in northeastern Kansas, Tectonics, 3, 367-384, 1984.

Smythe, D., Dipping reflectors at passive margins off Northwest Europe (abstract), in Programs and Abstracts, p. 77, European Geophysical Society and European Seismology Committee, London, 1982.

St. Julien, P., A. Slivitsky, and J. Feininger, A deep structural profile across the Appalachians of southern Quebec, in Contributions to the Tectonics and Geophysics of Mountain Chains, Mem. 158, edited by R. Hatcher, Jr., H. Williams, and I. Zietz, pp. 103-111, Geological Society of America, Boulder, Colo., 1983.

Thomas, W., Basement-cover relations in the Appalachian fold and thrust belt, Geol. J., 18, 267-276, 1983.

Thompson, B., A. Nekut, and A. Kuckes, A deep crustal electromagnetic sounding in the Georgia Piedmont, J. Geoph. Res., 88, 9461-9473, 1983.

Wehr, F., and L. Glover, III, Stratigraphy and tectonics of the Virginia-North Carolina Blue Ridge: Evolution of a late Proterozoic – early Paleozoic hinge zone, Geol. Soc. Amer. Bull., 96, 285-295, 1985.

Westbrook, G., M. Smith, J. Peacock, and M. Poulter, Extensive underthrusting of undeformed sediment beneath the accretionary complex of the Lesser Antilles subduction zone, Nature, 300, 625-628, 1982.

Wiltschko, D., and D. Eastman, Role of Basement warps and faults in localizing thrust fault ramps, in Contributions to the Tectonics and Geophysics of Mountain Chains, Mem. 158, edited by R. Hatcher, Jr., H. Williams, and I. Zietz, pp. 177-190, Geological Society of America, Boulder, Colo., 1983.

Woollard, G., and H. Joesting, Bouguer gravity anomaly map of the United States, Scale 1:2,500,000, American Geophysical Union, Washington, D.C., 1964.

Yeats, R., S. Hasan Khan, and M. Akhtar, Late Quaternary deformation of the Salt Range of Pakistan, Geol. Soc. Am. Bull., 95, 958-966, 1984.

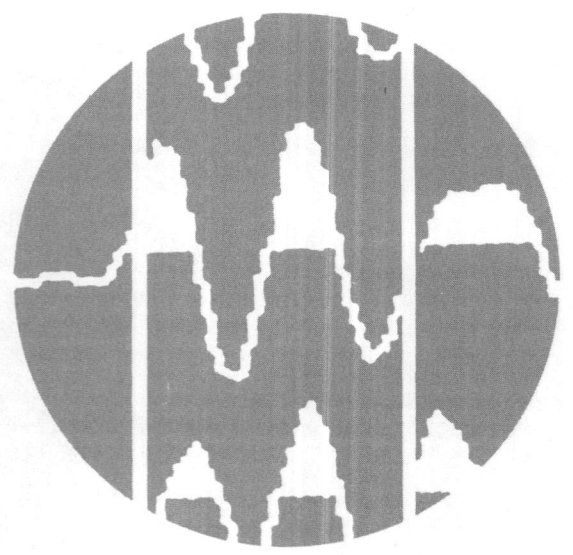

REFLECTIONS FROM THE SUBCRUSTAL LITHOSPHERE

Karl Fuchs

Geophysikalisches Institut
University Fridericiana, Karlsruhe, West-Germany

Abstract. Today there is a wide consensus that near-vertical seismic reflections out of the crust and from the crust-mantle boundary are not generated by isolated first-order discontinuities but rather by laminated zones which increase the reflectivity by constructive interference of multiple internal reflections in certain frequency bands. Although at present there are only a few observations of near-vertical reflections from the subcrustal lithosphere, in this region of the upper mantle the conditions appear to be favourable for the generation of such reflections. Probing of the upper mantle on long-range seismic profiles revealed properties of the subcrustal lithosphere which make it very likely that the same reflection mechanism could be effective here. The observation of unexpectedly high P-wave velocities of up to 8.5-8.6 km/s as shallow as 10-30 km below the crust-mantle boundary in layered zones embedded in regions of "normal" upper mantle material of around 8.0-8.2 km/s indicates that velocity contrasts in the upper mantle may become as large as at the crust-mantle boundary. A second property of the upper mantle recently discovered is an azimuth-dependent velocity distribution. This anisotropy starting at the Moho reaches into the upper mantle to a depth of about 100 km or even deeper. It may be generated by a preferred orientation of olivine in a more or less horizontal flow pattern. The same flow in a shear stress field orienting the crystals will also transform lateral heterogeneities with non-horizontal boundaries into flat horizontal layers in which the velocities depend mainly on depth. Such a shear-flow flattening is most likely the mechanism which produces laminated zones forming the reflectors within the crystalline earth.

From these properties of the subcrustal lithosphere there is good reason to expect that reflections from the upper mantle exist and can be found if a search is made for them as systematically and thoroughly as has been done during the past decade in the crust. In some cases there is already observational evidence for the existence of upper-mantle reflections.

Introduction

The first deep crustal reflections (Fig.1) were recorded as a chance discovery by Junger (1951) in Kern County. Even at that time the experts did not believe in what they saw. Steinhart and Meyer (1961) concluded: "The conflicting reports and the conflict of opinion concerning the interpretation to be attached to these reports means that the evidence for or against reflections at near-vertical angle of incidence is not yet conclusive. Since this matter is so important, future experiments should seek definitive evidence for or against the existence of these reflections". In a way that situation is similar today with regard to the exploration of the subcrustal lithosphere.

Structure of the Subcrustal Lithosphere from Seismic-Refraction Experiments

During the past twenty years seismic-refraction studies on long-range profiles have revealed heterogeneities in the subcrustal lithosphere which are in strong contrast to standard models of this part of the earth's interior. The observed velocity contrasts in the uppermost mantle are as large as those encountered at the crust-mantle boundary.

How can these heterogeneities be observed in record sections of long-range profiles? Figure 2 shows the record section from the well known experiment in France in 1972 (Hirn et al., 1973). It can be recognized that the Pn-phase travelling with a group velocity of around 8.1 km/s is in fact composed of a number of branches with velocities higher than 8.1 km/s. Using synthetic-seismogram techniques it was possible to show that these high-velocity branches are generated by overcritical reflections from high-velocity layers in the subcrustal lithosphere. The velocities range between 8.4-8.6 km/s, not only in France but also in Britain (LISPB; Faber, 1978), Southern Germany and in Scandinavia (FENNOLORA) (Fuchs, 1983), as revealed by similar experiments in these regions.

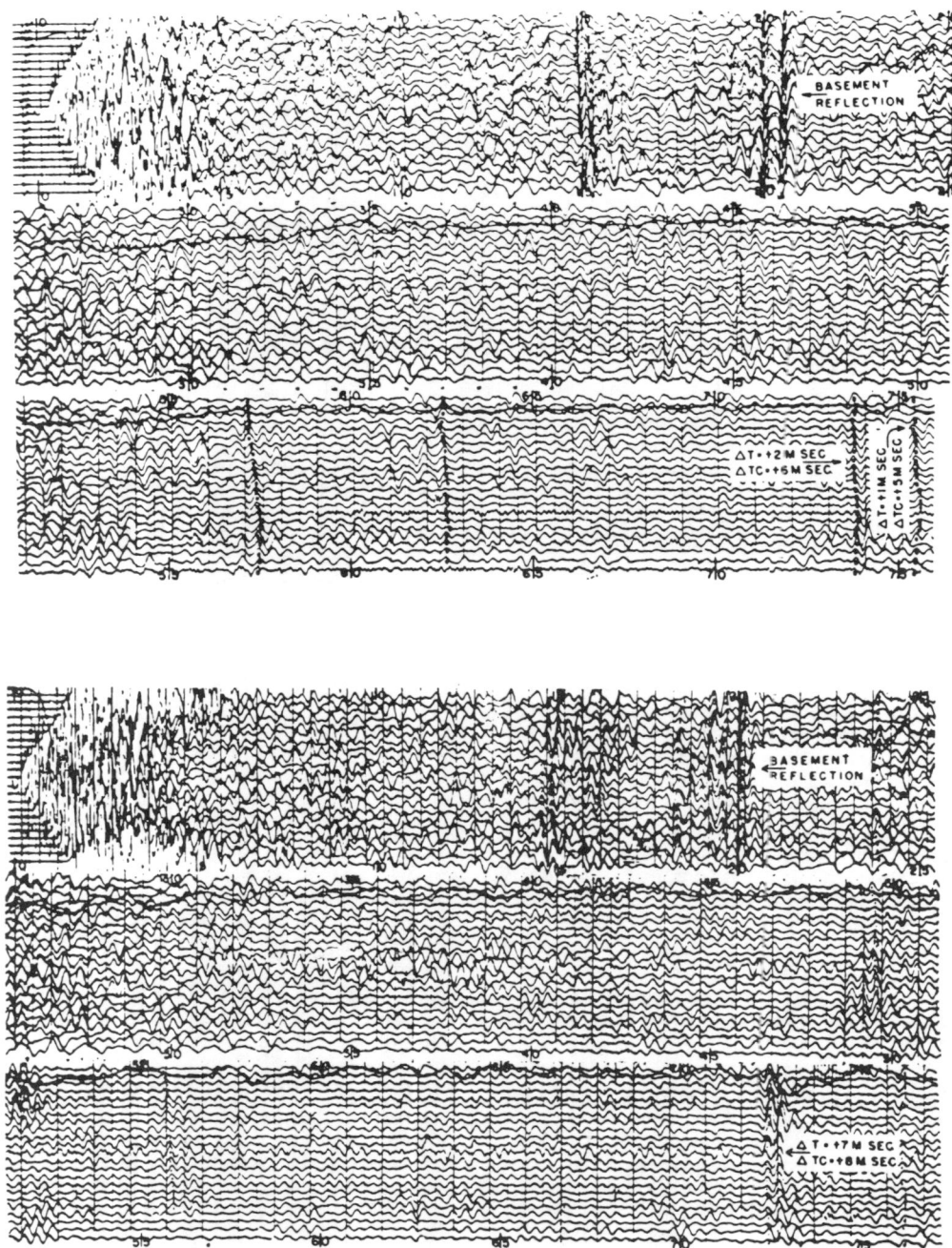

Fig. 1. Junger's (1951) observations of deep crustal reflections in Kern County/ California were the first documented in the literature. They still are among the most convincing recordings of reflections from the deep crystalline crust. The two examples display record sections in three parts. The deepest reflection in both cases is close to 7.5 sec echo-time.

The velocity-depth distributions inverted from these record sections are characterized by a sequence of high- and low-velocity layers. The presence of the low-velocity layers provides the delays of the various high-velocity branches, thereby lining them up with the effective "group velocity" of about 8.1 km/s. The velocity distribution from France (Fig. 3, Hirn et al., 1973)

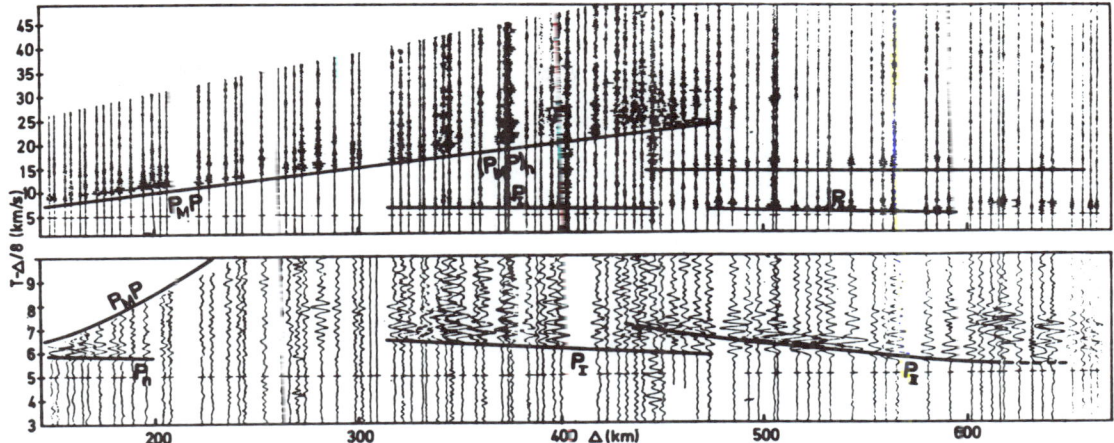

Fig. 2. Record section from the first west-European long-range profile in France (Hirn et al., 1973). - Reduction velocity 8 km/s. PmP is the reflection from the crust-mantle boundary. (PmP)n is PmP multiply reflected between the free surface and the crust-mantle transition. Pn is the refracted or headwave guided at the crust-mantle boundary (compare Fig. 3). PI: phase returned from the bottom of the low-velocity zone at about a depth of 50-60 km. PII: phase returned from the high velocity gradient at a depth of about 80-90 km.

serves as an example. All presently available velocity distributions in the upper mantle on continents have been compiled by Prodehl (1984); some examples from this compilation are to be seen in Figure 4.

Can we expect reflections from these zones at near-vertical incidence? If the velocity contrast is a measure of the likelihood to obtain reflections from a heterogeneous region, then there should be a good chance to observe reflected energy from the subcrustal lithosphere, in fact the chances should be as large as from the Moho. But a simple velocity contrast alone is certainly not sufficient to produce observable amplitudes. To explain the observed amplitudes of reflections from within the crust and especially from the transition zone at the crust-mantle boundary, reflections from fine velocity laminations have been proposed by various authors (Fuchs, 1968b; Meissner, 1966; Deichmann and Ansorge, 1983). These laminated zones may arise from magmatic injection, differentiation or from tectonic processes. Lamination enhances amplitudes in certain frequency bands and generates the reverberations of the observed interference.

It has been noted many times that the Moho appears differently in wide-angle refraction as compared to near-vertical incidence reflection surveys. In general, in a refraction experiment the Moho manifests itself by the overcritical PmP reflection with strong signal/noise ratio and dominant frequencies around 5 Hz. Also the refracted phase from the topmost mantle (Pn) is seen in most cases, although with much weaker amplitudes. Near-vertical incidence reflections from the crust-mantle transition are generally characterized by signals with a dominant frequency between 10-15 Hz. The reflection from the Moho are not observed everywhere and typically they can be correlated only over a few kilometers. In some cases the Moho reflection is even defined as the last reflection in a band of reflections from the lower crust. These differences indicate that it is very likely that seismic refraction and the reflection methods reveal different aspects of the same medium.

Synthetic seismograms for a finely layered

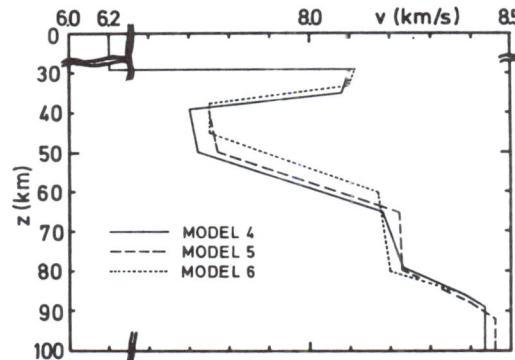

Fig. 3. Model of P-velocity distribution in the lower lithosphere in France derived by Hirn et al. (1973). Variations labeled models 4,5 and 6 indicate the estimate of the range of possible models compatible with travel time, phase velocity and amplitude data. Compare also Figure 4, where this model is reproduced as no. 1 and models no. 2 and 3 are interpretations by other authors (Kind, 1974; and Ansorge, 1975, respectively). Here the high-velocity layers are embedded in material of "normal" mantle velocity (i.e., 7.9-8.0 km/s).

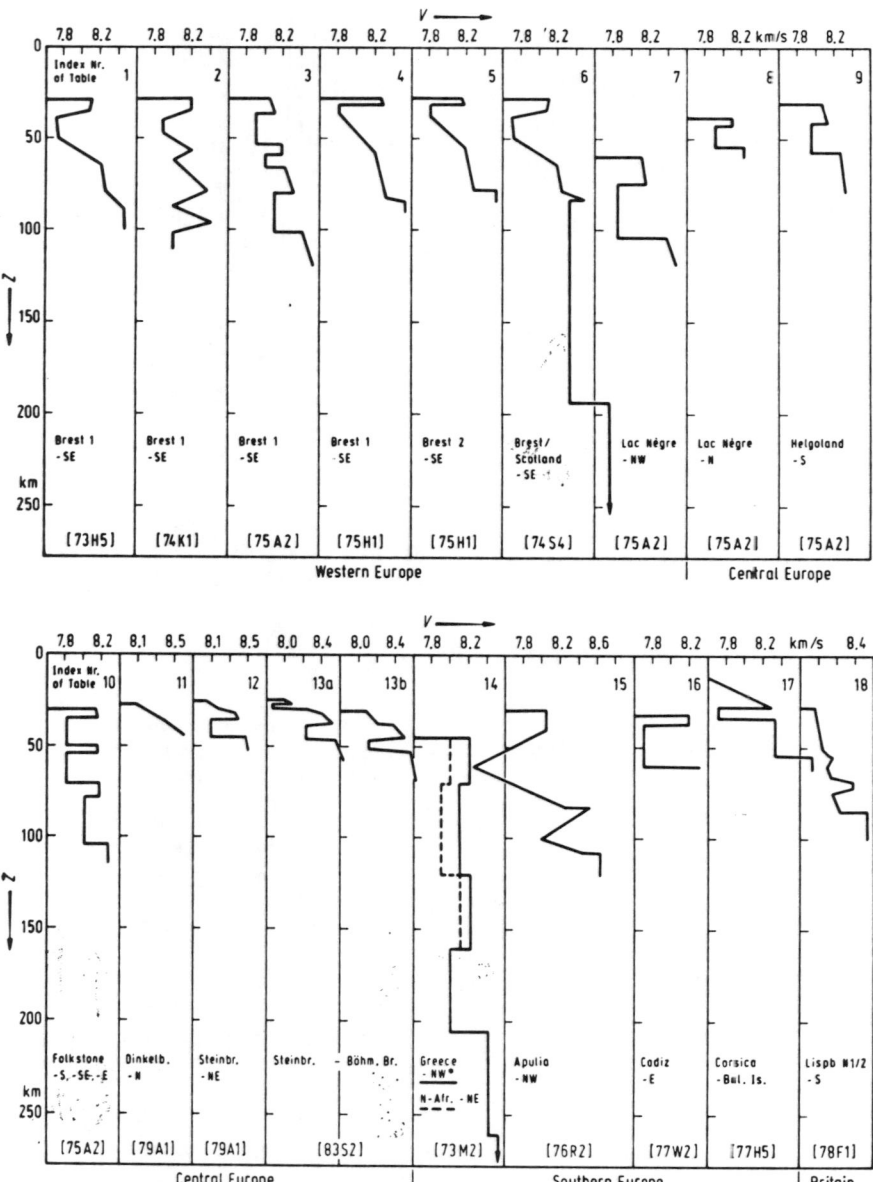

Fig. 4. A compilation of P-velocity distributions in the upper mantle from all parts of the world, mainly from controlled source experiments (from Prodehl, 1984; for references and tables see the cited paper). The occurence of velocities as high as 8.6 km/s within the top 100 km is the most important feature of these studies since it contradicts classical models of the upper mantle.

structure of the earth's crust (Fig. 5) have been computed on the CDC 205 using a fast version (Sandmeier, 1984; Sandmeier and Wenzel, 1985) of the reflectivity method (Fuchs, 1968a; Fuchs and Müller, 1971) at the Computer Center of Karlsruhe University. A complete record section extending from near-vertical incidence (15 km) to 190 km, i.e. to overcritical distances with regard to Moho, is shown in Figure 6 (Sandmeier, 1984). The amplitudes are normalized to the maximum of every trace. No automatic gain control (AGC) has been applied which explains the weak amplitudes of the later arrivals at short distances.

This section shows very clearly the strong amplitudes for the wide-angle data compared to those of near-vertical incidence. The same source signal was used for all distances. No Q was applied, and yet a noticeable change in the dominant frequency can be observed: the overcritical reflection from zone (4) starts to become a low-freqency signal from about 100 km onwards. This phase must have tunnelled through the thin high-velocity lamellas in zone (4) (Fuchs and Schulz, 1976). Only the low freqencies have penetrated this structure and were reflected from its bottom.

Thus the low-freqency content of refraction experiments appears to be not so much an effect of the source spectrum but rather of the transmission properties of the medium. The low-frequency arrivals in wide-angle observations are indirect evidence for the finely layered structure of the medium. In contrast, attempts to observe low-freqency crustal and Moho reflections at small distances (Fuchs and Kappelmeyer, 1962) during large explosions in Lago Lagorai failed: the dominant frequencies of the observed reflections were still between 10-15 Hz although special low-frequency geophones (4.5 Hz) had been used.

What reason is there to expect near-vertical reflections to be generated in the subcrustal lithosphere? The studies of Sn and Pn propagation in the upper mantle (Molnar and Oliver, 1969) revealed that high freqency waves of about 10 Hz can indeed propagate through the upper mantle. Reflections form the crystalline part of the lithosphere require some sort of lamination pattern with alternating velocities to enhance their amplitudes. What processes in the upper mantle could be producing such laminated zones similar to those proposed to enhance crustal reflections? What observations are available to indicate the nature of such processes?

The high-velocities of the subcrustal lithosphere observed on long-range profiles are in contrast to the seismic velocities predicted from petrological models for this depth range. Figure 7 displays, as an example, an "observed" velocity-depth distribution for Southern Germany (Stangl, 1984; Fuchs, 1983) together with predicted velocities for a number of petrological models for the upper mantle (see Table 1). The geotherm is that of a young continent (see Fuchs, 1983).

The predicted depth-distribution of velocities are clearly smaller than the observed distribution. It is practically impossible to match observed and predicted distribution either by a change in composition or in temperature. Since a pronounced azimuthal variation of Pn velocities has been detected in this region by Bamford (1973, 1977) in the topmost mantle close to the Moho, the most likely explanation for the discrepancy is the assumption that this anisotropy is not restricted to a thin veneer at the top of the mantle but rather extends into the subcrustal lithosphere and furthermore increases with depth (Fuchs, 1983). It is best understood by a preferred orientation of olivine crystals. The process leading to a preferred orientation requires a considerable horizontal shear deformation of the subcrustal lithosphere. In the context of this paper is is

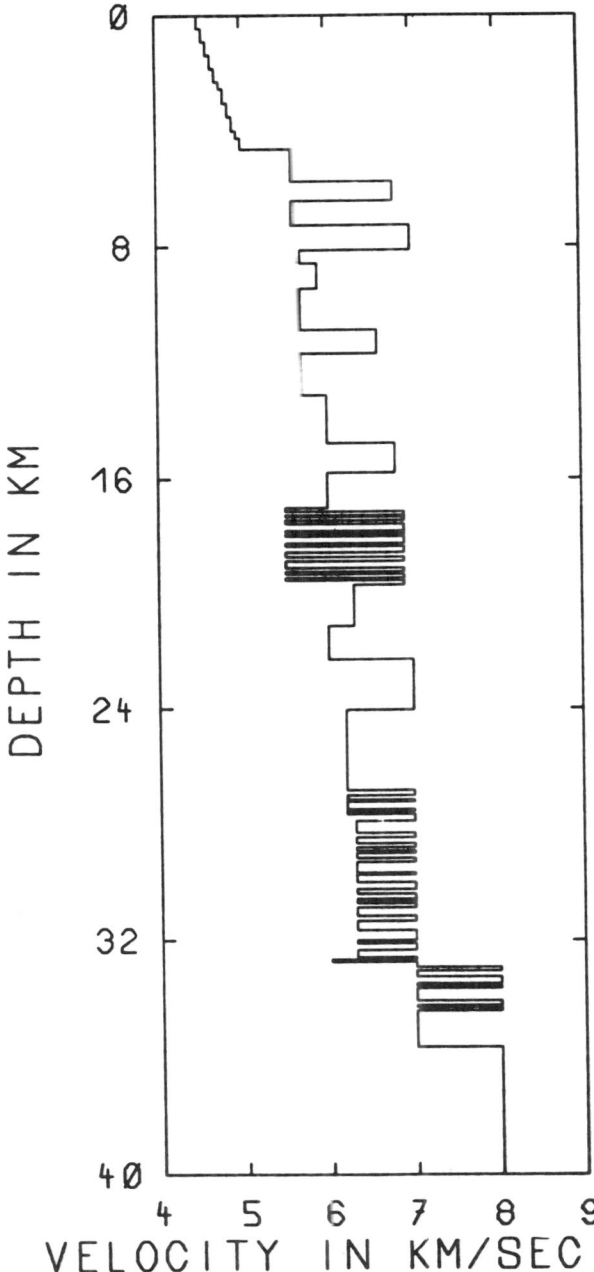

Fig. 5. Schematic crustal model with 4 laminated zones (after Smithson et al., 1977). Corresponding synthetic record section is displayed in Figure 6.

not important whether this deformation is caused by stresses from the lithosphere-asthenosphere boundary, by subduction or by other tectonic processes. The continental subcrustal lithosphere seems to be dominated by creep deformation since this region of the continental upper mantle appears to be mostly free of earthquakes (Meissner

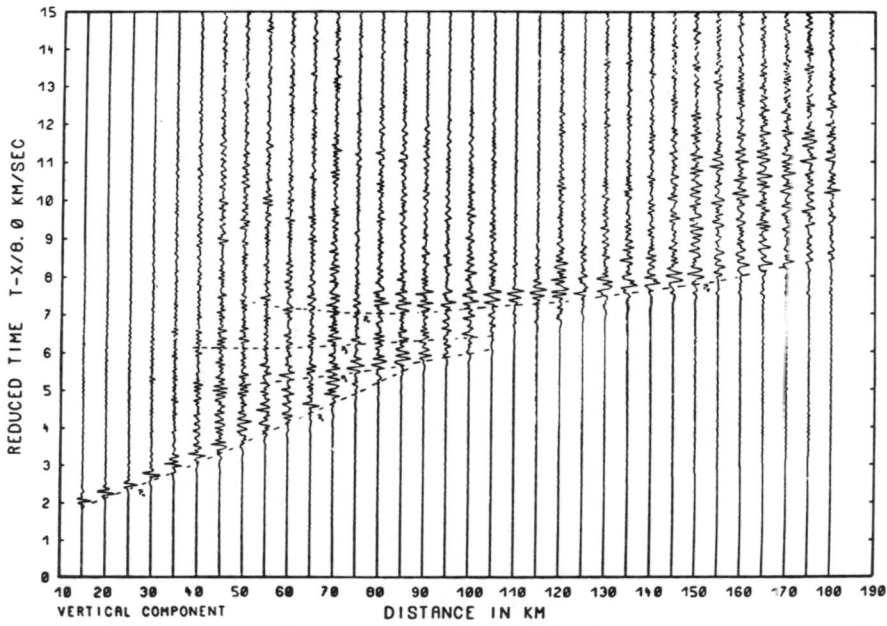

Fig. 6. Synthetic record section corresponding to laminated crustal model in Figure 5 computed with a fast version of the reflectivity program by Sandmeier (1984).

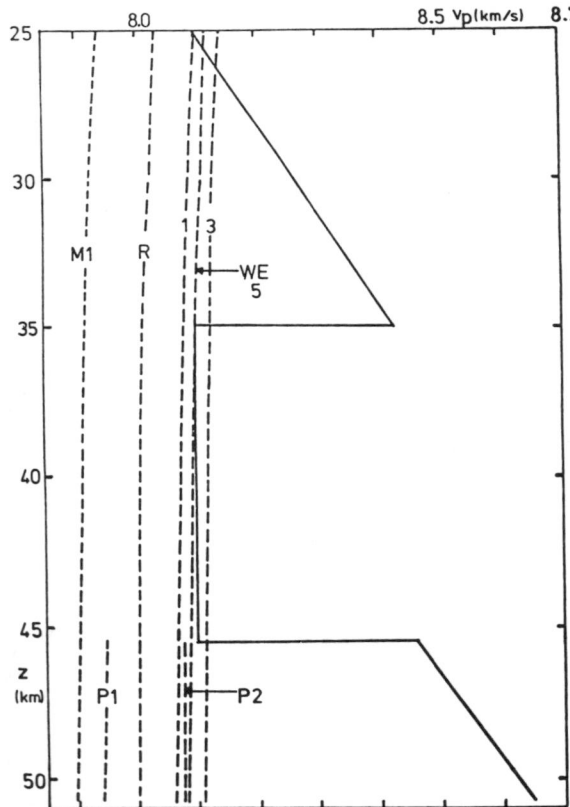

Fig. 7. Comparison of mantle P-velocity models inverted from refraction seismic observations (thick solid line) and velocities predicted from petrological models (dashed lines; M1, P1, P2, R, 1, 3, 5 = WE) in Table 1 (Fuchs, 1983).

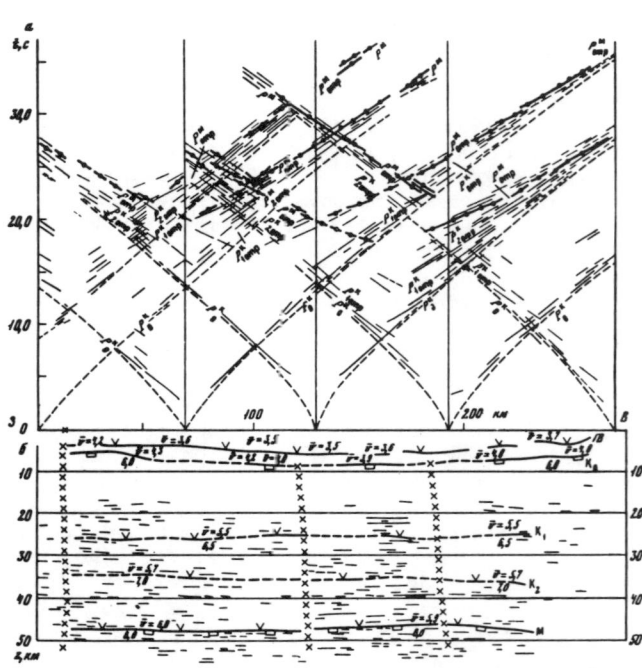

Fig. 8. Submoho reflections in the Tienshan/USSR (Zunnunov, 1980). Upper part: near- and wide-angle travel-time curves; lower part: depth section. The reflections at small distances extend to about 20 s, corresponding to a depth range well below Moho.

TABLE 1. Modal compositions (Vol%) of mantle xenoliths

Model/Location	Short Form	Olivine	Ortho-pyroxene	Clino-pyroxene	Spinell	Garnet
Spinell-Pyrolite	R	52	28	16	4	-
PHN-1569	P1	55.2	39.7	2	-	3.1
PHN-1611	P2	63.4	9.5	16.7	-	10.4
NE-Bavaria	1	65	23	9	3	-
Dreiser Weiher	3	75	17.5	8.4	0.9	-
North Hessian Depression	4	73.4	18.6	6.7	1.3	-
Western Eifel	5 = WE	73	19	7	1	-
Reference Model	M1	50	45	4	1	-

For references see Fuchs (1983).

and Strehlau, 1982). Only in very rare circumstances are hypocenters located in the continental upper mantle (Chen and Molnar, 1983). Any horizontal deformation necessary to produce the preferred orientation of olivine, if it lasts long enough and if its amplitude is large enough, transforms vertically elongated heterogeneities into thin horizontal lamellas.

Evidence for Reflections from below Moho

It was probably in the USSR where subcrustal reflections were first recorded during continuous crustal profiling. One typical example is shown in Figure 8 (Zunnunov, 1980). Although this is not a record section but a line drawing of the most prominent phases, there is no doubt that reflected energy may be followed back to more than 20 s at small distances from the shot point, i.e. depths well below the Moho.

A special deep crustal reflection survey by Dohr (1970) in the upper Rhinegraben proper in southwestern Germany revealed numerous reflections between 7-9 and 18 s (Fig. 9). The reflections between about 7-10 s are from the lower crust. At larger echo-times reflections can be recognized with amplitudes as large as from the crust. They are almost certainly related to the mantle diapir which developed with the rift formation.

An outstanding example of subcrustal reflectors came from the BIRPS experiment (Brewer et al., 1983; see also this volume). It deserves special attention since the pronounced reflector is dipping distinctly below Moho.

Conclusions

Refraction seismic probing of the subcrustal lithosphere provides evidence for unexpected strong vertical heterogeneities in the top 100 km of the upper mantle with velocity contrasts as large as encountered already at the crust-mantle boundary, where laminated zones are responsible for strong reflected signals. The observation of anisotropy in parts of the continental upper mantle caused by a preferred orientation of olivine crystals is indirect evidence for strong horizontal shear deformations in the subcrustal lithosphere which can produce sequences of nearly horizontal layers with alternating velocities. This is also indirect evidence that tectonic processes are not restricted to the crust but extend into the upper mantle. So far only a few examples of sub-Moho reflections are documented in the literature. However, most reflection surveys today stop near 12 s echo-time, i.e. near the crust-mantle boundary. So far no reflections have been expected from the upper mantle, and therefore nobody looked for them.

It took more than 40 years after the Moho had been discovered from refracted seismic waves to see this major boundary of the earth's interior also in near-vertical reflections. We should not again wait that long to attempt to observe reflections from the subcrustal lithosphere.

Reflection profiling is a key project of the International Lithosphere Program. With comparatively little effort it should be possible to obtain near-vertical reflections from the subcrustal lithosphere to a depth of 100-200 km during

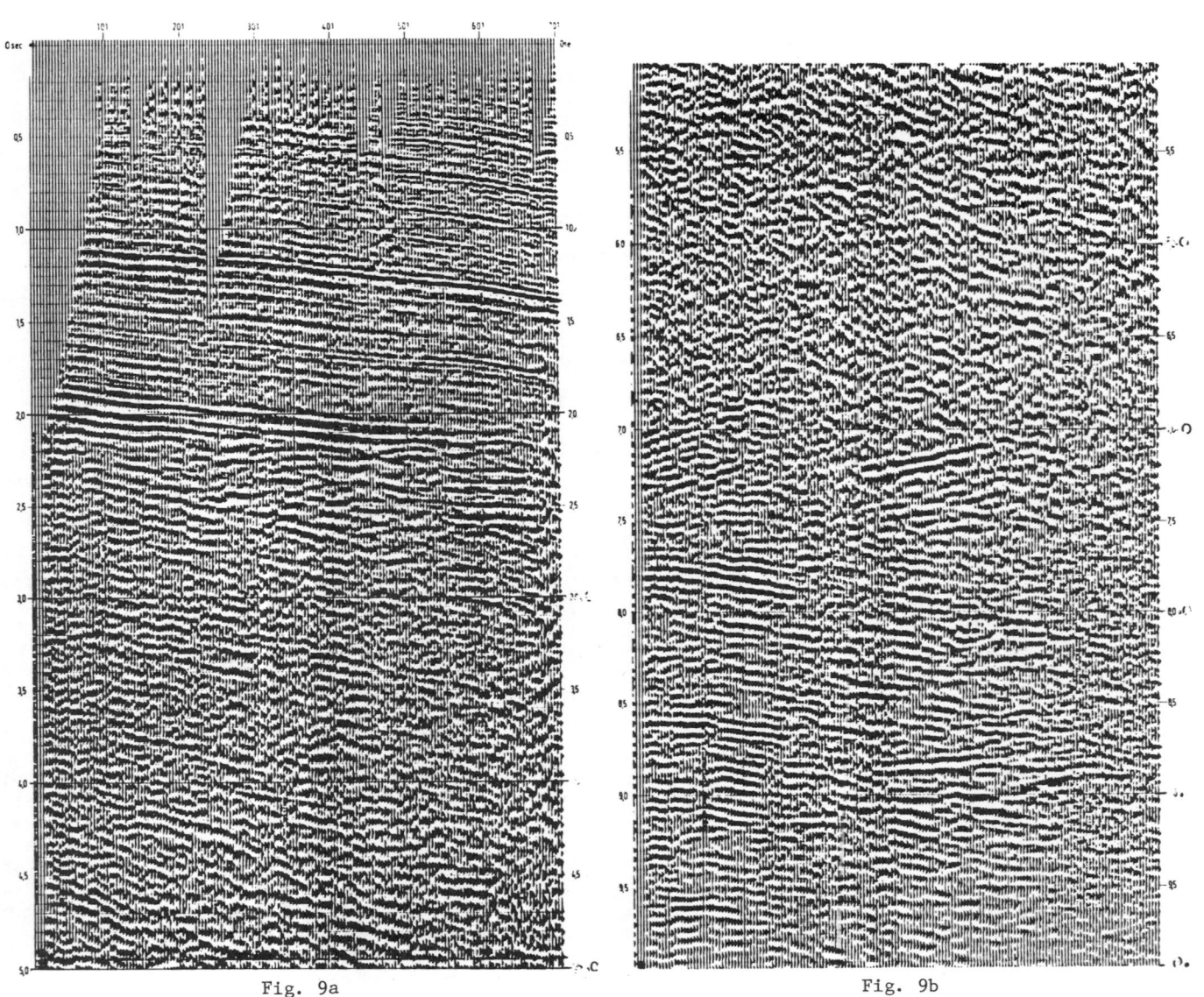

Fig. 9. Submoho reflections form the Upper Rhinegraben proper (Dohr, 1970). The reflections between about 7-10 s are from the lower crust. The reflections with larger echo-times are to be related to the mantle diapir beneath the southern part of the Upper Rhinegraben.

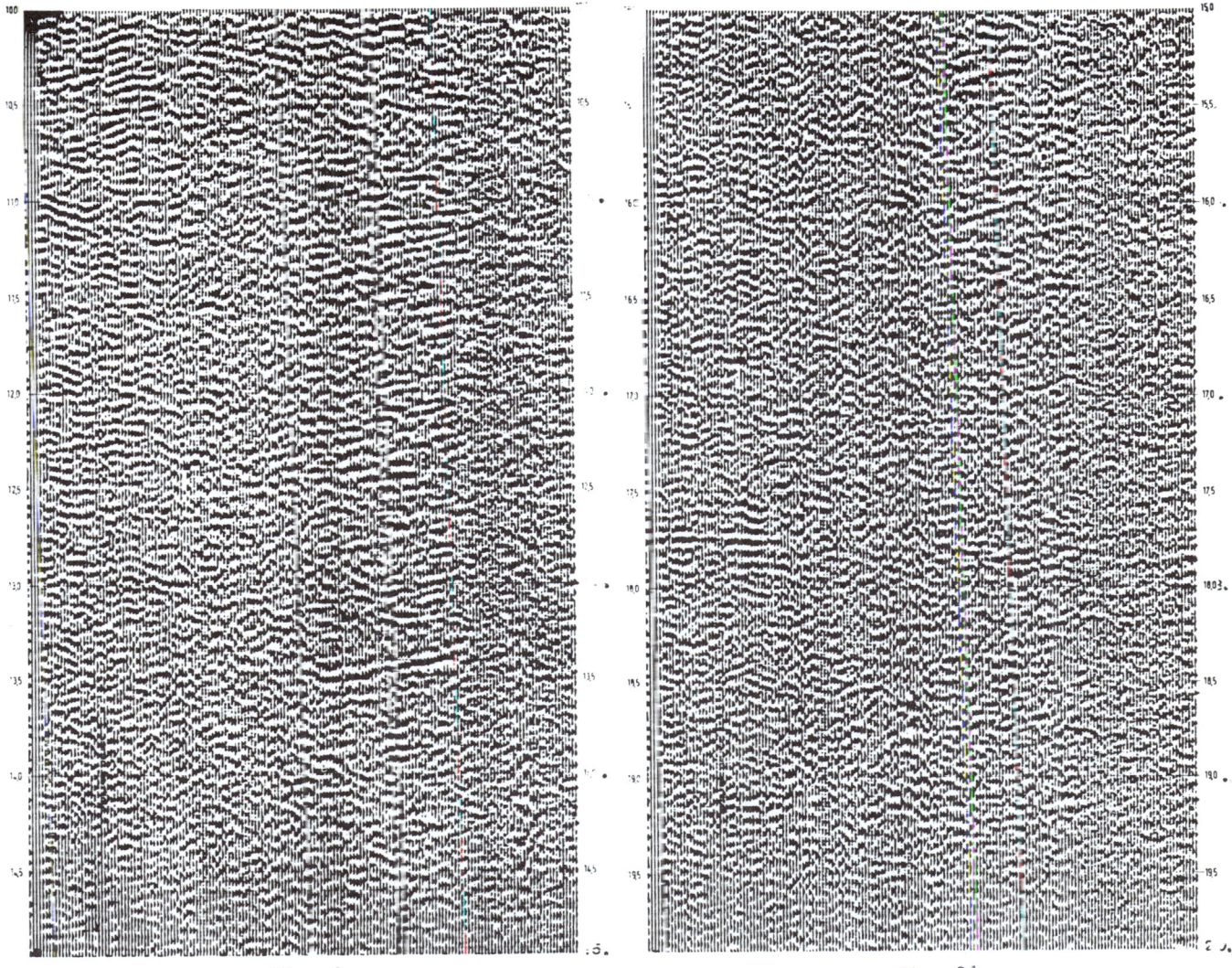

Fig. 9c Fig. 9d

Fig. 9. (continued)

reflection surveys. Long-range refraction surveys would not become superfluous. I have shown that near-vertical reflections and wide-angle refractions reveal different aspects of the structure of the earth's interior. Refraction probing, in addition and parallel to reflection surveys, provides the information on velocities required to understand the physical properties and the composition of the medium under investigation.

Acknowledgments. I am indepted to C. Prodehl, K.-J. Sandmeier and F. Wenzel for continuing discussions on the nature of seismic dicontinuities. C. Prodehl critically read the manuscript and helped to improve it. The ideas presented in this paper were developed within the special research program "Stress and stress release in the lithosphere (SFB 108) established by the Deutsche Forschungsgemeinschaft (German Research Society) at Karlsruhe University. K.-J. Sandmeier's synthetic seismograms were computed on the CDC CYBER 205 of the Computer Center at Karlsruhe University. I thank also the two reviewers for their kind suggestions to improve the manuscript. SFB 108-Contribution No. 080. Geophysical Institute Karlsruhe Contribution No. 290.

References

Ansorge, J., Die Feinstruktur des obersten Erdmantels unter Europa und dem mittleren Nordamerika, Dissertation, University of Karlsruhe, 1975.
Bamford, D., Refraction data in Western Germany -

a time-term interpretation, Z. Geophys., 39, 907-927, 1973.

Bamford, D., Pn-velocity anisotropy in a continental upper mantle, Geophys. J.R.A.S., 49, 29-48, 1977.

Brewer, J.A., Mathews, D.H., Warner, M.R., Hall, J., Smythe, D.K., Whittington, R.J., BIRPS deep seimic reflection studies of the British Caledonides, Nature, 15 September, 305, 206-210, 1983.

Chen, Wang-Ping, Molnar, P., Focal depths of intracontinental and intraplate earthquakes and their implications for the thermal and mechanical properties of the lithosphere. J. Geophys. Res., 88, 4183-4214, 1983.

Deichmann, N., J. Ansorge, Evidence for lamination of the lower continental crust beneath the Black Forest (Southwestern Germany). J. Geophys., 52, 108-118, 1983.

Dohr, G., Reflexionsseismische Messungen im Oberrheingraben mit digitaler Aufzeichnungstechnik und Bearbeitung. In: Graben Problems. Schweizerbart Stuttgart, 207-218, 1970.

Faber, S., Refraktionsseismische Untersuchung der Lithosphäre unter den britischen Inseln, PhD-Thesis, University of Karlsruhe, 132 p., 1978.

Fuchs, K., The reflection of spherical waves from transition zones with arbitrary depth-dependent elastic moduli and density. J. Physics earth, Special Issue, 16, 27-41, 1968a.

Fuchs, K., Das Reflexions- und Transmissionsvermögen eines geschichteten Mediums mit beliebiger Tiefen-Verteilung der elastischen Modulen und der Dichte für schrägen Einfall ebener Wellen, Z. Geophysik, 34, 389-413, 1968b.

Fuchs, K., Recently formed elastic anisotropy and petrological models for the continental subcrustal lithosphere in southern Germany, Phys. Earth Planet. Interior, 31, 93-118, 1983.

Fuchs, K., O. Kappelmeyer, Report on Reflection measurements in the Dolomites, Sept. 1961, Boll. Geof. Teor. Appl., 4, 133-141, 1962.

Fuchs, K., G. Mueller, Computation of synthetic seismograms with the reflectivity method and comparison with observations, Geophys. J.R.A.S., 23, 417-433, 1971.

Fuchs, K., K. Schulz, Tunneling of low-freqency waves through the subcrustal lithosphere, J. Geophysics, 42, 175-190, 1976.

Hirn, A., L. Steinmetz, R. Kind, K. Fuchs, Long-range profiles in Western Europe: II. Fine structure of the lower lithosphere in France (South. Bretagne), Z. Geophys., 39, 363-384, 1973.

Junger, A., Deep reflections in Big Horn County, Montana, Geophysics, 16, 499-505, 1951.

Kind, R., Propagation of seismic energy in the lower lithosphere, Z. Geophysik, 40, 188-202, 1974.

Meissner, R., An interpretation of the wide angle measurements in the Bavarian Molasse Basin, Geophys. Prosp., 14, 7-16, 1966.

Meissner, R., J. Strehlau, Limits of stresses in continental crusts and their relation to the depth-freqency distribution of shallow earthquakes, Tectonics, 1, 73-89, 1982.

Molnar, P., J. Oliver, Lateral variations of attenuation in the upper mantle and discontinuities in the lithosphere, J. Geophys. Res., 74, 2648-2682, 1969.

Prodehl, C., Structure of the earth's crust and upper mantle. In: Landolt Börnstein, New Series, Springer, Heidelberg, V/2a, 97-206, 1984.

Sandmeier, K.-J., Veränderung und Erweiterung des Reflektivitätsprogramms zur Rechnung synthetischer Seismogramme, Diploma Thesis, University of Karlsruhe, 182 p. & App., 1984.

Smithson, S.B., N.B. Shive, S.K. Brown, Seismic velocity reflections, and structure of the crystalline crust. In: The earth's crust, Heacock, J.G., editor, AGU; GEOPHYSICAL MONOGRAPH, 20, 254-270, 1977.

Stangl, R., Geschwindigkeitstiefen-Verteilungen von P-Wellen im oberen Erdmantel Süddeutschlands, Diploma Thesis, University of Karlsruhe, 154 p., 1983.

Steinhart, J.S., R.P. Meyer, Explosion studies of continental structure, Carnegie Institution Washington, Publ. No. 622, 409 p., 1961.

Zunnunov, F.Kh., The Turanskaya plate and adjoining geostructures of the Tien Shan. In: Seismic models of the lithosphere for the major geostructures on the territory of USSR, Publishing House NAUKA, Moscow, 78-82, 1980.

DEEP CRUSTAL SIGNATURES IN INDIA AND CONTIGUOUS REGIONS FROM SATELLITE AND GROUND GEOPHYSICAL DATA

M. N. Qureshy and R. K. Midha

Department of Science & Technology[1], Technology Bhavan, Government of India
New Mehrauli Road, New Delhi 110016, India

Abstract. Topography and Bouguer anomalies show an inverse relationship in the Indo-Ganga-Brahmaputra basin and Himalaya-Hindu Kush region. The Indo-Ganga-Brahmaputra basin, devoid of any prominent closure on the Bouguer anomaly map, exhibits a -100 mgal closure on the Airy-Heiskanan map. The positive isostatic anomaly over the Himalaya can be explained by two possible intracrustal mass distributions that could represent either a remnant of the Neo-Tethys floor or large-scale igneous intrusions into the crust from the mantle. We thus suggest that the isostatic compensation in the Himalaya and the Indian shield is nearly complete. The correlation of Moho depths derived from the empirical relationship between crustal thickness and elevation with the depths determined from Deep Seismic Sounding (DSS) provides support for this contention. The steep gradient on the regionalized isostatic anomaly map of the area indicates that the Main Mantle Thrust (MMT), which is located west of the Nanga Parbat, may represent a direct continuation to the Main Central Thrust (MCT), which is located farther east. Likewise, the northern gradient of the isostatic anomaly map is correlated with the Northern Main Suture (NMS) in the Kohistan and Himalaya-Tibet regon. That prominent lineaments, some dating back to the Precambrian, such as the Narmada-Son rift, Godavari graben, and Gandak-Karakorum, show up conspicuously on both ground- and satellite-based gravity and magnetic maps suggests a genetic relationship between the near-surface and deep controlling stuctures. The Magsat vertical field (Z) map shows a positive anomaly (4 to 8 nT) over the southern Indian shield and a negative ENE-trending anomaly (-4 to -8 nT) over the northern shield, with the Narmada-Son line marking the transition zone. The negative magnetic anomaly over the Ganga-Himalaya area suggests a possible rise of geotherms in this region. There is growing evidence to show that some blocks in the Indian continent, such as the Aravali ranges, Narmada rift, Shillong plateau, and Godavari graben, are associated with recent tectonic activity, suggesting that the Indian shield is not as stable as normally thought and that it may be passing through a phase of rejuvenation.

Introduction

The application of plate tectonics and flexure models to the evolution of the Himalaya [e.g., Molnar, 1984; Karner and Watts, 1983; among others] has lately focused attention of many geoscientists to the Indian subcontinent. Different types of geotectonic models are in vogue to explain the uplift of the Himalaya and the occurrence of the ophiolite suite of rocks in the Indus suture zone (Figure 1). To appreciate the tectonic evolution of the Indian subcontinent, it is necessary to have an overview of the existing geophysical data on the country. The present paper collates existing ground-based and satellite-derived geophysical and geological data and presents an analysis with a view to elucidating deep crustal and supra-crustal structures.

Most of India is made up of a number of plateaus or remnants of plateaus girdled by the Kirthar and Sulaiman ranges to the west, the Himalaya to the north, and the Burma arc to the east (Figure 1). The southern peninsula, jutting into the Indian Ocean, is flanked by the Arabian Sea and the Bay of Bengal. India's broad structural units are the Peninsula, the Extra-Peninsula, and the intervening Indo-Ganga-Brahmaputra basin (Figure 1). The Peninsula consists of Pre-cambrian formations, Gondwanas (Carboniferous to Cretaceous), Deccan traps (Cretaceous to Eocene), and Cenozoic sediments along parts of its coast and shelf. Major physiographic-structural units and basement fabric trend in the NS (Indo-Pamir), NW (Godavari-Gandak), NE (Aravalli), NNW (Dharwar), and ENE (Satpura) directions (inset in Figure 2). The Extra-Peninsula, comprising Himalaya, Sulaiman, and Kirtha ranges, and the Burma arc, is made up of Precambrian to Quaternary rocks. The Indus suture zone and the Indo-

[1] The views expressed are the authors' and do not reflect in any way those of the Department of Science and Technology.

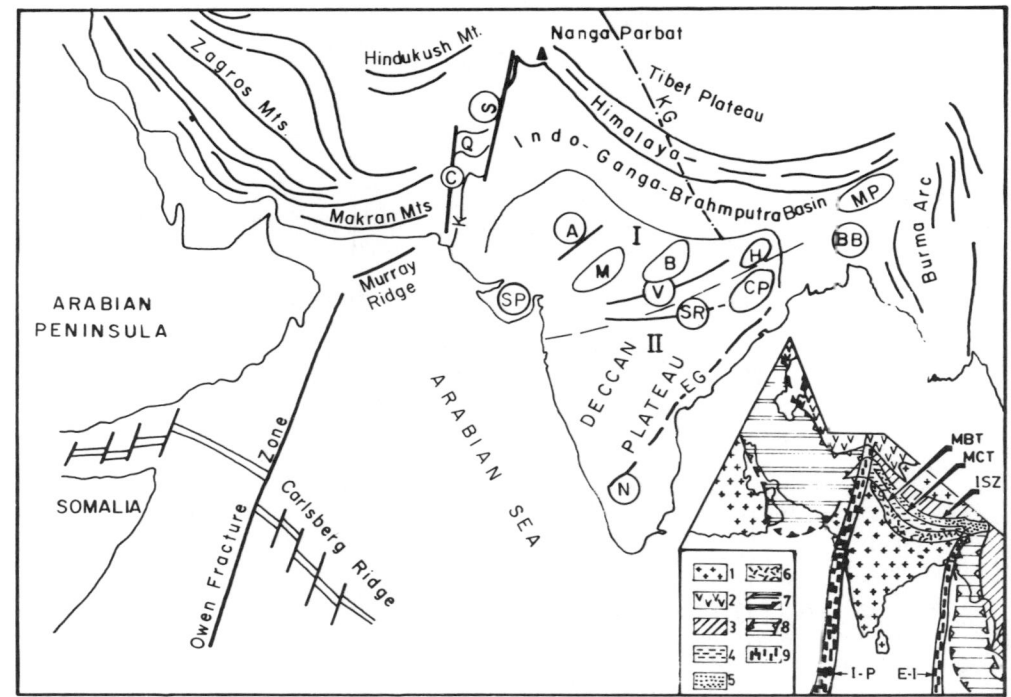

Fig. 1. Map of SE Asia showing major tectonic elements. I-North Indian shield; II-South Indian shield; B-Bundelkhand massif; BB-Bengal basin; C-Chaman fault; CP-Chota Nagpur plateau; EG- eastern Ghats; H-Hazaribagh plateau; K-Kirthar range; KG-Karakorum Gandak lineament; M-Malwa plateau; MP-Meghalaya (Shillong) plateau; N-Nilgiri hills; Q-Quetta; S-Sulaiman range; SP-Saurashtra Peninsula; SR-Satpura range; V-Vindhyan range. Inset - Zones of folding: 1-Precambrians; 2-Paleozoic; 3-Mesozoic; 4-Greater Himalaya, composed of Precambrians; 5-Tethys zone; 6-Indo-Ganga-Brahmaputra basin; 7-Mediter-ranean Alpine fold belt; 8-Circum-Pacific Alpine fold belt; 9-main transverse zones: IP-Indo- Pamir; EI-eastern Indian zone (Ninety East Ridge); MBT-Main Boundary Thrust; MCT-Main Central Thrust; ISZ-Indus suture zone.

Burmese arc are probable regions of plate subduction and continent-continent collision. The Chaman fault, on the west, is a transcurrent fault and is, presumably, connected to the Owen fracture zone in the Arabian Sea through the Murray ridge (Figure 1). The Cenozoic sediments filling the Indo-Ganga-Brahmaputra basin overlie Precambrian rocks.

Gravity and isostasy have evoked interest in India since the early 19th Century. It is postulated here that isostatic compensation in India and the Himalaya-Hindu Kush region is probably nearly complete. This follows from the moderate magnitude (compared to the Bouguer anomalies) of the isostatic anomalies; the correlation of Bouguer anomalies with elevation [Qureshy, 1971a]; the relationship of crustal thickness (T) to elevation (E) and the compatibility of Moho depths obtained from wide-angle reflection and refraction profiling (Figure 4) [Kaila, 1982] with those calculated from the empirically obtained equation, $T = 35 + 5.9E$ [Qureshy, 1970]; and the correlation of the isostatic anomalies with local geological features that can adequately explain the anomalies.

Peninsular India, the Indo-Ganga-Brahmaputra basin, and the Himalaya constitute a single unit. The tectonic processes therein are mostly intra-plate in nature and are represented by rifts like the Cambay and Gondwana basins, ancient sutures with recent reactivation like the Tapi-Narmada-Son lineament, and the reactivated shield blocks like the Himalaya and the Shillong (Meghalaya) plateau. Some transverse geophysical features cut across plate and continental boundaries and seem to show continuity in space and time. These aspects may provide some constraints on the flexure models [e.g., Karner and Watts, 1983; Lyon-Caen and Molnar, 1983]. It seems that the major geophysically mapped lineaments, which

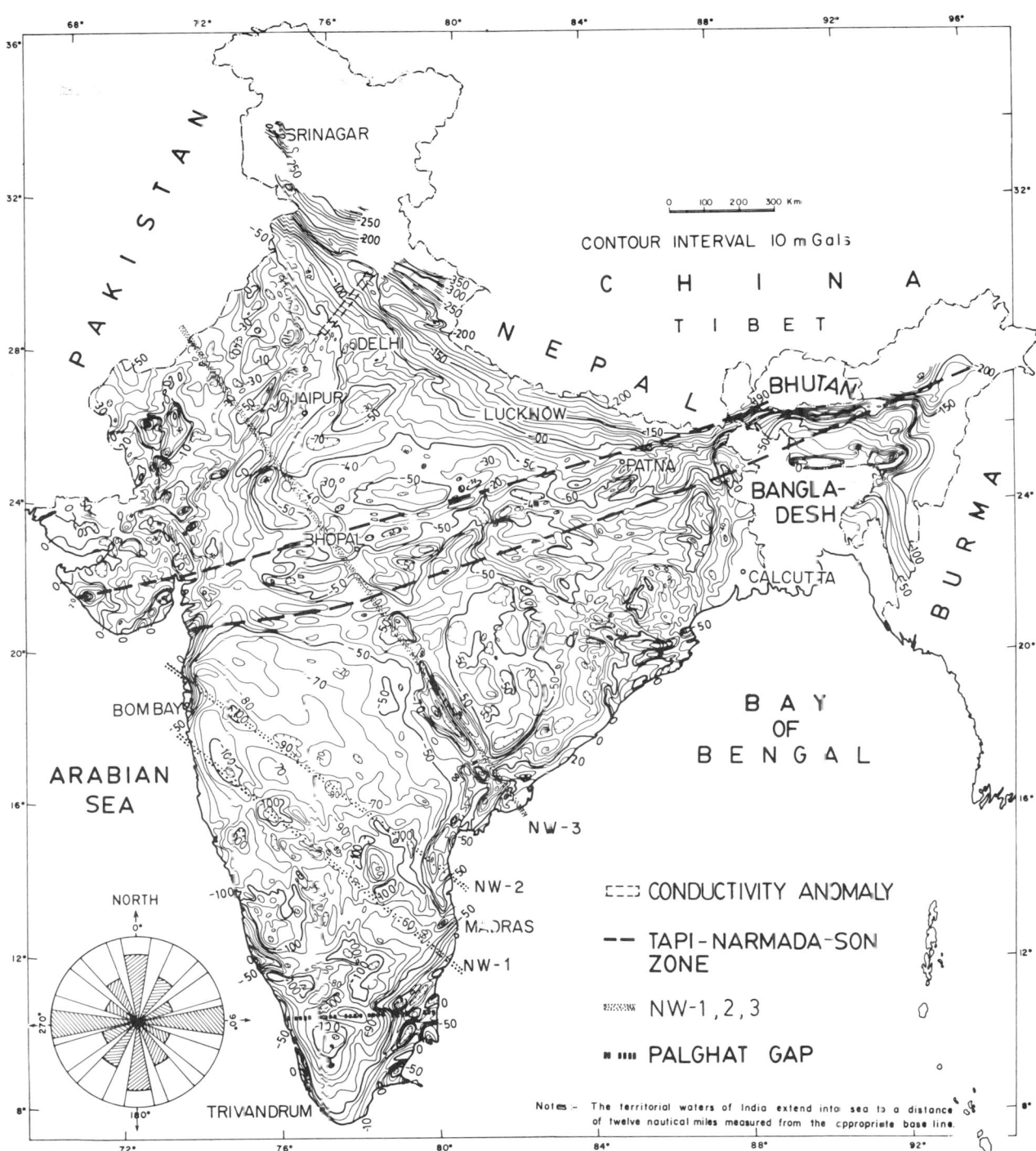

Fig. 2. Bouguer anomaly map of India. The dominant NE and ENE trend in the northern Indian shield, the NW trend in the western portion and somewhat oblate anomalous features in the northeastern segment of the southern Indian shield are noteworthy. NW1, NW2, and NW3 correspond to Bhima, Krishna, and Godavari trends, respectively. The Godavari lineament divides India into east-west segments and the Tapi-Narmada-Son zone into north-south segments. The inset (circle) shows the structural fabric in the sub-continent, as obtained by Eremencko [1969] from geomorphological, geological, and geophysical data.

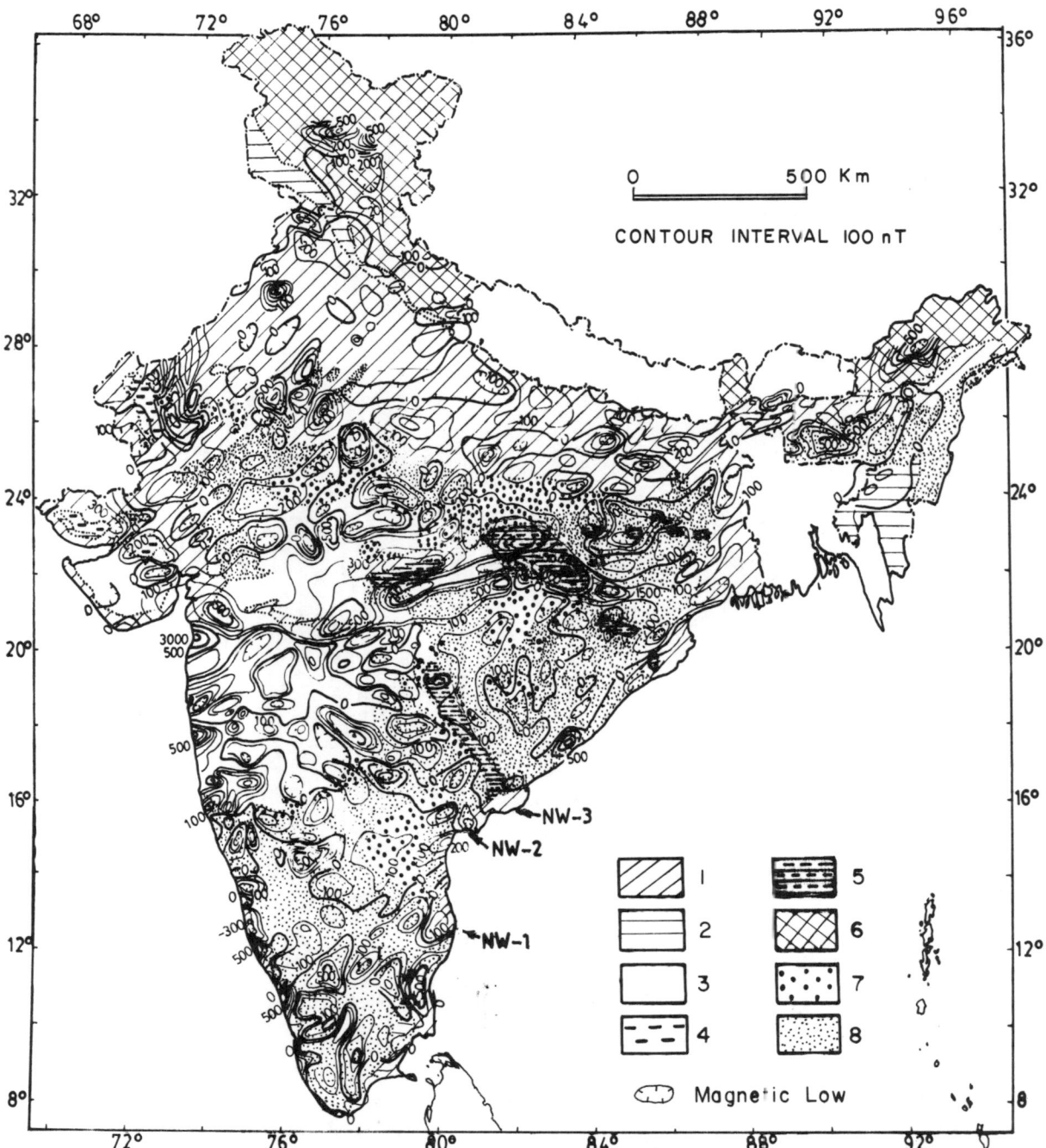

Fig. 3. Total intensity residual magnetic anomaly map superimposed on generalized geologic map of India. 1-Quaternary, 2-Tertiary, 3-Deccan trap, 4-Mesozoic; 5-Gondwana, 6-tectonized zone of Himalaya, 7-upper Proterozoic, 8-Precambrian tectonites of the Indian shield. Note apparent division of the Ganga basin by NW-trending magnetic zones with separate segments. An ENE-trending magnetic feature, coinciding with the Tapi-Narmada-Son zone (Figure 1), runs from the west coast (20°N) up to the NE extremity of India. This correlates with Magsat vertical (Z) field anomaly map (Figure 7). South of this feature the trends are mostly NW. A prominent low (22°N, 82°E) over the Mahanadi Gondwana basin (Figures 6 and 7) exceeding 1000 nT in magnitude, continues southwestward through the Satpura basin over the Deccan trap region, suggesting possible existence of Gondwanas beneath the traps. NW1, NW2, and NW3 are trends as in Figure 2.

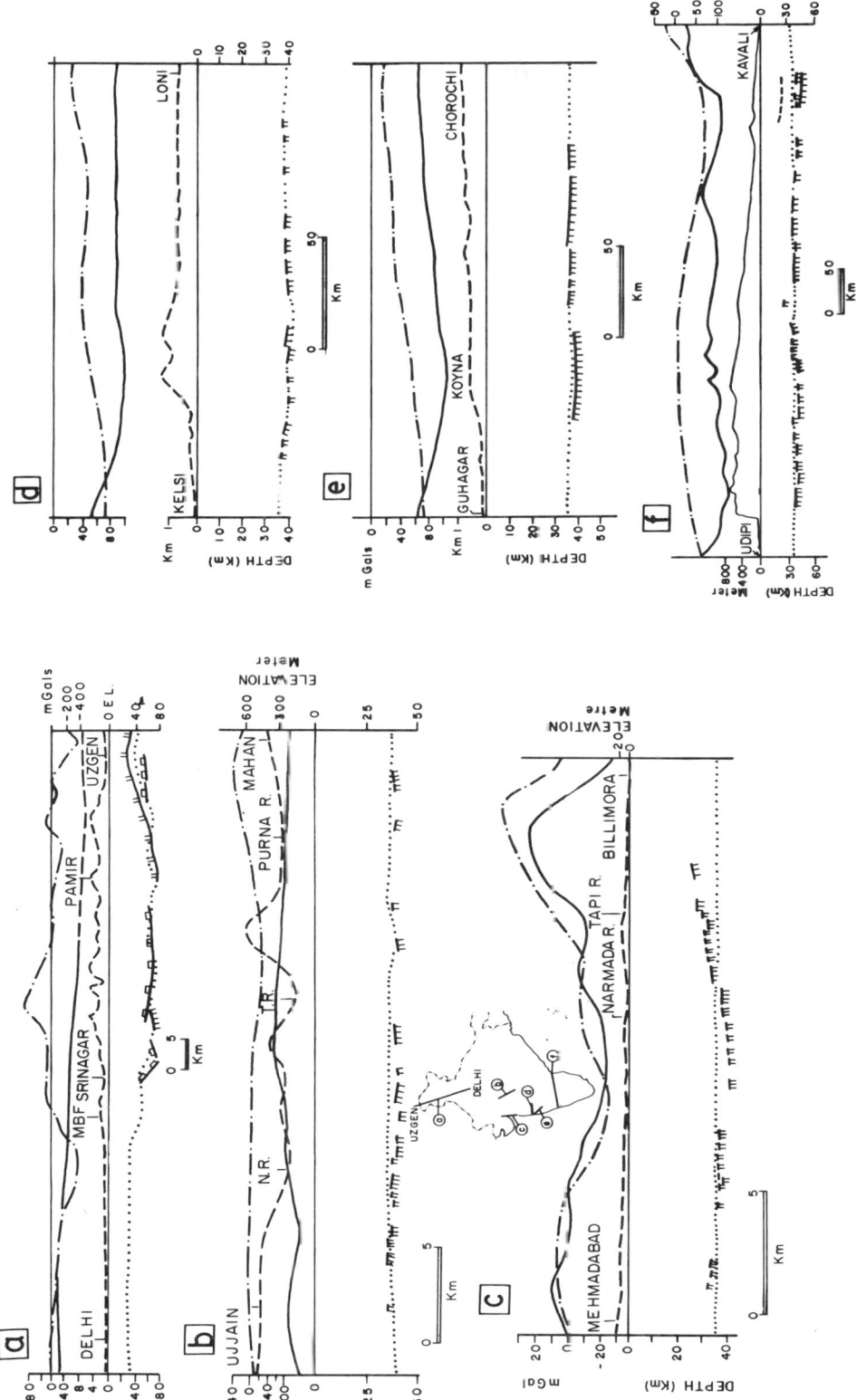

Fig. 4. Gravity anomaly, elevation and crustal depth profiles in India. (i) Elevation, (ii) Bouguer anomaly, (iii) isostatic anomaly, (iv) Moho from DSS, (v) crustal depth from isostatic consideration, (vi) and (vii) crustal depths on Delhi-Uzgen profile from Russian sources. Profiles: a-across Himalaya, b-across the Narmada, c-along the Cambay rift, d-across the south Indian shield, e and f-Koyna-Deccan trap region. Locations of profiles are shown in Figure 5.

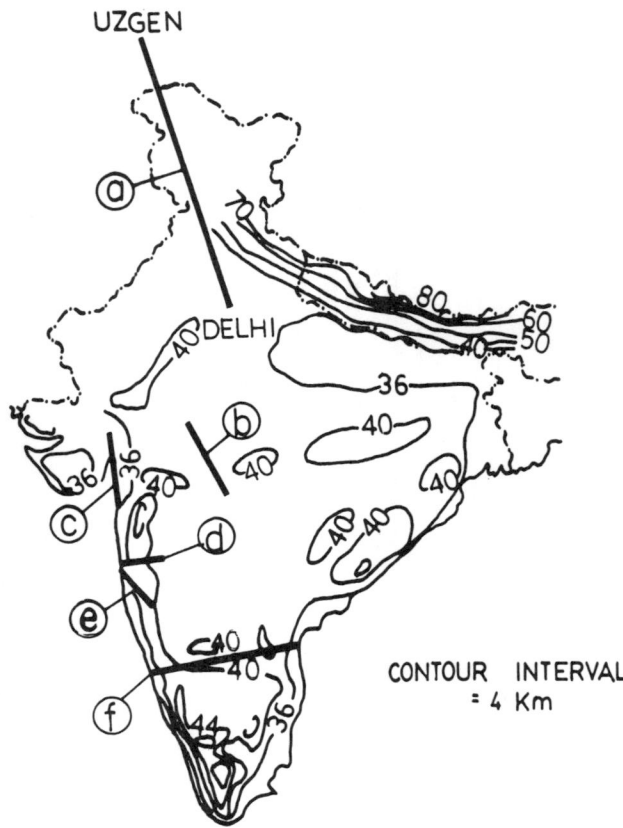

Fig. 5. Crustal thickness map of a part of India as calculated from the elevation: a – f show the location of DSS profiles discussed in text.

may be exogenic expressions of endogenic features, control the rejuvenation of blocks in the Indian plate and the rise of the Himalaya.

Data

Besides the ground geophysical surveys in the Himalaya and the Trans-Himalaya carried out by the Survey of India, Oil and Natural Gas Commission, National Geophysical Research Institute, satellites provide an additional data base. Most of these data, published as maps, form the basis for what follows [e.g., Gulatee, 1956; Qureshy, 1971a; Marussi, 1976; Chugh, 1976; Qureshy et al., 1981; Bhandari et al., 1984; Marsh, 1979].

Ground Gravity, Magnetic, and Array Data

The simple Bouguer anomaly map (Figure 2) of India [Qureshy and Warsi, 1980; Qureshy et al., 1981] incorporates some 31,000 point values. The observed gravity values refer to the value of 979,064 mgal at Dehradun, which corresponds to the value of 981,274 mgal at the International Base in Potsdam. The Bouguer anomalies are based on the International Gravity Formula (I.G.F.) of 1930 and a density of 2.67 g/cm^3 for the rock material above sea level. No terrain corrections were applied, though by simple calculations it was found to be negligible for most of India. Except for the hilly areas, where the terrain correction may be high, the accuracy of the Bouguer anomalies is estimated at 1.5 mgal. For details see Qureshy et al. [1981], who have given values reduced to the 1967 I.G.F.

The magnetic anomaly map [Figure 3, modified from Chugh, 1976] is based on some 900 point values. The total magnetic intensity anomaly (Figure 3) is the value at a point from which the mean of the values at the corners of a triangle has been subtracted. The results of geomagnetic induction studies are from Arora et al. [1982]. They operated an array of 24 Gough-Reitzel magnetometers in NW India and examined a variety of transient magnetic variations in the form of stacked profiles, contour maps of Fourier transform parameters and Parkinson induction arrows over the period range of 90 min to 20 min.

Seismic Data

The exploration seismic work conducted by the Oil and Natural Gas Commission [Bhandari et al., 1984] and refraction and wide-angle reflection profiles [Beloussov et al., 1980; Kaila, 1982] provide the seismic control for interpretation. The Moho depths from DSS and crustal thickness (T) calculated from elevation using the relation $T = 35 + 5.9E$ [Qureshy, 1970] are reproduced in Figures 4 and 5.

Satellite Gravity and Magnetic Data

Analysis of data from Landsat and Russian satellites (e.g., Meteor) of this region has been made by Ebblin [1976], Kazmi [1979], Powar [1981], Bhan [in Qureshy, 1982], and Bush et al. [1984], among others.

Satelllite-derived gravity data are from Marsh [1979], who presented free air total field anomalies and spherical harmonics of 0-12 and 13-22 degrees and order. Although satellite gravity data essentially show structures within the lower lithosphere and upper mantle, the correlation of some of the anomalies on satellite-derived gravity maps with the known surface geological features (Figure 6) and surface geophysical data, as in the Godavari graben, raises the possibility of a genetic relation between the deep-seated (endogenic) features indicated by the satellite-derived data and crustal (exogenic) structures. Even if the correlation is a mere coincidence, there is the question of the gravity effect of such unknown deep-seated features on the gravity effect of crustal features. These deep-seated effects thus assume importance in interpreting crustal anomalies and should be removed as "super regional."

The Magsat vertical field (Z) anomaly map (Figure 7) is prepared from the data processed by

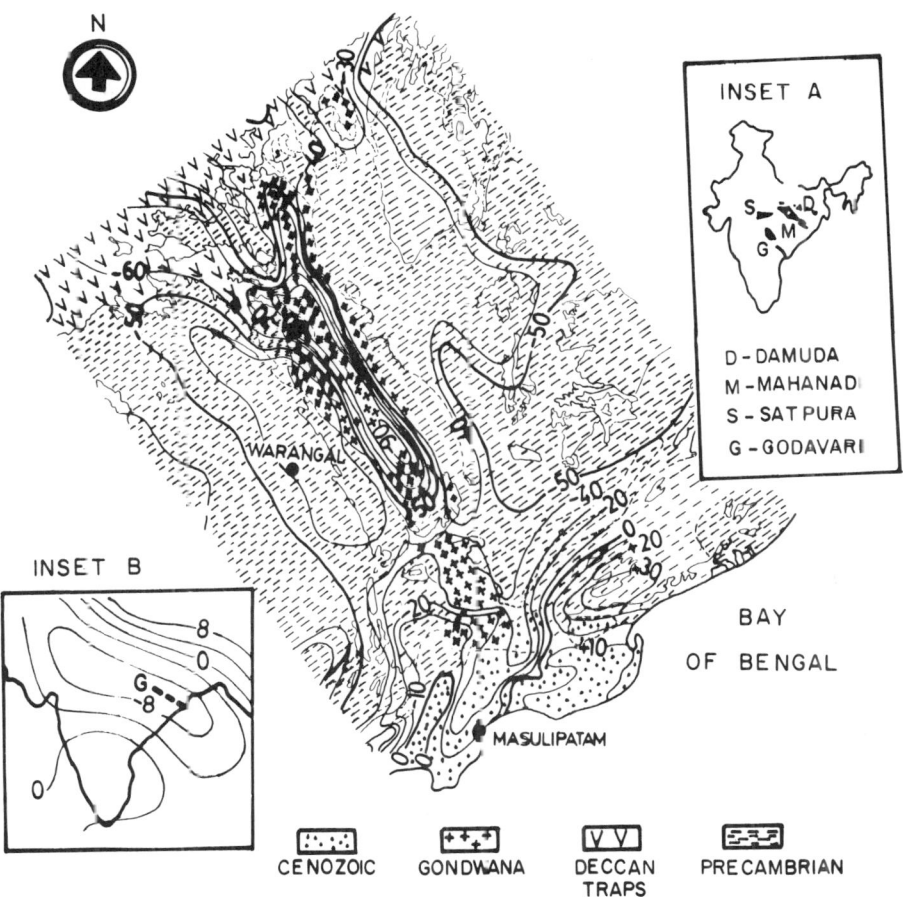

Fig. 6. Bouguer anomaly map of the Godavari graben from Qureshy et al. [1968b]. Note the 50 mgal low over the graben filled with low-density Gondwana sediments. Inset A shows the location of Godavari and other Gondwana basins in India. Inset B shows the gravity low on 13 - 22 harmonics field map from satellite [Marsh, 1979]. Note the general alignment of the satellite-derived low, which must be caused by deep-seated features, and the Godavari low from surface data.

Singh [1985] from Magsat investigator-B tapes supplied by NASA. Despite the long wave nature of the anomalies, some correlation with crustal features, and even surface lineations like the Narmada-Son, need not be a mere coincidence. Perhaps they reflect, in broad terms, rising of isogeotherms in the region of negative anomalies such as the northern Indian shield-Himalaya block.

Relationship of Geophysical Data to Broad Tectonic Elements

The Bouguer anomalies in India (Figure 2) show a variation of about 450 mgal. Positive anomalies up to +75 mgal occur near Bombay and in the Saurashtra peninsula (SP on Figure 1), and less than -375 mgal anomalies are observed in the vicinity of the India-China border. The east coast is associated with values higher than -50 mgal, becoming positive in places, a range of values that exhibits a normal continental margin effect. In contrast, the west coast is characterized by negative anomalies varying between -50 mgal and -100 mgal south of 18°N, and by near-zero and positive values north of 20°N. The magnetic anomaly map shows broad correlation (Figure 3) with the gravity anomalies and regional geology. Figure 4 shows that the DSS data largely substantiate the Airy-Heiskanen isostatic model as far as crustal thickness is concerned. The relationship of the geophysical data to broad tectonic elements of the subcontinent follows by region.

The South Indian Shield

The south Indian shield may be defined to be the region of India south of the Tapi-Narmada-Son zone (Figure 1), which in the west lies south of

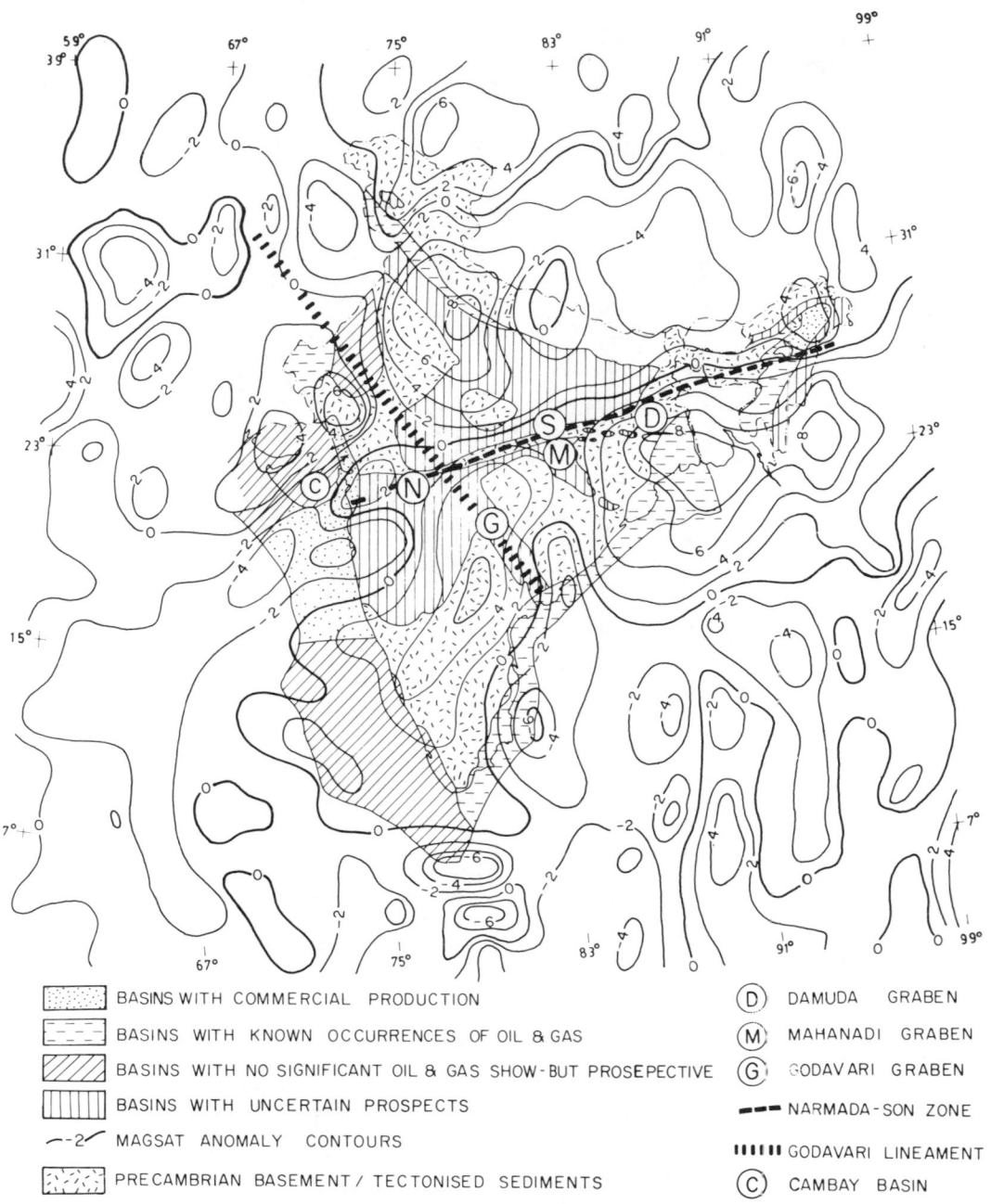

Fig. 7. Magsat-derived vertical field (Z) anomaly map of India [adopted from Singh, 1985] and the contiguous regions superimposed on the petroliferous basins in India [from Bhandari et al., 1984]. Noteworthy features are (i) the negative anomaly north of the Narmada-Son line on the north Indian shield-Himalaya block; (ii) positive anomaly over the southern, northwestern (Rajasthan-Punjab), and northeastern portion of India; and (iv) the association of a negative anomaly with a closure of -4 nT over commercially productive oil fields of the Bombay high-Cambay region and the northeast region in India and Potwai plateau (32°N, 72°E) of Pakistan. The Narmada-Son and Godavari lineaments demarcate zone of differing magnetic crust as is the case with the gravity field maps (Figure 2).

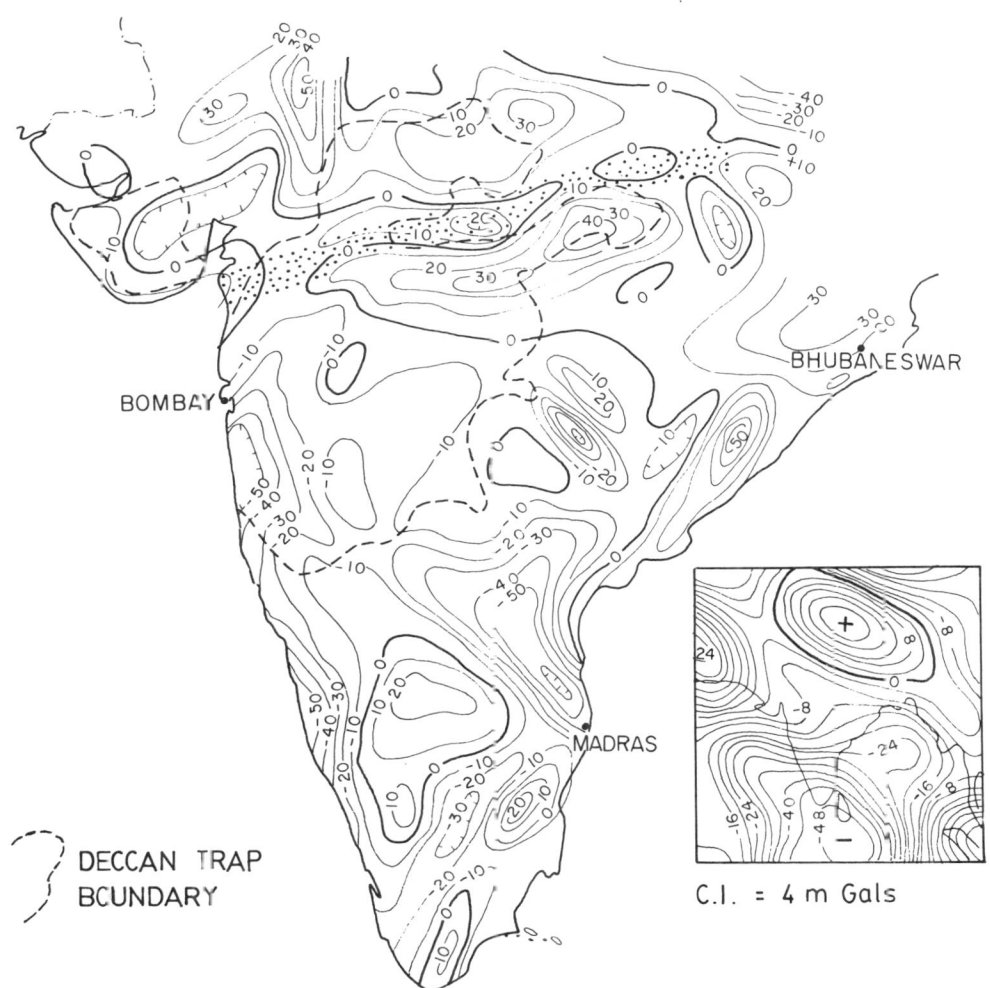

Fig. 8. Regionalized residual isostatic anomaly map of Peninsular India. The inset shows the total satellite-derived gravity (free air) anomaly field from Marsh [1979], which was removed from the Airy-Heiskanen isostatic anomaly maps as a "super regional." Note the low (stippled) over the Narmada-Son zone.

20°N. The one-degree average isostatic anomalies [Qureshy and Warsi, 1980] in the region are negative, with an average of -40 mgal and a minimum of -80 mgal. The question of negative anomalies over the Peninsula has been a matter of discussion for decades [see, e.g., Qureshy, 1970]. A factor in lowering the gravity field over this region may be the Indian Ocean gravity low, whose effect varies from near-zero over the Ganga basin to -48 mgal at the southern tip of India (inset B in Figure 6). The removal of this effect from the isostatic anomalies results in considerable reductions in their magnitude (Figure 8). The remaining anomalies can be accounted for by intra-crustal sources such as the low-density granite and gneissic bodies, upper Proterozoic (Purana) basins such as the Kurnool, Kaladgi, and Bhima [Qureshy and Warsi, 1978], and the Gondwana basins filled with subnormal density material (Qureshy et al., 1968a).

Besides these effects arising from the near-surface, low-density material, the gravity and magnetic anomaly maps (Figures 2 and 3) show northwest-trending zones commencing near the southeastern coast of India near 16°N and extending toward Bombay. Nearly two-thirds of this region is covered by the Deccan traps (Figure 3), which obscure the underlying structure. However, the Landsat-derived lineaments [Powar, 1981], earthquake epicenters [Guha and Padale, 1981], and the geohydrological map [Deshmukh et al., 1976] of this region all show correlation of the gravity and magnetic lineaments with the Godavari, Krishna, and Bhima rivers. The parallelism of the northwest-trending zone on the 13-22

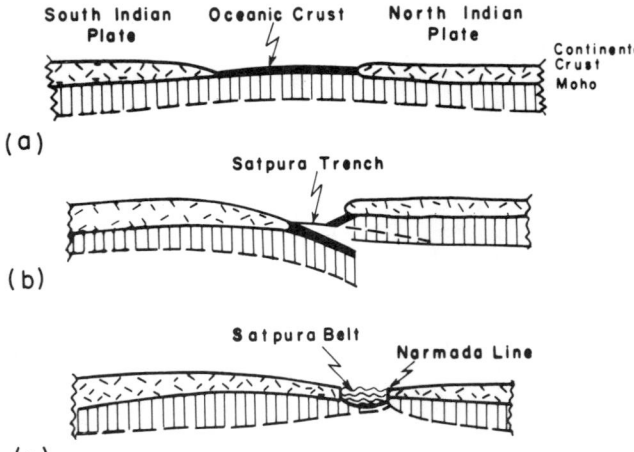

Fig. 9. Cartoon showing the possible evolution of the Precambrian Satpura belt resulting from the subduction of the south Indian plate beneath the north Indian plate some 900 Ma [after Iqbaluddin, 1984]. This paleo-suture is probably the locale of the Tapi-Narmada-Son zone and a series of ENE-trending gravity highs and lows. It also marks the transition zone north of which the Magsat field is negative.

spherical harmonics satellite-derived gravity field (inset B in Fig. 6) and the northwest-trending features (NW-1, NW-2, NW-3) on the ground-based gravity and magnetic anomaly maps (Figures 2 and 6) indicates a possible genetic relation of the northwest trend in India with phenomena occuring at great depths.

The North Indian Shield and the Tapi-Narmada-Son Zone

The northern margin of the south Indian shield is marked by ENE-trending lineaments between 21°N and 24°N as mapped from Landsat imagery and gravity and magnetic data [Qureshy, 1982]. This corresponds to the Tapi-Narmada-Son zone [see, e.g., Kaila, 1985], which stretches from near Bombay in the west to the northeastern extremity of the subcontinent (Figure 2). It is generally interpreted as a rift, although Iqbaluddin [1984] calls it a paleo-suture along which the older Precambrian plate to the south was subducted to give rise to the Satpura orogeny, dated around 900 Ma (Figure 9). A narrow gravity high, the Ratlam-Patna [Qureshy and Warsi, 1975], in the north may mark its northern boundary (Figure 2). Likewise, the southern boundary may also be an ENE-trending high south of the Tapi River (21°N, 74°E, Figure 2).

The residual isostatic anomaly map (Figure 8) shows the Tapi-Narmada-Son zone to be a region of broad gravity high with a narrow low in the middle. It encompasses parts of the Satpura hills, the Vindhyan ranges, and the Malwa plateau. The total intensity magnetic anomaly map (Figure 3) shows a set of narrow ENE-trending magnetic highs and lows corresponding to this zone. The north-south and vertical-component magnetic anomaly maps, compiled through rectangular harmonic analysis by Arora and Waghmare [1984], show this zone as a regional low embodying a feeble positive anomaly. This low value suggests a possible rise of isotherms in the region. This result is consistent with the negative magnetic anomalies on the Magsat anomaly map (Figure 7) of the region north of the Narmada-Son line. The narrow gravity low on Figure 8 indicates the Narmada-Son line to be a depression, perhaps on the crest of an upwarp. This view was earlier expressed by Ghosh [1976] from geological and geomorphological data analysis. This zone finds an expression, as a transition zone, on the satellite-derived spherical harmonics map (Figure 6), and even more prominently, on the Magsat anomaly map (Figure 7). The broad gravity high over the Tapi-Narmada-Son zone (Figure 8) requires about 1.5 km of upwarp on the normal crust. The high can also be explained by movement of mantle material into the crust through faults and deep fractures. Yet another explanation of the high is Iqbaluddin's [1984] model referred to above for the Precambrian Satpura orogeny (Figure 10). This model attributes the high as possibly due to the caught-up paleo-ocean floor at depth and thus may indicate an ancient suture zone with a reactivation history, the latest episode being the formation of the Narmada rift.

Apart from the anomalies discussed above, the low Bouguer anomaly values occur over the Archean Bundelkhand granite massif, the Proterozoic Vindhyan cratonic basin, and the rift-like Gondwana basins filled with low-density continental sediments. The rejuvenated shield blocks, such as the Aravallis in the west and the Meghalaya (Shillong) plateau in the northeast, are associated with relatively high Bouguer and near-zero or positive isostatic anomalies (Figures 2 and 8). Some of these, such as the Aravallis (Figure 1), are Precambrian mobile belts reactivated in Cretaceous and post-Neogene times [e.g., Sen and Sen, 1983], and others, such as the Meghalaya (Shillong) plateau, have basic material [e.g., Murthy, 1970] and high seismicity associated with them. Positive isostatic anomalies over these features suggest mobility of mantle material as the cause of their uplift [Qureshy, 1971b].

Gondwana and other Rift Basins

A northwest-trending gravity anomaly zone cuts through India from its southeastern coast to the festooning ranges of Pakistan and beyond. In India it falls over the Godavari rift (Figure 2), Deccan traps, Vindhyans, and alluvial and sand tracts of northwestern India [Qureshy, 1978, 1982]. Kazmi [1979] reports a large number of

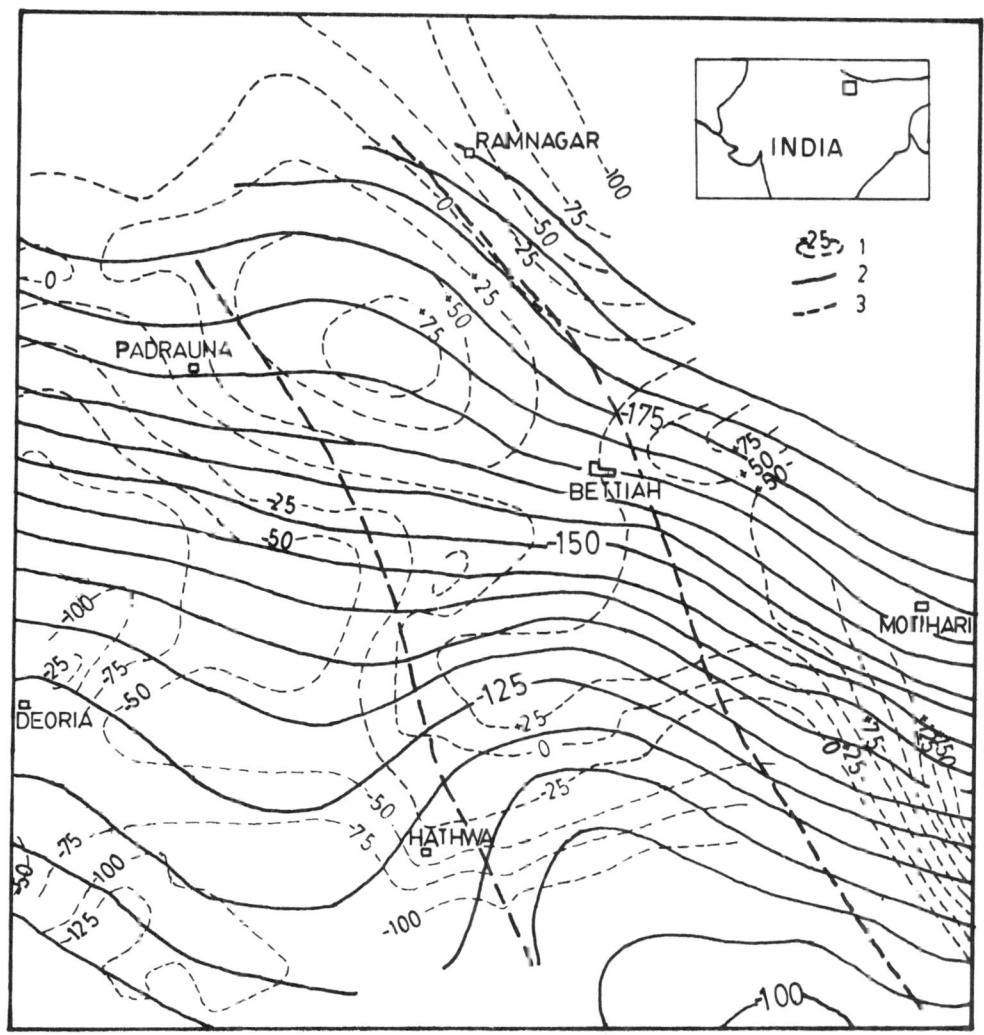

Fig. 10. Bouguer anomaly (simple) and magnetic intensity (vertical field) anomaly map of a part of eastern Ganga basin: 1-Magnetic values in nT, 2-Bouguer anomaly contours in mgal, and 3-approximate boundary of the Karakorum-Gandak lineament as expressed in this thick alluvial and sediment-filled Ganga basin. Note the NW-nosing of the gravity contours, which becomes less prominent in the north, probably due to the effect of sediments and the roots of the Himalaya, and the magnetic high values particularly between Padruna, Bettiah, and Ramnagar. This lineament is part of a megashear (Figure 1) mapped from satellite imagery by Bush et al. [1984].

NW-trending lineaments between 63°E and 25°N to 31°N in Pakistan from aerial photographs, Landsat imageries, and seismological data that are known to be active fault systems [Seeber and Armbruster, 1979]. This zone is a direct continuation of the Godavari lineament (Figure 2). A number of NW-trending lineaments from Landsat imageries have also been mapped in the central Indian part of this gravity feature [24°N, 76°E; Figure 8 of Qureshy, 1982] over the regions covered by the Deccan traps and Vindhyans. Its expression on the satellite-derived gravity map (Figure 6) as discussed above, suggests that the NW lineaments are probably controlled from great depths.

A NW-trending lineament, the Karakorum-Gandak, is parallel to the Godavari trend (Figures 1 and 10). This feature continues southeastward from the Karakorum region (where it is expressed in the Karakorum fault), through the Himalaya and Ganga basin, up to east-central India. This lineament is visible on the satellite imagery in the Himalaya and the shield portion. Over the Ganga basin, it is reflected on the gravity and magnetic field maps (Figure 12). Bush et al.

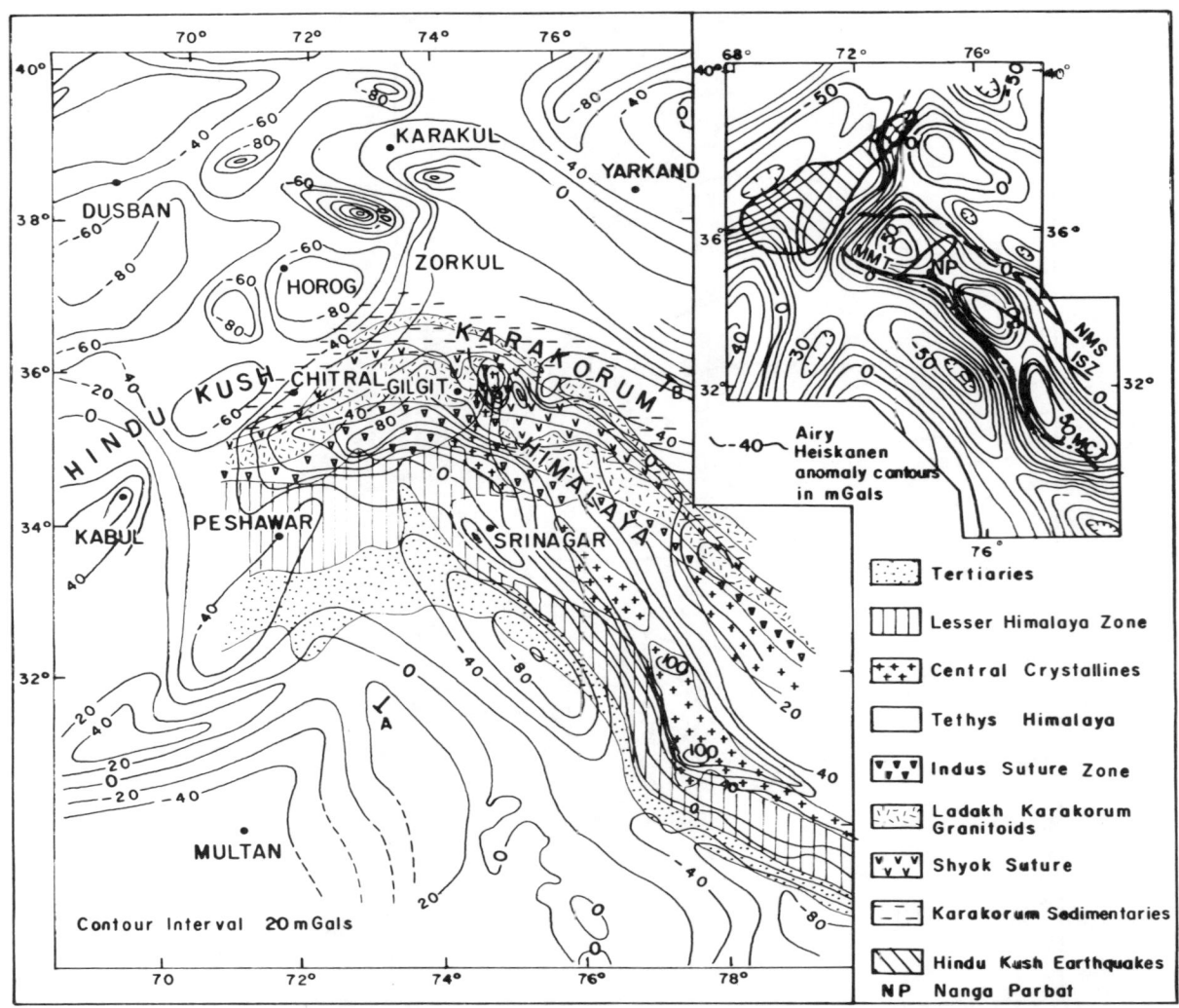

Fig. 11. Airy-Heiskanen isostatic anomaly map of Himalaya-Hindu Kush region. It is a composite map prepared from map published by Marussi [1976] and Qureshy [1971a] for a sea level crustal thickness of 30 km and crustal and upper mantle densities of 2.67 g/cm^3 and 3.27 g/cm^3. A prominent high with values reaching 100 mgal over the central crystallines in Himalaya and the Kohistan region is of particular interest. Note that, whereas MCT, MMT, and NMS show some correspondence with gravity gradients, ISZ does not have a clear correlation. The relationship of these tectonic elements is clearer on the one-degree average Airy-Heiskanen anomaly map in the inset. A characteristic feature of the gravity anomaly field is the NW-trending zone that intersects the NE-trending Hindu Kush zone. Thus the intersection zone demarcates the boundary of two differing tectonic zones as depicted in the inset in Figure 1 by the Indo-Pamir zone. A-B is the location of the Lyallpur-Karakorum Pass profile discussed in text.

[1984] call it the Turan Himalaya transcontinental lineament and show it to extend for 4500 km as a dextral shear (Figure 1). The existence of such features is enigmatic as they cut across plate and continental boundaries.

Like the Godavari basin, other Gondwana basins, namely the Satpura, Mahanadi, and Damuda (Figures 2 and 6), are also associated with gravity lows and are coal bearing and intruded by some peridotite dykes [Murthy, 1981]. A magnetic anomaly is also associated with the Damuda, Mahanadi, and Satpura basins (Figure 3). The southwestward extension of the magnetic anomaly over the Deccan trap-covered region indicates the possible extension of these rift-like basins beneath the traps.

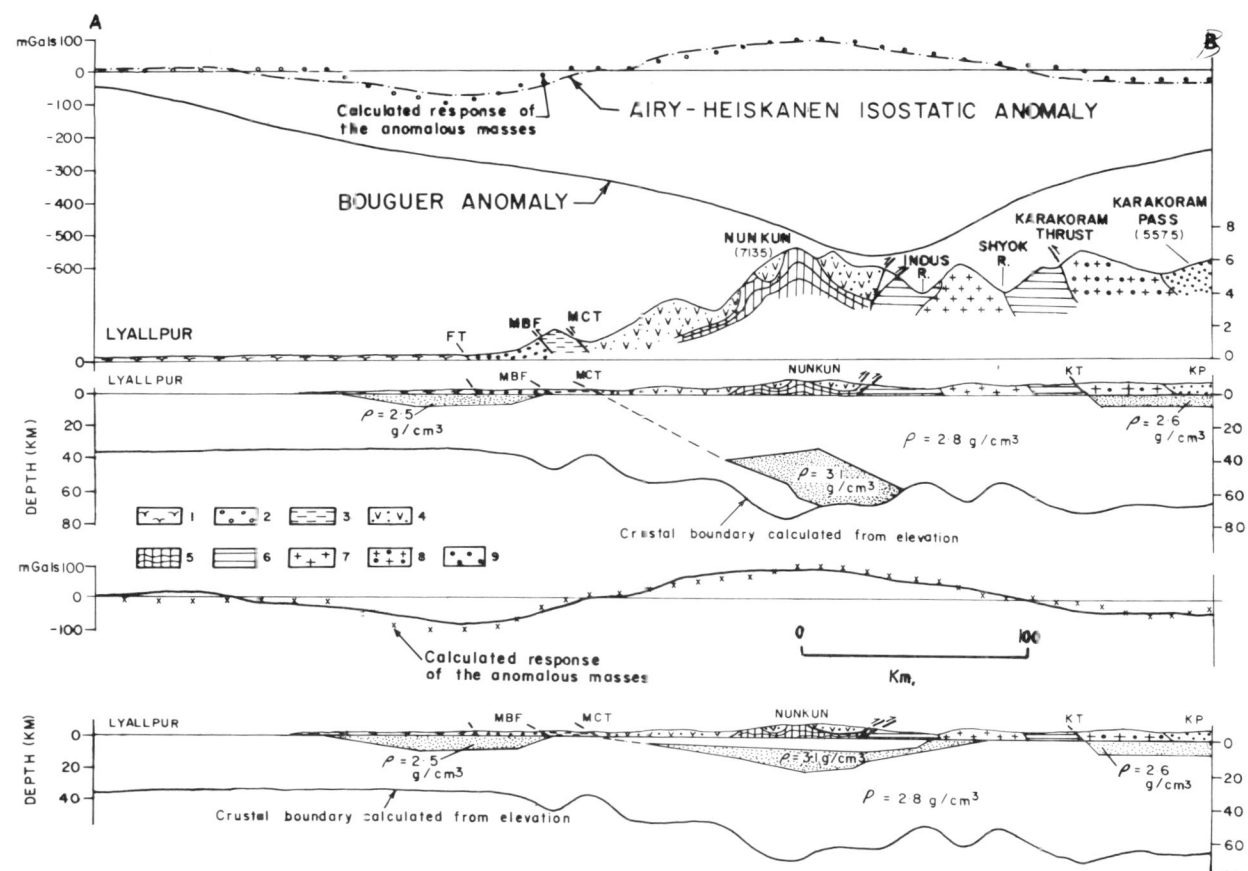

Fig. 12. Bouguer and isostatic anomaly profiles along AB section from Lyallpur to Karakorum Pass with surface geology and two possible anomalous mass distributions that can account for the isostatic anomaly. 1-Quaternary, 2-Lesser Himalaya, 3-Tethys zone, 4-central crystallines, 5-Indus and Shyok suture zones, 6-Ladakh batholith, 7-Paleozoics and Mesozoics, 8-Karakorum plutonic complex, 9-Karakorum granite. MBF-Main Boundary Fault; MCT-Main Central Thrust, FT-Frontal Thrust.

The oil-bearing Cambay basin (Figure 7) is known in some detail through exploration geophysics and drill hole data. These have revealed a 6-km thick, Tertiary sequence overlying the down-faulted traps [Kailasam and Qureshy, 1964; Ramanathan, 1981; Sen Gupta and Khattri, 1977]. A Bouguer anomaly high occurs along the axis of the basin. It is truncated by a NE-trending gravity high lying north of Bombay (Figure 2) that continues southwestward into the Arabian Sea. Takin [1966] inferred the Arabian Sea high to be caused by a relict magma chamber from which lighter acidic material had moved out. Similar gravity highs also occur in the Saurashtra peninsula (Figures 1 and 2) and a number of other areas in the Deccan trap region [Qureshy, 1981], where acid volcanism is known at a number of places. Takin's [1966] hypothesis of relict magma chambers may be extended to these areas as well. The west coast of India has a number of hot springs. Some mercury mineralization is also known. The negative Magsat anomaly (Figure 7), along with these features, suggests a rise in isogeotherms in this general area, which, incidentally, is also petroliferous [Bhandari et al., 1984].

The Magsat vertical field anomaly (Z) map (Figure 7) demarcates major divisions of the Indian shield. Significantly, negative anomalies occur over the northern Indian shield with an ENE trend that is oblique to the regional strike of the Himalaya. The transition zone is marked by the Narmada-Son line. The northern shield itself seems to be characterized by different magnetic characteristics. Positive anomalies occur west of the Godavari lineament, so this part of northwestern India, the Rajasthan shield, is more akin to the southern Indian shield, which is also characterized by positive anomalies. Note also that the commercially productive petroliferous basins (Figure 7) are associated with negative closures. Examples are the Cambay basin (-4 nT)

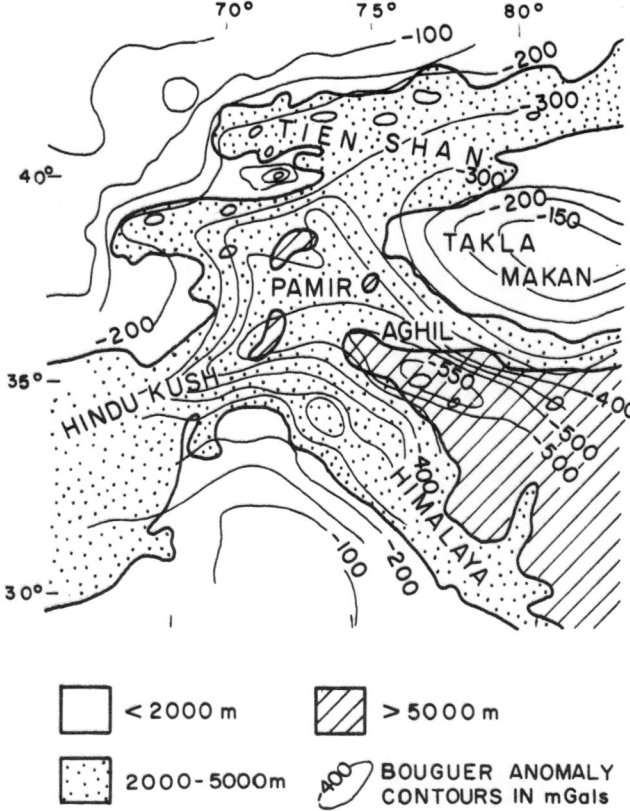

Fig. 13. A map of Hindu Kush region showing Bouguer anomalies superimposed on elevation [from Desio, 1964]. Note the correspondence of higher-than-5-km elevation region with the -400 to -500 mgal anomalies indicating the prevalence of isostasy.

and the Bombay off-shore region on the west coast, the northeastern region in Assam (-4 nT), and Attock (32°N, 72°E) in the Potwar basin. If the negative anomalies are due to the rise in isotherms, this correlation could be of some significance, particularly when we recall that anomalously high temperatures have been encountered in the Cambay oil field [Krishnaswamy, 1981].

The Ganga Basin-Himalaya-Hindu Kush Region

The Bouguer anomalies over the Ganga basin decrease from -50 mgal at its southern boundary to -200 mgal at its northern margin, where it abuts the Himalaya frontal thrust zone. The shield elements continue under the sediment cover of the basin as suggested by the extension of various gravity and magnetic anomaly zones, seismic data, and magnetic induction array data (Figures 2, 3, and 10). A sedimentary column of about 6 km is estimated from drill hole and geophysical data near the India-Nepal border [Agarwal, 1981], and a sedimentary thickness of the order of 20 km is reported in the western section from DSS data [Kaila et al., 1984], yet no significant Bouguer anomaly closures occur (Figure 2) over the basin. It may be that the isostatic root of the Himalaya obscures the effect of the sediments. After applying the Airy-Heiskanen isostatic correction to the Bouguer anomalies, a closure of -100 mgal results over the basin (Figure 11). The low can be accounted for by Cenozoic sediments filling the basin.

The magnetic anomaly map (Figure 3) shows variations in the magnetic characteristics over the Ganga basin, which seems to be divided into subbasins [Agarwal, 1981]. It can be seen on Figure 3 that the magnetic anomaly relief in the west (Punjab, 74°E-80°E) is low with near-zero anomalies; from 80°E to 82°E (West Uttar Pradesh), the relief is higher with negative values and an ENE trend. Farther east, it is followed by an ENE-trending broad high up to 85°E, which, in turn, gives way to a strong ENE-trending negative value up to 86°E. The eastern boundary of the Ganga basin is marked by a NE-trending high jutting into the Himalaya around 90°E.

Geomagnetic transient variation data [Arora et al., 1982] from magnetometer array studies in northern India indicate a major NE-trending conductivity structure under the sediments of the Ganga basin (Figure 2). Paralleling the Aravalli (NE) strike, it extends into the Himalaya and is coincident with one of the gravity and magnetic anomaly axes [Qureshy, 1982] as well as the Delhi-Dehradun ridge. These features appear to be the direct continuation of the northern Indian shield elements and may have influenced the basin evolution.

Figure 11 is an Airy-Heiskanen anomaly map compiled from Gulatee [1956], Marussi [1976], and Qureshy [1971a] and superimposed on regional geology from the Nepal-India border to the Hindu Kush region. It shows possible correlation with some geological features and lack of correlation with others. A NW-trending gravity high over the High Himalaya extends up to the vicinity of Chitral in the Kohistan region. It intersects the northeastward extension of the Chaman fault and its associated gravity contours northeast of Kabul. The gravity contours cut across the Nanga Parbat massif, which is oblique to the main structural grain of the NW-Himalaya. The northern boundary of the gravity high more or less follows the NMS of Bard et al. [1979]. Its southern boundary correlates with the MMT [the southern suture of Bard et al., 1979] west of Nanga Parbat, which is a region characterized by outcrops of the high-density Kohistan sequence [Tahirkheli et al., 1979]. The southeastern boundary of the gravity high seems to demarcate the MCT. The gravity high in this region may indicate the existence of high-density material, probably somewhat akin to the Kohistan sequence, at depth beneath the greater Himalaya. The gravity field pattern also suggests that MMT may be the westward continuation of MCT; that is, the two may join at depth beneath the Nanga Parbat (inset in Figure 11).

Figure 12 shows surface geology, the Bouguer anomaly, the Airy-Heiskanen anomaly, the crust-mantle boundary, and two permissible mass distributions that can account for the isostatic anomaly profile in the Himalaya from Lyallpur to the Karakorum Pass (see AB on Figure 11 for location). These calculations assume two-dimensional mass distribution. The positive isostatic anomaly is assumed to be caused by the presence of heavy material within the crust. The low over the Ganga basin is attributed to the sediments (density = 2.5 g/cm^3) that fill the basin. Similarly, the low northeast of the Karakorum Pass is attributed to the presence of granites (density = 2.6 g/cm^3). These assumptions are justified from the results of seismic studies, which indicate the uppermost 12 km of the Himalaya [Beloussov et al., 1984] to be comprised of Phanerozoic rocks overlying the Precambrian basement; that is, the bulk density up to this level would be near normal or lower than normal. Further, Beloussov et al. [1984] report low-velocity zones at depths of 10-15 km and 25-30 km. They also state that an interface occurs at a depth of 40 km, which, according to them, may be the upper surface of the basaltic layer.

Although there is little to choose between the two possible models, or any other alternative for that matter, both suggestions have merit. According to the plate tectonic model, in which the Asian and Indian land masses come in contact [Rai, 1983; Jiqing, 1984; Yanshin et al., 1984], the second model would seem to be preferred. If, on the other hand, the Indian plate gets subducted along the major mantle-tapping thrusts, such as MCT/MMT, the counter thrusting of the subducted slab will lead to the first model. This model can also be produced by the ascent of ab-normally heated mantle material as indicated by the negative magnetic anomalies over the northern Indian shield-Himalaya region (Figure 7).

The nature of the gravity data in the Himalaya is very regional, and therefore interpretation in terms of surface density variations due to the near-surface lithologic changes may not be warranted. Also, the geologic units in the Nunkun area [Vohra et al., 1982] include the Precambrian metamorphics, central crystallines, Tethyan quartzites, phyllites, limestone, Panjal trap, agglomerate, and slates. If the entire section from sea level and above of about 7 km is formed of a density higher than 2.67 g/cm^3, as used in the Bouguer anomaly reduction, the density of the rock material above sea level would have to be of the order of 2.94 g/cm^3 to explain the 80-100 mgal isostatic anomaly. This is considered much too high for the formations present in this area. Further, the gravity high does not pertain to a specific region. It extends along the entire Himalaya strip and therefore may not be caused by near-surface lithologic variations.

We prefer the simplest interpretation as presented above, a model that is consistent with the little that is known from geophysical and geological data. Despite some variation along the strike, cross sections at different parts of the Himalaya are similar enough that gross features of the geology can be depicted by one representative example. Thus, the Lyallpur-Karakorum Pass profile (Figure 12) may be illustrative for the whole Himalaya. Also, it is perhaps significant that the Hindu Kush seismic zone (stippled on inset on Figure 11) lies on the transition zone between the NW Himalaya gravity high and the NE Hindu Kush gravity low.

Figure 13 shows the relationship of Bouguer anomalies to elevation in the Hindu Kush region. The inverse relationship of Bouguer anomalies and elevation over the High Himalaya can also be seen on the Lyallpur-Karakorum Pass profile on Figure 12. This correlation, along with the relationship of crustal thickness and elevation reported by Desio [1964], demonstrates the prevalence of isostasy in this region.

Conclusions

The compatibility of Moho surface determined from DSS data and crustal thickness as calculated from the empirical relationship between elevation and crustal thickness in the Indian shield and the Himalaya signifies the prevalence of isostasy. It may mean that crustal thickening beneath the Himalaya, though a possible result of thin-skin tectonics and/or the subducting Indian plate, nevertheless is in keeping with isostasy.

Because of the strong influence of compensating masses in such highly elevated regions, the Bouguer anomalies may lead to misleading inferences. Heiskanen [1953] cautioned that, by basing inferences on Bouguer anomalies alone, we may be chopping off the wrong bark. The example of the Indo-Ganga-Brahmaputra basin is illustrative in this context. It may be recalled that over this basin, which has a sedimentary thickness of 6-8 km, no closures are observed on the Bouguer anomaly map (Figure 2). However, the removal of the isostatic effect (Figure 11) brings out a -100 mgal closure over this basin (32°15'N, 75°30'E). It is therefore necessary to remove the isostatic effect for a meaningful interpretation of gravity field in such regions.

We consider the Himalaya to be grossly in isostatic equilibrum because the positive isostatic anomalies there can be accounted for by the intracrustal masses as shown in Figure 12. From the inset in Figure 11, it also appears that MMT is the westward extension of MCT at depth; they may be connected beneath the Nanga Parbat, rather than through looping around it.

The association of major anomaly zones, as mapped from surface geopysical data and satellite-derived lineations, indicates that some of these lineations may extend to great depths. Lineaments such as the Godavari, Gandak-Karakoram, and Indo-Pamir cut across the plate and continental boundaries. The limited data on the Himalaya-Hindu Kush region and the Pakistan ranges inhibit appreciation of the full significance of these traverse megalineaments in terms of plate tectonics and the evolution of these mountain ranges.

The Magsat, satellite-derived gravity, surface magnetic and gravity data, and the occurrence of earthquakes of magnitudes up to 6.7 indicate that the Indian shield, despite its antiquity, is not as stable and rigid as is usually presumed. Whether the mobility of the shield (including the Himalaya) is the result of the continent-continent collision or the ascent of hot material from the mantle is yet to be resolved.

Acknowledgments. One of us (MNQ) is grateful to S. Varadarajan, Secretary, Department of Scientific & Industrial Research, and the then Secretary, Department of Science & Technology, for permission to participate in the International Symposium on Deep Structure of the Crust at Cornell University, Ithaca, NY, U.S.A. We thank Yash Pal, Secretary, Department of Science & Technology, for encouragement to continue these studies; Iqbaluddin, Surendar Kumar, K. R. Gupta, and H. Rai for helpful discussion, and C. V. Sitaram for help in model computations. We have greatly benefited from comments of one anonymous reviewer and editorial assistance of Judy Healey of Cornell University.

References

Agarwal, R. K., Structure and tectonics of Indo-Gangetic Plains, Geophysical case histories of India, Assn. Expl. Geophys., Hyderabad, India, I, 29-46, 1981.

Arora, B. R., and S. Y. Waghmare, Delineation of long wave length magnetic anomalies over central India by rectangular harmonic analysis, Proc. Indian Acad. Sci., 93, 4, 353-362, 1984.

Arora, B. R., F. E. M. Lilley, M. N. Sloane, B. P. Singh, B. J. Srivastava, and S. N. Prasad, Geomagnetic induction and conductive structures in north-west India, Geophys. J. R. Astr. Soc., 459-475, 1982.

Bard, J. R., M. Maluski, P. H. Mattle, and F. Proust, The Kohistan sequence: crust and mantle of an obducted island arc, Proc. Int. Comm. on Geodynamics, 87-94, 1979.

Beloussov, V. V., B. S. Volvovsky, I. S. Volvovsky, I. Kh. Khamrabaev, K. L. Kaila, and A. Marussi, Deep structure of central Asia along the Tien-Shan-Pamir Himalayas geotraverse, in Tectonics of Asia, 27th Int. Geol. Cong., Moscow, Colloquium 05, 5, 29-39, 1984.

Bhandari, L. L., B. S. Venkatachalam, P. Mitra, R. Kumar, D. C. Srivastava, and S. Nanjundaswamy (Eds.), Petroliferous basins of India, Petroleum Asia Journal, 7, 229 pp., 1984.

Bush, V. A., V. G. Trifonov, and S. S. Shulz, Systems of Eurasian active lineaments on the basis of space image interpretation, 27th Int. Geol. Cong., Moscow, 5, 51-64, 1984.

Chugh, R. S., Contribution of Geodetic and Research Branch (Survey of India) towards integrated national earth science policy-status report, presented at Symposium on Integrated Earth Science Policy, G.S.I., Calcutta, 1976.

Desio, A., Karakoram Mountains, in Mesozoic Cenozoic Orogenic Belts, edited by A. M. Spencer, Geol. Soc. London, Pub. 4, pp. 255-266, 1964.

Deshmukh, D. S., B. K. Banerjee, Raghava Rao, B. D. Pathak, P. G. Adyalkar, B. P. C. Sinha, N. C. Prasad, K. K. Sharma, S. P. Sinha, Roy, S. K. Sharma, and D. K. Dutt, Geohydrological Map of India, Central Ground Water Board, New Delhi, India, 1976.

Ebblin, C., Tectonic lineaments in Karakoram, Pamir and Hindukush from ERTS imageries, Rend. Acc. Naz. Lincei, S. VIII, LX, fasc. 3, 1976.

Ermenko, N. A., Some new ideas about the tectonics of India, in Selected Lectures on Petroleum Exploration, edited by S. N. Bhattacharya and V. V. Sastri, I.I.P.E., O.N.G.C., Dehradun, pp. 3-18, 1969.

Ghosh, D. B., The nature of Narmada-Son lineament, Geol. Surv. India Misc. Publ., 34, 119-132, 1976.

Guha, S. K., and J. G. Padale, Seismicity and structure of the Deccan trap region, in Deccan Volcanism and Related Basalt Provinces in Other Parts of the World, edited by K. V. Subba Rao and R. N. Sukeshwala, Geol. Soc. India, Bangalore, Mem. 3, pp. 153-164, 1981.

Gulatee, B. L., Gravity data in India, Survey of India Technical Publication, 10, 95 pp., 1956.

Heiskanen, W. A., The geophysical applications of gravity anomalies, Trans. AGU, 34, 11-15, 1953.

Iqbaluddin, Palaeo-seismic belt from the Vindhyan basin of India, 27th Int. Geol. Cong., Moscow, II, Section 04, 05, 310-311, 1984.

Jiqing, Huang, New research on the tectonic characteristics of India, in Tectonics of Asia, 27th Int. Geol. Cong., Moscow, Colloq. 05, 5, 13-28, 1984.

Kaila, K. L., Deep seismic sounding studies in India, Spec. Issue Geophys. Res. Bull., NGRI, Hyderabad, 309-328, 1982.

Kaila, K. L., Tectonic framework of Narmada-Son lineament - A continental rift system in central India from deep seismic soundings, this volume, 1985.

Kaila, K. L., K. M. Tripathi, and M. M. Dixit, Crustal structure along Wular Lake-Gulmarg-Neoshera profile across Pir Panjal range of the Himalayas from deep seismic soundings, J. Geol. Soc. of India, 25, 706-719, 1984.

Kailasam, L. N., and M. N. Qureshy, On some anomalous Bouguer gravity anomalies in India, in Advancing Frontiers in Geology and Geophysics, edited by A. P. Subramaniam and S. Balakrishna, Indian Geophysical Union, pp. 135-146, 1964.

Karner, G. D., and A. B. Watts, Gravity anomalies and flexure of the lithosphere at mountain ranges, J. Geophys. Res., 88, 10449-10477, 1983.

Kazmi, A. H., Active fault systems in Pakistan, in Geodynamics of Pakistan, edited by A. Farah and K. A. De Jong, Geol. Surv. Pakistan, pp. 285-294, 1979.

Krishnaswamy, V. S., The Deccan volcanic episode, related tectonics and geothermal manifestations, in Deccan Volcanism and Related Basalt Provinces in Other Parts of the World, edited by K. V. Subba Rao and R. N. Sukeshwala, pp. 1-7, 1981.

Lyon-Caen, H., and P. Molnar, Constraints on the structure of the Himalaya from an analysis of gravity anomalies and a flexural model of the lithosphere, J. Geophys. Res., 88, 8171-8191, 1983.

Marsh, J. G., Satellite-derived gravity maps, A Geophysical Atlas for Interpretation of Satellite-derived Data, edited by P. E. Lowman, Jr., and H. V. Frey, NASA, Greenbelt, MD., 54, 1979.

Marussi, A., Geotectonica Della zone Orogenich del Kashmir Himalaya, Karakoram-Hindu Kush-Pamir, Accademia Nazionale dei Lincei, 21, 17-25, 1976.

Molnar, P., Structure and tectonics of the Himalaya: constraints and implications of geopysical data, Ann. Rev. Earth Planet. Sci., 2, 489-518, 1984.

Murthy, A. V. N., Tectonic and mafic igneous activity in northeast India in relation to upper mantle, Proc. Second Symp. on Upper Mantle, Hyderabad, 287-304, 1970.

Murthy, A. V. N., Late Mesozoic-early Tertiary volcanism in the Trans-Deccan trap areas of the Indian shield: a synthesis, in Deccan Volcanism and Related Basalt Provinces in Other Parts of the World, edited by K. V. Subba Rao and R. N. Sukeshwala, Geol. Soc. India, Mem. 3, pp. 93-100, 1981.

Powar, K. B., Lineament fabric and dyke pattern in the western part of the Deccan volcanic province, in Deccan Volcanism and Related Basalt Provinces in Other Parts of the World, edited by K. V. Subba Rao and R. N. Sukeshwala, Geol. Soc. India, Mem. 3, pp. 45-57, 1981.

Qureshy, M. N., Relation of gravity to elevation, geology and tectonics in India, Proc. Second Upper Mantle Symposium, Hyderabad, India, 1-23, 1970.

Quereshy, M. N., Geophysical investigations in Himalayas, Himalayan Geology, I, 166-177, 1971a.

Qureshy, M. N., Relation of gravity to elevation and rejuvenation of blocks in India, J. Geophys. Res., 76, 545-557, 1971b.

Qureshy, M. N., Relationship of regional gravity to tectonics and possible zones of mineralization in western ESCAP region, Proc. Third Regional Conference on Geology and Mineral Resources of Southeast Asia, Bangkok, 157-164, 1978.

Qureshy, M. N., Geophysical and Landsat lineament mapping - an approach illustrated from west-central and south India, Photogrammetria, 37, 161-184, 1982.

Qureshy, M. N., and W. E. K. Warsi, Role of Regional gravity surveys in a concept oriented exploration programme: some inferences from a study in a shield area of central India, J. Geol. Soc. India, 16, 44-54, 1975.

Qureshy, M. N., and W. E. K. Warsi, Gravity anomalies over the Purana formations, Proc. The Purana Formations of Peninsular India, Univ. of Sagar, M. P., India, 1978.

Qureshy, M. N., and W. E. K. Warsi, A Bouguer anomaly map of India and its relation to broad tectonic elements of the sub-continent, Geophys. J. R. Astr. Soc., 61, 235-242, 1980.

Qureshy, M. N., N. Krishna Brahman, P. S. Aranamadhu, and S. M. Naqvi, Role of granite intrusions in reducing the density of the crust as illustrated from a study of the Cuddapah basin, Proc. Roy. Soc., London, 304, 449-464, 1968a.

Qureshy, M. N., N. Krishna Brahman, S. C. Garde, and B. K. Mathur, Gravity anomalies and Godavari rift, Geol. Soc. Am. Bull., 79, 1221-1230, 1968b.

Qureshy, M. N., N. Krishna Brahman, S. M. Nagvi, P. S. Aravanadhu, S. C. Garde, W. E. K. Warsi, S. C. Bhatia, D. V. Subba Rao, C. Subrahmanyam, Rafi Ahmad, and S. Venkatachalam, Gravity Data in India, Special Report (gravity data), 551 pp, NGRI-1, 1981.

Rai, H., Geology of the Nubra valley and its significance on the evolution of the Ladakh Himalaya, in Geology of Indus Suture Zone of Ladakh, Wadia Birth Centenary Commemorative Volume, WIHG, Dehradun, pp. 79-97, 1983.

Ramanathan, S., Some aspects of Deccan volcanism of western Indian shelf and Cambay basin, in Deccan Volcanism and Related Basalt Provinces in Other Parts of the World, edited by K. V. Subba Rao and R. N. Sukeshwala, Geol. Soc. India, Bangalore, 3, pp. 198-217, 1981.

Seeber, L., and J. Armbruster, Seismicity of the Hazara arc in northern Pakistan: decollment versus basement faulting, in Geodynamics of Pakistan, edited by A. Farah and K. A. DeJong, Geol. Surv. Pakistan, 131-142, 1979.

Sen Gupta, S., and K. N. Khattri, Some aspects of geodynamics of the Indian sub-continent, Proc. Ind. Nat. Sci. Acad., 41, 339-357, 1977.

Sen, D., and S. Sen, Post-Neogene tectonism along the Aravalli Range, Rajasthan, India, Tectonophysics, 93, 75-98, 1983.

Singh, B. P., "Magsat" Studies, paper presented at International Workshop of Data Processing, Pune, India, 1985.

Sukeshwala, R. N., Deccan basalt volcanism, in Deccan Volcanism and Related Basalt Provinces in Other Parts of the World, edited by K. V. Subba Rao and R. N. Sukeshwala, Geol. Soc. India, Bangalore, Mem. 3, pp. 8-18, 1981.

Tahirkheli, R. A. K., M. Mattauer, F. Proust, and P. Tapponnier, The India Eurasia suture zone in northern Pakistan, synthesis and interpretation of recent data at plate scale, in Geodynamics of Pakistan, edited by A. Farah and K. A. De Jong, Geol. Survey Pakistan, pp. 125-130, 1979.

Takin, M., An interpretation of the positive gravity anomaly over Bombay on the west coast of India, Geophys. J. R. Astr. Soc., 11, 527-538, 1966.

Vohra, C. P., B. S. Jangpangi, P. C. Mehrotra, D. P. Singh, V. M. K. Puri, and P. Mehta, Geology of the Warwan Nunkun area, J & K State, Proc. Himalaya Geology Seminar (1976), Misc. Pub. Geol. Surv. India, 41, 2, 56-63, 1982.

Yanshin, A. L., V. E. Khain, and Yu. G. Gatinsky, The principal problem of tectonics in Asia, in Tectonics of Asia, 27th Int. Geol. Cong. Moscow, Colloq. 05, 3-12, 1984.

SEISMIC REFLECTION PROFILES OF PRECAMBRIAN CRUST: A QUALITATIVE ASSESSMENT

Allan K. Gibbs

Institute for the Study of the Continents, Cornell University, Ithaca, New York 14853

Abstract. Seismic reflection profiles of Precambrian terranes reveal some structural features that were probably developed in Precambrian times. Seismically transparent zones are associated with granitoid rocks, anorthosites, and gneisses; zones of stratified, relatively continuous upper crustal reflection are typical of Precambrian continental sedimentary basins; relatively continuous, coherent, moderately-dipping reflection zones are associated with faults at Precambrian terrane boundaries; and zones of complex reflections and diffractions occur in gneiss terranes and basement to continental sedimentary basins. Features at Moho depths are typically weak or barely perceptible, discontinuous, and in some instances appear to have moderate dips. The base of the crust is revealed elsewhere by a decrease in the abundance of reflections. The similarity of structures in Archean, Proterozoic, and Phanerozoic crust may be due to qualitatively similar tectonic processes in all eras. The retention of Precambrian structures in middle and lower crust implies that such crust has remained coherent through subsequent time.

Introduction

Seismic reflection profiles of Precambrian terranes have yielded surprises, many enigmas, and some data that add to the understanding of Precambrian crustal development. These data may eventually help to answer some fundamental questions of Precambrian geology, including:
- Did tectonic styles evolve with time, leaving systematic differences in the structures and lithologies of Archean, Proterozoic, and Phanerozoic crust?
- Are particular kinds of Precambrian upper crust associated with specific kinds of deeper crust, or are upper and lower crust generally not related to one another?
- Is deeper crust typically older than shallow crust, as interpreted by Lowman [1984] and Sollogub and Chekunov [1983]?
- Are there Precambrian Mohos, where lower crust and adjacent uppermost mantle have been attached and intact since the Precambrian?
- Why does North American Proterozoic crust have a greater tendency than Archean crust to be covered with Phanerozoic sedimentary rocks?

Precambrian structures are most easily studied where they are not obscured by younger structures. Areas that have either little Phanerozoic deformation, or in which the Phanerozoic deformation can be discriminated from older features include gneiss and granite-greenstone terrane in Minnesota; gneiss terrane in New York; intracratonic rift basins in Kansas and Michigan; terrane boundary zones in Wyoming, Minnesota, and New York; and an intracratonic basin in Texas and adjacent Oklahoma.

Varieties of Precambrian Crust and their Seismic Reflection Characteristics

Most Precambrian upper crust can be classified into: greenstone-granite terranes; metasedimentary fold belts; gneiss and granulite terranes; anorogenic magmatic terranes; and continental sedimentary basins. Major zones of high strain that occur along terrane boundaries can also be considered as a category. Other categories might be added, and some crust overlaps several categories. Most Precambrian basement of the United States can be classified by this scheme, though some categories must be grouped together for small-scale maps (Figure 1).

Granite - Greenstone Terranes

Greenstone belts contain volcanic and sedimentary rocks of low to medium metamorphic grade, typically folded into synclinal belts between more abundant granitoid rocks and gneisses. The volcanic rocks are the oldest or nearly the oldest rocks in many shields. Many greenstone belts have randomly branching folds, but in other shields the belts and associated granitoid rocks are both highly elongated. Geological and geochemical evidence indicates that many greenstone belts were formed in unstable marine volcanic environments, not on or adjacent to continental crust. Sedimentary rocks of other belts were partially derived from continental sources,

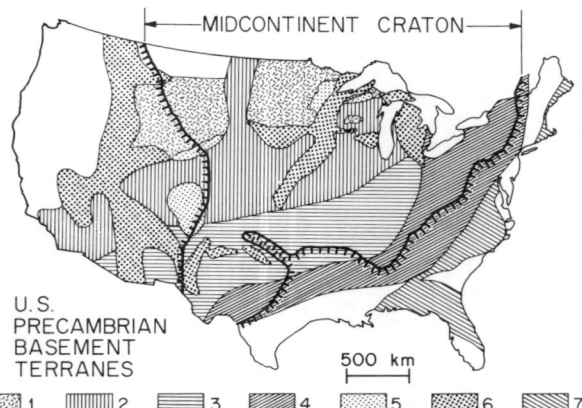

Fig. 1a. Precambrian basement terranes of the United States. Key: 1 - Archean rocks; Wyoming and Superior Provinces; 2 - Proterozoic granite-greenstone and high-grade terranes; 3 - areas with abundant Proterozoic anorogenic felsic rocks; 4 - Grenville terrane; 5 - platform cover or marginal sequences older than 1.7 Ga; 6 - platform cover or marginal sequences younger tha 1.7 Ga; 7 - other Precambrian basement terranes of Atlantic margin. Bold lines delimit the craton by the Laramide (west) and Appalachian (east) deformation belts. Compiled using data from King [1976], Hoffman et al. [1982], Van Schmus and Bickford [1981], Denison et al. [1984], Dutch [1983], Reed et al. [1982], Zietz et al. [1982], Hildenbrand et al. [1982], Anderson [1983], Condie [1981], Chowns and Williams [1983], Rankin [1983], Muehlburger et al. [1980], and others. The boundary between the areas with Proterozoic anorogenic felsic rocks and other varieties of Proterozoic basement is highly schematic: other anorogenic rocks also occur in the north and other basement types occur within the area shown.

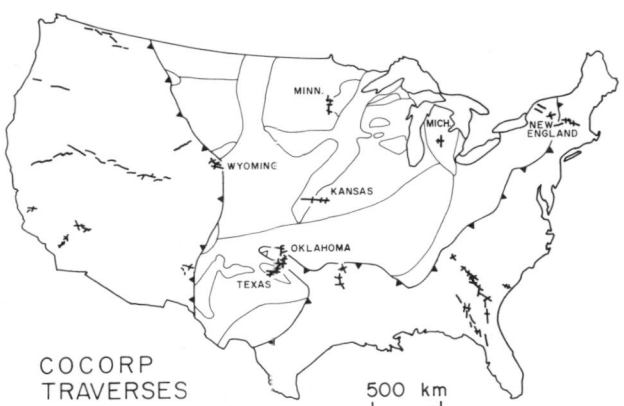

Fig. 1b. COCORP traverses to mid-1984.

and a few belts were clearly deposited on older continental crust.

The transformation from unstable, marine crust on which greenstone belts formed to the continental crust in which they now occur is not thoroughly understood. Accretion of various island arcs and associated fore- and back-arc basins into transitional crust as in the Philippines and Japan may be a useful analogy, [e.g. Condie, 1982; Dimroth et al., 1982; Rogers, 1984], though there are also other models.

Direct evidence of the nature of the middle crust beneath an Archean granite-greenstone terrane is provided by the Kapuskasing uplift in Ontario [Percival and Card, 1983]. Tabular batholiths of gneissic tonalite and granodiorite and an intrusive anorthosite and gabbro complex occur at inferred former crustal depths of 10 to over 25 km.

Greenstone belts and associated granitoid rocks in the western Superior province of Canada and Minnesota, India, western Australia, western Africa, and the eastern Amazonian craton are highly elongated. Strike-slip and thrust faults as well as penetrative elongation fabrics are present in some of these areas [e.g. Schwerdtner et al., 1979; Sims and Peterman, 1981]. In some instances elongation occurred before deposition of the first continental sedimentary rocks and intrusion of post-orogenic granites. In other cases the younger continental rocks were also deformed in a similar manner.

Archean granite-greenstone terranes are exposed in northeastern Minnesota and Wyoming, and Proterozoic ones are exposed in the fault blocks of Arizona, New Mexico, and Colorado, in the Penokean belt of northern Wisconsin and Michigan, and as highly deformed rocks in the Appalachian Mountains. Much of the basement of the central and southern midcontinent may consist of Proterozoic granite-greenstone terrane or its more intensely metamorphosed equivalents. Most of the crust of this region did not emerge from the mantle until the Early Proterozoic, according to Nd-Sm isotopic evidence [McCulloch and Wasserburg, 1978].

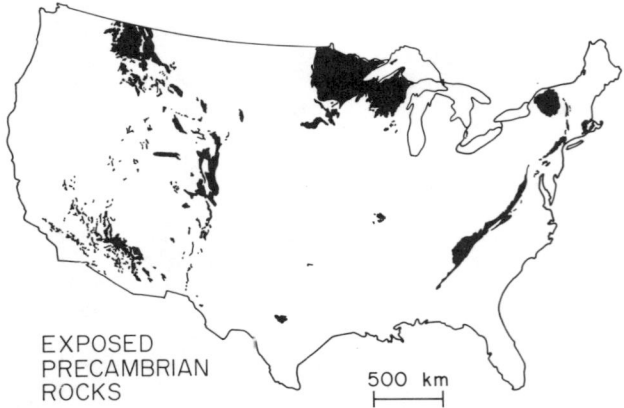

Fig. 1c. Areas with Precambrian rock exposures, after King [1976].

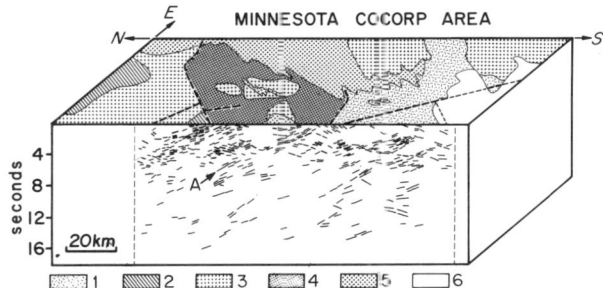

Fig. 2. Migrated reflections from the N-S lines of the Minnesota COCORP survey projected onto a single plane. Bold lines represent the stronger reflections. COCORP data from Gibbs et al. [1984]. Geology modified from Morey et al. [1981]. Key to geology: 1 - Archean gneiss terrane, known or inferred to be over 3 Ga; 2 - Superior Province greenstone belts; 3 - granitoid rocks: Archean in north and central portions, Proterozoic in the southeastern portions; 4 - Great Lakes tectonic zone: Archean gneisses and schists, part of the Superior Province; 5 - Proterozoic sedimentary and minor volcanic cover rocks; 6 - Cretaceous cover rocks. Dashed lines are faults. The N-dipping reflection zone at "A" is thought to correlate with a major, ENE-striking thrust fault that separates Superior Province crust [above] from higher-grade rocks, possibly correlatable with those of the Minnesota River Valley terrane [below]. Bold reflections about 1 s beneath the slate outlier in the center of the section may reveal the basement, possibly along a decollement. Reflections with migrated two-way travel times of about 13-15 s are thought to come from the base of the crust. Stronger, essentially horizontal reflections at these times on E-W cross lines may be correlative, indicating a segmented base of the crust, with moderately N-dipping segments.

The northern end of the COCORP line in Minnesota traversed a short segment of the Superior province granite-greenstone terrane (Figure 2). The data aquisition and processing were not optimized for resolution of the shallowest features. Thus, although the seismic features correlate with the geological map, observations about such terranes are tenuous and might not be broadly applicable.

The Giant's Range granite is relatively devoid of reflections in the top 2-3 s (6-8 km). Beneath this are numerous complex, variably-dipping reflections and diffractions. About 8 km of greenstone belt were traversed, showing complex, short reflections in the first 1 s of two-way travel time (roughly equivalent to the top 3 km of the crust) on both time and migrated sections. Steep fold structures, typical of such belts could produce this pattern. Numerous, strong, relatively coherent reflections wih moderate dips occur beneath the greenstone belt.

This change in seismic character might indicate that the greenstone belt rocks do not persist below about 3 km in this area. Relatively strong and continuous reflections at 2.5-5 s occur in the area just south of the greenstone belt. They may correspond with folds of comparable dimension in middle crustal rocks with high impedance contrast. Steeply-dipping faults are revealed by discontinuities in both the potential field maps and the seismic profiles.

The seismic characteristics of granite-greenstone terranes have also been examined by deep seismic sounding (DSS) surveys. These use explosive sources, far fewer shotpoints, and much longer geophone arrays than the COCORP surveys. Such surveys rely on near-critical-angle reflections, yielding data from great depths, but the long ray paths inevitably result in poorer resolution of small features. DSS traverses have crossed granite-greenstone terranes in western peninsular India [Kaila et al. 1979; Kaila and Bhatia, 1981], western Australia [Drummond et al., 1981], and Karelia in the Soviet Union [Kokorina et al., 1981]. The regional geology of the western Indian terrane was described by Naqvi et al. [1978] and Drury et al. [1984] and Karelian geology was described by Rybakov and Lobach-Zhuchenko [1981]. Greenstone belts and granitoid rocks in each of these regions are elongate, and the seismic traverses crossed the strikes and revealed seismic features associated with thrust faults and the development of the elongation patterns.

Many steeply-dipping faults were shown on the original Indian DSS data [Kaila et al., 1979]. These faults were interpreted on the basis of terminations of reflections, diffractions, steeply-dipping reflections, and lateral changes in seismic character of the crust. Detection of steeply-dipping faults from reflection data is

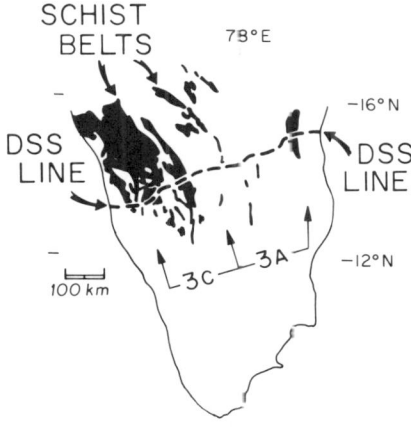

Fig. 3. Deep seismic sounding data from peninsular India, redrawn and modified from Kaila et al. [1979]. Sketch map shows the locations of the WSW-ENE traverse line and the greenstone and schist belts.

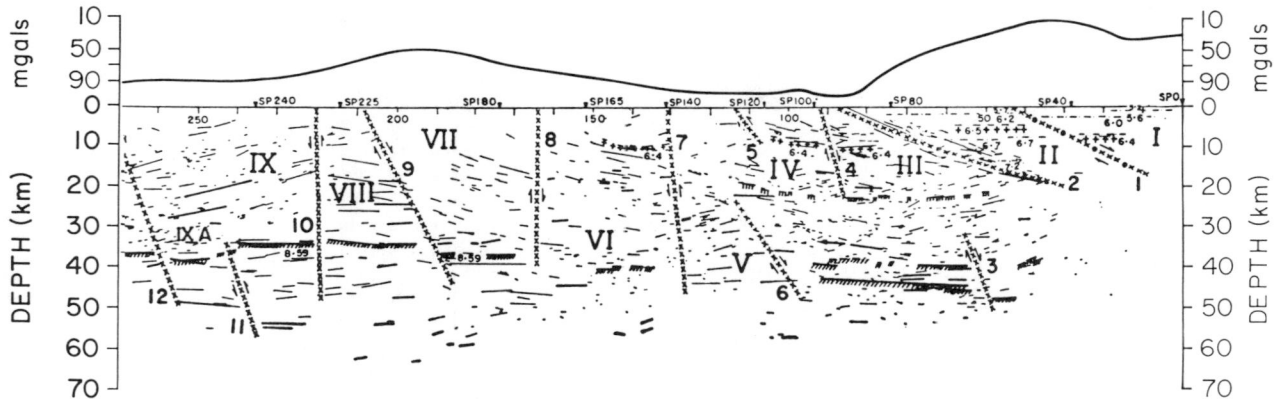

Fig. 3a. Portion of the Indian DSS data as shown by Kaila et al., [1979], with interpreted steeply dipping faults [lines of x's]. Reflector segments shown by lines, Moho discontinuity shown by hachured lines.

typically speculative, even with high-resolution data. Figure 3 shows these profiles both with and without these steep faults.

The western part of the Indian DSS traverse crossed an Archean granite-greenstone terrane and revealed an east-dipping, imbricate reflection fabric persisting to at least mid-crustal depths. Some of these east-dipping reflection zones correspond with Late Archean thrust faults [Kaila et al., 1979; Drury et al., 1984], and they may persist through the crust, and may even offset Moho reflections (Figures 3b and 3d).

The Karelian granite-greenstone terrane has a complexly and heterogeneously layered structure cut by moderately-dipping reflecting surfaces, interpreted as Archean thrust faults [Kokorina et al., 1981]. Some of the latter appear to persist into the mantle, offsetting the Moho.

Metasedimentary fold belts

Metasedimentary fold belts differ from greenstone belts by the relative scarcity of mafic volcanic rocks and the more common presence of metasedimentary rocks of continental provenance in them. At higher or lower metamorphic grades this category overlaps gneiss terranes and continental sedimentary basins, respectively. Proterozoic metasedimentary fold belts include the Labrador Trough in eastern Canada, the Early Proterozoic iron-bearing strata of the Lake Superior region, and part of the Grenville province of eastern North America. Archean examples include some of the folded metasupracrustal rocks of western India [Naqvi et al., 1978] and the English River and Quetico metasedimentary gneiss belts of the Superior province.

The Minnesota COCORP traverse crossed a thin upper crustal lens of foliated metapelites that are correlated with the Proterozoic Animikie basin [D. L. Southwick, A. K. Gibbs, and P. A. O'Day, unpublished work 1983]. The Animikie sedimentary rocks in adjacent eastern Minnesota have tight northward-verging folds with wavelengths of a few kilometers [Morey et al., 1981; Holst, 1984]. The shallowest COCORP data show complex reflections above a strong, 10-km long and upwardly concave reflection zone, interpreted respectively as folded Proterozoic metasedimentary rocks located above a more strongly reflective decollement.

COCORP studies of the Grenville metasedimen-

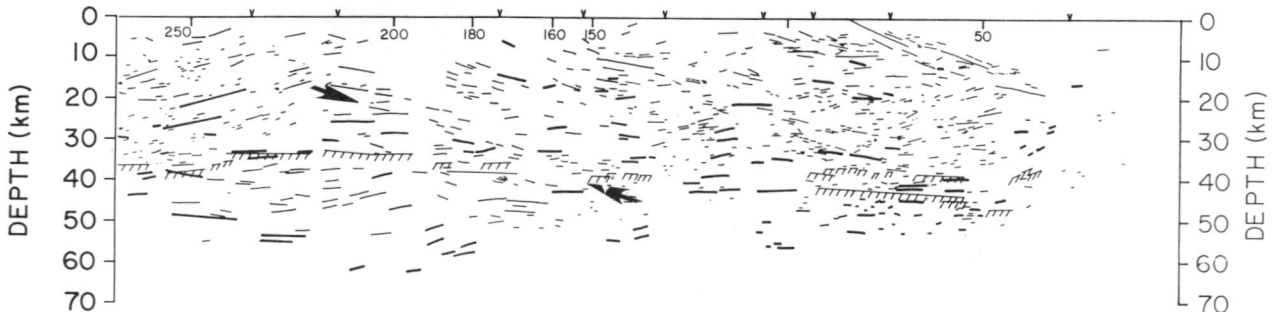

Fig. 3b. Indian DSS data of Fig. 3a without interpreted faults. Arrows point to possible moderately-dipping fault zones that might offset the Moho, and correlate with thrust faults that have been mapped at the surface.

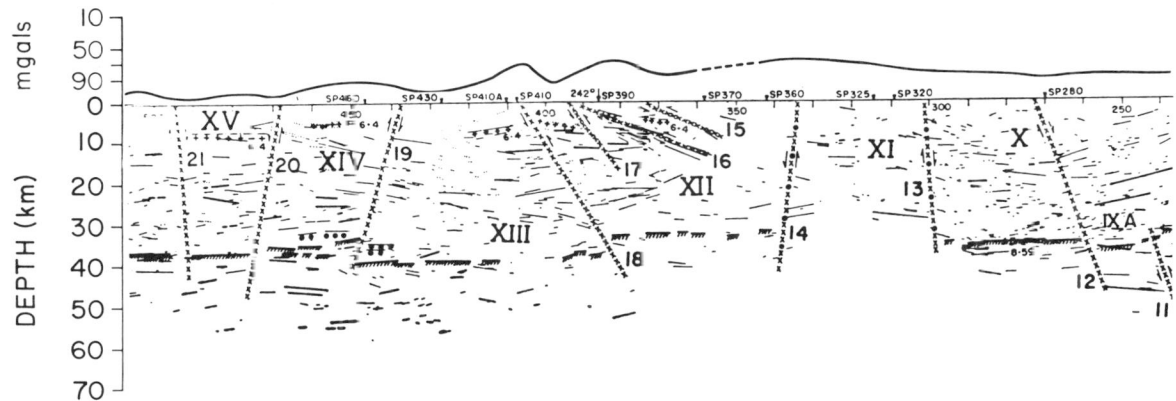

Fig. 3c. Another portion of the Indian DSS data, as in Fig. 3a.

tary fold belts of the northwest Adirondacks did not reveal distinct differences in seismic character between medium-grade metasedimentary rocks and the surrounding gneisses [Klemperer et al., 1983a].

The eastern half of the Indian DSS survey crossed the folded and faulted Proterozoic Cuddapah basin, and revealed features that Kaila et al. [1979] associated with basement to this basin, strata within it, and with the prominent thrust faults that cut the basin.

Granulite and Gneiss Terranes

Medium to high-grade ortho- and paragneisses and granulites are present in all Precambrian shields, both as discrete terranes and as smaller slivers in granite-greenstone terranes and metasedimentary fold belts. Continuous transitions occur between the latter, lower-grade categories and this granulite and gneiss category. The presence of granulites helps to discriminate between high- and medium-grade gneiss terranes. Temperatures and pressures estimated from mineral assemblages show that rocks of this category were metamorphosed and deformed in the middle and lower crust. Archean examples occur in the northern Wyoming province, the Minnesota River Valley, and in northern Minnesota. The Grenville province is the most extensive Proterozoic example; others occur in the Penokean gneisses of east-central Minnesota, and in the southwestern U.S. Drill hole evidence in the midcontinent has revealed vast regions of medium-grade rocks [King, 1976; Van Schmus and Bickford, 1981a; Denison et al., 1984].

Complex patterns of reflections and diffractions, distributed unevenly through the crust are common in gneiss terranes. In many instances distinct reflection groups occur in the middle crust, between about 3 and 8 s, but in the Adirondacks there are prominent groups both above and below that depth range. Many of the reflection zones in gneiss and granulite terranes correspond in dimension and style with folds that are visible at the surface.

Adirondack gneisses appear in diverse patterns in the COCORP data [Brown et al., 1983; Klemperer et al., 1983a, 1985a]. A zone of moderate numbers of reflections with variable dips is present in the upper 6 s beneath the medium-grade northwest Adirondacks, with few reflections in the underlying crust. The character of the reflections are consistent with, but do not fully reveal the broad-wavelength folds seen at the surface. High-grade orthogneisses and anorthosite in the central Adirondacks appear to be seismically transparent, and far beneath these

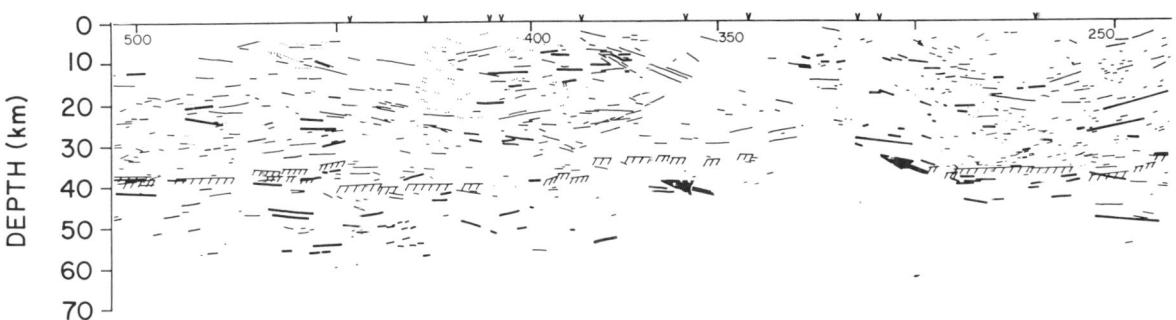

Fig. 3d. Data of Fig. 3c without interpreted faults, as in Fig. 3b.

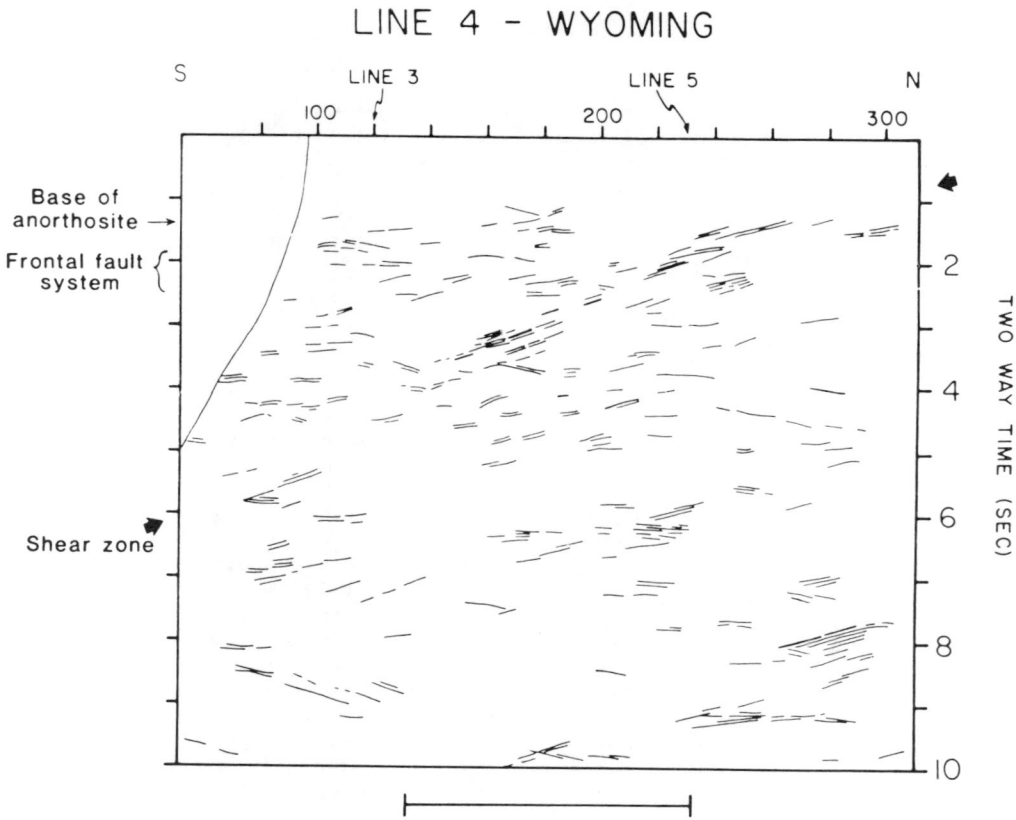

Fig. 4. Line 4 of the Laramie, Wyoming COCORP survey, after Allmendinger et al. [1982]. The S-dipping reflection zone is correlated with the Mullen Creek - Nash Fork shear zone. Note the presence of other, parallel features in other parts of the section.

are the remarkably strong, coherent, and extensive reflections of the Tahawus complex, extending from about 5.5 to 8.5 s. The enigmatic Tahawus complex is of the same approximate depth as an electrical conductivity anomaly [Nekut et al., 1977]. Gneissic, cumulate igneous, and metasedimentary layering have all been considered possible causes of the coherent reflections. In the Grenville crust beneath the Taconics, reflections are more abundant in the lower crust than in the upper part.

The eastern part of the COCORP Kansas survey crossed a zone with few reflections in the upper 3 s and relatively broad, variably-dipping and crossing, complex reflections and diffractions deeper in the sections [Brown et al., 1983].

Gneisses in the center and southern parts of the COCORP Minnesota traverse have more uniform seismic characteristics than in the northern granite-greenstone terrane. Groups of reflections and diffractions occur in a discontinuous zone that rises from about 6.5 s in the north to about 2 s in the south. The positions of these features correspond in general with the distribution of relatively dense and magnetic rocks, revealed by gravity and magnetic data. These rocks do not reach the surface, and thus their identity is obscure. They might be relatively mafic, lower crustal rocks from either the Superior province or Minnesota River Valley terrane, thrust up to middle and upper crustal levels.

Terrane Boundary Zones

Geological features that are common at the boundaries of different Precambrian basement terranes include a broad and imbricate array of faults and shear zones, juxtaposition of rocks with differing metamorphic grades, and contrasts in density of the lower crusts of adjacent terranes [e.g. Gibb et al., 1983; Coward and Daly, 1984]. The COCORP surveys in both Minnesota and Wyoming crossed parts of such terrane boundary zones. COCRUST work in Canada has also addressed terrane boundary problems along the western periphery of the Superior province.

The COCORP Minnesota survey crossed the Great Lakes tectonic zone, which separates the late Archean granite-greenstone terrane of the Superior province from the older Archean high-grade gneiss terrane of the Minnesota River Valley [Sims et al., 1980]. A relatively strong reflection zone with northerly dips of about 30° projects to the surface at the approximate position of a major fault mapped from gravity and magnetic data. This reflection zone is more nearly planar and continuous than any of the features in the Superior province crust above it. The zone of essentially parallel reflections about 0.4 to 2 s thick persists from the upper limit of data to more than 8 s. It is interpreted as one of a system of parallel fault zones that separate the two terranes. The gneisses south of this particular zone could be a detached part of the middle and lower Late Archean Superior province crust or part of the older Archean terrane of the Minnesota River Valley.

The COCORP Laramie, Wyoming survey crossed the Mullen Creek-Nash Fork shear zone, the boundary of the Archean Wyoming province and adjacent Proterozoic granite-greenstone terrane [Allmendinger et al., 1982]. The shear zone has both south-dipping thrusts and strike-slip faults within and adjacent to it. The Wyoming province has a Proterozoic continental marginal sequence along its southern edge, in thrust contact with the adjacent southern Proterozoic granite-greenstone terrane [Houston et al., 1979; Condie, 1984]. Where the COCORP survey crossed the fault zone it is cut by the younger Laramie Anorthosite and Laramide thrust faults, obscuring the Precambrian structures. An imbricate reflection zone that dips about 55° south is barely discernable beneath these younger features, in a position approximately coincident with the inferred Proterozoic collision zone (Figure 4).

Both the Minnesota and Wyoming surveys revealed very weak, discontinuous, events in the deepest crust, many with moderate dips in the same directions as the stronger reflections attributed to faults in the middle and upper crust. Although they are at the limits of perception, the patterns are consistent with the existence of deep structures of similar origin and age as those of the shallower crust.

The COCRUST program in Canada has included seismic reflection and refraction studies of the boundary of the Archean Superior province with the Proterozoic Churchill province [Green et al., 1980, 1981, 1985; Hajnal et al., 1984]. In both northern and southern Manitoba the boundary zones have west-dipping reflections with two-way travel times of about 5 to 6 s. Seismic refraction and potential field data also show the Archean Superior province dipping westward beneath the Churchill province. Refraction work indicates that the thrust and strike-slip fault zones along the contact probably persist throughout the entire crustal thickness [Hajnal et al., 1984].

Anorogenic magmatic terranes

These terranes consist of various felsic and mafic igneous rocks that erupted onto and into relatively stable or extending cratons. Either mafic or felsic rocks may predominate. Where abundant sedimentary rocks accompanied anorogenic extrusives, there is a continuum between this category and continental sedimentary basins. Some anorogenic terranes have been affected by later orogenesis, as in the Grenville province.

Anorogenic magmatic rocks include basaltic and rhyolitic dikes, sills, and volcanic rocks, layered intrusive complexes, and a characteristic association of rhyolites, epizonal granites, mangerites, and anorthosites. The mantle is considered the major source of the mafic rocks of the anorogenic suite. Intrusion of such material into the lower and middle crust may melt the surrounding rocks, generating more siliceous rocks [e.g. Anderson, 1983]. Layered intrusive complexes are likely to be present in the middle or lower crust in regions where felsic anorogenic igneous rocks have intruded the upper crust. Passively emplaced and undeformed layered intrusive complexes might appear as extensive, coherent reflections if there is sufficient cumulate layering, otherwise as seismically transparent zones.

The distribution of the anorogenic magmatic association in the United States has been discussed by Van Schmus and Bickford [1981] and Anderson [1983]. Good examples are exposed in SE Missouri, central Wisconsin, and the Stillwater Igneous Complex of Montana. Middle to Late Proterozoic anorogenic granites and rhyolites are common in the subsurface in a wide belt from eastern California through the southern midcontinent and as far north as central Wisconsin and the Grenville terrane. Late Precambrian felsic volcanic rocks are also present in Florida. The Canadian Precambrian has relatively few examples of such rocks, aside from the Grenville and Nain provinces, where mid-crustal level anorthosites and layered complexes occur.

The boundaries of anorogenic magmatic terranes are not well-exposed in North America, but in the Amazonian craton the anorogenic rocks end along an irregular boundary zone that cuts across most structures in the granite-greenstone and gneiss basement [Gibbs and Barron, 1983]. In Kansas, a distinct boundary between a southern area with abundant Proterozoic felsic volcanic rocks in the subsurface and a northern area with older metamorphic rocks locally cut by anorogenic granites was traced aeromagnetically [Yarger, 1981]. The COCORP Kansas traverse in the northern terrane found a seismically transparent zone, attributed to possibly anorogenic granitoid plutons, beneath the Paleozoic sedimentary rocks [Brown et al., 1983]. The crust beneath the transparent zone in Kansas is seismically complex, without evidence of any large, structurally simple mafic intrusive bodies such as might be expected in areas of substantial crustal melting.

Proterozoic anorthosites profiled in Wyoming and the Adirondack Mountains of New York show relatively few reflections. A much more reflective zone beneath the Laramie, Wyoming anorthosite might arise from cumulate rocks at the anorthosite base [Allmendinger et al., 1982]. The much thicker reflection zone far beneath the Adirondack anorthosite could also be due to cumulate rocks, though it might be unrelated to the anorthosite [Klemperer et al., 1985].

Continental Sedimentary Basins

This broad category includes deposits formed in epicontinental seas, continental margins, aulacogens, rifts, and other intracratonic basins. Their seismic reflection characteristics are broadly similar to those of Phanerozoic basins.

Some of these basins are almost entirely sedimentary; others contain volcanic and hypabyssal igneous rocks, characteristically either mafic, felsic, or both. Reflections are produced where volcanic rocks, sills, or sedimentary rocks with sufficient seismic velocity contrasts are interstratified. The Precambrian continental crust underlying such basins may belong to any of the other categories described above. Basins that contain igneous rocks might be underlain by deep crustal magma chambers.

Examples of Precambrian sedimentary and low-grade metasedimentary rocks of continental provenance include the Early Proterozoic iron-bearing basins of the Lake Superior district; the Middle Proterozoic Belt Supergroup in Montana and Idaho; the Unita Mountain Group in Colorado and Utah; the Sioux and Baraboo quartzites of the north-central states; the Late Proterozoic Windermere Group in Montana; various groups in southeastern California and adjacent regions; and the Keweenawan volcanics and clastic sedimentary strata along the Central North American Rift System, exposed in the Lake Superior region. Most Archean continental sedimentary rocks have higher metamorphic grades and more intense deformation.

Part of a previously unknown, 7-10 km thick, cratonic basin was profiled in Texas and Oklahoma [Oliver et al., 1976; Brewer et al., 1981]. The profile shows strong and continuous reflections that reveal undeformed layering. Stratified upper crustal reflections in the eastern part of the Abo Pass, New Mexico line may also arise from a Proterozoic cratonic basin [Klemperer and Oliver, 1982; De Voogd et al., in press].

Assymetric rift basins, with rotated sedimentary and volcanic fill were profiled in Kansas [Serpa et al., 1984] and Michigan [Brown et al., 1982]. Seismic sections of these basins are similar to those of the Tertiary Rio Grande rift [Brown et al., 1980]. The lower parts of these basins tend to have a larger number of strong reflections, which may arise from interstratified volcanic and sedimentary rocks. The upper parts tend to have fewer such reflections, perhaps because of their dominantly sedimentary character [Serpa et al., 1984; Brown et al., 1982].

Beneath the stratified crust in these basin profiles the reflection and diffraction patterns are complex and probably arise from structures formed before the basins developed.

The Deepest Precambrian Crust and Mohos

Were continental lower crust and upper mantle typically so weak that they could not retain structures generated during craton formation? Detachment of upper and middle crust from lower crust and upper mantle is thought to occur in Phanerozoic orogens [Meissner, 1974; Meissner and Weaver, 1985; Hirn, 1984; Matthews and Cheadle, 1985] and may have also occurred in Precambrian orogens. Roy Chowdhury and Hargraves [1981] further suggested that because of higher temperatures at the base of the crust in the Precambrian, the base of the crust might actually have been the base of the lithosphere. Are there any instances where the lower crust and uppermost mantle have remained intact and attached since the Precambrian? The great stability of many Archean cratonic regions indicates the strength of Precambrian continental lithosphere, but does not rule out the possibility of relatively passive modification of the deep crust. Lower or basal crustal weakness might show up as a lack of correlation between the features of the upper and lower crust and uppermost mantle in seismic profiles of Precambrian terranes, and an absence of deep structures that are demonstrably Precambrian.

The base of the cratonic crust in the U.S. appears in seismic reflection profiles as a decrease in the number of reflections (in Kansas and the Adirondacks) and as discontinuous, weak reflections at Moho depths (in Minnesota, Wyoming, and Texas). Tenuous evidence of structural integrity of Precambrian lower crust and uppermost mantle can be seen in some of these weak reflections, and in the stronger, though less adequately resolved reflections in DSS traverses.

North-dipping reflection zones on the COCORP Minnesota traverse appear to penetrate to the deepest crust. Discontinuous Moho-level reflections remain at the same approximate depths in spite of their apparent dips, and are possibly offset by faults or folds. There is little geological evidence of younger deformation of the region that could have produced these deep structures, and they are thought to have originated during Late Archean collision of the Minnesota River Valley and Superior terranes.

Roy Chowdhury and Hargraves [1981] noted that Moho reflections in the Indian and Ukraine DSS surveys, although locally offset, are typically nearly horizontal, while the reflections in the overlying crust have a dipping fabric of reflections. They attributed the widespread horizontality of the Moho to ductile smearing, evidence of basal crustal weakness. However, the redrawn version of the Indian data (Figures 3b and 3d) shows that the discordance in reflection dips is not restricted to Moho levels, but is also visible in the lower crust. This discordance together with some of the east-dipping reflections and even some of the apparent offsets in the Moho are probably associated with the episodic, west-verging thrusting in this region between the Late Archean and Middle Proterozoic [Drury et al., 1984].

Some deep Precambrian crust has clearly been modified by younger events. Precambrian lower crust in the Basin and Range traverse in Utah and Nevada was apparently weakened and detached from the uppermost mantle during Tertiary extension, and the resulting strong, coherent Moho-level reflections truncate structures in the overlying Precambrian crust [Klemperer et al., 1985].

Conclusions

Two major inferences about Precambrian structures are supported, though certainly not categorically proven, by seismic reflection studies:

Interpretations that are qualitatively consistent with modern tectonic processes can explain most of the Precambrian features that have been resolved: no very different, uniquely Precambrian tectonic processes are required. Terrane boundary zones with deeply-penetrating, moderately dipping and imbricate reflection zones can be interpreted as collisional orogens. Precambrian extensional, continental sedimentary basins yield excellent images that are entirely comparable to their younger counterparts.

The apparent persistence of some Precambrian structures throughout the crust in several regions implies that this crust, and possibly also the attached upper mantle, has remained coherent and refractory through subsequent time. The compositions and thermal histories of the lower crust and upper mantle of such regions did not result in thorough annealing, destruction of Precambrian structures, or detachment of crust and mantle.

Reflection seismology greatly enlarges the accessible Precambrian geological record. It adds much more resolution to crustal sections than has been possible previously, and, in conjunction with potential-field data, it permits access to the vast regions covered by epicontinental sedimentary basins.

We need to learn the typical seismic reflection characteristics of all of the varieties of Precambrian crust. Many traverses across each type are needed in order to separate typical and circumstantial features. Granite-greenstone terranes and anorogenic magmatic terranes are some of the most fundamental categories of continental crust, yet they have barely been examined by reflection seismology.

With much more deep crustal data from many more continents we might detect quantitative differences in the geometry and relative abundances of seismic reflection features of Archean, Proterozoic, and Phanerozoic crust. Higher Precambrian heat flows and the progressive differentiation of the crust and mantle suggest that such differences might exist.

Acknowledgments. This work has benefited from many discussions and suggestions by other present and former participants in the COCORP program, including J. Oliver, L. Brown, S,. Kaufman, R. Allmendinger, D. Nelson, C. Ando, S. Klemperer, J. Brewer, L. Serpa, and B. Payne. Figure 1a was prepared in cooperation with Karl Wirth. COCORP work is supported by NSF Grants EAR-82-12445, EAR-83-13378 and EAR-84-18157. The author's work in the Guiana Shield is supported by NSF Grant EAR 8207422. Jack Oliver, Larry Brown, and Simon Klemperer provided valuable reviews of an earlier draft of this paper. Institute for the Study of the Continents Contribution No. 20.

References

Allmendinger, R. W., J. A. Brewer, L. D. Brown, S. Kaufman, J. E. Oliver, and R. S. Houston, COCORP profiling across the Rocky Mountain Front in southern Wyoming, Part 2: Precambrian basement structure and its influence on Laramide deformation, Bull. Geol. Soc. Am. 93, 1253-1263, 1982.

Anderson, J. L., Proterozoic anorogenic granite plutonism, in Proterozoic Geology: Selected Papers from an International Proterozoic Symposium, edited by L. G. Medaris et al., Geol. Soc. Amer. Mem. 161, 133-154, 1983.

Brewer, J. A., L. D. Brown, D. Steiner, J. E. Oliver, S. Kaufman, and R. E. Denison, Proterozoic basin in the southern midcontinent of the United States revealed by COCORP deep seismic reflection profiling, Geology 9, 569-575, 1981.

Brown, L., C. Ando, S. Klemperer, J. Oliver, S. Kaufman, B. Czuchra, T. Walsh, and Y. W. Isachsen, Adirondack-Appalachian crustal structure: the COCORP northeast traverse, Bull. Geol. Soc. Am., 94, 1173-1184, 1983.

Brown, L., C. E. Chapin, A. R. Sanford, S. Kaufman, and J. Oliver, Deep structure of the Rio Grande Rift from seismic reflection profiling, Jour. Geophys. Res., 85, 4773-4800, 1980.

Brown, L., L. Jensen, J. Oliver, S. Kaufman, and D. Steiner, Rift structure beneath the Michigan Basin from COCORP profiling, Geology, 10, 645-649, 1982.

Brown, L., L. Serpa, T. Setzer, J. Oliver, S. Kaufman, R. Lillie, D. Steiner, and D. Steeples, Intracrustal complexity in the United States midcontinent: preliminary results from COCORP surveys in northeastern Kansas, Geology, 11, 25-30, 1983.

Chowns T., and Williams, C., Pre-Cretaceous rocks beneath the Georgia Coastal Plain - Regional implications, in Studies related to the Charleston, South Carolina Earthquake of 1886 - Tectonics and Seismicity, edited by G. Gohn, Prof. Paper 1313, U.S. Geol. Surv., pp. L1-L42, 1983.

Condie, K. C., Precambrian rocks of the southwestern United States and adjacent areas of Mexico, Resources Map 13, New Mexico Bur. Mines Min. Res., 1981.

Condie, K. C., Plate-tectonics model for Proterozoic continental accretion in the southwestern United States, Geology 10, 37-42, 1982.

Condie, K. C. and C. A. Shadel, An early Proterozoic volcanic arc succession in southeastern Wyoming, Can. Jour. Earth Sci., 21, 428-436, 1984.

Coward, M. P., and M. C. Daly, Crustal lineaments and shear zones in Africa: Their relationship to plate movements, Precamb. Res., 24, 27-45, 1984.

Denison, R. E., E. G. Lidiak, M. E. Bickford, and E. B. Kisvarsanyi, Geology and geochronology of Precambrian rocks in the central interior region of the United States, U.S. Geol. Surv. Prof. Paper 1241-C, C1-C20, 1984.

Dimroth, E., L. Imreh, N. Goulet, and M. Rocheleau, Evolution of the south-central part of the Archean Abitibi Belt, Quebec. Part I: Stratigraphy and palaeogeographic model, Can. Jour. Earth Sci., 19, 1729-1758, 1982.

Drummond, B. J., R. E. Smith, and R. C. Horwitz, Crustal structure in the Pilbara and northern Yilgarn blocks from deep seismic sounding, in Archean Geology, edited by J. E. Glover and D. I. Groves, Spec. Publ. No. 7, Geol. Society of Austral. 33-42, 1981.

Drury, S. A., N. B. W. Harris, R. W. Holt, G. J. Reeves-Smith, and R. T. Wightman, Precambrian tectonics and crustal evolution in south India, Jour. Geol., 92, 3-20, 1984.

Dutch, S. I., Proterozoic structural provinces in the north-central United States, Geology, 11, 478-481, 1983.

Gibb, R. A., M. D. Thomas, P. L. Lapointe, and M. Mukhopadhyay, Geophysics of proposed Proterozoic sutures in Canada, Precamb. Res., 19, 349-384, 1983.

Gibbs, A. K., and C. N. Barron, The Guiana Shield reviewed, Episodes, 6, no. 2, 7-14, 1983.

Gibbs, A. K., B. Payne, T. Setzer, L. D. Brown, J. E. Oliver, and S. Kaufman, Seismic-reflection study of the Precambrian crust of central Minnesota, Bull. Geol. Soc. Am. 95, 280-294, 1984.

Green, A. G., Results of a seismic reflection survey across the fault zone between the Thompson nickel belt and the Churchill tectonic province, northern Manitoba, Can. Jour. Earth Sci., 18, 13-25, 1981.

Green, A. G., O. G. Stephenson, G. D. Mann, E. R. Kanasewich, G. L. Cumming, Z. Hajnal, J. A. Mair, and G. F. West, Cooperative seismic surveys across the Superior-Churchill boundary zone in southern Canada, Can. Jour. Earth Sci., 17, 617-632, 1980.

Green, A. G., Z. Hajnal, and W. Weber, An evolutionary model of the western Churchill Province and Western margin of the Superior Province in Canada and the north-central United States, Tectonophysics, in press, 1985.

Green, A. G., M. J. Berry, C. P. Spencer, E. R. Kanasewich, S. Shiu, R. M. Clowes, C. J. Yorath, D. B. Stewart, J. D. Unger, and W. H. Poole, Recent seismic reflection studies in Canada, this volume.

Hajnal, Z., C. M. R. Fowler, R. F. Mereu, E. R. Kanasweich, G. L. Cumming, A. G. Green, and A. Mair, An initial analysis of the earth's crust under the Williston Basin: 1979 COCRUST experiment, Jour. Geophys. Res., 89, 9381-9400, 1984.

Hildenbrand, T. G., R. W. Simpson, R. H. Godson, and M. F. Kane, Digital colored residual and regional Bouguer gravity maps of the conterminous United States, with cut-off wavelengths of 250 km and 1000 km, scale 1:7,500,000, Map GP-953-A, U.S. Geol. Surv. 1982.

Hirn, A., J-C. Lepine, G. Jobert, M. Sapin, G. Wittlinger, X. Zhong Xin, G. En Yuan, W., Xiang Jing, T. Ji Wen, X. Saho Bai, M. R. Pandey, and J. M. Tater, Crustal structure and variability of the Himalayan border of Tibet, Nature, 307, 23-25, 1984.

Hoffman, P. F., K. D. Card, and A. Davidson, The Precambrian: Canada and Greenland, in Perspectives in Regional Geological Syntheses, edited by A. R. Palmer, D-NAG Spec. Publ. 1, Geol. Soc. Amer., pp. 3-5, 1982.

Holst, T. B., Evidence for nappe development during the early Proterozoic Penokean orogeny, Minnesota. Geology, 12, 135-138, 1984.

Houston, R. S., K. E. Karlstrom, F. A. Hills, and S. B. Smithson, The Cheyenne Belt: The major Precambrian crustal boundary in the western United States, (abstract) Geol. Soc. Am. Abstracts with Programs 11, 446, 1979.

Kaila, K. L., and S. C. Bhatia, Gravity study along the Kavali-Udipi deep seismic sounding profile in the Indian peninsular shield: some inferences about the origin of anorthosites and the Eastern Ghats Orogeny, Tectonophysics, 79, 129-143, 1981.

Kaila, K. L., K. Roy Chowdhury, P. R. Reddy, V. G. Krishna, Hair Narain, S. I. Subbotin, V. B. Sollogub, A. V. Chekunov, G. E. Kharetchko, M. A. Lararenko, and T. V. Ilchenko, Crustal structure along Kavali-Udipi profile in the Indian peninsular shield from deep seismic sounding, Jour. Geol. Soc. India, 20, 307-333, 1979.

King, P. B., Precambrian geology of the United States; an explanatory text to accompany the geologic map of the United States, Prof. Paper 902, U.S. Geol. Surv., 85 pp., 1976.

Klemperer, S. L., L. Brown, J. E. Oliver, C. Ando, and S. Kaufman, Crustal structure in the Adirondacks, in Seismic Expression of Structural Styles, edited by A. W. Bally, Studies in Geology, 15, Vol. 1, Am. Assoc. Petrol. Geol., 5.12-5.16, 1983a.

Klemperer, S. L., and J. E. Oliver, The advantage of length in deep crustal reflection profiles, First Break, 20-27, April 1983b.

Klemperer, S. L., L. D. Brown, J. E. Oliver, C. J. Ando, B. L. Czuchra, and S. Kaufman, Some results of COCORP seismic reflection profiling in the Grenville-age Adirondack Mountains, New York State, Can. Jour. Earth Sci. 22, 141-153, 1985a.

Klemperer, S. L., T. A. Hauge, J. E. Oliver, and C. J. Potter, The Moho in the northern Basin and Range province, Nevada, along the COCORP 40°N seismic reflection transect, Bull. Geol. Soc. Am., in press 1985b.

Kokorina, L. K., E. A. Miagkova, and V. V. Yakovleva, The results of complex geological and geophysical studies of the earth crust in central Karelia, in Geological, Geochemical, and Geophysical Investigations in the Eastern Part of the Baltic Shield, edited by K. Puustinen, Papers issued to the 10th General Meeting of the Finnish-Soviet Joint Geological Working Group, Rovaniemi, 7-11 September 1981, Comm. Sci. Tech. Coop. Finl. Sov. Un., Helsinki, 59-70, 1981.

Lowman, P. D., Jr., Formation of the earliest continental crust: Inferences from the Scourian Complex of northwest Scotland and geophysical models of the lower continental crust, Precamb. Res., 24, 199-215, 1984, .

McCulloch, M. T., and G. J. Wasserburg, Sm-Nd and Rb-Sr chronology of continental crust formation, Science, 200, 1003-1011, 1978.

Matthews, D., and Cheadle, M., Deep reflections from the Caledonides and Variscides west of Britain and comparison with the Himalayas, this volume.

Meissner, R., Viscosity-depth-structure of different tectonic units and possible consequences for the upper part of converging plates, Jour. Geophys., 40, 57-73, 1974.

Meissner, R., and Weaver, T., Reflection data from the Variscides, this volume.

Morey, G. B., B. M., Olsen, and D. L. Southwick, Geologic map of Minnesota: East central Minnesota, bedrock geology, scale 1:250,000, Minnesota Geol. Surv., 1981.

Muehlburger, W. R., The shape of North America during the Precambrian, in Continental Geology, edited by B. C. Burchfiel, J. E. Oliver, and L. T. Silver, Nat. Acad. Sci., Wash., D.C., pp. 175-183, 1980.

Naqvi, S. M., V. Divakara Rao, and Hari Narain, The primitive crust - evidence from the Indian shield, Precamb. Res., 6, 323-345, 1978.

Nekut, A., J. E. P. Connerney, and A. F. Kuckes, Deep crustal electrical conductivity: Evidence for water in the lower crust, Geophys. Res. Lett., 4, 239-242.

Oliver, J. E., M. Dobrin, S. Kaufman, R. Meyer, and R. Phinney, Continuous reflection profiling of the deep basement, Hardeman County, Texas, Bull. Geol. Soc. Am., 87, 1537-1546, 1976.

Percival, J. A., and K. D. Card, Archean crust as revealed in the Kapuskasing uplift, Superior province, Canada, Geology, 11, 323-326, 1983.

Rankin, D. W., T. W. Stern, J. McLelland, R. E. Zartman, and A. L. Odom, Correlation chart for Precambrian rocks of the eastern United States, Prof. Paper 1241-E, U.S. Geol. Surv., E1-E18, 1983.

Reed, J.C., Jr., P. K. Sims, F. S. Houston, L. T. Silver, D. W. Rankin, and M. W. Reynolds, Precambrian of the conterminous United States, in Perspectives in Regional Geological Syntheses, edited by A. R. Palmer, D-NAG Spec. Pub. 1, Geol. Soc. Amer., p. 8-13, 1982.

Rogers, J. J. W., Evolution of continents, Tectonophysics, 105, 55-69, 1984.

Roy Chowdhury, K., and R. B. Hargraves, Deep seismic soundings in India and the origin of continental crust, Nature, 291, 648-650, 1981.

Rybakov, S. I., and S. B. Lobach-Zhuchenko, Greenstone belts of the Fenno-Karelian Craton, in Geological, Geochemical, and Geophysical Investigations in the Eastern Part of the Baltic Shield, edited by K. Puustinen, Papers issued to the 10th General Meeting of the Finnish-Soviet Joint Geological Working Group, Rovaniemi, 7-11 September 1981, Comm. Sci. Tech. Coop. between Finl. Sov. Un., Helsinki, 19-41, 1981.

Schwerdtner, W. M., D. Stone, K. Osadetz, J. Morgan, and G. M. Stout, Granitoid complexes and the Archean tectonic record in the southern part of northwestern Ontario, Can. Jour. Earth Sci., 16, 1965-1977, 1979.

Serpa, L.,. T. Setzer, H. Farmer, L. Brown, J. Oliver, S. Kaufman, J. Sharp, and D. W. Steeples, Structure of the southern Keweenawan rift from COCORP surveys across the midcontinent geophysical anomaly in northeastern Kansas, Tectonics, 3, 367-384, 1984.

Sims P. K., and Z. Peterman, Archean rocks in the southern part of the Can. Shield - A review, Spec. Publ. No. 7, Geol. Soc. Austral., pp. 85-98, 1981.

Sims, P. K., K. D. Card, G. B. Morey, and Z. E. Peterman, The Great Lakes tectonic zone - a major crustal structure in central North America, Bull. Geol. Soc. Am., 91, 690-698, 1980.

Sollogub, V. B., and A. V. Chekunov, The lithosphere of the Ukraine, First Break, June 1983, 9-17, 1983.

Van Schmus, R., and M. E. Bickford, 1981, Proterozoic chronology and evolution of the midcontinent region, North America, in Precambrian

Plate Tectonics, edited by A. Kroner, Elsevier Sci. Pub. Co., Amsterdam, pp. 261-296.

Voogd, B. de, L. Brown, and C. Merey, Nature of the eastern boundary of the Rio Grande rift from COCORP surveys in the Albuquerque basin, New Mexico, Jour. Geophys. Res. in press, 1985.

Yarger, H. L., Aeromagnetic survey of Kansas, Eos Trans. AGU, 62, 173-178, 1981.

Zietz, I., compiler, Composite magnetic anomaly map of the United States: Part A, Conterminous United States, scale 1:2,500,000, Map GP-954-A, U.S. Geol. Surv. 1982.

COMPOSITION, STRUCTURE AND EVOLUTION OF THE EARLY PRECAMBRIAN LOWER CONTINENTAL CRUST: CONSTRAINTS FROM GEOLOGICAL OBSERVATIONS AND AGE RELATIONSHIPS

Alfred Kröner

Institut für Geowissenschaften, Johannes Gutenberg-Universität
Postfach 3980, 6500 Mainz, West Germany

Abstract. Ancient granulite terranes have been interpreted as segments of juvenile crust, added to the continents during accretion along active plate margins. I review the geology and age relationships in the major early Precambrian granulite provinces in the light of this model and conclude that the majority of high-grade terranes display rock assemblages and structures that are unlike those found in modern accretion belts. In Archean terranes age data and isotopic systematics also show that granulite formation preceded greenstone belt formation significantly in the majority of cases examined and that many high-grade complexes have long deformational histories extending over hundreds of Ma. Shallow-water metasediments of continental character predominate in many granulite terranes and display a surprising continuity of lithological layering as exemplified by the supracrustal assemblages in Sri Lanka. I suggest that many of these layers and associated recumbent structures and thrusts constitute near-horizontal seismic reflectors as seen in lower crustal seismic profiles. Granulites may form in a variety of tectonic settings that are all compatible with the plate tectonic concept. However, oversimplification and gross generalization of granulite genesis do not contribute to a better understanding of the evolution of the ancient lower continental crust.

Introduction

Although there is now general agreement that the Earth's continental crust grew more rapidly in the early Precambrian than in later times (for review see Reymer and Schubert, 1984), the mechanism of this growth is still poorly understood. A variety of models are presently discussed that involve juvenile continental additions along active plate margins and through magmatic underplating (for review see Kröner, 1984a) and that increasingly recognize the role of crustal recycling in the process of continent formation and crustal differentiation (Armstrong, 1981; DePaolo, 1983).

Most presently popular scenarios relate early continent formation to voluminous lateral arc accretion, resulting from fast plate turnover, but new models for the thermal evolution of the Earth (Christensen, 1985), calculations on crust-addition rates (Reymer and Schubert, 1984) and paleomagnetic data (Kröner et al., 1984) do not favor such situations that result in unrealistically high growth rates. Although some form of plate tectonics seems to have operated since at least 3.5 Ga ago (Kröner, 1984a), we have no record of the style of plate interaction other than evidence for a predominance of horizontal deformation, both in the upper and lower ancient continental crust.

Early Precambrian high-grade gneiss terranes are exposed in several continents and are commonly regarded as segments of the ancient lower continental crust, brought up to the surface by tectonic processes (e.g. Windley, 1984). They contain diagnostic rock assemblages, structures, geochemical patterns and isotopic systematics that constrain models of crustal growth and differentiation and the form of plate interaction that may have been responsible for their formation and present setting. This paper tries to relate surface field geology of ancient granulite terranes to recent advances in geochemistry, isotope geology and reflection seismology and suggests that there is no unique setting for the evolution of these crustal segments and, perhaps, for the ancient lower continental crust.

Geological Characteristics of Ancient Granulite Terranes

High-grade gneiss terranes make up a significant proportion of the Precambrian shields, and there seems little doubt that the majority of these are of Archean age (Windley, 1984). Two tectonic environments of their occurrence are presently recognized:
- areally extensive or localized regions within or around Archean granite-greenstone terranes that commonly lack a distinct structural linearity and that were stabilized early in the cratonic development, often constituting the sialic basement of the greenstone belts. Examples are the gneisses in the Rhodesian (Zimbabwe) craton (Nisbet et al., 1981), the Ancient Gneiss Com-

plex of Swaziland (Hunter, 1974; Jackson, 1984), the granulites in the core of the Vredefort structure in South Africa (Hart et al., 1981), the West Nile Complex of central Africa (Cahen et al., 1984), the West Yilgarn Gneiss Domain of Western Australia (Gee, 1979; Groves and Batt, 1984; Myers and Williams, 1985), the granulite terrane of southern India (Drury et al., 1984), the Qianxi granulites of Hebei Province, northern China (Jahn and Zhang, 1984a, 1984b), Enderby Land of Antarctica (Kamenev, 1982; Ravich, 1982; Black and James, 1983) and the Pikwitonei granulite domain in central Manitoba, Canada (Weber, 1983).

linear "mobile belts" of high-grade assemblages that commonly display a structural grain distinctly different from neighboring crustal segments and often apparently "transect" or border older cratonic blocks. Examples are the Archean Limpopo belt of southern Africa (Barton and Key, 1981; Van Biljon and Legg, 1983), the Eastern Ghats belt of India (Drury et al., 1984), the Highland Series of Sri Lanka (Cooray, 1978), the Lewisian-West Greenland belt of the North Atlantic craton (Bridgwater et al., 1978), the Kapuskasing structural zone in the central Superior Province of Canada (Percival and Card, 1983), the Inari-Belomoride granulite belt of Finnish Lapland (Barbey et al., 1984), as well as several middle to late Proterozoic gneiss-granulite belts in Africa (e.g. Namaqua-Natal belt), Australia (Musgrave and Arunta Blocks) and in the North Atlantic Province (e.g. Grenville-South Norwegian domain).

The linear mobile belts of the latter category frequently contain rock assemblages derived from metasediments, metavolcanics and granitoid intrusives that are also found in Phanerozoic orogenic belts, and they are thus generally ascribed to collision tectonics in analogy with the evolution of the Himalaya orogen (Dewey and Burke, 1973; Windley, 1981). In some cases, however, such simple models are difficult to reconcile with the complex polymetamorphic and deformational history of granulite belts that demonstrably extend over several hundred million years (e.g. Limpopo belt of southern Africa, Barton and Key, 1981) and where high-grade events significantly postdate crust formation of the affected rocks by up to 1000 Ma (e.g. Fyfe Hills, East Antarctica, DePaolo et al., 1982). In addition, the prevalence of shallow-water, clastic, continent-derived sedimentary assemblages in almost all granulite terranes remains unexplained by most collision models.

Tarney and Windley (1977) and Weaver and Tarney (1981) have noted a close chemical and structural similarity of several Archean grey gneiss granulite terranes with the root zones of modern accretionary complexes such as the Rocas Verdes Complex in the Andes of southern Chile and have proposed that these gneisses represent early Precambrian juvenile crust that accreted or subcreted during subduction processes in intraoceanic or continental margin settings. The Scourian Complex of NW Scotland has become one of the often-cited model examples for the formation of juvenile lower crust, the more so since the overall geochemistry of its gneisses resembles modern calc-alkaline igneous assemblages, and isotopic data convincingly show that crust-formation ages are virtually indistinguishable from the age of high-grade metamorphism (Hamilton et al., 1979).

Accepting the accretion-subcretion growth model for the Lewisian gneisses the question remains whether these rocks are representative of the ancient lower continental crust in general (Lowman, 1984) and of exposed high-grade terranes in particular, and whether the rock types and structural settings found in modern accretionary plate margins are also comparable to those in other ancient granulite complexes.

The Lewisian Complex consists of interlayered bimodal gneisses with a bulk composition broadly equivalent to andesite and are interpreted as a predominantly volcanogenic supracrustal assemblage, locally interlayered with undoubted metaquartzites and marbles. These rocks were intruded by tonalitic to trondhjemitic plutons, now interbanded or tectonically interleaved with the above supracrustal succession (for details see Lowman, 1984 and literature cited therein). An important constraint on the origin and evolution of this complex is that no pre-Scourian basement has been recognized and that isotopic systematics preclude both the supracrustals and the intrusive rocks to be derived from much older preexisting continental crust (Hamilton et al., 1979). Even such metasediments as quartzites, marbles and pelitic gneisses are apparently of local derivation and are seen as the erosion products of emerging arc complexes or as deposits in intra-arc or marginal basins (Weaver and Tarney, 1981). In this regard the Scourian resembles some of the Ketilidian and older gneisses of southern West Greenland (Karlsbeek, 1984; Taylor et al., 1980) and the early Archean Isua-Amitsoq assemblage farther north (Nutman et al., 1984).

Many other Archean high-grade gneiss terranes, however, neither show the compositional characteristics nor the age relationships as found in the Lewisian and Greenland gneisses. In the Limpopo belt of southern Africa, for example, the complexly deformed Sand River gneisses of apparently supracrustal origin were subjected to granulite metamorphism ca. 3.8 Ga ago (Fripp, 1983), significantly earlier than the formation of the granite-greenstone terranes in the adjacent Kaapvaal and Rhodesian cratons. Subsequent consolidation, uplift and stretching of this early continental crust led to deposition of platform-type supracrustal lithologies on this basement some 3.3 to 3.5 Ga ago in which metasediments predominate over possible metavolcanics (Watkeys et al., 1983). These rocks were subjected to a further granulite event at ca. 3.1 Ga ago and apparently remained at deep crustal levels until about 2.7 Ga ago when they were rapidly uplifted during a tectonic episode that also imprinted the presently observed linear geometry on the marginal zones of the Limpopo belt (Barton and Key, 1981, see also papers in Van Biljon and Legg, 1983). A gradual, uninterrupted metamorphic transition from the low-grade granite-greenstone terrane of the northern Kaapvaal craton to the high-grade gneisses of the Limpopo belt (Du Toit et al., 1983) may represent an almost complete crustal section placed on edge by extensive thrusting (Coward and Fairhead, 1980) and suggesting that the pre-greenstone grey gneisses may underlie most, if not all, of the cratonic upper crustal low-grade rocks (Du Toit et al., 1983; Nisbet et al., 1981).

Another crust-on-edge profile appears to be exposed in the Vredefort structure south of Johannesburg, South Africa, where amphibolite-facies granite gneiss grades downwards into a granulitic complex composed of supracrustal rocks interleaved with quartzofeldspathic leuco-gneisses of probable igneous origin (Hart et al., 1981). The granulitic supracrustals contain metaquartzites, calc-silicate gneisses, BIF and metavolcanics and may have been generated about 3.8 Ga ago (Hart et al., 1981), at approximately the same time as the Sand River gneisses of the Limpopo belt farther north. They were metamorphosed and isotopically homogenized at about 3.5 Ga ago, i.e. they already constituted part of the lower crust of the Kaapvaal craton when the 3.5-3.0 Ga old granite-greenstone complexes formed at upper crustal levels and, together with the associated leuco-gneisses, they were the anatectic source for upper crustal granitoids (Hart et al., 1981).

Also in southern Africa, the Ancient Gneiss Complex consists of bimodal grey gneisses that contain numerous high-grade metasedimentary relicts of apparently shallow-water provenance (Hunter, 1974; Tankard et al., 1982) and with a metamorphic U-Pb zircon age of 3.5 Ga (Tegtmeyer et al., 1985). Here again these rocks were already at deep crustal levels when the nearby Barberton greenstone belt began to form, and rapid uplift of the Ancient Gneiss Complex-type lower crust is revealed by the preservation of 3.5 Ga old detrital zircons in the greenstone belt clastic sediments (Tegtmeyer and Kröner, unpubl. data).

The remarkable similarity in rock types, chronology and apparent crustal structure of the three regions of southern Africa described above suggests that large portions of the granite-greenstone upper crust of the Kaapvaal and, perhaps, the Rhodesian cratons are underlain by older high-grade gneiss assemblages that consist of granulitic orthogneisses intrusive into widespread banded successions of predominantly metasedimentary, supracrustal origin (Nisbet, 1984). This type of crust is significantly different from the Scourian Complex of Scotland and, as shown above, was already in existence when the overlying granite-greenstone terranes formed. Clearly, the accretion-subcretion model of Weaver and Tarney (1981) does not seem applicable to this extensive segment of the lower continental crust.

The Western Gneiss Terrane of the Yilgarn Block of Western Australia has many similarities with the occurrences described above and consists of tonalitic-trondhjemitic-adamellitic orthogneisses with abundant inclusions of a metaanorthosite-gabbro-ultramafic suite and with tectonically concordant layers of supracrustal rocks that are largely of sedimentary origin (Myers and Williams, 1985). The polymetamorphic and complexly deformed terrane is considered to be an extensively exposed example of sialic basement upon which the 2.6-3.0 Ga old greenstone sequences of the Yilgarn block were deposited (Gee et al., 1981). Geophysical data (Archibald et al., 1981) and xenocrystic zircons in greenstone volcanics (Compston et al., 1985) support the contention that these granulitic gneisses constitute the entire lower crust of the Yilgarn Block. Granulite facies conditions were apparently attained some 3.35 Ga ago (De Laeter et al., 1981), and the primary ages of the metasediments and granitoids may be between 3.5 Ga and 3.6 Ga. Detrital zircons from cross-bedded metaquartzites have yielded the oldest mineral age so far known on Earth at 4.2 Ga (Froude et al., 1983) and suggest that granitoid sialic crust must have existed well before the now preserved Archean greenstone belts were formed.

The extent of the high-grade crust in the Western Gneiss Terrane and underneath the Yilgarn granite-greenstone layer, the significant time interval between rock formation and granulite metamorphism, and the fact that granite-greenstone evolution is at least 300 Ma younger than consolidation and uplift of the gneisses all make it impossible to explain the evolution of the entire Yilgarn crust in terms of the juvenile accretion model. It is particularly noteworthy that metavolcanic rocks have not been recognized in the supracrustal successions of the Western Gneiss Terrane (Myers and Williams, 1985; Myers, in press), in marked contrast to the Scourian rocks, and reworking of assemblages possibly as old as 4.2 Ga is indicated rather than juvenile addition of new crust during or shortly before the granulite event.

A particularly well preserved section through the continental crust is exposed in southern India where an unbroken transition from (upper crustal) amphibolite facies to (lower crustal) granulite facies has been the subject of detailed studies (Hansen et al., 1984; Condie and Allen, 1984). Field relationships show the late Archean to early Proterozoic Dharwar greenstone assemblages to rest unconformably on the older Peninsular gneiss that contains numerous supracrustal remnants and grades into the extensive granulite terrane of southern Karnataka (Viswanatha et al., 1982).

The gneisses and granitoids have yielded ages between 3.4 Ga and 3.0 Ga, based on modern isotopic work by various dating methods on amphibolite-grade rocks, and at least some of them are of anatectic origin (Taylor et al., 1984), while the supracrustal enclaves are undated. Granulite-facies metamorphism was attained at about 2.5-2.7 Ga ago (Ramiengar et al., 1978), thus significantly after crust-formation as in the case of the Western Gneiss Terrane of Australia and the Fyfe Hills of Antarctica. Virtually all investigators agree that the Indian section represents a crustal profile, and there is little doubt from field relationships that granulitization of the gneisses took place after a complex and polyphase structural evolution. Newton et al. (1980) have suggested that charnockitization has occurred without a significant change in temperature and pressure but merely as a result of dehydration effected through a CO_2-rich vapor phase that migrated upwards from below, a process that these authors named "carbonic metamorphism".

One of the key regions to understand the evolution of metasediment-dominated granulite terranes appears to be the island of Sri Lanka (Ceylon) where the (?Archean to early Proterozoic?) Highland Series constitutes an assemblage of typically shallow-water deposits, apparently devoid of any appreciable volcanic components (Cooray, 1978). The succession is tightly folded with flat, recumbent structures and thrusts dominating so that the overall attitude of the main regional foliation and lithological layering shows gentle dips to the NW. In spite of the intense deformation individual layers can be followed for long distances in the field and display complicated structural interference patterns (Geological Map of Sri Lanka, 1982; see also Fig. 1). Metaquartzites and marbles are particularly good marker beds that show the above structures and, in some cases, can be followed for distances of up to 100 km (Fig. 1).

Spectacular quarry outcrops near the village of Kurunegala reveal the same relationship as in southern India, namely that charnockitization postdates and overprints a polyphase deformation pattern. Field relationships also suggest that the granulite-grade Highland Series were thrust over the generally lower grade orthogneisses of the adjacent Vijayan Complex (Vitanage, 1985 and own observations, 1984).

The Sri Lankan, Indian, Western Australian and African examples of extensive quartzitic, pelitic and calcareous supracrustals in high-grade terranes as well as numerous similar assemblages from other granulite complexes (e.g. Enderby Land, Antarctica, Ravich, 1982; NE China, Wang et al., 1985; Guayana shield, Venezuela, Dougan, 1974; southern Norway, Andreae, 1974) suggest deposition on or near relatively stable and mature continental crust (Condie, 1984) rather than accretion in or under arc complexes as required by the juvenile growth model.

Tectonics of Accretionary Terranes

The proposal for the evolution of high-grade terranes in accretionary complexes of convergence zones was based on the accretion models of Karig (1974) and others where subduction required accretion which, in turn, required constant seaward growth of a plate margin. Modern studies, including multichannel seismic data and IPOD drilling revealed, however, that tectonic processes at active margins are more complex than originally expected, and such features as tectonic erosion, extensional structures and subduction of large amounts of sediments were discovered. Also, many convergent margins were found to lack accretionary prisms (von Huene, 1984).

All accretionary models agree in that the involved oceanic and trench sediments are strongly disrupted during the convergence process, with imbrication, décollement and "tectonic kneading" (Scholl et al., 1980) being the dominant deformation processes (Fig. 2). Von Huene (1984) showed that the style of deformation largely depends on the pore-fluid pressure and, during fast convergence, high pressures will weaken the material, cause it to fail and thus strongly disrupt original stratigraphic successions to form tectonic mélanges. The Franciscan mélange at the coast of California (Hsü, 1971) is a good example of this process and of tectonic mixing of plate margin sediments with remnants of oceanic crust. Disruption down to scales of several cm is characteristic of such assemblages (Fig. 3).

These tectonic features are in marked contrast to the well preserved and extensive lithological units found in many granulite terranes as described above and make it unlikely that such terranes accreted in convergence zones comparable to modern active margins. It is also difficult to visualize how typical shelf or continental basin-type sediments such as cross-bedded coarse quartzites and carbonates could be subducted and underplated and still maintain lithological continuity over such large distances as shown by the granulites of Sri Lanka (Fig. 1).

Discussion and Models

Ancient granulite belts in which there is a large proportion of calc-alkaline metavolcanic and intrusive material and where high-grade metamorphism occurred shortly after protolith formation (e.g. Scourie Complex, Weaver and Tarney, 1981; Qianxi gneisses of China, Jahn and Zhang, 1984a) may have formed in ancient convergence zones during lateral juvenile continental growth as proposed by Weaver and Tarney (1981). It is argued here, however, that these crustal segments are not typical and representative of the entire lower continental crust. Granulite terranes with a predominance of metasedimentary supracrustals and tectonically and magmatically interleaved granitoids are areally much more extensive, and their formation is not genetically related to greenstone belt formation. Rather, they predate the greenstones, often seem to underlie them (as revealed by crustal contamination of originally mantle-derived mafic magmas, e.g. Hegner et al., 1984; Chauvel et al., 1985 and in press; Compston et al., 1985) and experienced their granulitization significantly after protolith formation.

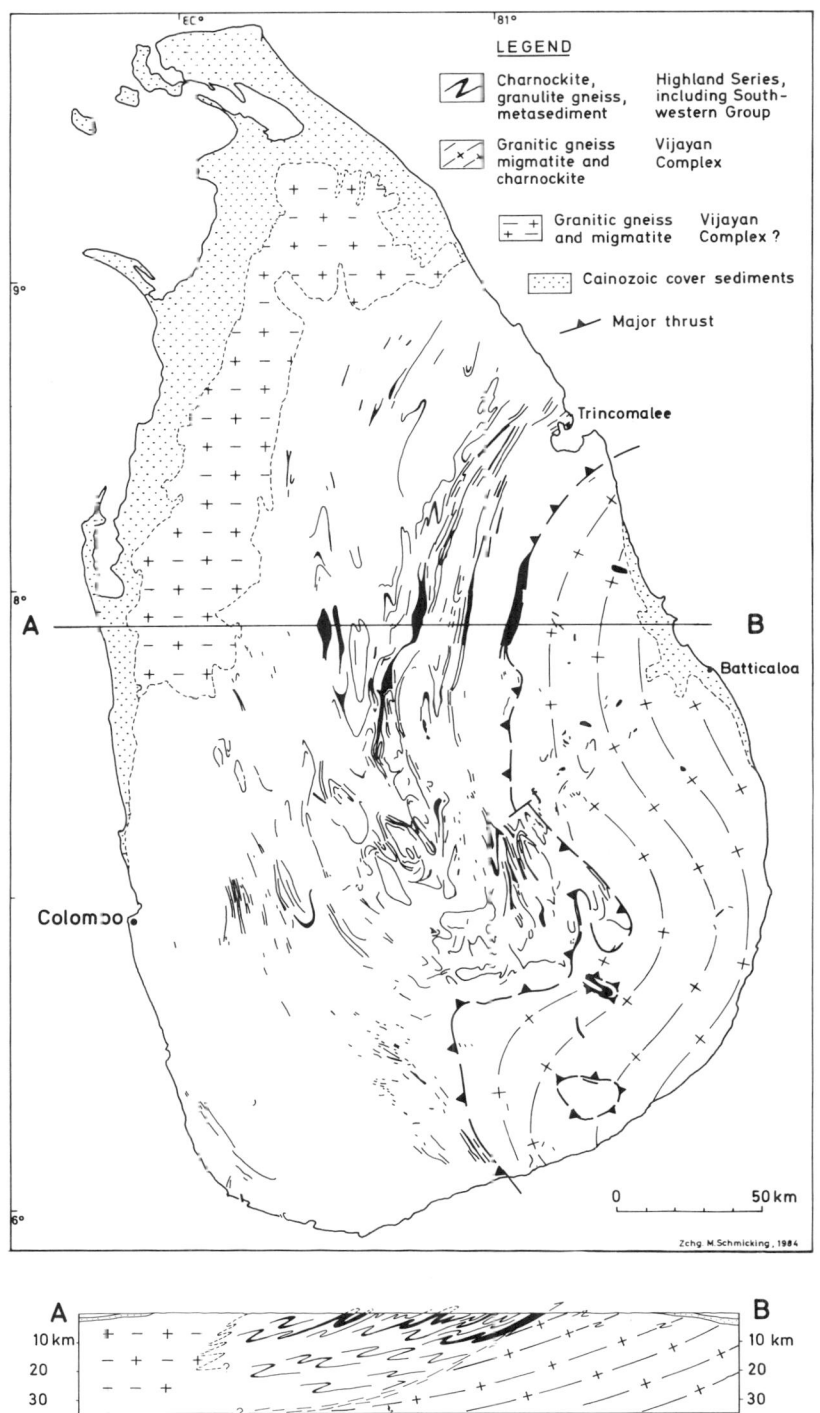

Fig. 1. Simplified and schematic geological map (top) and speculative cross-section (bottom) of the basement complex of Sri Lanka, showing the supracrustal Highland Series with individual layers of metaquartzite and marble (black) and thrust contact with Vijayan orthogneisses (map based on Geol. Map of Sri Lanka, 1982, thrust modified after Vitanage, 1985). Section shows simplified deformation pattern as revealed by surface exposures and suggests tilting of (lower crustal) Highland Series resulting from movement along (listric?) thrust.

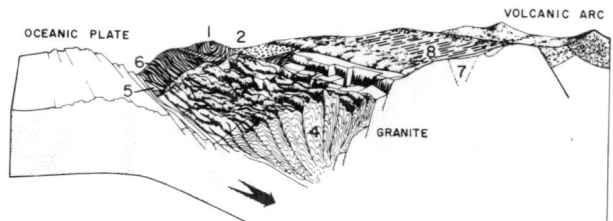

Fig. 2. Schematic view of tectonic disruption during sediment accretion along active plate margin (from Moberly et al., 1982).

Ancient granulite terranes are frequently separated from the (overlying) lower grade granite-greenstone assemblages by transitional zones over which quartzofeldspathic gneisses commonly become migmatized and give way to charnockites while other lithologies develop a granulite-facies mineralogy (Newton and Hansen, 1983). This may suggest that the gneissic basement was transferred, almost vertically, to the lower crust while granitoid intrusions considerably thickened the upper crust. However, the ubiquitous evidence for horizontal shortening in gneiss terranes clearly shows that vertical motion alone was not responsible for the transfer to the lower crust.

The mounting evidence for deep fault and thrust zones in and on the margins of high-grade terranes suggest that plane-strain thrust restacking of previously thinned or mechanically weakened continental crust may have been the predominant mechanism in the Archean to produce high-grade rocks. Such faults may extend to the base of the crust where they gradually flatten out (Bally, 1981; Oliver et al., 1983). The thrusting mechanism, coupled with intense deformation and granitoid intrusions, may have transferred the once shallow-level sediments to deep crustal regions where they recrystallized and experienced chemical modification, perhaps related to CO_2-streaming that is now seen as the major agency of metamorphic dehydration (charnockitization) in the lower crust (Newton et al., 1980).

Katz (1981) ascribed the generation of Archean granulite terranes to large-scale intracratonic transform motion along zones of weakness demarcated by sediment-filled aulacogens, while Jahn and Zhang (1984a) relate the close association of tonalite-trondhjemite-granodiorite gneisses with metavolcanics and metasediments in granulite complexes to magmatic underplating combined with intracrustal thrusting and stacking of crustal slices. The marked structural linearity that commonly defines the narrow Precambrian high-grade "mobile belts" may also be due to such thrusting whereby large segments of the crust are placed on edge, as suggested for the Archean Limpopo belt of southern Africa (Coward, 1976) and the Proterozoic Nagssugtoqidian belt of West Greenland (Bak et al., 1975). Fig. 4 illustrates this model for a hypothetical Archean crustal section and implies that the high-grade gneisses underlie the greenstone belts and only reach the surface after crustal tilting and thickening through thrusting, followed by erosion. Coward (1983) suggested that the forces required to drive segments of the lower crust for large distances on to upper crustal rocks may have operated at plate collision zones or some distance away in the continental interior due to propagation of structures away from the collision zone.

One of the most significant new results from reflection seismology is the recognition of shallow-dipping or near-horizontal reflectors in almost all sections through the lower continental crust (Smithson et al., 1977; Oliver, 1978). Although it is generally agreed that these reflections indicate crustal heterogeneities, their true nature remains obscure since it has only been possible in a few instances to directly relate deep crustal reflectors to surface geology (e.g. the Wind River Mts., USA; NW Scotland).

Several recent studies show that ductile deformation zones are good reflectors (Smithson et al., this volume; Jones and Nur, 1984), and if this is true the model of tectonic interthrusting as proposed above may receive considerable support from seismic profiles. However, it is also possible that many of the reflections recognized in the lower crust arise from compositional heterogeneities (Phinney, 1978) such as found in granulite terranes dominated by supracrustal assemblages as described from Sri Lanka. Kern and Schenk (1985) have shown from laboratory experiments on granulites that variations in compressional and shear wave velocities are strongly related to mineralogical composition; e.g. garnetiferous metapelitic gneisses have velocities in the lower crust that are equivalent to those in mafic rocks or significantly higher (up to 7.6 km/s) while felsic lithologies have velocities of only 6.2-6.4 km/s.

Fig. 3. Mixed sedimentary/ophiolitic mélange of the Franciscan complex south of San Francisco, California.

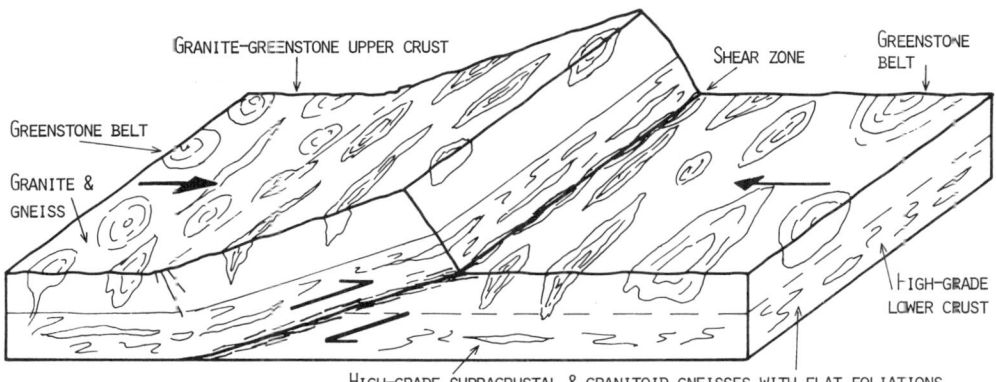

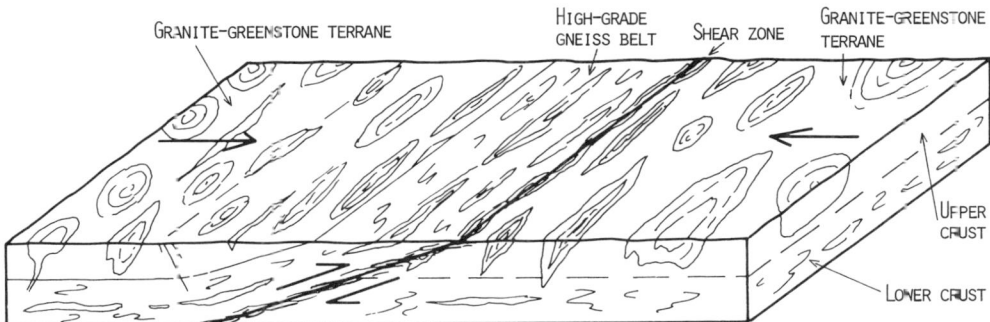

Fig. 4. Schematic and idealized section through the Archean continental crust to show suggested relationship between upper crustal granite-greenstone terranes and lower crustal high-grade gneiss terranes. Upper section displays crustal interstacking by low-angle thrust that results in tectonic juxtaposition of lower crustal gneisses with upper crustal granite-greenstone rocks. A wide listric shear zone develops along the thrust. Lower section shows the same situation after removal of the overthrust block by erosion. Tilting of the crust has thus resulted in a transition zone from upper to lower crust on exposed erosion surface. Model is not to scale and is drawn to explain the possible origin of Archean high-grade gneiss belts such as the Limpopo belt in southern Africa. Note that this model assumes continuity of high-grade gneiss under the granite-greenstone terrane (from Kröner, 1984b).

Structural studies of surface outcrops in Sri Lanka confirm a dominance of subhorizontal or gently NW-dipping foliation surfaces and thrusts that are also axial planar to tight isoclinal folds. These folds are particularly well displayed by marker beds such as metaquartzite and marble, with thicknesses between several tens of meters and several hundred meters and fold limbs that are tens of kilometers long. The fact that this structural style predominates over a (tectonic) thickness of at least 5 km in the Sri Lankan Highland Series supracrustals strongly suggests that such structures also dominate at greater crustal depth as speculated in the cross-section of Fig. 1.

If compositional layering in granulites of supracrustal origin contributes significantly to seismic reflections as proposed here, then most lower crustal segments so far analyzed may contain a higher proportion of such rocks as is hitherto assumed. This conclusion reiterates the contention made earlier that ancient lower crustal types such as exposed in Western Australia, India and Sri Lanka are perhaps more representative of the lower continental crust in general than the Scourie-type assemblage.

Several Archean and Proterozoic linear granulite belts contain supracrustal sequences that resemble continental margin type deposits and ocean floor volcanics and have thus been interpreted in terms of continental collision of the modern Wilson cycle (Windley, 1981; Light, 1983) One well documented case may be the Harts Range area of the Arunta Block in central Australia where Ding and James (1985) showed that mid-Proterozoic granulite formation in reworked Archean basement was due to repeated semi-ductile to brittle thrusting of the basement under a tightly folded overthrust pile of relatively thin nappes

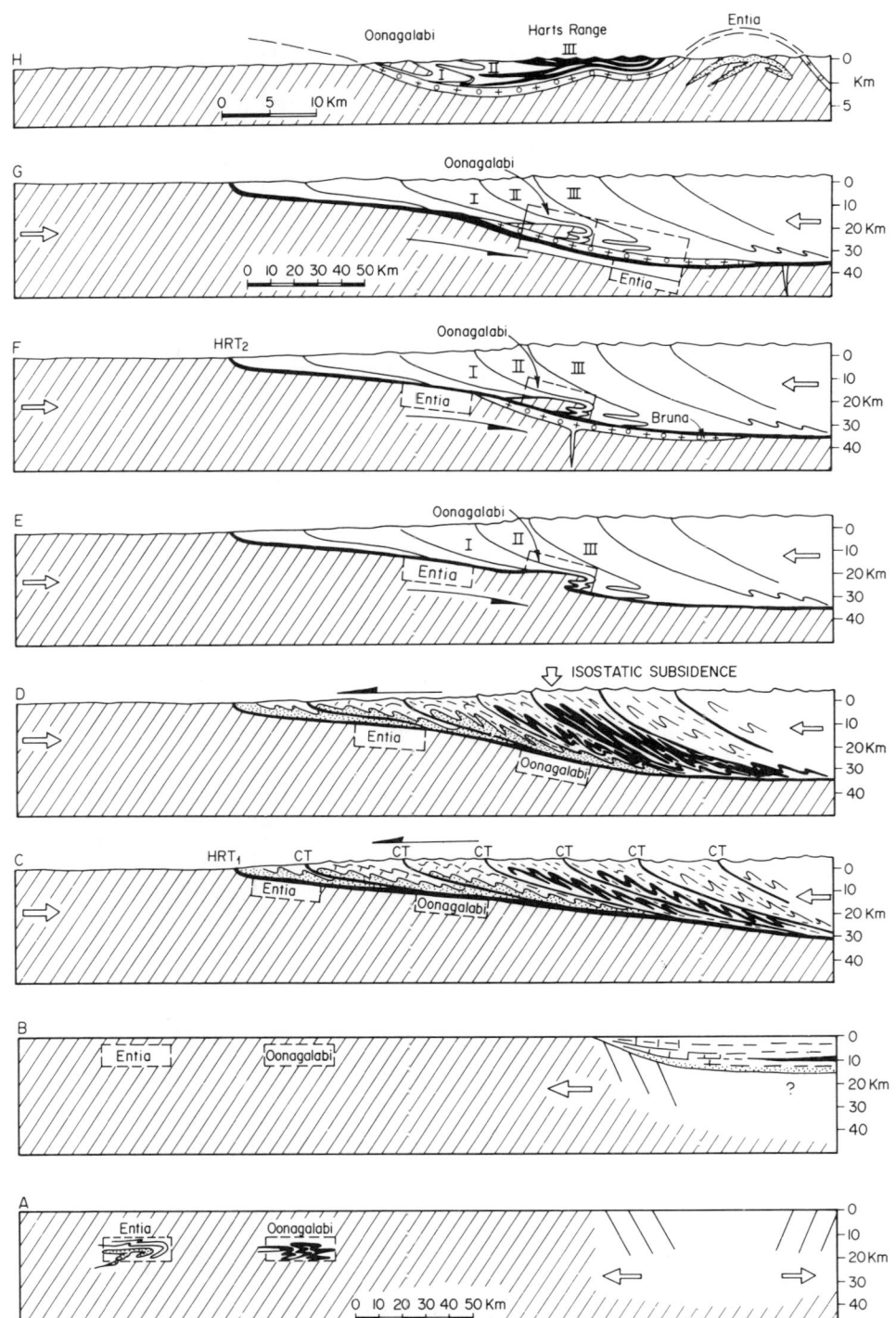

Fig. 5. Schematic sections across Harts Range area, central Australia, showing suggested evolution from rift basin (bottom) to thrust belt (top). Sections A-G are S-N, section H is W-E. Oblique ruling = reworked Archean crystalline basement, stippled and other rulings = Proterozoic Irindina supracrustal cover sequence. Model suggests that basement and cover were juxtaposed by progressive underthrusting of basement beneath stacked overthrust pile of thin cover nappes (from Ding and James, 1985).

consisting of early Proterozoic supracrustal assemblages (Fig. 5).

In other cases, however, the simple collision model is incompatible with field geology, geochemical data and geochronology as, for example, shown for the Inari-Belomoride granulite belt of Finnish Lapland (Barbey et al., 1985) and for the Namaqua mobile belt of southwestern Africa (Barton, 1983). In the Lapland belt Barbey et al. (1985) relate early Proterozoic supracrustal deposition to an intracontinental rift that develops into a wide ocean basin during crustal stretching of an Archean platform. Basin closure is facilitated by subcrustal delamination of mantle lithosphere, and A-subduction combined with crustal interstacking led to a foldbelt whose roots are in the lower crust where the supracrustal rocks are transformed into granulite assemblages. This scenario essentially follows the model of Kröner (1981, 1983) for "intracontinental" foldbelts and is entirely compatible with the plate tectonic concept. The South African Namaqua belt with well documented evidence for extensive crustal reworking during a ca. 1000-1100 Ma old granulite event has been interpreted in terms of collision of an "exotic crustal fragment" with the Archean Kaapvaal craton (Barton and Burger, 1983).

Conclusions

Exposures of high-grade gneiss terranes that are interpreted as segments of the ancient lower continental crust are so variable in composition, structural style and internal age relationships that no uniform model for their evolution can be suggested. However, some general patterns seem to emerge that may help to constrain speculations on lower crustal genesis.

First, crustal segments where crust-formation and granulitization occurred in a single episode of crustal differentiation such as proposed by Moorbath (1975) and as exemplified by the Scourian gneisses of Scotland appear to be less voluminous than complexes that exhibit protracted and polyphase tectono-metamorphic histories and where crustal differentiation significantly postdates primary crust-formation. This conclusion is in accord with field geology, isotopic data (Ben Othman et al., 1984) and dynamically feasible crust-addition rates (Reymer and Schubert, 1984).

Second, contemporaneous development of adjacent granite-greenstone and high-grade gneiss terranes as advocated by the active margin/marginal basin model of Tarney et al. (1982) can only be demonstrated in a few instances such as the Superior Province of Canada (Goodwin, 1981) and West Greenland (Nutman et al., 1984). The general setting, at least for the late Archean, seems to be that many greenstones developed on significantly older granitoid crust and that greenstone evolution was unrelated to granulite formation.

Third, the predominance of clastic, shallow-water, continent-derived metasediments in almost all granulite terranes with well preserved internal lithology is in marked contrast to the rock types and their tectonic disintegration as observed in most modern accretionary belts. Furthermore, most high-grade supracrustal successions demonstrate that tight isoclinal folding predates granulite formation, i.e. the structures had already formed well before the folded complex reached deep crustal levels. These relationships suggest that few, if any, ancient granulite terranes represent former accretion complexes of active margins. It is more likely that sedimentary basins developed on attenuated crust, either in rift-dominated settings or on passive margins, and that compressive deformation resulted in significant crustal thickening through thrusting, bringing the sedimentary assemblages to deep crustal levels as exemplified by the Limpopo belt, the Inari-Belomoride belt and the Harts Range of central Australia.

The formation of sedimentary basins and the crustal weakness to facilitate extensive thrusting can be visualized in settings where continental fragmentation occurs in wide, transcurrent zones with the formation of pull-apart basins and microcontinental crustal fragments. Such settings, as proposed for the early history of the Alpine System (Dewey et al., 1973), rarely lead to full ocean basins and the development of evolved active plate margins but generate numerous segments of fractured and/or attenuated crust that facilitate thrusting during subsequent periods of compression related to plate convergence. Relatively small changes in the direction of plate motion and significant shifts in the location of rotation poles (Euler poles) can cause attenuation and compression as suggested above (Dewey, 1975). It is therefore not surprising that exotic or suspect terranes should occur in areas of high-grade rocks (e.g. Barton and Burger, 1983) or be part of Archean crustal provinces as speculated by Hubert et al. (1984), and evolutionary models will have to consider this added complexity.

Lastly, it is emphasized that plate tectonic theory allows for a variety of dynamic settings in which high-grade rocks may develop, just as it presents us with a great diversity of settings that we commonly describe as "orogen". To conclude with a recent comment of Moorbath (1984): "Just as there are granites and granites there most certainly are granulites and granulites", and we should rather look at the differences between granulite terranes in order to understand lower crustal evolution than oversimplify and generalize their characteristics with the purpose to fit them to a particular and currently popular global setting.

Acknowledgments. I thank the West German Science Foundation (DFG) for travel grants to visit several granulite terranes, in particular Sri Lanka.

References

Andreae, M.O., Chemical and stable isotope composition of the high-grade metamorphic rocks from the Arendal area, southern Norway, Contrib. Mineral. Petrol., 46, 169-188, 1974.

Archibald, N.J., L.F. Bettenay, M.J. Bickle and D.I. Groves, Evolution of Archaean crust in the eastern Goldfields Province of the Yilgarn Block, Western Australia, in Archaean geology, edited by J.E. Glover and D.I. Groves, pp. 491-504, Geol. Soc. Australia Spec. Publ. 7, 1981.

Armstrong, R.L., Radiogenic isotopes: the case for crustal recycling on a near-steady-state-no-continental growth earth. Phil. Trans. R. Soc. Lond., A 301, 443-472, 1981.

Bak, J., K. Sorensen, J. Gnocott, J.A. Korstgard, D. Nash and J. Watterson, Tectonic implications of Precambrian shear belts in western Greenland, Nature, 254, 566-569, 1975.

Bally, A.W., Thoughts on the tectonics of folded belts, in Thrust and nappe tectonics, edited by K. McClay and N.J. Price, pp. 13-32, Geol. Soc. London Spec. Publ. 9, 1981.

Barbey, P., J. Convert, B. Moreau, R. Capdevila and J. Hameurt, Petrogenesis and evolution of an early Proterozoic collisional orogenic belt: the granulite belt of Lapland and the Belomorides (Fennoscandia), Geol. Soc. Finland Bull., 56, in press, 1985.

Barton, E.S., Reconnaissance isotopic investigations in the Namaqua mobile belt and implications for Proterozoic crustal evolution - Namaqualand geotraverse, in Namaqualand metamorphic complex, edited by B.J.V. Botha, pp. 45-66, Geol. Soc. S. Afr., Spec. Publ. 10, 1983.

Barton, E.S. and A.J. Burger, Reconnaissance isotopic investigations in the Namaqua mobile belt and implications for Proterozoic crustal evolution - Upington geotraverse, in Namaqualand metamorphic complex, edited by B.J.V. Botha, pp. 173-192, Geol. Soc. S. Afr., Spec. Publ. 10, 1983.

Barton Jr., J.M. and R. Key, The tectonic development of the Limpopo mobile belt and the evolution of the Archaean cratons of southern Africa, in Precambrian plate tectonics, edited by A. Kröner, pp. 185-212, Elsevier, Amsterdam, 1981.

Ben Othman, D., M. Polvé and C.J. Allègre, Nd-Sr isotopic composition of granulites and constraints on the evolution of the lower continental crust, Nature, 307, 510-515, 1984.

Black, L.P. and P.R. James, Geological history of the Archaean Napier complex of Enderby Land, in Antarctic earth sciences, edited by R.L. Oliver P.R. James and J.B. Jago, pp. 11-15, Austr. Acad. Sci., Canberra, 1983.

Bridgwater, D., K.D. Collerson and J.S. Myers, The development of the Archaean gneiss complex of the North Atlantic region, in Evolution of the Earth's crust, edited by D.H. Tarling, pp. 19-69, Academic Press, London, 1978.

Cahen, L., N.J. Snelling, J. Delhal and J.R. Vail, The geochronology and evolution of Africa, 512 pp., Clarendon Press, Oxford, 1984.

Chauvel, C., B. Dupré, N.T. Arndt and G.A. Jenner, Isotopic heterogeneities in Archean greenstone belts (abstract), Terra Cognita, 5, 206, 1985.

Chauvel, C., B. Dupré and G.A. Jenner, The Sm-Nd age of Kambalda volcanics is 500 Ma too old! Earth Planet. Sci. Lett., in press.

Christensen, U.R., Thermal evolution models for the Earth, J. geophys. Res., 90, 2995-3007, 1985.

Compston, W., I.S. Williams, I.H. Cambell and J. Gresham, Xenocrystic zircons in the Kambalda volcanics, Research School of Earth Sciences, Annual Report 1984, Australian National University, 94-95, 1985.

Condie, K.C., Archean geotherms and supracrustal assemblages, Tectonophysics, 105, 29-41, 1984.

Condie, K.C. and P. Allen, Origin of Archaean charnockites from southern India, in Archaean geochemistry, edited by A. Kröner, G.N. Hanson and A.M. Goodwin, pp. 182-203, Springer-Verlag, Berlin/Heidelberg, 1984.

Cooray, P.G., Geology of Sri Lanka, in Proceedings of 3rd regional conference on geology and mineral resources of southeast Asia, edited by P. Nayla, pp. 701-710, Asian Inst. Technol., Bangkok, Thailand, 1978.

Coward, M.P., Archaean deformation patterns in southern Africa. Phil. Trans. R. Soc. Lond., A 283, 313-331, 1976.

Coward, M.P., Some thoughts on the tectonics of the Limpopo belt, in The Limpopo belt, edited by W.V. Van Biljon and J.H. Legg, pp. 175-180, Geol. Soc. S. Afr., Spec. Publ. 8, 1983.

Coward, M.P. and J.D. Fairhead, Gravity and structural evidence for the deep structure of the Limpopo belt, southern Africa, Tectonophysics, 68, 31-43, 1980.

De Laeter, J.R., W.G. Libby and A.F. Trendall, The older Precambrian geochronology of Western Australia, in Archaean geology, edited by J.E. Glover and D.I. Groves, pp. 145-158, Geol. Soc. Australia, Spec. Publ. 7, 1981.

DePaolo, D.J., The mean life of continents: estimates of continent recycling rates from Nd and Hf isotopic data and implications for mantle structure, Geophys. Res. Lett., 10, 705-708, 1983.

DePaolo, D.J., W.I. Manton, E.S. Grew and M. Halpern, Sm-Nd, Rb-Sr and U-Th-Pb systematics of granulite facies rocks from Fyfe Hills, Enderby Land, Antarctica, Nature, 298, 614-619, 1982.

Dewey, J.F., Finite plate evolution: some implications for the evolution of rock masses at plate margins, Am. J. Sci., 275-A, 260-284, 1975.

Dewey, J.F. and K.C.A. Burke, Tibetan, Variscan and Precambrian basement reactivation: products of continental collision, J. Geol., 81, 683-692, 1973.

Dewey, J.F., W.C. Pitman III, W.B.F. Ryan and J. Bonnin, Plate tectonics and the evolution of the Alpine System, Geol. Soc. America Bull., 84, 3137-3180, 1973.

Ding, P. and P.R. James, Structural evolution of the Harts Range area and its implication for the development of the Arunta Block, central Australia, Precambrian Res., 27, 251-276, 1985.

Dougan, T.W., Cordierite gneisses and associated lithologies of the Guri area, northwest Guayana shield, Venezuela, Contrib. Mineral. Petrol., 46, 169-188, 1974.

Drury, S.A., N.B.W. Harris, R.W. Holt, G.J. Reeves-Smith and R.T. Wightman, Precambrian tectonics and crustal evolution in southern India. J. Geol., 92, 3-20, 1984.

Du Toit, M.C., D.D. Van Reenen and C. Roering, Some aspects of the geology, structure and metamorphism of the southern marginal zone of the Limpopo metamorphic complex, in The Limpopo belt, edited by W.J. Van Biljon and J.H. Legg, pp. 121-142, Geol. Soc. S. Afr., Spec. Publ. 8, 1983.

Fripp, R.E.P., The Precambrian geology of the area around the Sand River near Messina, central zone, Limpopo mobile belt, in The Limpopo belt, edited by W.J. Van Biljon and J.H. Legg, pp. 89-102, Geol. Soc. S. Afr., Spec. Publ. 8, 1983.

Froude, D.O., T.R. Ireland, P.D. Kinny, I.S. Williams, W. Compston, I.R. Williams and J.S. Myers, Ion microprobe identification of 4,100-4,200 Myr-old terrestrial zircons, Nature, 304, 616-618, 1983.

Gee, R.D., Structure and tectonic style of the western Australian shield, Tectonophysics, 58, 327-369, 1979.

Gee, R.D., J.L. Baxter, S.A. Wilde and I.R. Williams, Crustal development on the Archaean Yilgarn Block, Western Australia, in Archaean geology, edited by J.E. Glover and D.I. Groves, pp. 43-56, Geol. Soc. Australia, Spec. Publ. 7, 1981.

Geological Map of Sri Lanka, Scale: 8 miles to one inch, Geol. Survey Dept., Colombo, Sri Lanka, 1982.

Goodwin, A.M., Archaean plates and greenstone belts, in Precambrian plate tectonics, edited by A. Kröner, pp. 105-135, Elsevier, Amsterdam, 1981.

Groves, D.I. and W.D. Batt, Spatial and temporal variations of Archaean metallogenetic associations in terms of evolution of granitoid-greenstone terrains with particular emphasis on the western Australian shield, in Archaean geochemistry, edited by A. Kröner, G.N. Hanson and A.M. Goodwin, pp. 73-100, Springer-Verlag, Berlin/Heidelberg, 1984.

Hamilton, P.J., N.M. Evensen, R.K. O'Nions and J. Tarney, Sm-Nd systematics of Lewisian gneisses: implications for the origin of granulites, Nature, 277, 25-28, 1979.

Hansen, E.C., R.C. Newton and A. Janardhan, Pressures, temperatures and metamorphic fluids across an unbroken amphibolite facies to granulite facies transition in southern Karnataka, India, in Archaean geochemistry, edited by A. Kröner, G.N. Hanson and A.M. Goodwin, pp. 161-181, Springer-Verlag, Berlin/Heidelberg, 1984.

Hart, R.J., H.J. Welke and L.O. Nicolaysen, Geochronology of the deep profile through the Archean basement at Vredefort, with implications for early crustal evolution, J. geophys. Res., 86, 10663-10680, 1981.

Hegner, E., A. Kröner and A.W. Hofmann, Age and isotope geochemistry of the Archaean Pongola and Usushwana suites, Swaziland, southern Africa: a case for crustal contamination of mantle derived magma, Earth Planet. Sci. Lett., 70, 267-279, 1984.

Hsü, K.J., Franciscan mélanges as a model for eugeosynclinal sedimentation and underthrusting tectonics, J. geophys. Res., 76, 1162-1170, 1971.

Hubert, C., P. Trudel and L. Gélinas, Archean wrench fault tectonics and structural evolution of the Blake River Group, Abitibi belt, Quebec, Can. J. Earth Sci., 21, 1024-1032, 1984.

Hunter, D.R., Crustal development in the Kaapvaal craton. I. The Archaean, Precambrian Res., 1, 259-294, 1974.

Jackson, M.P.A., Archaean structural styles in the Ancient Gneiss Complex of Swaziland, southern Africa, in Precambrian tectonics illustrated, edited by A. Kröner and R. Greiling, pp. 1-18, E. Schweizerbart'sche Verlagsbuchhandlung, Stuttgart, 1984.

Jahn, B.-M. and Z.-Q. Zhang, Archean granulite gneisses from eastern Hebei Province, China: rare earth geochemistry and tectonic implications, Contrib. Mineral. Petrol., 85, 224-243, 1984a.

Jahn, B.-M. and Z.-Q. Zhang, Radiometric ages (Rb-Sr, Sm-Nd, U-Pb) and REE geochemistry of Archaean granulite gneisses from eastern Hebei Province, China, in Archaean geochemistry, edited by A. Kröner, G.N. Hanson and A.M. Goodwin, pp. 204-234, Springer-Verlag, Berlin/Heidelberg 1984b.

Jones, T.D. and A. Nur, The nature of seismic reflections from deep crustal fault zones, J. geophys. Res., 89, 3153-3171.

Katz, M.B., A shear-mobile transform belt in the Precambrian Gondwanaland of Africa-South America, Geol. Rundschau, 70, 1012-1019, 1981.

Kamenev, E.N., Regional metamorphism in Antarctica, in Antarctic geoscience, edited by C. Craddock, pp. 429-433, University of Wisconsin Press, Madison, 1982.

Karig, D.E., Evolution of arc systems in the western Pacific, Ann. Rev. Earth Planet. Sci., 2, 51-75, 1974.

Karlsbeek, F., Relations between Archaean rocks and Proterozoic mobile belts in the Precambrian shield of Greenland (abstract), NATO Advanced

Study Institute, Norway, The deep Proterozoic crust in the North Atlantic Provinces, p. 18, 1984.

Kern, H. and V. Schenk, Elastic wave velocities in rocks from a lower crustal section in southern Calabria (Italy), Phys. Earth Planet. Int., in press.

Kröner, A., Precambrian plate tectonics, in Precambrian plate tectonics, edited by A. Kröner, pp. 57-90, Elsevier, Amsterdam, 1981.

Kröner, A., Proterozoic mobile belts compatible with the plate tectonic concept, in Proterozoic geology, edited by L.G. Medaris Jr., C.W. Byers D.M. Mickelson and W.C. Shanks, pp. 69-74, Geol. Soc. America, Mem. 161, 1983.

Kröner, A., Evolution, growth and stabilization of the Precambrian lithosphere, Phys. Chem. Earth, 15, 69-106, 1984a.

Kröner, A., Fold belts and plate tectonics in the Precambrian, Proc. 27th Int. Geol. Congr., 5, 247-280, VNU Science Press, Utrecht, 1984b.

Kröner, A., P.W. Layer and M.O. McWilliams, Archaean palaeomagnetism: evidence for continental drift and the existence of a dipolar magnetic field since ca. 3.5 billion years ago (abstract) Terra Cognita, 4, 78, 1984.

Light, M.P.R., The Limpopo mobile belt: a result of continental collision, Tectonics, 1, 325-342, 1982.

Lowman Jr., P.D., Formation of the earliest continental crust: inferences from the Scourian complex of northwest Scotland and geophysical models of the lower continental crust, Precambrian Res., 24, 199-215, 1984.

Moberly, R., G.L. Shepherd and W.T. Coulbourn, Forearc and other basins, continental margin of northern and southern Peru and adjacent Ecuador and Chile, in Trench-forearc geology, edited by J.K. Leggett, pp. 171-189, Geol. Soc. London, Spec. Publ., 10, 1982.

Moorbath, S., Geological interpretation of whole-rock isochron dates from high grade gneiss terrains, Nature, 255, 391-395, 1975.

Moorbath, S., Origin of granulites, Nature, 312, 290, 1984.

Myers, J.S., High grade terrains in and around the Yilgarn Block of Western Australia, J. geol. Soc. London, in press.

Myers, J.S. and I.R. Williams, Early Precambrian crustal evolution at Mt Narryer, Western Australia, Precambrian Res., 27,

Newton, R.C. and E.C. Hansen, The origin of Proterozoic and late Archean charnockites - evidence from field relations and experimental petrology, in Proterozoic geology, edited by L.G. Medaris Jr., C.W. Byers, D.M. Michelson and W.C. Shanks, pp. 167-178, Geol. Soc. America, Mem., 161, 1983.

Newton, R.C., J.V. Smith and B.F. Windley, Carbonic metamorphism, granulites and crustal growth, Nature, 288, 45-50, 1980.

Nisbet, E.G., The continental and oceanic crust and lithosphere in the Archaean: isostatic, thermal, and tectonic models, Can. J. Earth Sci. 21, 1426-1441, 1984.

Nisbet, E.G., J.F. Wilson and M.J. Bickle, The evolution of the Rhodesian craton and adjacent terain: tectonic models, in Precambrian plate tectonics, edited by A. Kröner, pp. 161-184, Elsevier, Amsterdam, 1981.

Nutman, A.P., J.H. Allaart, D. Bridgwater, E. Dimroth and M. Rosing, Stratigraphic and geochemical evidence for the depositional environment of the early Archaean Isua supracrustal belt, southern West Greenland, Precambrian Res., 25, 365-396, 1984.

Oliver, J., Exploration of the continental basement by seismic reflection profiling, Nature, 275, 485-488, 1978.

Oliver, J., I. Cook and L. Brown, COCORP and the continental crust, J. geophys. Res., 88, 3329-3347, 1983.

Percival, J.A. and K.D. Card, Archean crust as revealed by the Kapuskasing uplift, Superior Province, Canada, Geology, 11, 323-326, 1983.

Phinney, R.A., Interpretation of reflection seismic images of the lower continental crust (abstract), Eos Trans. AGU, 59, 389, 1978.

Ramiengar, A.S., M. Ramakrishnan and M.N. Viswanatha, Charnockite-gneiss complex relationships in southern Karnataka, J. geol. Soc. India, 19, 411-419, 1978.

Ravich, M.G., The Lower Precambrian of Antarctica, in Antarctic Geoscience, edited by C. Craddock, pp. 421-427, University of Wisconsin Press, Madison, 1982.

Reymer, A. and G. Schubert, Phanerozoic addition rates to the continental crust and crustal growth, Tectonics, 3, 63-77, 1984.

Scholl, D.W., R. von Huene, T.L. Vallier and D.G. Howell, Sedimentary masses and concepts about tectonic processes at underthrust ocean margins, Geology, 8, 564-568, 1980.

Smithson, S.B., P.N. Shive and S.K. Brown, Seismic velocities, reflections and structure of the crystalline basement, in The Earth's crust, edited by J.F. Heacock, Am. geophys. Union, Geophys. Monogr. Ser. 20, 254-270, 1977.

Smithson, S.B., R.A. Johnson, C.A. Hurich and D.M. Fountain, Crustal reflections and crustal structure, this volume, 1985.

Tankard, A.J., D.K. Hobday, M.P.A. Jackson, D.R. Hunter, K.A. Eriksson and W.E.L. Minter, Crustal evolution of southern Africa, 3.8 billion years of earth history, 523 pp., Springer-Verlag, New York, 1982.

Tarney, J. and B.F. Windley, Chemistry, thermal gradients and evolution of the lower continental crust, J. geol. Soc. London, 134, 153-172, 1977.

Tarney, J., B.L. Weaver and B.F. Windley, Geological and geochemical evolution of the Archaean continental crust, Rev. Bras. Geociênc., 12, 53-59, 1982.

Taylor, P.N., S. Moorbath, R. Goodwin and A.C. Petrykowski, Crustal contamination as an indicator of the extent of early Archaean continental crust: Pb isotopic evidence from the late Archaean gneisses of West Greenland, Geochim. Cosmochim. Acta, 44, 1437-1453, 1980.

Taylor, P,N., B. Chadwick, S. Moorbath, M. Ramakrishnan and M.N. Viswanatha, Petrography, chemistry and isotopic ages of Peninsular gneiss, Dharwar acid volcanic rocks and the Chitradurga granite with special reference to the late Archaean evolution of the Karnataka craton, southern India, Precambrian Res., 23, 349-375, 1984.

Tegtmeyer, A.R., A. Kröner and A.W. Hofmann, Isotopic systematics of early Archaean high-grade greenstones from Swaziland, southern Africa (abstract), Terra Cognita, 5, 205, 1985.

Van Biljon, W.J. and J.H. Legg (Eds.), The Limpopo belt, 203 pp., Geol. Soc. S. Afr., Spec. Publ., 8, 1983.

Viswanatha, M.N., M. Ramakrishnan and J. Swami Nath, Angular unconformity between Sargur and Dharwar supracrustals in Sigegudda, Karnataka craton, south India, J. geol. Soc. India, 23, 85-89, 1982.

Vitanage, P.W., Tectonics and mineralization in Sri Lanka, Geol. Soc. Finland Bull. (Sahama-volume), in press, 1985.

von Huene, R., Tectonic processes along the front of modern convergent margins - Research of the past decade, Ann. Rev. Earth Planet. Sci., 12, 359-381, 1984.

Wang, R.-M., S. He, Z. Chen, P. Li and F. Dai, Geochemical evolution and metamorphic development of the early Precambrian in eastern Hebei, China, Precambrian Res., 27, 111-130, 1985.

Watkeys, M.K., M.P.R. Light and T.J. Broderick, A retrospective view of the central zone of the Limpopo belt, Zimbabwe, in The Limpopo belt, edited by W.J. Van Biljon and J.H. Legg, pp. 65-80, Geol. Soc. S. Afr., Spec. Publ., 8, 1983.

Weaver, B.L. and J. Tarney, Lewisian gneiss geochemistry and Archaean crustal development models, Earth Planet. Sci. Lett., 51, 171-180, 1981.

Weber, W., The Pikwitonei granulite domain: a lower crustal level along the Churchill-Superior boundary in central Manitoba (abstract), in Workshop on a cross section of Archean crust, edited by L.D. Ashwal and K.D. Card, pp. 95-97, Lunar Planet. Sci. Inst., Houston, Tech. Rpt. 83-03, 1983.

Windley, B.F., Precambrian rocks in the light of the plate tectonic concept, in Precambrian plate tectonics, edited by A. Kröner, pp. 1-20, Elsevier, Amsterdam, 1981.

Windley, B.F., 1984, The evolving continents, 2nd ed., 399 pp., John Wiley & Sons, London, 1984.

PRECAMBRIAN CRUSTAL STRUCTURE OF THE NORTHERN BALTIC SHIELD FROM THE FENNOLORA PROFILE: EVIDENCE FOR UPPER CRUSTAL ANISOTROPIC LAMINATIONS[1]

Kenneth H. Olsen

Earth and Space Sciences Division, MS C335, Los Alamos National Laboratory, Los Alamos, New Mexico 87545, U.S.A.

Carl-Erik Lund

Institute of Solid Earth Physics, University of Uppsala, S-75122, Uppsala, SWEDEN

Abstract. Because Archean heat generation was two to four times its present value, the rates and style of crustal evolution and global tectonic mechanisms during the Archean and Proterozoic (3900-600 Ma) were possibly quite different than those familiar from Phanerozoic plate tectonics. In particular, the lateral scales and depths of convection and lithospheric subduction elements may have been smaller than contemporary analogs. Structures preserved in the upper and midcrustal levels of the cratonic area of the northern Baltic Shield therefore may be very useful in formulating more detailed models of Precambrian lithospheric tectonics. The northern part of the NNE-SSW-trending FENNOLORA profile traverses Precambrian basement complexes ranging in age from 1800-2800 Ma, which are adjacent to the 2800 Ma Kola nucleus. We use reflectivity method synthetic seismogram modeling to assist interpretation of a 700-km-long segment of the FENNOLORA line running from Northcape, across portions of Norway, Finland, and Sweden, to about the Arctic circle in the south. Our Finnish colleagues also provided, for comparison, a record section approximately perpendicular to FENNOLORA running southeastward for approximately 300 km across Finnish Lappland (FINLAP). FENNOLORA record stations reveal an *en echelon* pattern of P-wave first arrivals with apparent velocities between 5.0 and 6.8 km/s. The *en echelon* pattern is observable in both north- and south-trending directions from shotpoint G. This suggests a fine structure of the upper crust to depths approximately 20 km consisting of several alternating high- and low-velocity layers, each about 1 or 2 km thick. On the other hand, the *en echelon* pattern cannot be clearly seen on the perpendicular FINLAP profile from shotpoint G, which implies some of the laminations are anisotropic with the high speed axis trending approximately north-south. One speculative interpretation is that the anisotropic layers are basaltic fragments of Archean or Proterozoic oceanic crust that were "stranded" beneath thin sialic crust by very shallow angle subduction zones.

INTRODUCTION

Seismic reflection and refraction/wide-angle-reflection studies of crustal structure of shields or cratons are of great importance to two fundamental topics concerning the nature and evolution of the Earth: (1) understanding early stages of continental growth and evolution during the Archean (before 2500 Ma) and Proterozoic (2500-600 Ma) and how these may have differed from more familiar plate tectonic processes in the Phanerozoic; and (2) near surface (a few km depth) high-resolution observations of such complex metamorphosed and reworked rocks may provide valuable insights in interpreting the geophysical nature of middle and lower levels of continental crust elsewhere in the world. Heretofore, these subjects have been dominated by geochemical, geochronological, and petrologic data; but advances in both near-vertical reflection and refraction/wide-angle-reflection techniques and in geographical coverage now are beginning to provide high-resolution geophysical constraints for tectonic modeling.

Because of greater abundance of long-lived radionuclides, bulk earth heat production rates at the end of the Archean (2500 Ma) were two times the present value and three to four times greater for crustal rocks such as gneiss and granulites (Lambert, 1976). The consequent higher thermal gradients thus imply that the primitive continental plates were thinner, ocean-continent subduction processes were faster and took place at shallower angles, and indeed that subduction of thin continental crust (A-subduction) could have played an important role in orogenic processes (Kröner, 1981). Although some researchers postulate an even greater variety of crustal tectonic models (e.g., dominantly vertical movements, massive planetesimal impacts—Witschard, 1984) in the early Precambrian, there is general agreement that, while rates and styles of plate interactions differed significantly, the same basic plate tectonic processes probably operated during the late Archean and Proterozoic time (2700-600 Ma) when most of the crustal development and evolution of the Baltic Shield took place (Kröner, 1981; Berthelsen, 1985). Structures preserved in the upper and midcrustal levels of the Baltic Shield and observable by seismic reflection and refraction techniques may therefore be very useful in understanding Precambrian lithospheric tectonics and evolution.

The middle and lower continental crust in most parts of the world probably consists of mylonites and of granulite facies metamorphic rocks of andesitic or dioritic composition (Smithson, 1978). Since shields often contain extensive granulite belts at the surface, seismic measurements may also assist interpretation of lower crustal seismic data elsewhere (Chroston and Evans, 1983).

[1] FENNOLORA Contribution No. 6

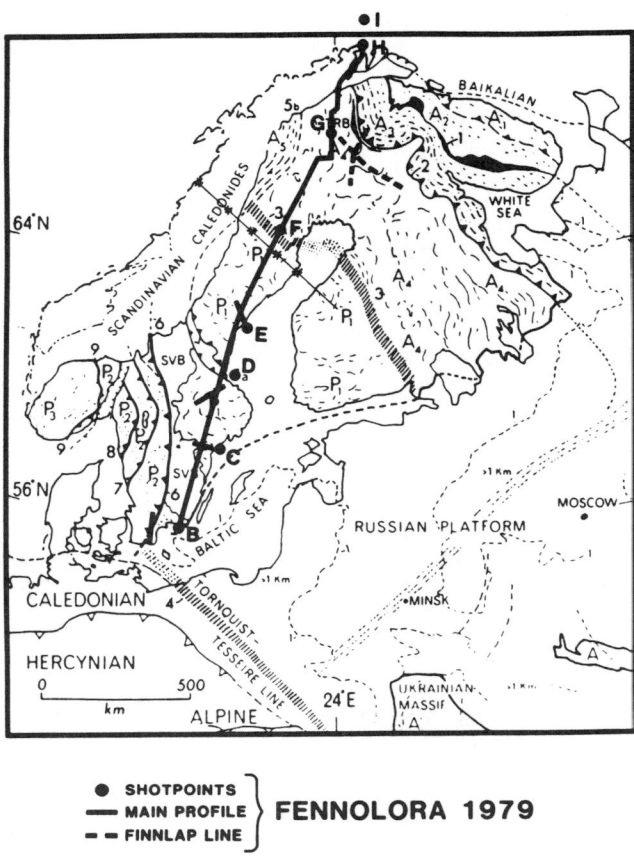

Fig. 1. Tectonic map of Scandanavia with the shotpoints (B, C, D, E, F, G, H, and I) and recording lines for the FENNOLORA and FINLAP profiles superimposed. Also shown is the 1972 Blue Road profile (*-*-*). Tectonic base map adapted from Berthelsen (1985), by permission. Archean and Proterozoic tectonic provinces within the Baltic Shield are labeled as A_1, A_2, ..., P_1, P_2, ..., etc. Major tectonic boundaries numbered from 1 to 9 (see text for discussion).

FENNOLORA PROFILE

Figure 1 shows locations of shotpoints and lines along which recording stations were deployed for the FENNOscandian LOng-RAnge profile (FENNOLORA) during August, 1979. The experiment was primarily designed to study the fine structure of the upper mantle including the lithosphere/astenosphere transitional region (Lund, 1979b, 1983) and, hence, a very long profile (approximately 1900 km) over a stable and relatively homogeneous part of the crust was necessary. Because of the emphasis on lower lithospheric depths (>50 km), rather large shotpoint spacing (170 to 400 km) was acceptable. For good upper crustal structure resolution using reversed profiling, a smaller separation (70-100 km) is preferable; but the large intershot intervals are somewhat compensated for by the relatively close average spacing of recording stations. Twenty-eight explosions (0.6-8.0 tons) were detonated at eight shotpoints along the main line from Bornholm (shotpoint B) to Northcape (shotpoints H and I). Ninety-six, matched, three-component portable seismograph stations were deployed at 539 sites along the main profile to yield the average station separation of 3.5 km. To supplement the main profile, an international team recorded a 300-km-long line running SE from shot-point G, across Finnish Lapland, approximately perpendicular to the main FENNOLORA line (Figure 1). The 300 km SE-trending line together with a shorter (150 km) line subparallel to the main FENNOLORA line are designated as the FINLAP profiles (Luosto et al., 1983). The entire FENNOLORA project was coordinated by a working group of the European Seismological Commission, and seismologists from fourteen European countries participated in the field work (Lund, 1983).

In this paper we discuss only the preliminary interpretation of the seismic record sections from shotpoint G in the northern part of the profile. We confine our analysis here mainly to P-wave arrivals at ranges less than 300 km; additional interpretation and modeling using S-wave arrivals and crustal guided-wave phases ($\bar{P}$ and Lg) is in progress. We employ standard techniques of phase correlation and forward modeling with one-dimensional travel-time calculations. After acceptable velocity depth models are found from travel-times, we refine them by modeling amplitude and waveform characteristics with one-dimensional, reflectivity method synthetic seismogram computations (Kind, 1978).

SEISMIC RECORD SECTIONS

Vertical component record sections out to ranges of 300 km from shotpoint G-north, G-south, and G-east (FINLAP), are respectively displayed in Figures 2, 3, 4, together with derived velocity-depth models, travel-time correlations and synthetic seismogram sections. The crustal model for FINLAP is taken directly from the travel-time analysis of Luosto et al. (1983) and was not independently rederived by us since it appears to adequately fit the observations. General features of the crustal models (compared in Figure 5) are: (1) Pn velocities significantly greater than 8.0 km/s (8.2-8.4 km/s), (2) Moho depths in the range between 44 and 49 km, (3) evidence of steepening velocity gradients in the lower crust, (4) P-wave velocities exceeding 7.0 km/s at depths between approximately 34 km and the Moho, and (5) alternating velocity maxima and reversals in the middle and upper crust except for the FINLAP profile where distinct maxima and minima are not clearly observed.

Although our interpreted Moho depths may be uncertain by 3 or 4 km due to the effects of the upper crustal velocity reversals, they are in good agreement with values of total crustal thickness of 45-46 km derived by the spectral ratio technique from teleseismic data measured at permanent seismic observatories in northern Sweden, Finland, and Norway (Bungum et al., 1980). P-wave velocities above 7.0 km/s and generally steepening velocity gradients in the lower part of the crust seem to be common characteristics of the deep crust in other parts of the Baltic Shield; such properties are also observed at more southerly FENNOLORA shotpoints (Prodehl and Kaminski, 1984) and in the NW-SE Blue Road profile across central Scandanavia (Lund, 1979a). We speculate that such higher velocities and gradients may be a consequence of thickening of initially thin (10-20 km?) primitive crust by one or more of several hypothosized lithospheric evolutionary processes occurring in early Precambrian times—such as magmatic underplating, ensialic orogeny, intracontinental subduction, etc.—as discussed by Kröner (1981).

The most noteworthy aspect of record section and travel-time interpretations of Figure 2, 3, and 4 are the *en echelon* patterns of first P-wave arrivals out to ranges of approximately 150 km from the shotpoints. At least three *en echelon* branches having apparent velocities between 6.0 and 6.8 km/s—which are rather high for shallow crustal material—are clearly correlatable on both the north- and south-trending sections from shotpoint G. Such *en echelon*

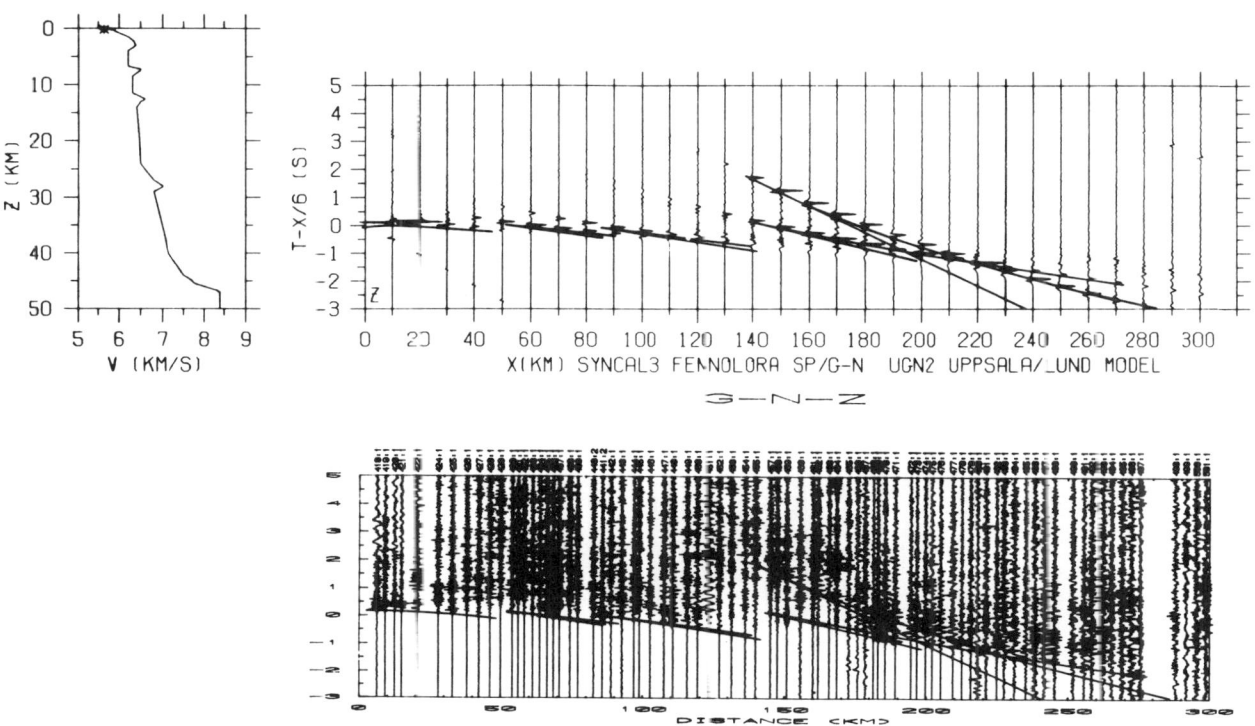

Fig. 2. Vertical component record section for FENNOLORA shotpoint G-north (G-N) below, synthetic section above, and travel-time branches calculated from the P-wave velocity-vs-depth model shown at left.

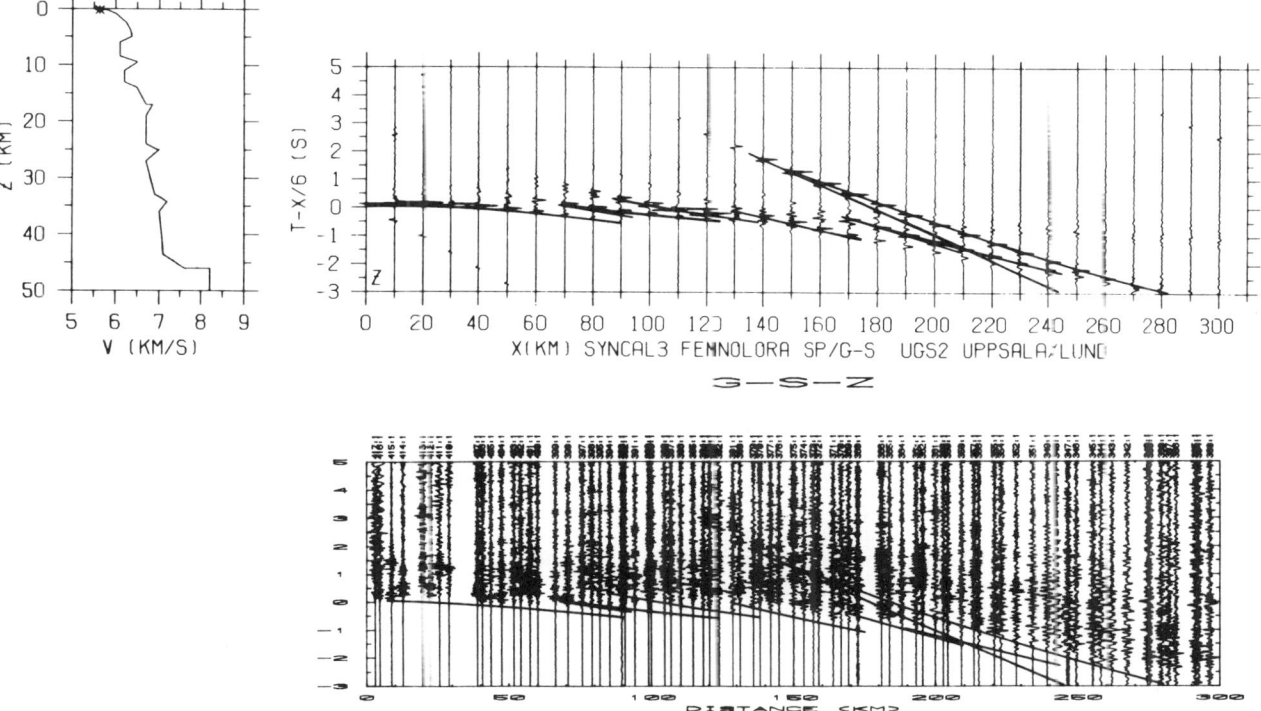

Fig. 3. Observed (below) and synthetic (above) vertical component record section for shotpoint G-south (G-S).

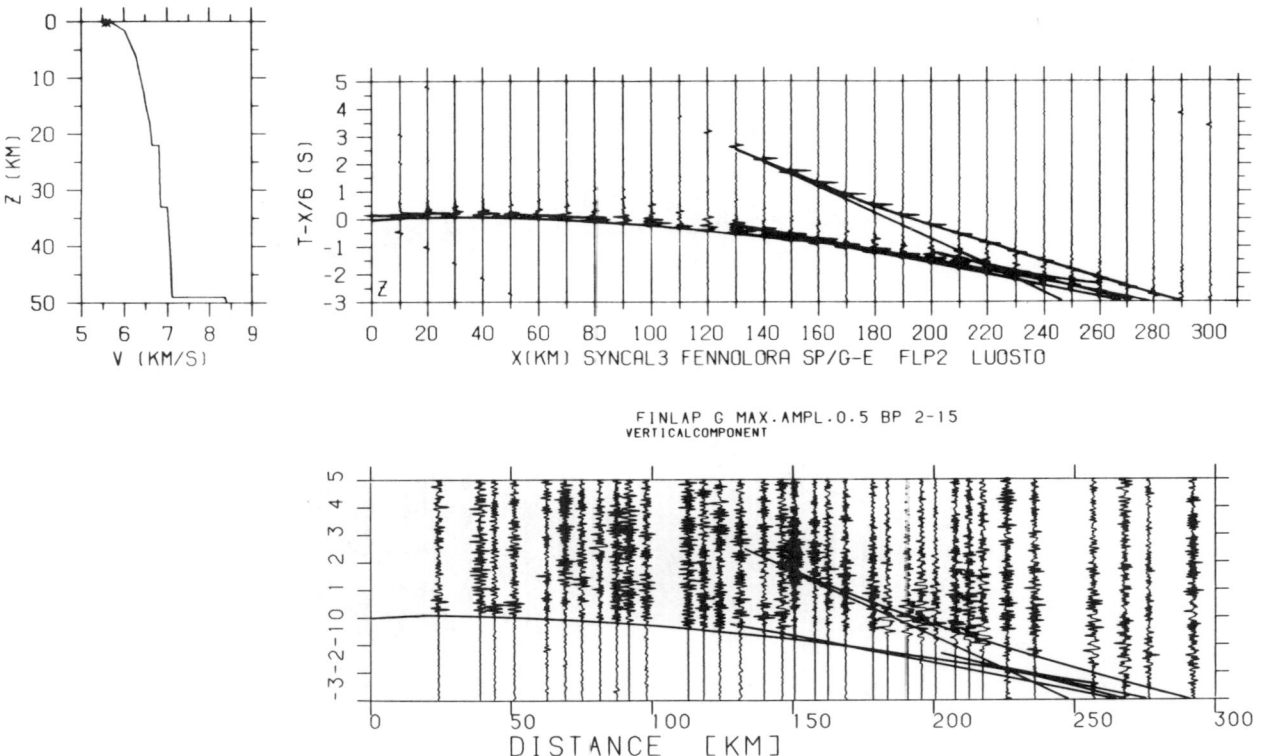

Fig. 4. Observed (below) and synthetic (above) vertical component record section for shotpoint G-east (G-E), FINLAP.

travel-time patterns are most straightforwardly interpreted as due to a fine structure of a predominantly horizontally stratified upper crust. As shown in the figures, the models consist of several high- and low-velocity layers, each only 1-2 km thick down to depths of about 20 km. The gradients shown for the high velocity "teeth" are necessary to properly match the growth and decay of amplitudes within each travel-time branch. *En echelon* branches are also clearly seen on radial component and somewhat less clearly on transverse component record sections (not shown). Alternative interpretations of the multiple *en echelon* branches are that they may be caused by (1) lateral variations in near-surface rock types or formation dips, or (2) an "interference-like" effect due to source effects near the shotpoint. We have examined geological maps of the region and find no obvious correlation with rock type in this very complex metamorphic terrain. Another argument against the variable surface geology interpretation is that such *en echelon* patterns are also found near the more southerly shotpoints along the FENNOLORA lines, although they are not as clearly developed as for shotpoint G. An "interference" shotpoint cause could be disproved by moving a shotpoint about 10 km to see if the pattern remains stationary. Although the precise coordinates of the repeated FENNOLORA shotpoints in the lakes and ocean may have varied by about 1 km, we believe the existence of these types of patterns at other shotpoints also argues against this interpretation.

Similar kinds of *en echelon* travel-time branches have been observed on very long European profiles, except that the previous observations concern apparent velocities near 8 km/s and ranges from approximately 200 km to over 1000 km (e.g., Hirn et al., 1973; Lund, 1979c). These have been interpreted in terms of laminated structures in the subcrustal lithosphere. Deichmann and Ansorge (1983) discuss evidence for lamination in the lower continental crust in southwestern Germany but, to our knowledge, these northern FENNOLORA sections are the first to show evidence for such shallow crustal laminations so clearly. Although velocity-depth profiles for shotpoint G towards the north and towards the south show some similarities in number, thicknesses, and approximate depths of the velocity minima, they are not identical. This suggests the sub-horizonal stratifications in the basement are not continuous over distances of the order of more than 20 to 50 km. This leads to the picture that they may be more like "flakes" distributed more or less randomly horizontally and in depth rather than extensive bedding as might be expected in tectonically undisturbed young sedimentary basins elsewhere in the world.

As can be seen from Figure 4, the record section from the nearly perpendicular FINLAP profile (shotpoint G-E) shows no indication whatsoever of *en echelon* branches in the 0-150 km range as do the FENNOLORA lines from the same shotpoint. Although the average station spacing for FINLAP is about twice that for FENNOLORA, inspection of a "decimated" SP-G-N or SP-G-S record section shows the lamination effect should be easily seen on the FINLAP line if the velocity contrasts between laminations were as strong as for the NNE-SSW direction. Near shotpoint G, both FINLAP and FENNOLORA traverse much the same complex granitoid/greenstone belt. We are thus lead to the conclusion that some of the shallow crustal laminations or "flakes" exhibit a velocity anisotropy (approximately 5%) with the fast axis oriented NNE-SSW approximately along the strike of the main FENNOLORA profile. We shall discuss below that this also turns out to be the trend of main age progression of cratonization, thrusting and other predominantly horizontal tectonic movements in the Baltic Shield. We also note

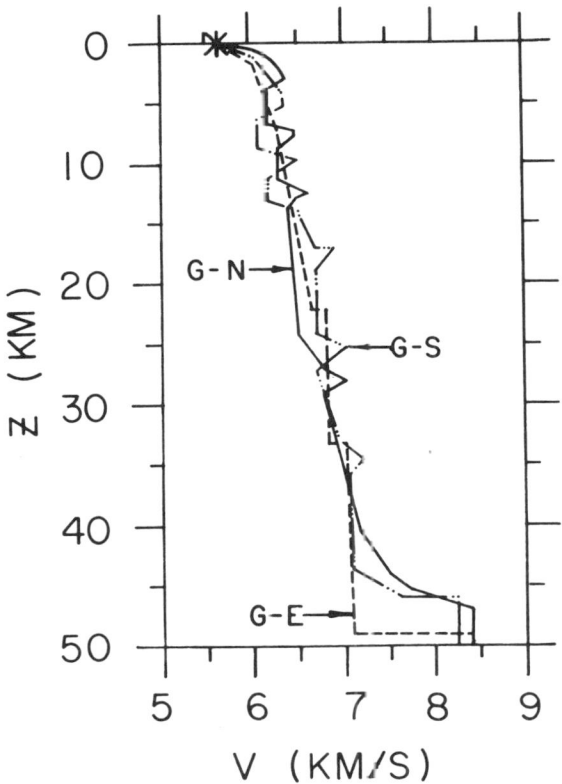

Fig. 5. Comparison of crustal velocity models for FENNOLORA profiles G-N, G-S, G-E, and H-S.

that the earlier Blue Road profile in central Norway and Sweden (Figure 1) had a trend nearly parallel to FINLAP and, while shallow crustal travel-time branches can be broken into segments, an *en echelon* pattern cannot be clearly discerned. We therefore suggest that this basic shallow crustal "flake" anisotropy possibly persists beneath much of the northeastern and eastern Baltic Shield.

DISCUSSION

Witschard (1984, with extensive references therein) recently presented a detailed review of geological, geochemical, and geophysical investigations in northernmost Sweden across which the G-S part of the FENNOLORA profile runs. He suggests these data support a picture of ensialic tectonic evolution for this region, but the large, violent meteorite impacts in the early Archean created circular and arcuate structures which may have exerted significant control over subsequent tectonic evolutionary patterns.

Berthelsen (1985) has proposed a tectonic model for the crustal evolution of the entire Baltic Shield area, including the northern part which has lacked extensive analysis heretofore. We attempt to give highlights of his model in Figure 1. Berthelsen (1985) characterizes Precambrian crustal rocks of the Baltic Shield by a dualistic classification: (1) crustal age provinces (capital letters in Figure 1, e.g., A = Archean, P = Proterozoic) which are defined as to when the first/oldest continental (mantle derived?) crust was formed, and (2) structural age provinces (numerical subscripts to letters, e.g. A_1, A_2, etc.) defined according to the age when the youngest large-scale orogenic patterns were formed. In this way, for example, where the first-formed crust is of Archean age, it is divided into tectonic units (A_1-A_5) according to the degree and style of Proterozoic reworking. Several important tectonic boundaries are recognized; all were developed during the Proterozoic and are numbered 1-9 in Figure 1. They are interpreted variously as (a) continent-continent collision sutures, such as the approximately 2100 Ma Kola suture (No. 1); (b) healed subduction sutures related to primitive shallow angle oceanic subduction beneath thin Archean crust (No. 3 and 4) and several intracrustal sutures formed by "tectonic underplating" and crustal doubling (A-type subduction) similar to contemporary Himalayan collisional tectonics (No. 2, 5, 6, 7, 8, 9). The picture thus developed is that of a progressive cratonization spreading in a general southwestward trend from the approximately 2800 Ma Kola nucleus (units A_1-A_3). Strikes of principal orogenic foldbelts are NW-SE perpendicular to the cratonization trends. For interpretation of the anisotropic seismic velocities observed from shotpoint G, we are mainly concerned with the Karelian tectonic unit, A_4.

Perhaps one of, or a combination of, three general types of hypothesized Precambrian tectonic processes may be able to account for our observation of shallow anisotropic seismic layering in the northern Baltic Shield: (1) shallow angle subduction processes near a (thin) continental-oceanic destructive margin, (2) dominantly horizontal imbricated faulting or older supracrustals due to thrusting, (3) sheet-like ultramafic intrusives emplaced in the shallow crust

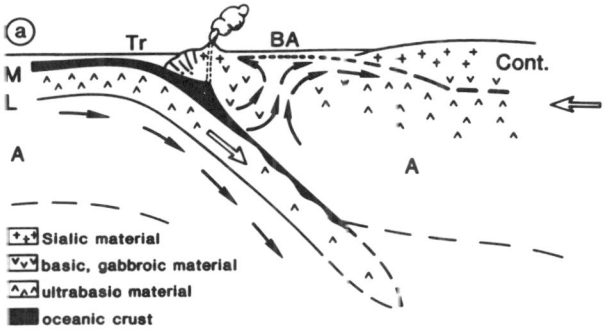

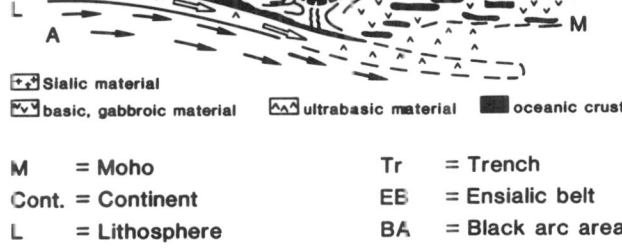

Fig. 6. Generalized cross sections contrasting the present day (above) and hypothesized (below) Archean and early Proterozoic patterns of crustal subduction and mantle convection at destructive ocean/continent plate boundaries. Secondary convection cells behind shallowly descending early Precambrian subduction zones give rise to ensialic basins and isolation of "flakes" of oceanic crust in the thin, warm primitive continental crust. Adapted and redrawn from Meissner (1979).

as a result of the proximity of hot mantle. We illustrate only the first of these suggestions in the following discussion.

Figure 6 is a cartoon showing hypothesized differences between early Precambrian and contemporary subduction processes at continental-oceanic destructive margins. In late Archean-early Proterozoic time, numerous small sialic plates accreted, the destructive margin rocks were metamorphosed to high grade terrains, and the ensialic back-arc basins were reworked to form greenstone belts. Gabbroic and/or basaltic oceanic crust (corresponding approximately to "layer 3" of modern-day oceanic crust) were underthrust beneath the thin primitive continental crust at shallow-angle subduction zones and later fragmented and isolated when the continental margin shifted outward from the growing shield nucleus.

If indeed such fragments of early oceanic crust became "stranded" as suggested above, their gabbroic composition means that they contain significant amounts of olivine. The olivine crystals could have been highly oriented as a result of flow and thrusting stresses associated with the subduction processes and thus would cause an anisotropy in the P-wave velocity much as is today observed throughout the oceans (Bott, 1982). It is well known (Bott, 1982) that the fast direction of P-wave anisotropy occurs parallel to the oceanic fracture zones, i.e. perpendicular to the ridge axes and in the direction of the spreading vectors.

We thus suggest that one explanation of our observation of anisotropic shallow crustal "fragments" from the FENNOLORA-FINLAP data could be a Precambrian crustal evolution scenario similar to that sketched above. Tectonic boundary 3 of Figure 1 is best interpreted as an approximately 2000-Ma-old subduction zone off the SW margin of the Karelian proto-shield (Berthelsen, 1985), and the directions of stresses during subduction and folding of the Karelian unit (A_4) are consistent with the fast axes of the anisotropic shallow crustal layers.

Acknowledgments. We are indebted to U. Luosto of the University of Helsinki for providing us with large scale record sections of the FINLAP profile. A. Berthelsen kindly provided us with material prior to publication, and we greatly appreciate discussions with him and with A. Kröner concerning the mysterious world of Precambrian tectonics. We hope we have not unduly misrepresented their views in our own tectonic speculations. S. Mykkeltveit and T. Kvaerna of NORSAR contributed substantially to the initial computer processing of data and construction of record sections.

REFERENCES

Berthelsen, A., A tectonic model for the crustal evolution of the Baltic Shield, in Shaer, J. P., and Rodgers, J., (eds.): *The Anatomy of Mountain Belts*, Princeton University Press (in press), 1985.

Bott, M. H. P., *The Interior of the Earth, its Structure, Constitution and Evolution*, 2nd Ed., 403 p., Elsevier, New York, 1982.

Bungum, H., S. E. Pirhonen, and E. S. Husebye, Crustal thicknesses in Scandanavia, *Geophs. J. R. Astron. Soc.*, 63, 759-774, 1980.

Chroston, P. N., and C. J. Evans, Seismic velocities of granulites from the Seiland petrographic province, north Norway: Implications of Scandinavian lower continental crust, *J. Geophys.*, 52, 14-21, 1983.

Deichmann, N., and J. Ansorge, Evidence for lamination in the lower continental crust beneath the Black forest (Southwestern Germany), *J. Geophys.*, 52, 109-118, 1983.

Hirn, A., L. Steinmetz, R. Kind, and K. Fuchs, Long range profiles in Western Europe: II. Fine structure of the lower lithosphere in France (Southern Bretagne), *Z. Geophysik*, 39, 363-384, 1973.

Kind, R., The reflectivity method for a buried source, *J. Geophys.*, 44, 603-612, 1978.

Kröner, A., Precambrian plate tectonics, in *Precambrian Plate Tectonics*, edited by A. Kröner, 57-90, Elsevier, Amsterdam, 1981.

Lambert, R. St. J., Archean thermal regimes crustal and upper mantle temperatures, and a progressive evolutionary model for the Earth, in *The Early History of the Earth*, edited by B. F. Windley, 3363-373, Wiley, London, 1976.

Lund, C. E., Crustal structure along the Blue Road profile in northern Scandanavia, *Geologiska Föreningens i Stockholm Förhandlingar*, 101, 191-294, 1979a.

Lund, C. E., Fennoscandian Long-Range Profile, *Geologiska Förenigens i Stockholm Förhandlingar*, 101, 342, 1979b.

Lund, C. E., The fine structure of the lower lithosphere underneath the Blue Road profile in northern Scandanavia, *Tectonophysics*, 56, 111-122, 1979c.

Lund, C. E., Fennoscandian Long-Range Project 1979 (FENNOLORA), in *Proc. 17th Assembly of the European Seismol. Commission, Budapest, 1980*, edited by E. Bisztricsany and G. Szeidovitz, 511-515, Elsevier, Amsterdam, 1983.

Luosto, U., S. M. Zverev, I. P Kosminskaya, and H. Korhonen, Observations of FENNOLORA shots on additional lines in Finnish Lapland, in *Proc. 17th Assembly of the European Seismol. Commission, Budapest, 1980*, edited by E. Bisztricsany and G. Szeidovitz, 517-521, Elsevier, Amsterdam, 1983.

Meissner, R., Fennoscandia - A short outline of its geodynamical development, *GeoJournal*, 3, No. 3, 227-233, 1979.

Prodehl, C., and W. Kaminski, Crustal structure under the FENNOLORA profile, in *Proceedings of the First Workshop on the European Geotraverse (EGT) - the northern segment*, edited by D. A. Galson and St. Mueller, 43048, European Science Foundation, Strasbourg, 1984.

Smithson, S. B., Modeling continental crust: structural and chemical constraints, *Geophys. Res. Lett.*, 5, 749-752, 1978.

Witschard, F., The geological and tectonic evolution of the Precambrian of northern Sweden—A case for basement reactivation?, *Precambrian Res.* 23, 273-315, 1984.

EVIDENCE FOR AN INACTIVE RIFT IN THE PRECAMBRIAN FROM A WIDE-ANGLE REFLECTION SURVEY ACROSS THE OTTAWA-BONNECHERE GRABEN

Robert Mereu, Dapeng Wang[1], and Oliver Kuhn[2]

Department of Geophysics, University of Western Ontario
London, Ontario, Canada N6A 5B7

Abstract. During the summer of 1982 the Canadian Consortium for Crustal Reconnaissance using Seismic Techniques (COCRUST) conducted a major long range seismic refraction and wide angle reflection experiment across the Grenville province of the Canadian Shield. One of the main aims of the experiment was to investigate the structure and origin of the Ottawa-Bonnechere graben from a set of in-line and fan-type profiles recorded both along the length of the Graben and in directions perpendicular to it. Wide-angle reflection observations from the Central Gneiss Belt to the north of the graben revealed a very simple one layered crust with a sharp Moho. Large amplitude wide-angle PmP reflected waves were clearly identified. This was in sharp contrast to the poor or non-existent PmP signals associated with the profiles obtained along the Graben. The data supports the theory that the Graben was part of the St. Lawrence rift system and that the Moho was disrupted with the entry of upper mantle material into the crust. The results of the first arrival direct wave observations showed that the near-surface velocities of rocks of the Grenville province varied from 5.8 to 6.4 km/s. There was no evidence for any intermediate discontinuity, however, regional differences in velocity gradients within the upper crust were pronounced and had a major influence on the appearance of the record sections. The complexity of the energy in the coda supported geological observations that the rocks of the Central Gneiss belt are more homogeneous than those associated with the Central Metasedimentary belt south of the Graben. Amplitude and direct wave arrival time fluctuations all indicated that the seismic velocity function is a fractal quantity. Seismic waves on travelling through the crust smooth out the small scale variations to create regions of high and low velocity each separated from the other by both lateral and vertical velocity gradients.

Introduction

During the summer of 1982, The Canadian Consortium for Crustal Reconnaissance using Seismic Techniques (COCRUST) conducted two major long range seismic refraction and wide-angle reflection experiments across the Grenville and Superior provinces of the Canadian Shield in Ontario and Quebec. Major tectonic features of the region under study were the Central Metasedimentary belt, the Ottawa-Bonnechere graben, the Central Gneiss belt, the Grenville Front and the Abitibi Greenstone belt. Four seismic lines AO, BC, CD, and DE each approximately 300 km in length as well as a number of fan profiles were obtained from a total of 16 shots as shown in the map in Figure 1. Geological and geophysical summaries of the tectonic features shown on this map were given by Wynne-Edwards (1972), Forsyth (1981) and Goodwin and Ridler (1970). Detailed descriptions, complete record sections from all 16 shot points and preliminary interpretations of all the results were given by Mereu (1983), Mereu et al (1985) and Crossley (oral comm., 1985.). In this paper the major results of the wide-angle reflection observations along the Ottawa-Bonnechere graben are reviewed and compared with the results obtained from the more geologically homogeneous Gneiss terrane to the north of the graben. See lines AO and CD respectively.

The Ottawa-Bonnechere graben is a northwesterly trending structurally depressed zone running roughly parallel to the Ottawa river. It is approximately 60 km wide and is bounded on the north-east by the Laurentian highlands and on the south-west by the Madawaska Highlands both of which have a maximum relief of about 400 meters. The zone is marked by numerous block faults and dykes. The geology of the graben was given by Kay (1942). Kumarapelli and Saul (1966) have postulated that the graben and

---

[1] Now at the Department of Geology and Geophysics, University of Wyoming.

[2] Now at Geo-X, Calgary, Alberta.

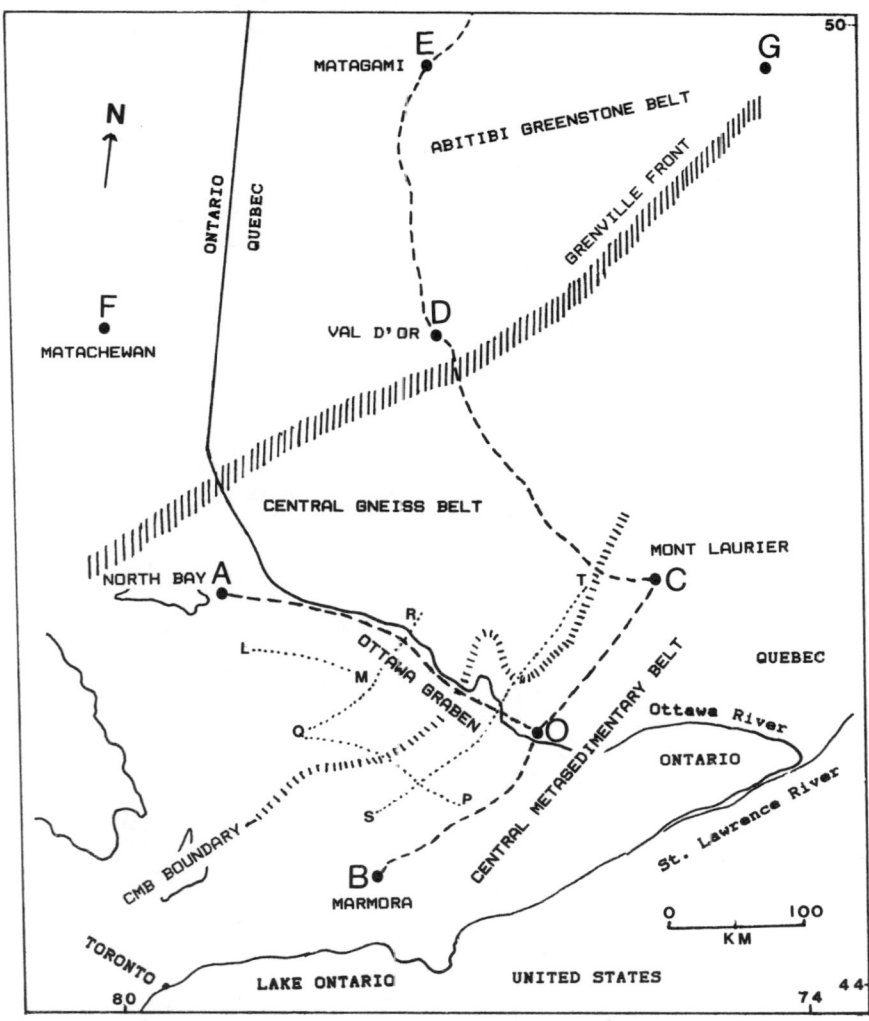

Fig. 1. Shot points and seismic lines for the 1982 COCRUST seismic experiment. Shot points (A,B,O,C, and D) and lines (AO,BO,OC, and CD) belong to the Ottawa-Bonnechere graben - Grenville Front experiment (see Mereu et al. 1985). Shot points (D,E,F, and G) and seismic line DE belong to the Abitibi Greenstone belt experiment. Dotted lines QR and QP indicate the approximate positions where the PmP wide angle reflected wave sample the Moho from the fan shots at A and B respectively. Dotted lines ST and LM indicate the approximate offset positions where the Pn waves sample the Moho from the fan shots at A and B respectively.

the St. Lawrence River valley form part of an ancient rift system that runs from Newfoundland to Lake Superior. Burke and Dewey (1973) have argued that the graben could be the failed arm of a triple junction system. Geological evidence, based mainly on the ages of dykes (Doig and Barton 1968), indicates that the graben is a feature which developed at a much later time after the Grenville orogeny - probably as late as Lower Cretaceous time.

The Central Gneiss belt lies in the region of the Grenville province just to the south of the Grenville Front. The rocks of this belt have been observed by geologists to be much more homogeneous in character than most other rocks of the Canadian shield such as those of the Central Metasedimentary belt.

Experimental Results

The direct wave observations

An interpretation of the structure of the upper crust was made from the analysis of the onset times and amplitudes of the direct waves. The most significant feature of these

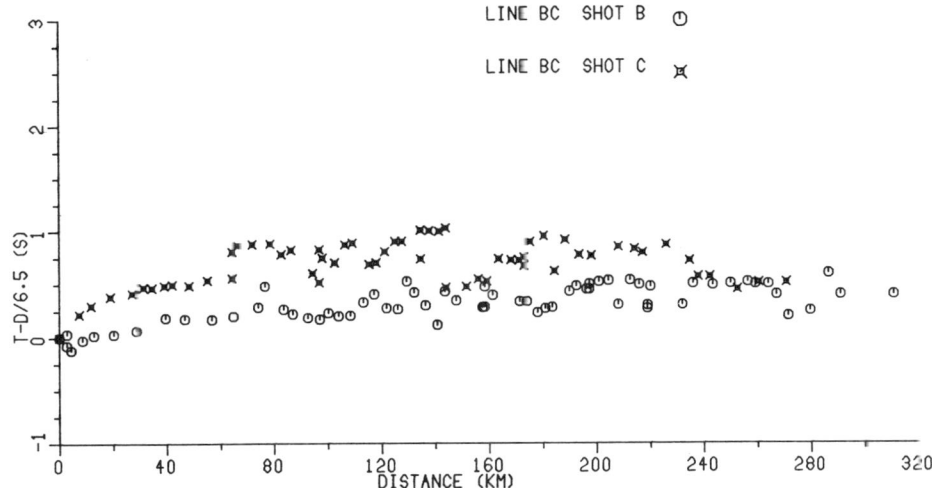

Fig. 2. Direct wave travel-time observations for line BC. Note how the arrival times for shot C are delayed compared to those for shot B.

observations was the fairly large variations which occurred in the near surface velocities as determined from examinations of traces obtained a few kilometres from the shot points. Relatively low values of 5.8 km/s were recorded near A and C and much larger than average values of 6.4 km/s were recorded near B and O. There was little evidence in any of the profiles of the whole experiment for any regional intermediate discontinuity such as the Conrad. Instead the characteristics of the direct wave record sections indicated that they were mainly dominated by the effects of strong lateral and vertical velocity gradients within the crust. Figure 2 shows as an example of the direct wave travel-time observations for line BC which runs roughly perpendicular to the Graben. The scatter in the observations was typical of what was seen for all the lines. The fact that almost all the observations from shot point C are delayed by about 0.5 seconds compared to shot B at the other end of the line is interesting and is a direct result of the fact that at C, there is a region of relatively low near surface velocities (5.8 km/s) compared to the region near B where the values were closer to 6.4 km/s. For short distances, the rays spend much more time in the near surface low velocity regions. The effects of the near surface structure are minimized for larger distances as the more deeply penetrating rays approach the surface at a steeper angle.

Although there was a rough correlation of the near surface velocity lows with nearby gravity lows, the direct correlation of the velocity values with detailed complex geological features at the surface was not clear. This is not surprising as an examination of the very small scale variations in the composition of rocks and minerals would indicate that the seismic velocity would also vary significantly from point to point in a medium. In fact, the seismic velocity function could easily be described as a fractal function, or a function with random variations and no derivative. The property of fractals and numerous examples were given by Mandelbrot (1977). Seismic waves on travelling through the earth will act as a low-pass filter and smooth out the small scale local variations in a manner analogous to that used by map-makers who smooth out fractal coastlines when drawing a map. The smoothing effect does not reduce the fluctuations to zero but rather broadens their effects such that there will be regions of higher than average velocity as well as regions of lower than average velocity. This can easily be seen when a set of random numbers is filtered with a low-pass filter. The smoothing effect is obviously dependent on the frequency of the signals. The instantaneous seismic velocity of a rock will have a deterministic component equal to the average value plus an indeterministic component which will in general vary from point to point in the medium. In the three dimensional case the waves act as three dimensional low pass spacial filters and the correlation distance of the inceterministic component will equal that of the filter. Furthermore the velocity fluctuations will give rise to fluctuations in travel-path directions. These in turn create paths that are always longer than the paths for the equivalent homogeneous earth. The path lengthening effects create fluctuated delays in the arrival times of the direct waves similar to those shown in Figure 2. In a numerical study of this problem, Mereu and Ojo (1981) and Ojo (1981) confirmed that lateral and vertical heterogeneities in rocks not only cause travel-time fluctuations typical of those seen in crustal refraction experiments, but also that a smooth continuous travel-time curve can easily be broken into one with short line

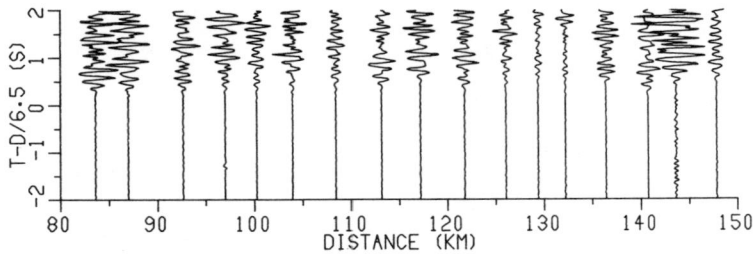

Fig. 3. First arrival amplitude fluctuations along line BC.

segments. The numerical experiments were repeated such that different sets of two-dimensionally filtered velocity variations were superimposed on a crust with a smooth vertical velocity gradient. It was found that there was a direct relationship between the correlation distance of the smoothing filter and the depth to the apparent discontinuities that were computed using conventional simple inversion techniques. The experiments also showed that as a result of the effects of the gradients, the energy would be focussed and defocussed such that significant amplitude variations would take place over short distances along the profile. A selection of traces from line BC is presented in Figure 3 to illustrate these amplitude fluctuations.

"Wide-angle reflections from the Moho"

Two in-line record sections (lines CD and OA) and one fan section for a shot from shot-point A to line BC were selected for illustration purposes for this paper. Two methods were used to plot the in-line sections. In the first manner (see Figure 4a and 4b), the displays are conventional in that the traces were band-pass filtered (4 to 12 Hz), normalized and then plotted at their respective station positions with a reducing velocity of 6.5 km/s. In the second method (see Figure 5a and 5b), the data from all the shots to a line were combined into a composite plot. A normal moveout correction was applied to all the traces and then the traces were plotted at the approximate locations where the PmP waves would sample the Moho (ie half way between the the shot and the recording stations). These traces were also plotted with their time axis downwards such that the resultant record sections were of the same form as used in near vertical reflection work. The NMO corrections were computed from a simple earth crustal model with a Moho depth of 37 km and velocity increasing from 6.2 km/s at the surface to 6.8 km/s just above the Moho. This second method of displaying the traces enables one to more easily map Moho topography from wide-angle reflection information. In the fan shot experiment see

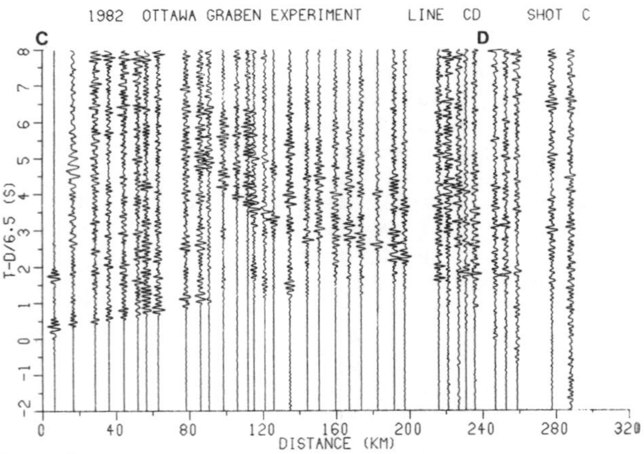

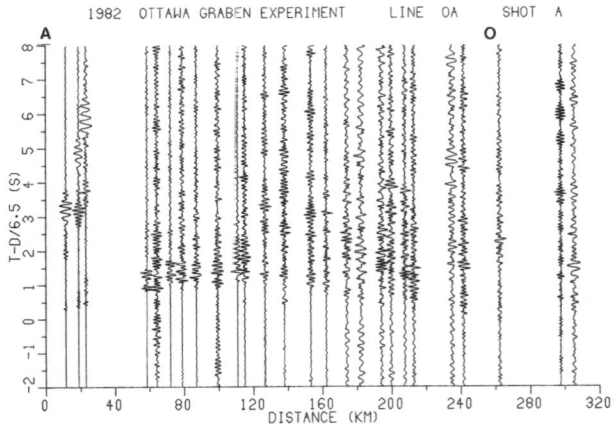

Fig. 4a. Record section for line CD, shot C. (Band-pass filtered 4 to 12 Hz)

Fig. 4b. Record section for line OA, shot A. (Band-pass filtered 4 to 12 Hz)

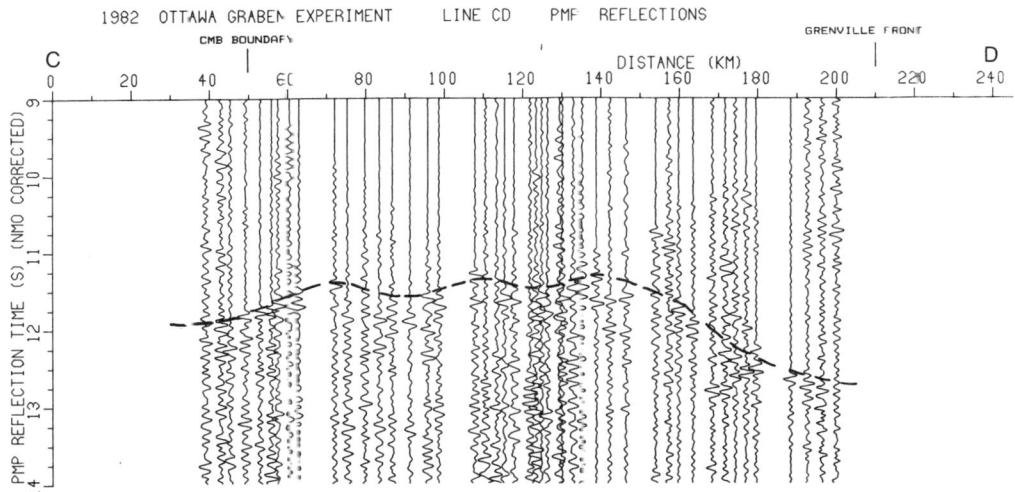

Fig. 5a. Composite record section for line CD, shots C and D. This section was corrected for the PmP normal moveout. Traces for both shots are plotted at the points along line CD where the PmP ray reflects off the Moho.

Figure 5b recorded along line BC from shot point A, the PmP waves sampled the Moho along line QR (see Figure 1).

One of the most significant observations one can make on the record sections is the great differences that exist in in the character of the PmP amplitudes. The pattern of traces for line CD (Figure 4a) is a good example of a section for a one layered crust overlying a very sharp Moho. The PmP waves are very pronounced and the weak amplitude Pn waves are easily visible. The section Figure 5a shows that topography is present on the Moho under line CD with a significant depression of at least 5 km near the Grenville Front. This section does however give us a picture of a rather continuous Moho. The section from line OA (Figure 4b) makes a very sharp contrast to the CD section. Very few PmP or Pn arrivals can be identified and where present are very much weaker and more scattered. It must be concluded that either the Moho does not exist or if it does ,the crust-mantle transition zone must be very thick. The fan section given in Figure 5b shows that a rather sharp fault-like structure is present. These results are consistent with the hypothesis that the graben originated as a rift-like structure in which the Moho was ripped open and faulted by the

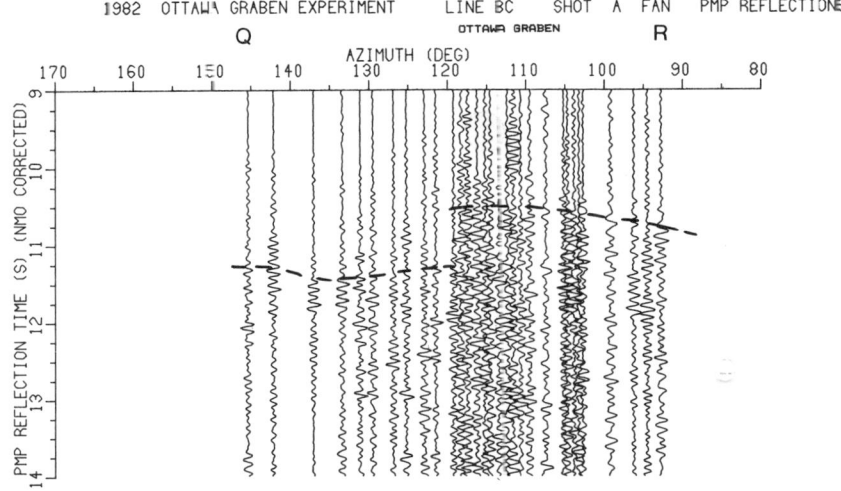

Fig. 5b. Record section for fan shot from A, recorded on line BC. This section was corrected for the PmP normal moveout. The traces are plotted at the reflection points along line QR at various azimuths from shot point A.

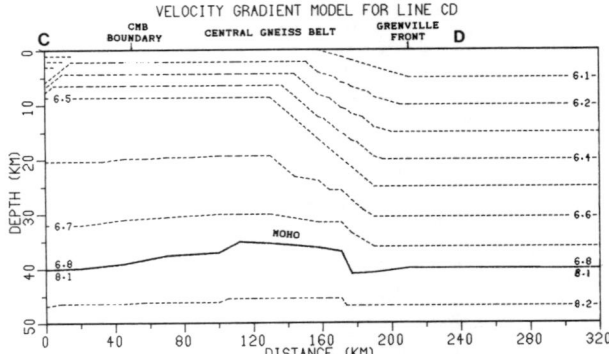

Fig. 6a. Velocity gradient Model for line CD. This line lies across the Central Gneiss belt. Dotted lines indicate 0.1 km/s velocity contour lines.

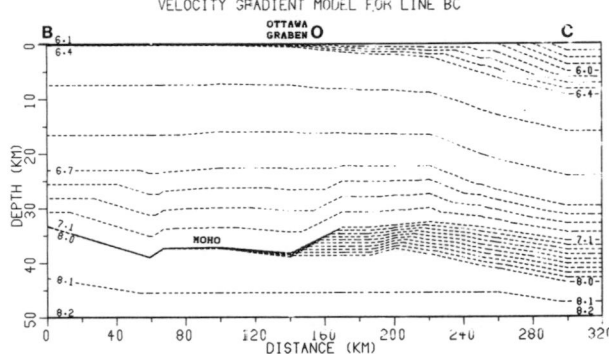

Fig. 6c. Velocity gradient Model for line BC. This line lies perpendicular to the Ottawa-Bonnechere graben. Dotted lines indicate 0.1 km/s contour lines.

entry of mantle material into the lower crust. Such action could easily ruin a good reflecting surface and the resulting heterogeneity disrupt the weak Pn signals so as to make them unobservable.

Interpretations

As a result of the conclusions drawn from the numerical experiments described above, the interpretations of the crustal profiles were made under the assumption that only lateral and vertical velocity gradients were present in the crust. It is felt that any attempt to fit line segments to the observations with the aim of obtaining depths to subsurface intermediate discontinuities would not lead to the most representative solution to the problem. The final models for lines AO, BC, and CD which were obtained using an iterative iteractive procedure, are given in Figure 6. These models are not precisely unique, however, any significant departure from the illustrated gradients will seriously affect the arrival-times and amplitudes. The velocity gradient model for line CD is of particular interest as it shows quite clearly that the Grenville Frontal zone is deep-seated and extends right down to a depressed Moho. Figure 7 shows synthetic seismograms for line CD shot C and line OA shot A. These should be compared with the observed sections given in Figure 4. A complete set of synthetics for the other lines and shot points and further details on the method of computation were given by Mereu et al (1985). There is very good agreement between the positions of the main arrivals on the synthetics and the observed sections. Small scale heterogeneities were not incorporated into the synthetics and this accounts for the fact that the observed signals were much more complex than the computed ones.

A comparison of the various profiles of the field experiments also showed that the coda energy following the main onsets was much more pronounced across the Central Metasedimentary belt than that across the Gneiss belt. This

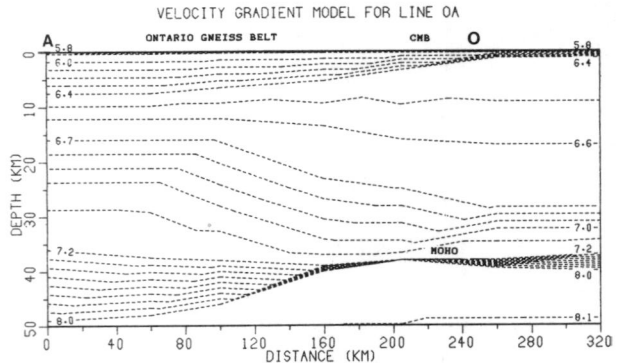

Fig. 6b. Velocity gradient Model for line OA. This line lies along the Ottawa-Bonnechere graben. Dotted lines indicate 0.1 km/s velocity contour lines.

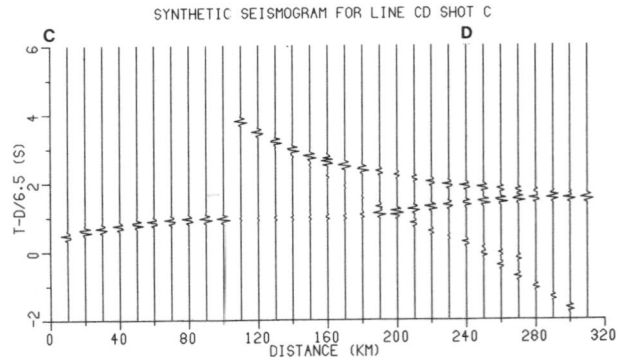

Fig. 7a. Synthetic record section for the line CD velocity model with shot point at C. Compare with Figure 4a.

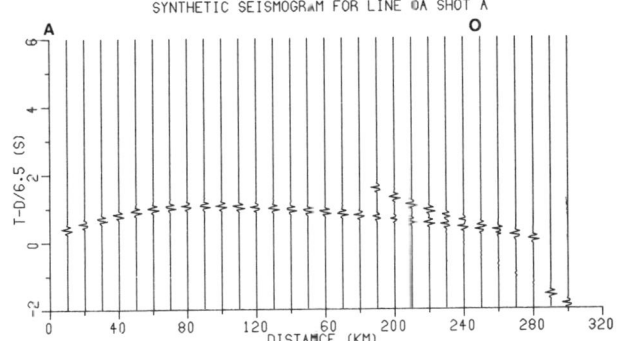

Fig. 7b. Synthetic record section for line OA velocity model with shot point at A. Compare with Figure 4b.

observation supports the geological observation that the Gneiss belt rocks are more homogeneous than other shield rocks.

Conclusions

Near-surface seismic velocities varied from 5.8 to 6.4 km/s. There was no evidence for any intermediate discontinuity, however, regional differences in velocity gradients within the upper crust were pronounced and had a major influence on the appearance of the record sections. Some of these effects probably had their origin as a result of the fact that the seismic velocity function is a fractal function.

The Moho along the Ottawa-Bonnechere graben is poorly defined. The seismic evidence indicates that upper mantle material may have disrupted the Moho. This also is in keeping with suggestions that the graben may be part of an ancient rift-like structure. The sharp step-like structure on the Moho on a line perpendicular to the graben indicates that an inactive fault is present beneath various portions of the graben.

Observations of the complexity of the seismic codas indicates that the Central Gneiss terrane is much more homogeneous than that of the Central Metasedimentary belt. This is in agreement with observations of surface geology, and supports the hypothesis that during the Grenville orogeny, Central Metasedimentary material was pushed northwestward against the more rigid Gneiss terrane.

The Grenville Front at depth is marked by a change in the character of the velocity gradient within the crust as well as a significant thickening of the crust by over 5 km along the boundary zone. This observation adds credence to the hypothesis that the Grenville front may have formed from a continent-continent collision mechanism. The geological evidence suggests the suture to be farther southeast within the Grenville Province. If that is the case, then the Grenville Front would mark the metamorphic boundary between the softer Grenville rocks and the cooler more rigid Superior rocks. The counterpart of deformations and shear zones which geologists have mapped at the surface could easily be a marked depression on the Moho.

Acknowledgments. The authors would like to thank the following members of the COCRUST group and field participants for their cooperation and assistance in making this experiment a success: A.G.Green, D.Forsyth, P.Morel, G. Buchbinder, M.Berry, F.Anderson, D.Hoy, C.Michaud, L.Parenteau (Earth Physics Branch), C.Brooks, C.Gagner (University of Montreal), E.Schwartz, E.Poterlot (Ecole Polytechnique), D.Crossley, C.Parker, C.Tsingas, W.Wang (McGill University), R.Clowes, R.Meldrum (University of British Columbia), R.duBerger, J.Villineuve (Universite du Quebec a Chicoutimi), G.West (University of Toronto), J.Baerg, M.Jeffered, S.Pamidi, J. Brunet, B.Price (University of Western Ontario).

We are also indebted to the following companies who gave us assistance with respect to locating the shot points: A.J.N.Drilling of Callendar, Ontario, Ambro Material and Construction of Brampton,Ontario, Lamarsh-McGuinty Inc Quebec, Poisson Granite Transport of Mont Laurier Quebec, Dennison Mines of Val d'Or and the Ontario Ministry of Natural Resources at North Bay. This research was funded by an EMR contract, an EMR research grant, and an NSERC grant (#A1793).

References

Burke, K. and J.F.Dewey, Plume-generated triple junctions: Key indicators in applying plate tectonics to old rocks, Journal of Geology, 81, 406-433, 1973.

Doig, R.and J.M.Barton ,Ages of carbonatites and other alkaline rocks in Quebec. Canadian Journal of Earth Sciences, 5, 1401-1407, 1968.

Forsyth, D., Characteristics of the Western Quebec seismic zone, Canadian Journal of Earth Sciences, 18, 103-119, 1981.

Goodwin, A.M. and R.H.Ridler, The Abitibi orogenic belt. In: Precambrian basins and geosynclines of the Canadian Shield. A.J. Baer, (ed),Geological Survey of Canada, Paper, 70-40, 1-30, 1970.

Kay, G.M., Ottawa-Bonnechere graben and Lake Ontario homocline. Bulletin of the Geological Society of America, 53, 585-646, 1942.

Kumarapelli, P.S. and V.A.Saull, The St.Lawrence valley system. Canadian Journal of Earth Sciences, 3, 639-658, 1966.

Mandelbrot, B.B, Fractals,form,chance and dimension, W.H.Freeman and Co. San Francisco, 1977.

Mereu, R.F. and S.B.Ojo, The scattering of seismic waves through a crust and upper mantle with random lateral and vertical inhomogeneities. Physics of the Earth and Planetary Interiors, 26, 233-240, 1981.

Mereu, R.F., The 1982 COCRUST seismic refraction experiment: Acquisition and interpretation of crustal refraction profiles across the Ottawa graben, the Central Metasedimentary Belt and the Grenville Front, Ontario and Quebec. Earth Physics Branch Open File Report Number 83-28, 1983.

Mereu, R.F., D.Wang, O.Kuhn, D.A.Forsyth, A.G.Green, P.Morel, G.Buchbinder, D.Crossley, E.Schwartz, R.duBerger, C.Brooks, and R.Clowes, The 1982 COCRUST seismic experiment across the Ottawa-Bonnechere Graben and Grenville Front in Ontario and Quebec. Geophysical Journal of the Royal Astronomical Society, In press. 1985.

Ojo, S.B., An investigation of P-wave scattering in the crust and upper mantle using travel-time fluctuations and array signal coherence, Ph.D. Thesis, University of Western Ontario, 1981.

Wynne-Edwards, H.R., The Grenville Province. In Variations tectonic styles in Canada. Edited by R.A.Price and R.J.Douglas. Geological Association of Canada, Special Paper 11, 263-334, 1972.

A POSSIBLE EXPOSED CONRAD DISCONTINUITY IN THE KAPUSKASING UPLIFT, ONTARIO

John A. Percival

Precambrian Geology Division, Geological Survey of Canada, Ottawa, Ontario K1A 0E4

Abstract. Mid-crustal seismic velocity discontinuities have been recorded at several localities in the Archean Superior Province of the Canadian Shield. In the central part of the province, the Kapuskasing uplift exposes a relatively complete oblique section through the upper two-thirds of the crust. At a structural depth within the uplift of approximately 20 km, based on geobarometry of metamorphic assemblages, a complex transition occurs between an upper region consisting dominantly of tonalitic gneiss in the amphibolite facies (average density ~2.70 g·cm^{-3}) and a lower level of interlayered tonalite, diorite, anorthosite, paragneiss and mafic gneiss in the upper amphibolite and granulite facies (average density ~2.82 g·cm^{-3}). The transition zone, thought to be analogous to a mid-crustal velocity discontinuity, separates upper crust, in which felsic plutonic compositions predominate, from lower crust, where intrusive rocks include mafic tonalite, diorite and anorthosite.

Introduction

The appplication of high-resolution seismic reflection techniques to crustal-scale problems has raised many questions concerning the deep crust. In this note some of the characteristics of the lower continental crust as deduced by geophysical techniques are compared with those of rocks exposed at the base of a cross-section through the upper two-thirds of the crust in the Archean Superior Province of the Canadian Shield (Percival and Card, 1983). At least 70 per cent of the present volume of continental crust existed by the end of the Archean (Taylor and McLennan, 1981; McCulloch and Wasserburg, 1978), emphasizing the need for an understanding of Archean crustal genesis in programs of crustal reflection profiling.

Limited seismic refraction and reflection studies have been carried out in the Superior Province, mainly in the west. In that area, the crust beneath the Wabigoon greenstone-granite terrane consists of an upper layer ~19 km thick of P-wave velocity 6.2 km·sec^{-1}, overlying a middle layer 3-5 km thick of 6.9 km·sec^{-1} and a lower layer ~16 km thick of 7.2 km·sec^{-1}, bounded below by the Moho at 38 km (Green et al., 1979; Hall and Brisbin, 1982). A mid-crustal seismic velocity discontinuity of ~0.3 km·sec^{-1} at about 20 km is also present in the Superior Province farther east (Berry and Fuchs, 1973). Results of a refraction experiment in the Kapuskasing area (Northey et al., 1985) are not yet published.

Explanations offered for mid-crustal velocity discontinuities at many localities in the continental crust include: (1) a compositional change from felsic to mafic with increasing depth, (2) a body of mafic or intermediate rock at mid-crustal depth (Smithson, 1978; Brown et al., 1980), (3) layered sill complexes (Clowes and Kanasewich, 1970), or (4) amphibolite overlying granulite (metamorphic boundary) (Mueller, 1977). Well-defined Conrad discontinuities do not necessarily produce seismic reflections (Kay and Kay, 1981; Oliver et al., 1983), suggesting a gradational or interlayered interface rather than a sharp contact.

The Kapuskasing uplift in the central Superior Province provides a unique view of the crust beneath the Abitibi-Wawa greenstone-granite terrane. Metamorphic pressure varies over a lateral distance of 100 km across the uplift, from a low of 2-3 kbar in metavolcanic rocks near Wawa (Studemeister, 1983) to a high of >8 kbar in the Kapuskasing zone proper (Percival, 1983), implying structural relief on the order of 20 km. Within the oblique section is a zone across which changes in structural style, metamorphic grade, bulk chemical composition and density provide a possible analog for the mid-crustal seismic velocity discontinuity noted elsewhere in the region. A review of the geological characteristics of the Kapuskasing uplift is necessary in order to develop the analogy.

Geology

The Kapuskasing uplift can be divided into 4 zones based on distinctive regional lithological, metamorphic and structural characteristics. From west to east (Fig. 1) these are: (1) the Michipicoten belt, (2) the Wawa gneiss terrane, (3) the Kapuskasing zone, and (4) the Ivanhoe Lake cataclastic zone. To the east is the Abitibi greenstone belt, constituting a fifth zone. The Michipicoten belt consists mainly of metavolcanic rocks of ultramafic, mafic and felsic composition, with intercalated greywacke, conglomerate, iron formation and chert (Goodwin, 1962; Sage, 1983), all metamorphosed from subgreenschist to amphibolite

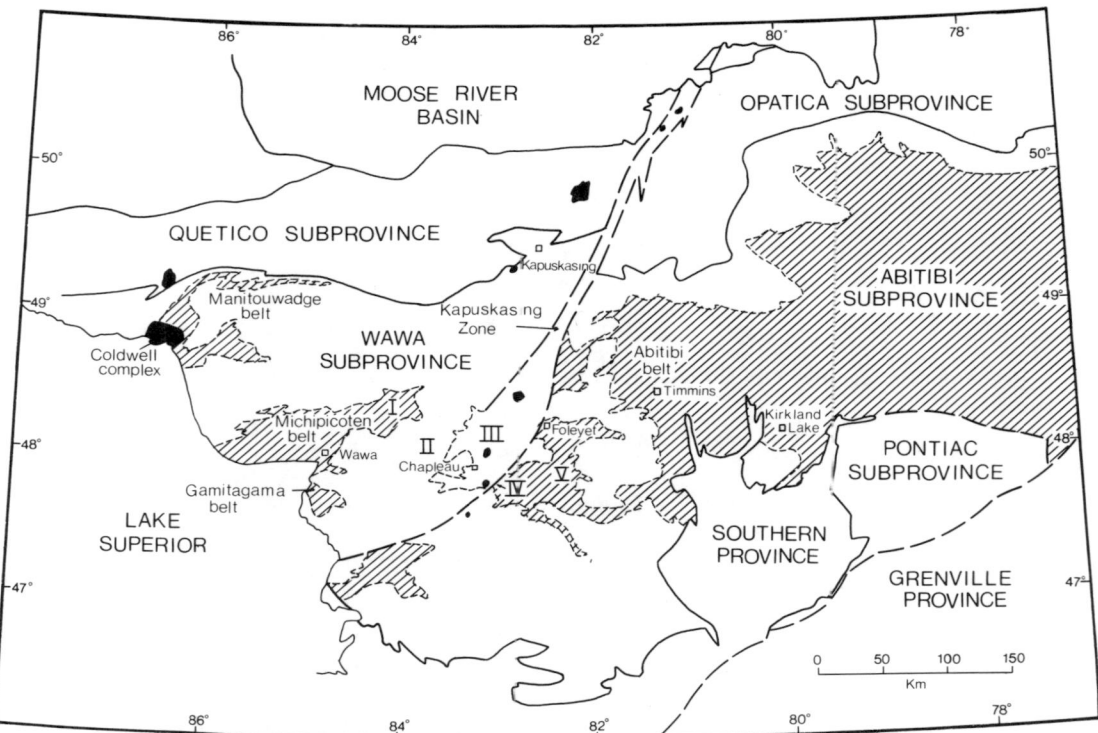

Fig. 1. Location of major lithologic and geographic features in the central Superior Province. I = Michipicoten belt; II = Wawa gneiss terrane; III = Kapuskasing zone; IV = Ivanhoe Lake cataclastic zone; V = Abitibi belt. Zones I to IV make up the Kapuskasing uplift. Diagonal ruling indicates greenstone belts. Dashed line between zones II and III in the Chapleau area represents the compositional, structural and metamorphic discontinuity thought to correspond to the mid-crustal discontinuity in seismic velocity (see detail in Fig. 3).

facies (Studemeister, 1983; Fraser et al., 1978). Large-scale dome and basin structures (Goodwin, 1962) as well as downward-facing strata (Attoh, 1980) have been recognized. Intrusive rocks include synvolcanic peridotite to granodiorite, younger granodiorite batholiths and still younger granite and syenite plutons (Card, 1982).

Supracrustal rocks of the Michipicoten belt are intruded to the southeast by a variety of igneous rocks in the Wawa gneiss terrane. This consists of variably xenolithic hornblende-biotite tonalitic gneiss cut by granodiorite and later granite and pegmatite. Several domal structures on the 20-25 km scale are defined by trends of foliation and fold axial surfaces. One dome (Robson Lake dome) adjacent to the Kapuskasing zone has a core of mafic gneiss, paragneiss and tonalitic gneiss, similar to the lithologic assemblage of the Kapuskasing zone to the east.

Units of paragneiss, mafic gneiss, tonalitic, dioritic and anorthositic rocks form northeast-striking, northwest-dipping belts in the Kapuskasing zone. Metamorphic grade is uniformly high, in the upper amphibolite and granulite facies, with PT conditions in the range 700-800°C, 6-8 kbar. Paragneiss, of probable supracrustal origin, contains the assemblage biotite-plagioclase-quartz$\pm$garnet$\pm$orthopyroxene$\pm$graphite and may have had a greywacke precursor. Migmatitic mafic gneiss contains the assemblage garnet-clinopyroxene-hornblende-plagioclase-quartz-ilmenite$\pm$orthopyroxene and is generally associated with paragneiss, although its origin is not known. Units of tonalite to diorite consist of hornblende-plagioclase$\pm$quartz$\pm$clinopyroxene$\pm$orthopyroxene$\pm$garnet. Igneous layering and minerals are preserved locally. The Shawmere anorthosite complex has a layered outer zone and homogeneous gabbroic anorthosite inner zone. Igneous textures and minerals are commonly preserved in the interior.

High-grade rocks of the Kapuskasing zone are separated from low-grade rocks and granite of the Abitibi belt to the east by the Ivanhoe Lake cataclastic zone, thought to be the surface expression of a northwest-dipping thrust fault (Percival and Card, 1983; Cook, 1985). Rocks of the Abitibi belt are mainly mafic metavolcanics, with minor felsic and ultramafic metavolcanics and metasediments, metamorphosed from subgreenschist to amphibolite facies (Jolly, 1978). Lithologic, structural and geochronological (Turek et al., 1982; Nunes and Pyke, 1980) similarities between the Michipicoten and Abitibi belts (Percival and Card, 1983, 1985) suggest that the two are parts of a once-continuous belt, now interrupted by the Kapuskasing uplift.

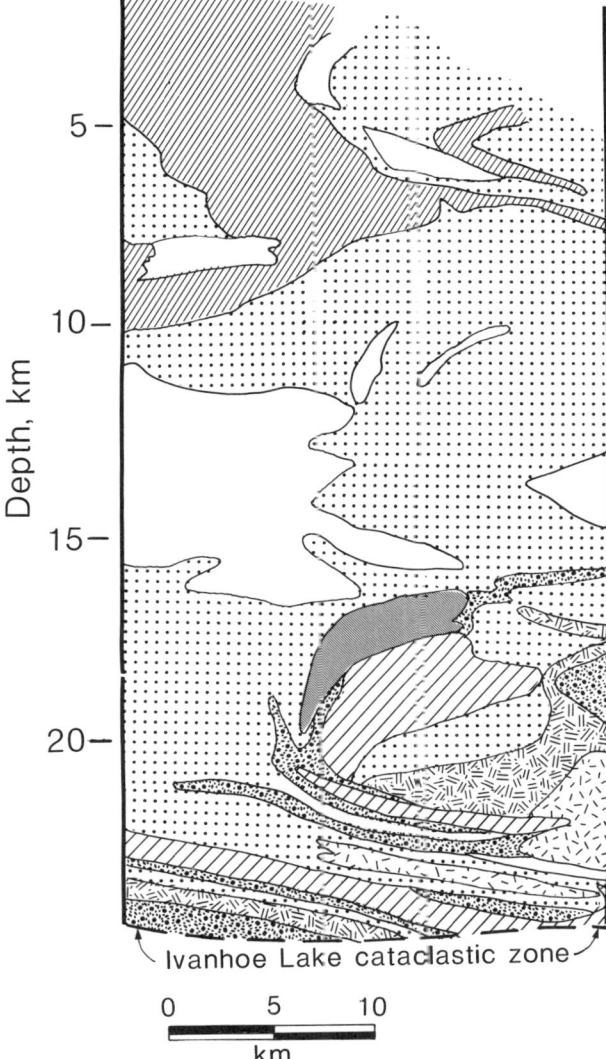

Fig. 2. Cross-section of Archean crust in the Wawa-Chapleau area, based on down-plunge projection technique (Percival and Card, 1985). Note 2:1 vertical exaggeration. Symbols as in Fig. 3.

Based on regional gravity, structural, metamorphic and geochronological data, Percival and Card (1983) suggested that the Michipicoten belt, Wawa gneiss terrane and Kapuskasing zone represent an oblique section through a slab of upper and middle crust that was thrust southeastward along the Ivanhoe Lake cataclastic zone onto the Abitibi belt, possibly during the Proterozoic. Although similar structures have been postulated for other crustal cross-sections at plate margins (Fountain and Salisbury, 1981), the Kapuskasing uplift is particularly informative in that large-scale lithological units can be correlated across this intracratonic structure. The reconstructed crustal section of Figure 2 shows upper crust, to a depth of 20 km, to consist mainly of tonalitic gneiss, with thin (<5 km) keels of supracrustal rock. Beneath the upper crust, extending to unknown depth (but possibly to the Moho), is the interlayered high-grade gneiss sequence of the Kapuskasing zone. Average densities for rocks of the three terranes are listed in Table 1.

A Mid-crustal Discontinuity

A difference in average rock density is evident between the relatively felsic Wawa gneiss terrane, with an average density of 2.70 g·cm^{-3} and the more mafic rocks of the Kapuskasing zone (average density ~2.82 g·cm^{-3}). The Kapuskasing zone has an associated 50 mGal positive gravity anomaly. Because seismic velocity is a function of density (Birch, 1961; Christensen and Fountain, 1975), it is likely that a seismic velocity discontinuity exists between the upper and lower parts of the section. As a first approximation, the magnitude of this discontinuity may be calculated by using the relationship $V_p = 0.31 + 2.27$ (Christensen and Fountain, 1975) and amounts to a value of 0.27 km·sec^{-1}. While this relationship is strictly valid only for granulites, more detailed calculations taking account of rock composition and mineralogy are outside the scope of this note and would at best only approximate laboratory-measured velocities or those derived from in situ seismic surveys. However, the magnitude of the approximation is remarkably similar to the measured mid-crustal discontinuity of 0.3 km·sec^{-1} reported for the Superior and Grenville provinces by Berry and Fuchs (1973). A velocity of 6.2–6.5 km·sec^{-1} was calculated for part of the Kapuskasing zone by Cook (1985) on the basis of a short reflection survey over tonalitic and anorthositic rocks. Preliminary results of a regional refraction survey indicate high seismic velocities in the upper crust over the Kapuskasing zone relative to those in the surrounding region (Northey et al., 1985; pers. comm., 1985).

Geobarometry of assemblages of garnet-clinopyroxene-plagioclase-quartz (Newton and Perkins, 1982) indicates pressure on the order of 6 kbar

TABLE 1: Summary of density determinations (g·cm^{-3})

	Average	Range	Proportion
Wawa Gneiss Terrane			
[1]Tonalitic gneiss (n=7)	2.70	2.66–2.75	100%
Aggregate density	2.70		
Kapuskasing Zone			
Anorthosite (7)	2.82	2.72–2.92	20%
Paragneiss (9)	2.77	2.70–2.88	20%
Mafic gneiss (16)	3.10	3.01–3.31	15%
Tonalite (7)	2.71	2.65–2.79	20%
Diorite (5)	2.80	2.76–2.83	25%
Aggregate density	2.82		

[1]Most outcrops of tonalitic gneiss contain 0–5% amphibolite inclusions and 0–15% granite, aplite and pegmatite dykes. The effect on density of these heterogeneities has not been taken into account.

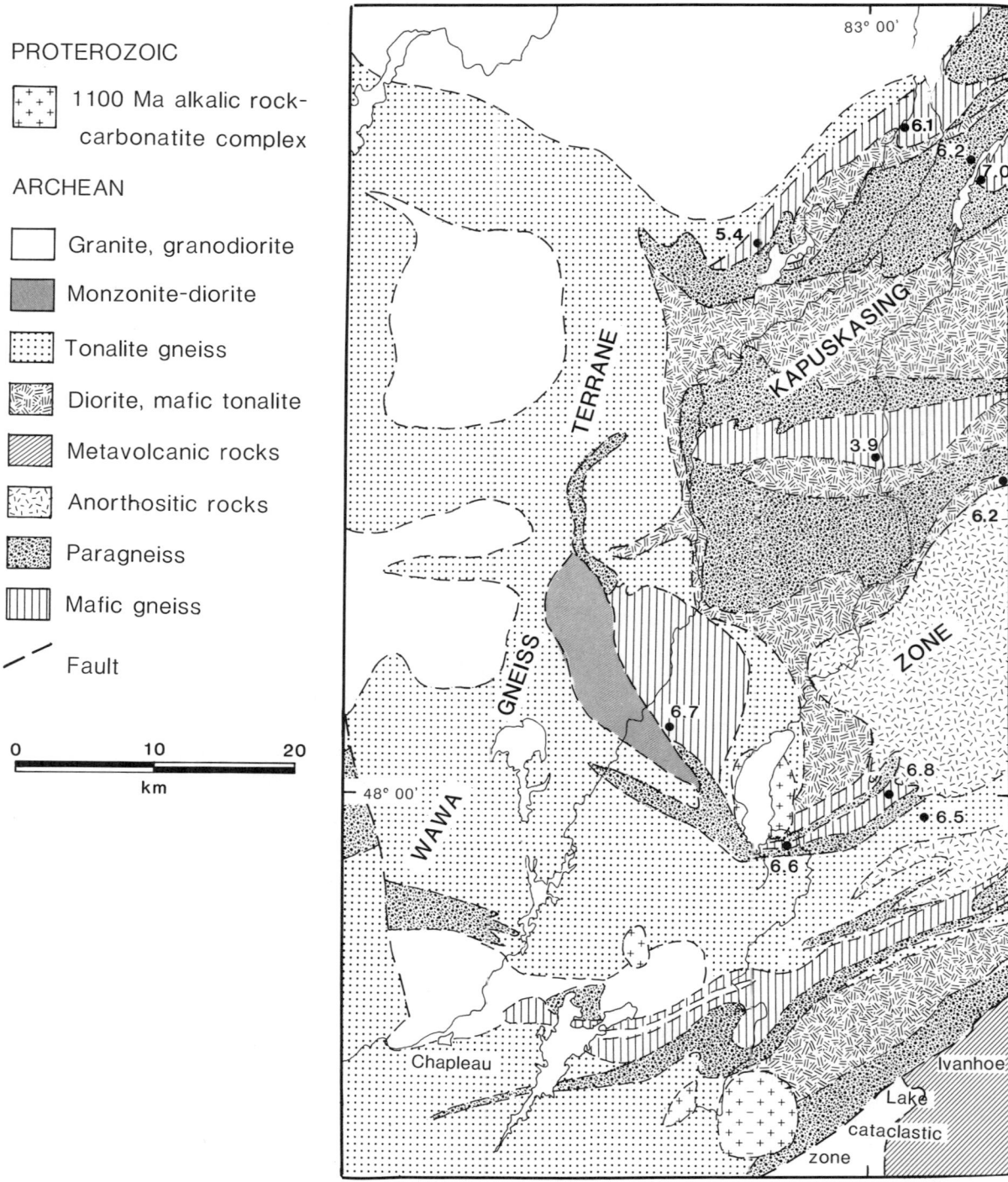

Fig. 3. Detailed map showing the transition between the Wawa gneiss terrane and Kapuskasing zone. Numbers beside large dots represent pressure, in kbar, at the time of metamorphic quenching, based on the assemblage garnet-clinopyroxene-plagioclase-quartz (Percival, 1983; Newton and Perkins, 1982).

at the boundary zone (Percival, 1983), corresponding to a depth of equilibration of ~20 km. This crustal level is similar to that at which mid-crustal discontinuities are recorded in many refraction surveys over crystalline terrane (Berry and Mair, 1980).

Mid-crustal seismic velocity discontinuities are commonly not zones that produce coherent reflections (Brown et al., 1983a; Oliver et al., 1983). This suggests that the discontinuity is not a sharp contact, but rather, occurs over some vertical interval.

The mid-crustal region in many localities corresponds to a zone of change from less reflective upper crust to more reflective lower crust (Brown et al., 1983a; Meissner, 1984). Suggested reasons for the lack of reflectivity in the upper crust are a homogeneous nature or a chaotic or vertical internal structure.

A change from dominantly upright structures to more horizontal attitudes takes place over an 80 km west-east section within the Wawa gneiss terrane. In the west, gneissosity in felsic plutonic rocks adjacent to steeply-dipping metavolcanics is also steep. To the east, gneissic layering is chaotic at the outcrop scale and domal on the regional scale. Near the Kapuskasing zone, gneissic layering is locally transposed into horizontal to gently-dipping shear zones. In the Kapuskasing zone, gneissic layering is parallel to lithologic contacts, dipping moderately and consistently northwest. Thus, within the Wawa gneiss terrane there may be an analog for many upper crustal reflection profiles.

Abundant horizontal to arcuate reflective horizons characterize the lower crust in many regions (Oliver et al., 1983; Brown et al., 1983a,b). Possible reasons for the seismic impedance constrasts include lithological layering (Smithson, 1978; Smithson et al., 1984), horizontal detachment zones (Allmendinger et al., 1983), mylonites (Fountain et al., 1984), sill-like intrusions (Clowes and Kanasewich, 1970) and fluid-bearing horizons (Jones and Nur, 1982; Fyfe, 1984). If the Kapuskasing zone is analogous to a reflective lower crust, then the probable cause is the 1/2 to 8 km-scale lithological layering in the Kapuskasing zone, with density contrasts up to 0.34 g·cm^{-3} and potential velocity contrasts on the order of 0.75 km·sec^{-1}.

An examination of the 10 km-wide transition zone between the Wawa gneiss terrane and Kapuskasing zone (Fig. 3) provides a possible explanation for the lack of seismic reflections at depths corresponding to mid-crustal velocity discontinuities. Along with the obvious but gradational compositional changes, the transition zone encompasses changes in metamorphic grade, from amphibolite to upper amphibolite and granulite facies, and in structural style, from domal to consistently northwest-dipping gneissosity. Refraction techniques would average the velocity change over this interval and thus a discrete discontinuity would be inferred. It would be difficult to define this zone on a reflection profile except as a zone of change, because contacts are gradational over several kilometres and vary in orientation from sub-horizontal to sub-vertical.

Discussion

In a comprehensive summary of the characteristics of the lower crust, Kay and Kay (1981) emphasized the extreme variability of rock composition, structure and mineralogy. Similarly, Smithson and Brown (1977) postulated that the deep crust consists of a structurally complex assemblage of both igneous and metamorphic rocks. In general, however, the overall composition of the lower crust corresponds to that of an intermediate igneous rock (Kay and Kay, 1981; Weaver and Tarney, 1984; Taylor and McLennan, 1981; Smithson et al., 1981) and the metamorphic rocks of the lower crust are at high grade. As noted by Smithson and Brown (1977) the intermediate composition is an average for interlayered mafic and felsic rock of both supracrustal and magmatic origin.

In the crustal cross-section based on the Kapuskasing uplift, most of the rocks are intrusive; less than 20 per cent of the Kapuskasing zone is of definite supracrustal origin and the percentage is lower in the Wawa gneiss terrane. Bulk compositional differences between the upper, more felsic part of the section and the lower, more mafic part may be accounted for by two factors: (1) intrusions in the basal part of the section are more mafic, including anorthosite, diorite and mafic tonalite, than the tonalites of the Wawa gneiss terrane, and (2) mafic rocks are both more abundant and at higher metamorphic grade and therefore denser in the lower part of the section. Magmatic additions in the form of sills of variable thickness and composition account for much of the present volume of the crust.

Although there is a general correlation between increasing depth in the crust and higher seismic velocity, the increase in velocity occurs across a discontinuity only in some areas. A continuous increase in velocity could be attributed to a gradual change in bulk composition, possibly controlled by buoyancy differences among various magma compositions. For example, lighter granitic rocks should generally rise to higher levels in the crust than denser gabbroic rocks. In the Kapuskasing section, the postulated discontinuity corresponds to the break between the Wawa gneiss terrane of tonalitic composition and the Kapuskasing zone containing supracrustal rocks and relatively mafic intrusions. Thick felsic plutons were intruded above the Kapuskasing supracrustal succession and thinner, more mafic sills were intruded within, suggesting that early lithological differences had a mechanical effect on pluton emplacement, in conjunction with gravity.

Metasedimentary granulites of supracrustal origin are common constituents of the lower crust in cratons, based on studies of xenolith suites in intrusive pipes (Kay and Kay, 1981) and exposed crustal cross-setions (Fountain and Salisbury, 1981). Several mechanisms have been proposed to account for the presence of supracrustal rocks in the lower crust, including, 1) tectonic burial by an overthrust slab (Newton et al., 1980), 2) incorporation of ocean-floor sediments into continental lower crust during subduction at Andean margins (Tarney and Windley, 1977), and 3) magmatic overaccretion (Wells, 1980; Percival and Card, 1983). The third mechanism accounts for the observed intrusive relationships between paragneiss and tonalites in the Kapuskasing zone. Precise U-Pb zircon dates on intrusive rocks from high and low structural levels across the Kapuskasing uplift indicate contemporary intrusion in a major crust-building episode between 2750 and

2665 Ma (Percival and Krogh, 1983; Turek et al., 1982). Layering of rock types on the km-scale in the Kapuskasing zone may be the product of the style of igneous intrusion or a modification caused by strain. Because the lower crust is ductile relative to the upper crust and upper mantle (Chen and Molnar, 1983), horizontal stresses may give rise to gently-dipping ductile shear zones in the lower crust. The effect would be to thin and accentuate layering.

Conclusions

The Kapuskasing uplift reveals an oblique cross-section through the upper 2/3 of the continental crust beneath an Archean greenstone belt. In this region the upper crust consists mainly of tonalitic gneiss in the amphibolite facies and granite, capped by thin metavolcanic-metasedimentary belts in the greenschist facies. The lower crust consists of units, on the 1/2 to 8 km scale, of paragneiss, mafic gneiss, and dioritic, tonalitic and anorthositic rocks, all metamorphosed to upper amphibolite and granulite facies at T=700-800°C, P=6-8 kbar. The interface between the two, with a 6 kbar metamorphic signature, encompasses changes, with increasing depth, of (1) composition, from tonalitic to interlayered felsic, intermediate, mafic and anorthositic rocks, (2) structural style, from large-scale domal geometry to northwest-dipping belts, and (3) metamorphic grade, from amphibolite to upper amphibolite and granulite facies. The interface may be analogous to discontinuities in seismic velocity detected at mid-crustal depths in many regions of the continental crust.

Acknowledgments. K.D. Card, A. Davidson and D. Forsyth are thanked for discussions and early reviews of the manuscript. Two anonymous reviews led to improved presentation.

References

Allmendinger, R.W., J.W. Sharp, D. Von Tish, L. Serpa, L. Brown, S. Kaufman, J. Oliver, and R.B. Smith, Cenozoic and Mesozoic structure of the Eastern Basin and Range, Utah, from COCORP seismic reflection data, Geology, 11, 532-536, 1983.

Berry, M.J., and K. Fuchs, Crustal structure of the Superior and Grenville Provinces of the northeastern Canadian Shield, Seismol. Soc. Amer. Bull., 63, 1393-1432, 1973.

Berry, M.J., and J.A. Mair, Structure of the continental crust: a reconciliation of seismic reflection and refraction studies, in The Continental Crust and its Mineral Deposits, edited by D.W. Strangway, Geol. Assoc. Can. Sp. Paper 20, 195-213, 1980.

Birch, F., The velocity of compressional waves in rocks to 10 kilobars, Part 2, J. Geophys. Res., 66, 2199-2224, 1961.

Brown, L., C. Chapin, A. Sanford, S. Kaufman, and J. Oliver, Deep structure of the Rio Grande rift from seismic reflection profiling, J. Geophys. Res., 85, 4773-4800, 1980.

Brown, L., L. Serpa, T. Setzer, J. Oliver, S. Kaufman, R. Lillie, D. Steiner, and D.W. Steeples, Intracrustal complexity in the United States midcontinent: preliminary results from COCORP surveys in northeastern Kansas, Geology 11, 25-30, 1983a.

Brown, L., C. Ando, S. Klemperer, J. Oliver, S. Kaufman, B. Czuchra, T. Walsh and Y.W. Isachsen, Adirondack-Appalachian crustal structure: The COCORP northeast traverse, Geol. Soc. Am. Bull., 94, 1173-1184, 1983b.

Card, K.D., Progress report on regional geological synthesis, central Superior Province, in Current Research, Part A, Geol. Surv. Can. Paper 82-1A, 23-28, 1982.

Chen, W-P., and P. Molnar, Focal depths of intracontinental and intraplate earthquakes and their implications for the thermal and mechanical properties of the lithosphere, J. Geophys. Res., 88, 4183-4214, 1983.

Christensen, N.I., and D.M. Fountain, Constitution of the lower continental crust based on experimental studies of seismic velocities in granulite, Geol. Soc. Am. Bull., 86, 227-236, 1975.

Clowes, R.M., and E.R. Kanasewich, Seismic attenuation and the nature of reflecting horizons with the crust, J. Geophys. Res., 75, 6693-6705, 1970.

Cook, F.A., Geometry of the Kapuskasing structure from a Lithoprobe pilot reflection survey, Geology, 13, 368-371, 1985.

Fountain, D.M., and M.H. Salisbury, Exposed cross-sections through the continental crust: implications for crustal structure, petrology and evolution, Earth Planet. Sci. Lett., 56, 263-277, 1981.

Fountain, D.M., C.A. Hurich, and S.B. Smithson, Seismic reflectivity of mylonite zones in the crust, Geology, 12, 195-198, 1984.

Fraser, J.A., W.W. Heywood, and M.A. Mazurski, Metamorphic map of the Canadian Shield, Geol. Surv. Can. Map 1475-A, 1978.

Fyfe, W.S., Fluid generation in deep continental crust (abstract) in International Symposium on Deep Structure of the Continental Crust: Results from Reflection Seismology, Cornell University, Ithaca, N.Y., 29-30, 1984.

Goodwin, A.M., Structure, stratigraphy and origin of iron formation, Michipicoten area, Algoma District, Ontario, Geol. Soc. Am. Bull., 73, 561-586, 1962.

Green, A.G., N.L. Anderson and O.G. Stephenson, An expanding spread seismic reflection survey across the Snake Bay-Kakagi Lake greenstone belt, northwestern Ontario, Can. J. Earth Sci. 16, 1599-1612, 1979.

Hall, D.H., and W.C. Brisbin, Overview of regional geophysical studies in Manitoba and northwestern Ontario, Can. J. Earth Sci., 19, 2049-2059, 1982.

Jolly, W.T., Metamorphic history of the Archean Abitibi belt, in Metamorphism of the Canadian Shield, Geol. Surv. Can. Paper 78-10, 63-78, 1978.

Jones, T., and A. Nur, Seismic velocity and anisotropy in mylonites and the reflectivity of deep crustal fault zones, Geology, 10, 260-263, 1982.

Kay, R.W., and S.M. Kay, The nature of the lower continental crust: inferences from geophysics, surface geology and crustal xenoliths, Rev. Geophys. Space Phys. 19, 271-297, 1981.

McCulloch, M., and G. Wasserburg, Sm-Nd and Rb-Sr

chronology of continental crust formation, Science, 200, 1003-1011, 1978.

Meissner, R., The continental crust in central Europe as based on data from reflection seismology (abstract) in International Symposium on Deep Structure of the Continental Crust: Results from Reflection Seismology, Cornell University, Ithaca, N.Y., 57-58, 1984.

Mueller, S., A new model of the continental crust, in The Earth's Crust: Its Nature and Physical Properties, edited by J.G. Heacock, Amer. Geophys. Union Geophys. Monograph 20, 289-317, 1977.

Newton, R.C., J.V. Smith and B.F. Windley, Carbonic metamorphism, granulites and crustal growth, Nature, 288, 45-50, 1980.

Newton, R.C., and D. Perkins, Thermodynamic calibration of geobarometers based on the assemblages garnet-plagioclase-orthopyroxene (clinopyroxene)-quartz, Am. Mineral. 67, 203-222, 1982.

Northey, D.J., G.F. West and F.A. Cook, Crustal scale seismic refraction and reflection studies of the Kapuskasing structural zone, a Lithoprobe project, (abstract) Geol. Assoc. Can., Prog. Abs., 10, A44, 1985.

Nunes, P.D., and D.R. Pyke, Geochronology of the Abitibi metavolcanic belt, Timmins-Matachewan area -progress report, Ont. Geol. Surv. Misc. Pap. 92, 34-39, 1980.

Oliver, J., F. Cook, and L. Brown, COCORP and the continental crust, J. Geophys. Res., 88, 3329-3347, 1983.

Percival, J.A., High-grade metamorphism in the Chapleau-Foleyet area, Ontario, Am. Mineral. 68, 667-686, 1983.

Percival, J.A., and K.D. Card, Archean crust as revealed in the Kapuskasing uplift, Superior Province, Canada, Geology, 11, 323-326, 1983.

Percival, J.A., and K.D. Card, Structure and evolution of Archean crust in central Superior Province, Canada, in Archean Supracrustal Sequences, edited by L.D. Ayres, P.C. Thurston, K.D. Card and W. Weber, Geol. Assoc. Can. Spec. Paper, 28, 179-192, 1985.

Percival, J.A., and T.E. Krogh, U-Pb geochronology of the Kapuskasing structural zone and vicinity in the Chapleau-Foleyet area, Ontario, Can. J. Earth Sci., 20, 830-843, 1983.

Sage, R.P., Josephine area, District of Algoma, in Summary of Field Work, 1983, Ont. Geol. Surv. Misc. Paper 116, 45-49, 1983.

Smithson, S.B., Modeling continental crust: structural and chemical constraints, Geophys. Res. Lett., 5, 749-752, 1978.

Smithson, S.B., and S. Brown, A model for lower continental crust, Earth Planet. Sci. Lett., 35, 134-144, 1977.

Smithson, S.B., R.A. Johnson and Y.K. Wong, Mean crustal velocity: a critical parameter for interpreting crustal structure and crustal growth, Earth. Planet. Sci. Let., 53, 323-332, 1981.

Smithson, S.B., R.A. Johnson, C.A. Hurich, and D.M. Fountain, Crustal reflections and crustal structure, (abstract) in International Symposium on Deep Structure of the Continental Crust: Results from Reflection Seismology, Cornell University, Ithaca, N.Y., 80-81, 1984.

Studemeister, P., The greenschist facies of an Archean assemblage near Wawa, Ontario, Can. J. Earth Sci., 20, 1409-1420, 1983.

Tarney, J., and B.F. Windley, Chemistry, thermal gradients and evolution of the lower continental crust, J. Geol. Soc. Lond., 134, 153-172, 1977.

Taylor, S.R., and S.M. McLennan, The composition and evolution of the continental crust: rare earth element evidence from sedimentary rocks, Phil. Trans. R. Soc. Lond. A301, 381-399, 1981.

Turek, A., P.E. Smith and W.R. Van Schmus, Rb-Sr and U-Pb ages of volcanism and granite emplacement in the Michipicoten belt, Wawa, Ontario, Can. J. Earth Sci., 19, 1608-1626, 1982.

Weaver, B.L., and J. Tarney, Empirical approach to estimating the composition of the continental crust, Nature, 310, 575-577, 1984.

Wells, P.R.A., Thermal models for the magmatic accretion and subsequent metamorphism of continental crust, Earth Planet. Sci. Lett., 40, 253-265, 1980.

SEISMIC CRUSTAL STRUCTURE NORTHWEST OF THUNDER BAY, ONTARIO

Roger A. Young[1], Jeffrey Wright[2], G.F. West

Geophysics Laboratory, University of Toronto, Toronto, Ontario, Canada M5S 1A7

Abstract. A seismic refraction-wide angle reflection survey of limited scope was carried out in the Shebandowan-Atikokan-Savant Lake region of western Ontario using open pit mine blasts as the principal energy sources. Conclusions drawn from the data are: 1) The seismic crustal structure varies laterally from one part of the area to another, but the variations are not substantial, nor is the crustal structure anywhere very unusual. The interpreted depth of the M discontinuity lies everywhere in the range 37-44 km. P wave velocity everywhere rises in the lower crust, reaching at least 7 km/s near the M discontinuity, whereas the upper crust displays a relatively more uniform velocity of less than 6.3 km/s. In three of the four interpreted sections, velocity jumps rapidly at an intermediate depth to define a distinct lower crustal layer. The depth to the top of this lower crustal layer, however, is quite variable (13-21 km). 2) An 8 km thick, near-surface capping of slightly higher velocity is interpreted for the more northerly profile which passes through the Sturgeon Lake-Savant Lake greenstone belt. It appears to be a manifestation of the more frequent occurrence of mafic metavolcanic material in crust which otherwise exhibits granitoid velocities (6.0 km/s). The result lends support to the view that greenstone belts of the Superior Province are generally of limited depth extent and are underlain by granitoids. 3) The M discontinuity at least beneath the northerly profile is a complex zone about 5 km thick, seemingly with a lamellar structure. Upper mantle velocity is not well defined by the surveys, but is consistent with previous regional estimates of about 8.1 km/s. On the northerly profile, there is evidence for a rise in mantle velocity to 8.3 km/s at about 50 km depth.

---

[1]Present address: Phillips Research Center, 169 GB, Bartlesville, OK 74004
[2]Present address: Chevron U.S.A., Inc., 1111 Tulane Avenue, New Orleans, LA 70112

Introduction

During the summers of 1973-75 and 1977, a series of crustal scale seismic refraction surveys was carried out on the Superior Province of the Precambrian Shield in an area about 200 km northwest of Thunder Bay, Ontario. The project was part of the Superior Province Geotraverse Project (Goodwin and West, 1974; Schwerdtner and West, 1979). This paper reports the principal results and conclusions of the work. Additional detail is given in theses by Young, 1979, and Wright, 1976.

Regional Geology

The Geotraverse area (48-52°N, 90-92°W) is a typical cross section of the Archean Superior Province which in turn is very similar to many Archean volcanic-plutonic terrains found elsewhere in shield areas of the world. Much of the Geotraverse area consists of varieties of granitoid gneiss in which a number of large and small metavolcanic assemblages (greenstone belts) are found. (Figure 1.)

Volcanic-plutonic terrain is cut by two roughly east-west striking belts of metasedimentary and granitoid gneissic rocks. These belts are known as the English River and the Quetico Metasedimentary Gneiss Belts or Subprovinces. They have a strongly banded gross structure, steep dips, a regular but sinuous east-west strike direction, and are bounded on their north sides by major east-west trending faults. In the Geotraverse, metamorphic grade is highest in the English River Belt--commonly upper amphibolite facies--while most of the volcanic-plutonic terrain is in the greenschist or lower amphibolite facies.

Although the study area is predominantly an Archean terrain, the southeast end of the Geotraverse area includes early Proterozoic Animikie and Keweenawan sedimentary rocks at the southwest end of Lake Superior.

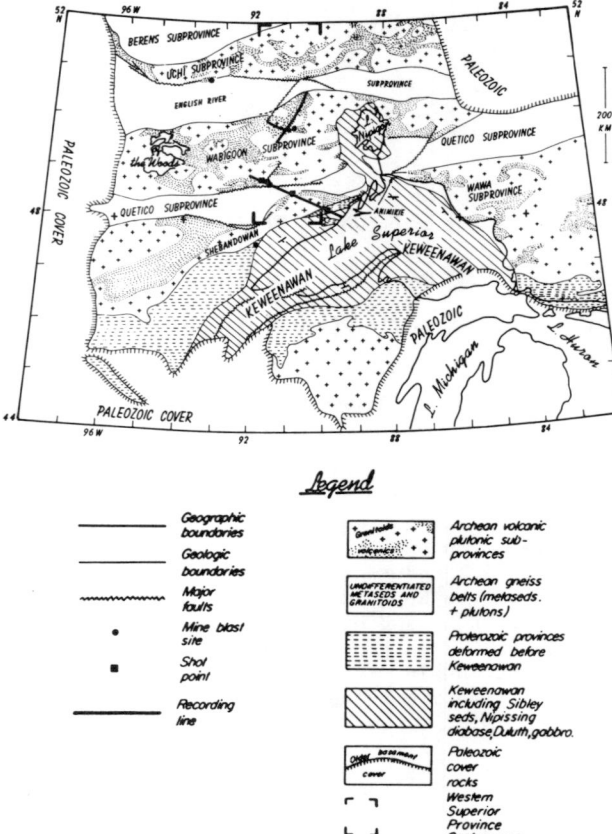

Fig. 1. Regional geological setting of the seismic survey.

Previous Work

The seismic structure of the Keweenawan Rift System was studied by the 1963 Lake Superior experiment (Smith et al., 1966; Berry and West, 1966) who found a deep basin (>50 km to M) under the lake. In the 1965 Project Early Rise, several very long radial profiles were recorded from a fixed shot point in Lake Superior (Iyer, 1969). One of the profiles (Mereu and Hunter, 1969) crossed the northwestern part of the Geotraverse area. Pn time-terms in the range 2.94 to 3.48 s were observed (based on a 8.05 km/s mantle velocity) and they were interpreted as indicative of crustal thicknesses from 30-35 km.

Hall and Hajnal (1969) and (1973) carried out explosion refraction surveys to the west of the Geotraverse in the Kenora-Red Lake area at about 94°W longitude crossing the English River and Uchi Subprovinces. They interpreted depths of the M discontinuity in the range 30-38 km, with a distinct Conrad discontinuity in the crust. Appreciable topography was found on both C and M horizons with an approximately inverse correlation between their depths. Godlewski and West (1977) analyzed earthquake group velocity dispersion from Flin Flon, Manitoba, to Thunder Bay, Ontario. They found an average crustal structure with M at 40 km and a shear wave velocity in the lower crust which is clearly greater than in the upper crust. Green et al. (1978 and 1979) conducted reflection experiments on the east side of Lake of the Woods which defined crustal structure there. Hall and Brisbin (1982) have reviewed work in the Kenora-Red Lake area. Berry and Mair (1977 and 1980) have compiled useful general reviews of reflection and refraction sounding of continental crust, which provide a background for comparison with our Geotraverse results.

Survey Methods

The original intent of our seismic field program was to record a series of explosion refraction profiles. Because of the logistical difficulty and the cost of drilling shot holes in the area and because of restrictions on shooting in lakes, we began in summer 1973 by employing old abandoned mine shafts as shot points. However, due to their limited depth extent and to their slow refilling with water after use, we obtained very limited low frequency energy from these shot points. This experience led to the utilization of mine blasts from various open pit mines in the region for the remainder of the survey.

Recordings were made with up to 11 pack-portable, three channel, digitally recording, chronometer-programmed, seismic recorders. Seismometers were 2 Hz horizontal or vertical instruments positioned on bedrock in almost all cases. Timing was by simultaneous recording of WWVB, with crystal chronometer back-up. Controlled explosions were timed electrically against WWVB, and all mine blasts were recorded by continuously running seismic instruments at one of two fixed base stations (Nym Lake or Sturgeon Lake). Also, one or more blasts at each mine site was also timed with locally placed recorders to establish relative blast-site to base-station timing. Local velocities were established at each mine site to assist in computing corrections for variations in blast position within the mine pits. The mines assisted by providing advance notice of blasting schedules, precise locations, charge size and patterns, etc. Advance knowledge of probable blast times enabled the field crew to program the seismic recorders. Base station recordings made it possible to compare seismograms from different blasts at the same pit. In most cases, the seismograms repeated quite well, insofar as gross features were concerned. In some cases, good replication was obtained on a cycle by cycle basis. However, occasional blasts (which were not in general used for the survey) gave very different seismograms, especially if delays totalling more than 200 ms were employed between parts of the blast pattern, or

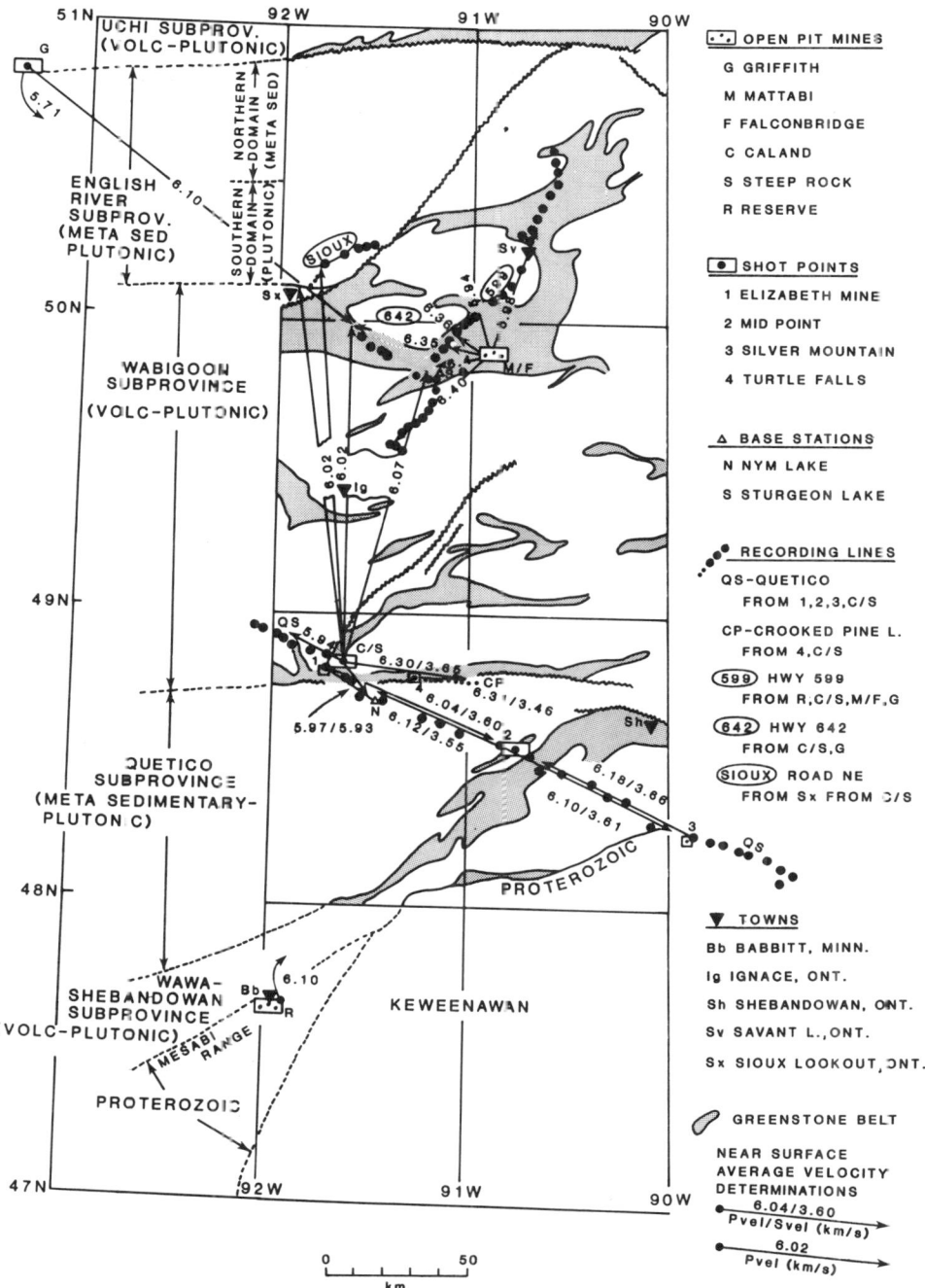

Fig. 2. Location of seismic sources and recordings in the Geotraverse and results of near-surface velocity surveys.

different parts of the pit were blasted at the same time.

Data Sets

Figure 2 shows the recording sites and sources used for the surveys. The most complete profile was a 135 km spread of 34 sites along Highway 599 which recorded blasts from the Caland and Steep Rock mines at Atikokan, Ontario, and the Reserve mine at Babbitt, Minnesota (ranges of 80-225 km and 205-340 km, respectively). Another profile (the first one recorded) crossed in a west-northwest direction from the

SEISMIC CRUSTAL STRUCTURE, CANADA   145

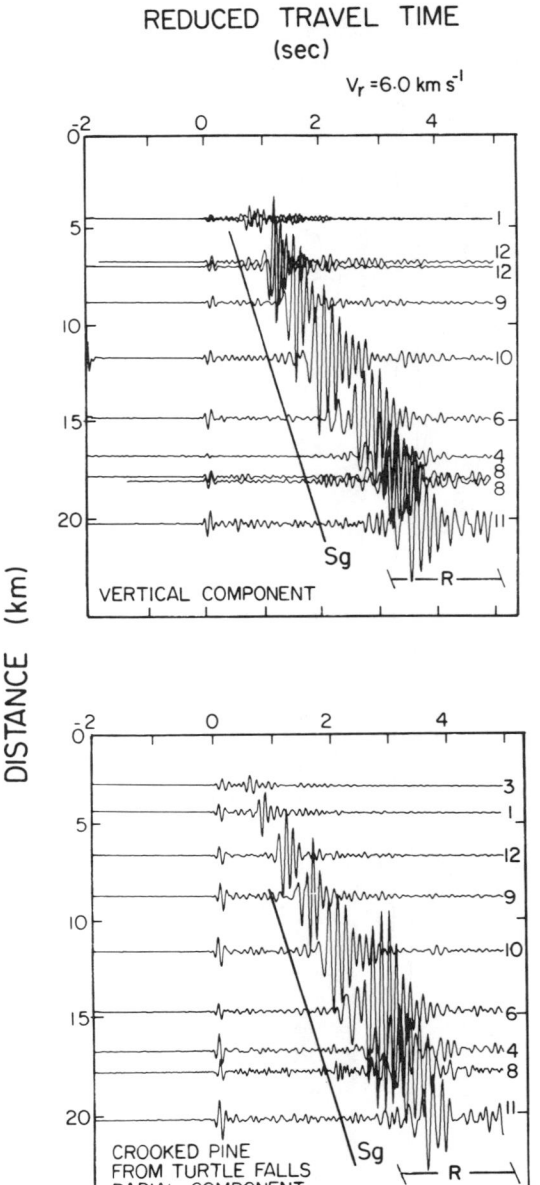

Fig. 3. The local velocity survey at Crooked Pine Lake. Note the simple character of the records. The surface low velocity layer is less than a kilometer deep. The velocity structure appears to be homogeneous beneath it.

ranges usually in the 20 km range and not exceeding 80 km to determine near surface velocity variation. Several fan profiles were observed at various ranges and a short in-line profile covering the range 155-195 km was observed along Highway 642 approximately perpendicular to the Highway 599 recording line, using a blast at the Griffith Mine south of Red Lake, Ontario. Although not highly significant in themselves, the supplementary surveys aid in supporting and extending the interpretation of the main surveys.

Data Analysis and Interpretation

The data are divided into four categories for convenient discussion: 1) local velocity data;

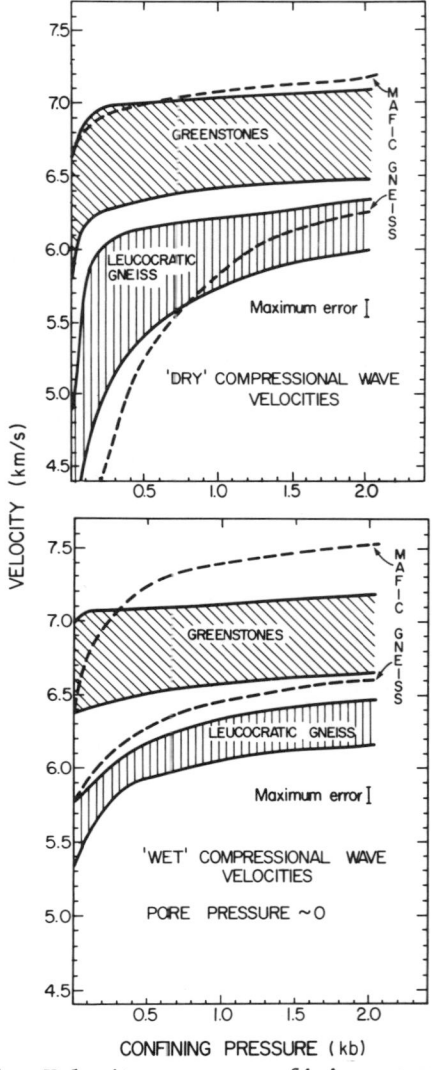

Fig. 4. Velocity versus confining pressure for 23 samples from selected parts of the Geotraverse (after Karson, 1979). Velocity was measured at 17 confining pressures.

Lake Superior shore for 185 km to a point about 70 km WNW of Atikokan. This line was recorded mainly with our own shots, but augmented by mine blasts from Caland and Steep Rock for the eastern end of the profile.

The main profiles were supplemented with a variety of other observations. Short profiles were recorded at several sites with maximum

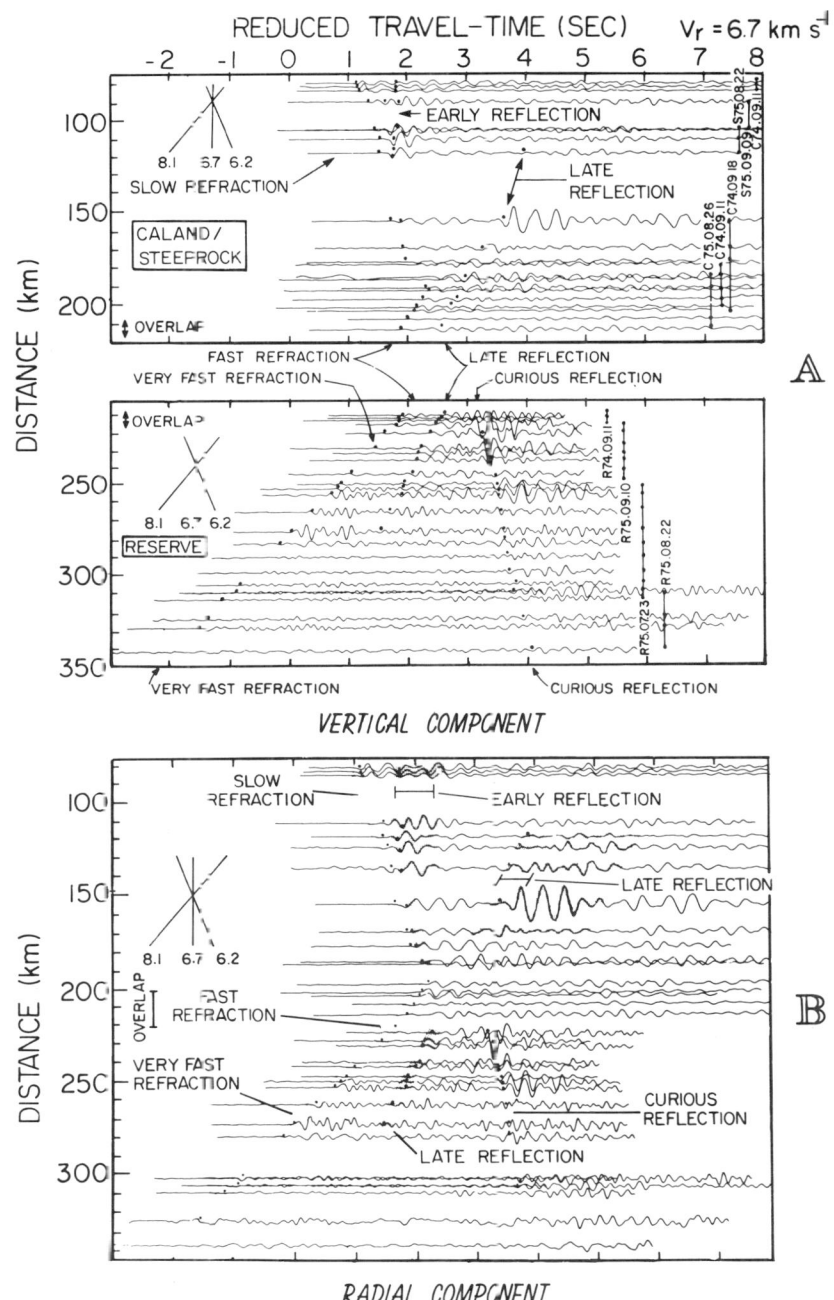

Fig. 5. Highway 599 seismic sections. A shows separately the vertical component reduced time seismic sections from the Caland/Steep Rock (C/S) and Reserve (R) shot points. Events are named. B shows the composite seismic section (radial component).

2) Highway 599 profile; 3) Quetico-Shebandowan profile; 4) additional data.

Local velocity data

As indicated in Berry and Mair (1980) and measured directly by Hajnal and Stauffer (1975), the P velocity in an exposed Shield usually rises rapidly with depth through a thin surface layer, reaching a rather uniform velocity near 6 km/s in much less than one kilometer. The velocity then does not usually rise much before a depth of 10-15 km is reached. In some cases, zones of somewhat lower velocity have been found

TABLE 1. P-wave Ray and Event Notation

Event name	Where observed[+]	Layer or - Interface -	Lateral waves (direct, cont. refractions, head waves)	Reflections
Slow refraction	1,2,3	Upper crust	Pg	
Early reflection	1	- UC -		PUP
(no event)	-	Mid-crust (may be LVZ)	-	
Intermediate reflection	3	- MC -		PMcP
Curious reflection	1	- MC -		PCP(LVZ)
Intermediate refraction	1,2,3	Lower crust	P*(headwave) PL(cont. refr.)	
Late reflection	1,2,3	- M -		PMP
Fast refraction	1,2,3	Sub-M transition layer	Pn	
(Very late reflection)	1	-SM -		PSMP
Very fast refraction	1	Uppermost mantle	PSM	

[+] 1 = 599 - C/S - R; 2 = QS; 3 = 642 - G

Table 1. Nomenclature for events observed on record sections and their ray-modelled counterparts.

in the 2 to 10-15 km depth range. As a result-- and because deeper crustal velocities are not usually more than 20% greater than the upper crustal values--the seismic time-distance curve usually has a long, constant velocity segment running from a very few to more than 100 km of source-detector separation.

The low velocity surface layer in the Geotraverse area is present but so thin as to be virtually unobservable in our surveys. Hammer seismic measurements to distances of 30 m gave velocities of 4.5-5.5 km/s; but in the explosion surveys, the intercept time of the approximately 6 km/s arrivals was certainly less than 0.1 s, and nearly constant apparent velocities were observed to distances of more than 100 km. Figure 2 shows the paths along which surficial velocity was measured and the average velocities so obtained. Figure 3 presents a survey along Crooked Pine Lake where very simple seismograms were obtained, indicating a lack of any significant velocity heterogeneity in that region even though the profile runs in metasedimentary rocks paralleled at distance of only about a kilometer by metavolcanic rocks on the north side of the Quetico fault. As indicated above, P wave delay due to the surface low velocity layer is not distinguishable (i.e., depth << 1 km), but some indication of an S wave surface layer is seen in the surface wave dispersion. A group velocity minimum of about 2.6 km/s is observed at about 8 Hz corresponding to a layer of order 0.3 km thick. The very simple character of the P wave arrivals in these and other short range seismograms shows that the wave form of the mine blasts can well be monitored at about 20 km range.

The observed near-surface P velocities shown in Figure 2 range from 5.71 to 6.40 km/s. The longer distance profiles (>50 km) show less variation (6.02-6.18 km/s) than the shorter ones, presumably due to more averaging on the longer paths. No systematic relationship of velocity to regional or local geology is obvious, even though higher velocities are expected in the mafic volcanics than in the granitoids and metasediments. It must be noted, however, that almost all paths traverse a variety of lithologies. Although the data is very limited, a case could be made for higher velocities being observed along the local strike of formations than across. Such anisotropy is to be expected if seismic velocity is heterogeneous on a local scale and the heterogeneities are elongated and systematically oriented.

Figure 4 shows laboratory velocity measurements on a suite of samples from the northern half of the Geotraverse from the English River gneisses and the Sturgeon Lake metavolcanic assemblage (Karson, 1979). One sees immediately that the sample velocities cover a much wider range than the field velocities and the "greenstone" (mafic metavolcanic) samples all have velocities well above 6.0 km/s at all pressures where appreciable crack closing has taken place (>0.2 kb wet (=20 MPa)).

To reconcile the laboratory and field observations, we should first note that the mafic gneiss samples were of high metamorphic grade and typical only of a few parts of the English River Subprovince. The "greenstone" specimens were typical of the lower amphibolite grade mafic metavolcanics commonly found in most of the greenstone belts of the Geotraverse. Generally, however, the greenstone belts also contain lower grade mafic volcanics, felsic volcanics and volcanogenic fragmentals, volcanogenic sediments and minor intrusions, all of

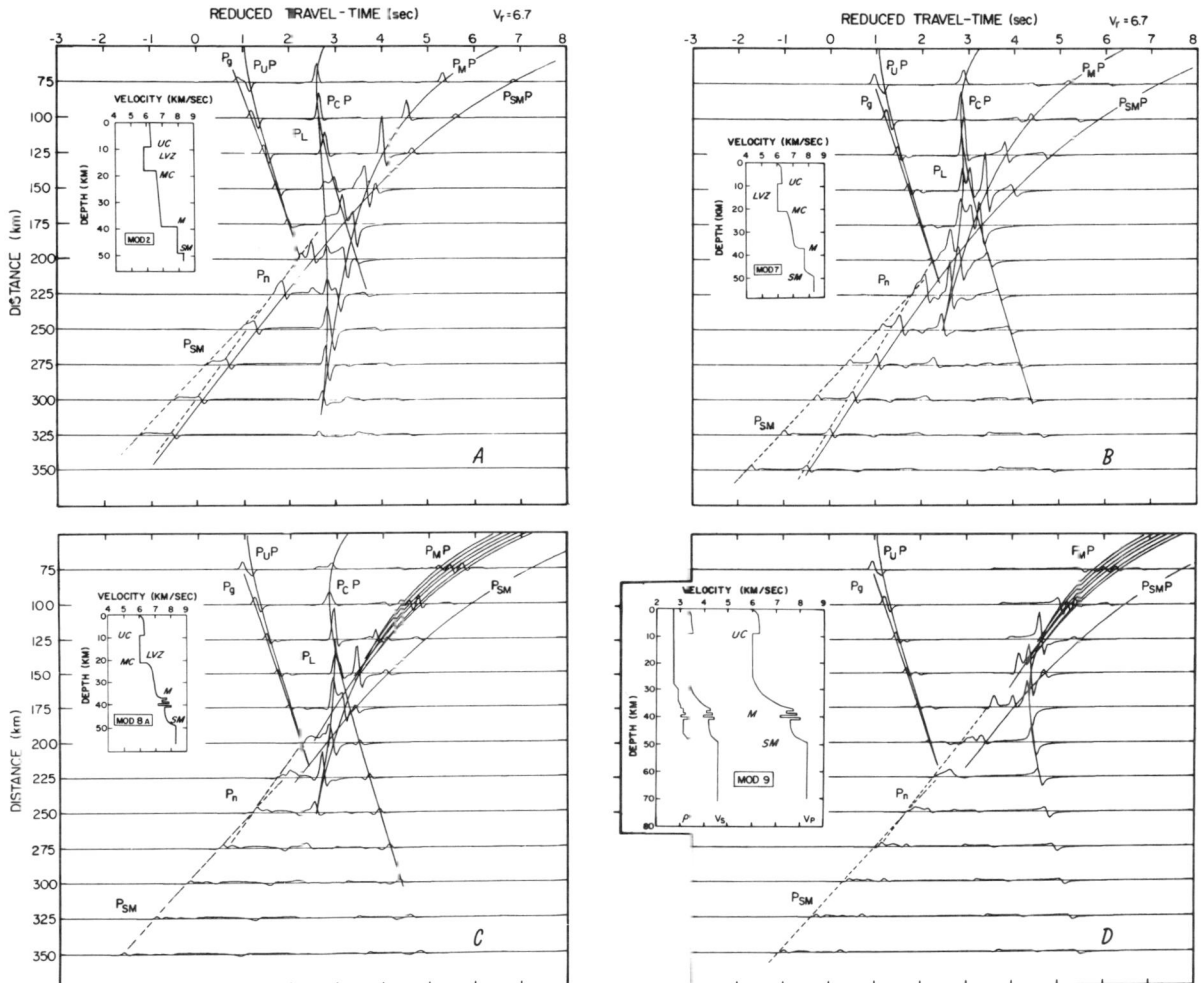

Fig. 6. Velocity models and synthetic seismograms for comparison with the 599-C/S-R composite section. A is a basic model (MOD2). B incorporates a more gradational lower crustal velocity structure to weaken and relocate PL (MOD7). C incorporates a lamellar M structure (MOD8A). D is an alternative to C which does not have a mid-crustal discontinuity (MOD9). C and D are better fits to the C/S and R seismic sections, respectively.

which are likely to be of lower velocity. Furthermore, the specimen velocity measurements are done at ultrasonic frequencies, higher by 10^4 to 10^5 than the field frequencies, and some dispersion inevitably accompanies whatever attenuation is present. Also, the specimens are free of the macroscopic cracks and imperfections that are present in the field and which will lower bulk velocity. Nevertheless, the in situ measurement of velocity in a greenstone belt by Hajnal and Stauffer (1975) gave about 6.6 km s^{-1}, and the velocity of the greenstone samples is consistently so much higher than the average velocities observed in the upper crust that it is very difficult to believe that the greenstone belts can become more extensive at depth. It is much more likely that they are relatively rare and shallow remnants in a granitoid host.

Highway 599 Profile

Figure 5 displays the vertical and radial component record sections from the Highway 599 profile. Three distinct first-arrival branches and three prominent later arrivals are named. These are the "slow," "fast," and "very fast" refractions (apparent velocities of 6.34 ±0.5, 7.72 ±.49, 8.30 ±.08 km/s, respectively) and the "early," "late," and "curious" reflections. The curious reflection is so named because it arrives later, has a low apparent velocity and is surprisingly strong. The various events are tabulated in Table 1, Column 1. The fact that the two record sections are so similar in their overlap region suggests that regional dip along the whole profile is minimal and that to a first approximation the structure can be treated as horizontally layered.

Ray tracing through a simple horizontal four-layer model indicates that a velocity reversal must be included to account for the late arrival times of the curious reflection. The crustal layers corresponding to this basic model are identified in Column 2 of Table 1, where the P wave notation for lateral waves (Column 3) and reflections (Column 4) is also given.

Figure 6A shows synthetic WKBJ seismograms (Chapman, 1978) for the basic model (insert). Note how the bottom of the upper crustal LVZ results in a late, slow reflection (PCP). Velocity gradients in the upper crust and especially in the lower crust are needed to generate curvature of Pg and to increase the apparent velocity of PMP, respectively. Note that the continuous refraction from the lower crust (PL) is not a first-arrival, in agreement with the absence of an intermediate velocity first-arrival in the data. The fast refraction and very fast refraction correspond, respectively, to Pn from the M boundary and to PSM from a sub-M boundary. A Pn apparent velocity of 7.7 km/s is appreciably lower than expected for a shield area (McConnell and McTaggert-Cowan, 1965); however, M boundary dip of only 2° increases Pn by .2 km/s.

A refined model (Figure 6B) incorporates a more gradational lower crustal structure which weakens PL and restricts the range of PMP. A velocity gradient above the SM boundary weakens PSMP and strengthens PSM. Incorporating a lamellar M structure (Figure 6C) focuses PMP around 150 km and provides more rapid amplitude fall-off at shorter ranges. Interference within this packet of reflections may account in part for the low frequency, multicycle appearance of the late reflection at 155 km (Figure 5). The minimum number of lamelli required, their thicknesses, and the maximum velocity contrast are not resolvable with the present data set, but five lamelli, each 0.8 km thick, with a maximum velocity contrast of 0.8 km/s yield a good match with the data. Convolution of the impulse response in Figure 6C with a 0.25-0.5 s duration wavelet approximating the source function of the large, extended mine blasts which generated the record sections gives an even closer correspondence between data and synthetics.

Turning to differences between the sections recorded from Caland/Steep Rock and Reserve, there is an important mismatch between the models and the C/S observations which involves the mid-crustal phases PL and PCP. Neither of these events is present on the shorter range portion of the data generated by blasts at C/S (Figure 5, top). Hence, the mid-crust illuminated from C/S must not have the midcrustal velocity jump. An alternative velocity model with a completely featureless mid-crust is presented in Figure 6D. The increase from crustal to mantle velocity is accommodated by a very steep gradient in the lowermost crust. The PCP-PL and PMP-Pn cusps have been collapsed into the vicinity of critical PMP, producing synthetics without events between PUP and PMP. Figure 6D is an elastic rather than an acoustic model and Vs was estimated from the laboratory measurement (Karson, 1979) on representative samples at appropriate confining pressure, and density was obtained from Vp following Birch (1960/61). Although synthetic seismograms could not be computed for dipping layering in the models, ray tracing was carried out for major events. Giving local dips of a few degrees to horizons improved the correspondence of the models to the data but had only minor effect on interface depths near the ray reflection points. Other forms of velocity inhomogeneity could provide similar effects. Thus, the interpreted dips were not considered to have great significance and only the horizontal models are given here.

Quetico-Shebandowan Profile

The Quetico-Shebandowan profile runs at a right angle to the Highway 599 profile and lies 100 km south of it (Figure 2). The east-west trend of alternating metasedimentary-plutonic and volcanic-plutonic subprovinces is traversed obliquely by the southeast-northwest QS profile.

Reversed vertical and radial component record sections are shown in Figure 7, with synthetic seismograms computed by the method of Wiggins and Madrid (1974) shown for comparison. Unfortunately, it was not logistically possible to extend these profiles to greater distance, and the cross-over to first arrival of Pn is seen only on one section where the Pn apparent velocity is poorly defined. The outstanding characteristic of these seismic sections is the absence of strong, correlatable, late arriving events such as are seen in the 599-C/S section (Figure 5). Although higher amplitudes do follow the early-arriving phases, indicating considerable scattering in the crust, only three early arriving events can be positioned on a travel time graph. They are denoted as Pg (apparent velocities 6.18 ±.03 and 6.04 ±.02 km/s); P* (6.70 ±.16 and 6.66 ±.26) and Pn (8.11 ±.28). These events were seen on the 599 profile, but with the intermediate refraction very differently positioned, i.e., at least one second behind the first arrivals. It is therefore named P* on the QS section rather than PL to correspond with classical nomenclature. Modelling P* requires a significant increase in velocity at midcrustal depth, but position of the event on the travel time curve requires that

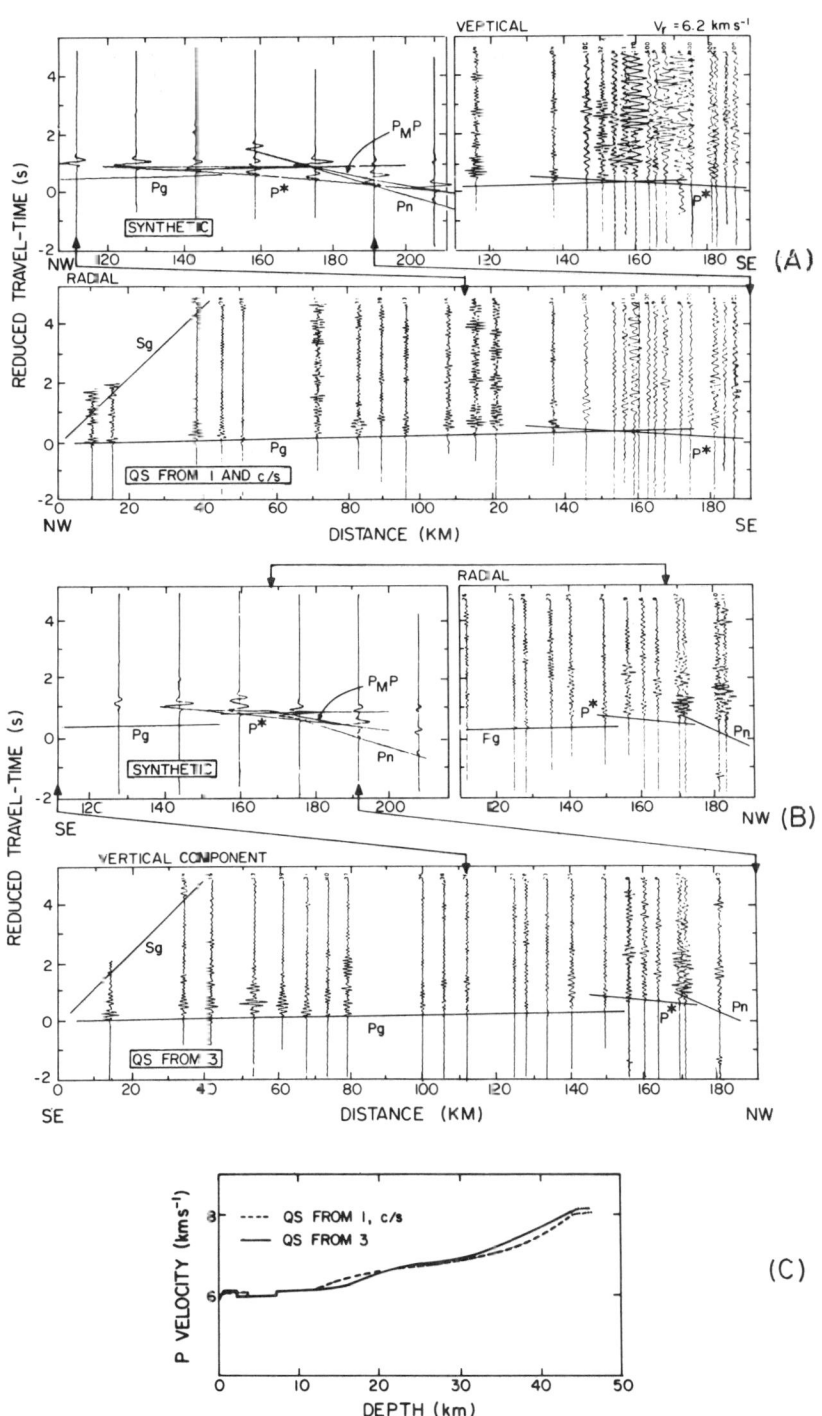

Fig. 7. The reversed Quetico-Shebandowan profile: (A) shot east from Caland/Steep Rock and Elizabeth mine; (B) shot west from Silver Mountain; (C) horizontally-layered models which generate synthetics in A and B.

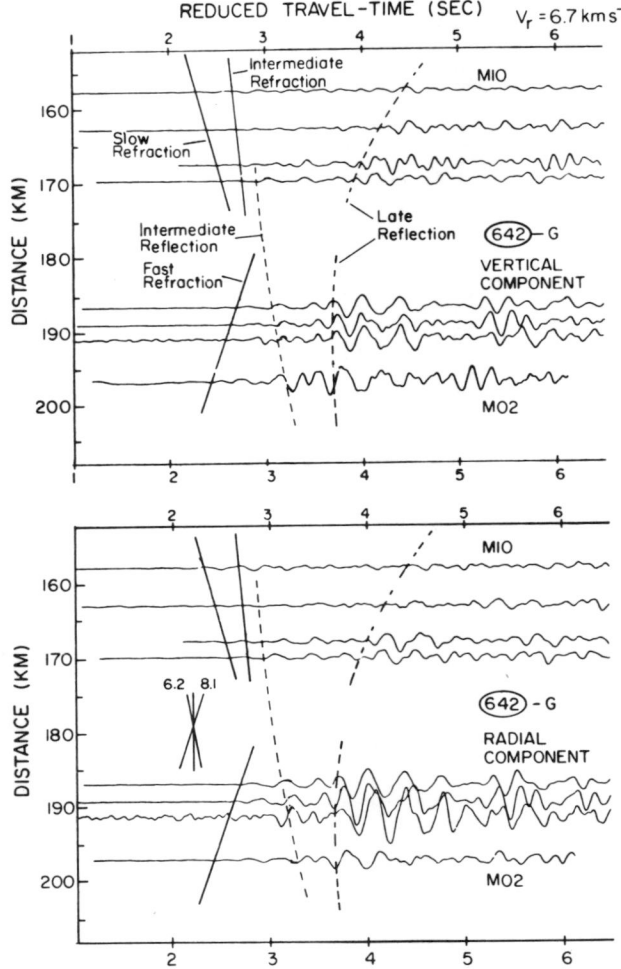

Fig. 8. The Highway 642-Griffith Mine section, with a simple layered mode. The event markings are observed coherent arrivals not identified events from a model.

the upper crustal average velocity not drop significantly below the velocity of Pg. In order to concentrate all triplications of the travel time curves into the first second following first arrivals, gradational velocity structures have been employed in the modelling. The best fitting horizontal structures for each profile are shown in Figure 7. The main predicted late events are shown in the synthetics of Figure 7. To produce the very slightly delayed onset of large amplitudes seen just behind Pg at some distances, a very minor upper crustal low velocity zone has been included in each model.

Additional Data

Although only 40 km long, the Highway 642-G profile spans the PMP critical reflection and helps to define the velocity structure in the southern domain of the English River Subprovince lying northwest of the previous two (Figure 2). The presence of three refraction branches (6.02 ±.19, 6.71 ±.06, and 7.91 ±.06 km/s, respectively) and two reflection branches (Figure 8) indicates, to a first approximation, a two-layer crust with boundaries at about 19 and 37 km. The presence of local topography on MC and M 200 km west of the present location (Hall and Hajnal, 1973) is certainly not inconsistent with the present limited data.

Discussion

Figure 9 presents a summary of the derived velocity models showing location of approximate subsurface coverage to which the models are most relevant. A crustal thickness of about 40 km (-3, +4 km) is common to all models despite different locations. This relative constancy in crustal thickness contrasts with the Ukrainian Shield block structure, described by Sollogub et al. (1973) where the M boundary is frequently offset by faults which slice through the entire crust. In the present study, a regional dislocation such as the Quetico Fault (northern boundary of the Quetico Subprovince) fails noticeably to disrupt velocity layering or to juxtapose a velocity contrast across the fault trace. The local surveys also show that the fault zone itself is not a low velocity zone, reinforcing petrologic evidence that fault healing has occurred (Bau, 1975).

The velocity structure of the crust does not become more simple with increasing depth. Lamellar structure at the crust/mantle boundary is suggested in the Wabigoon Subprovince by the present study to explain the weakness of the late M reflection at subcritical ranges and, in part, the low frequency content of the critical reflection around 155 km. C. Bois, et al. (this volume) present an example of a similar low frequency M reflection observed in France. K. Fuchs (personal communication, 1984) has shown by full wave equation synthetics that a sequence of lamelli illuminated obliquely behaves as a low pass filter. In addition, many deep reflection profiling surveys--thanks to higher frequency content and lesser lateral averaging--show evidence of such laminations: in Western Canada (Clowes and Kanasewich, 1970; Berry and Mair, 1977); in the Rio Grande rift (Brewer and Oliver, 1980); and near the Gabilan Mountains, California (Hale and Thompson, 1982). However, similar seismic events were not observed in the Quetico-Shebandowan profile, indicating either that the lamellar structure is not universal, or that the reflections were suppressed by crustal inhomogeneity or the relatively high frequency of most of the QS records.

An uppermost average crustal velocity of about 6.0 km/s is ubiquitous in the study area. This correlates with the widespread outcropping of

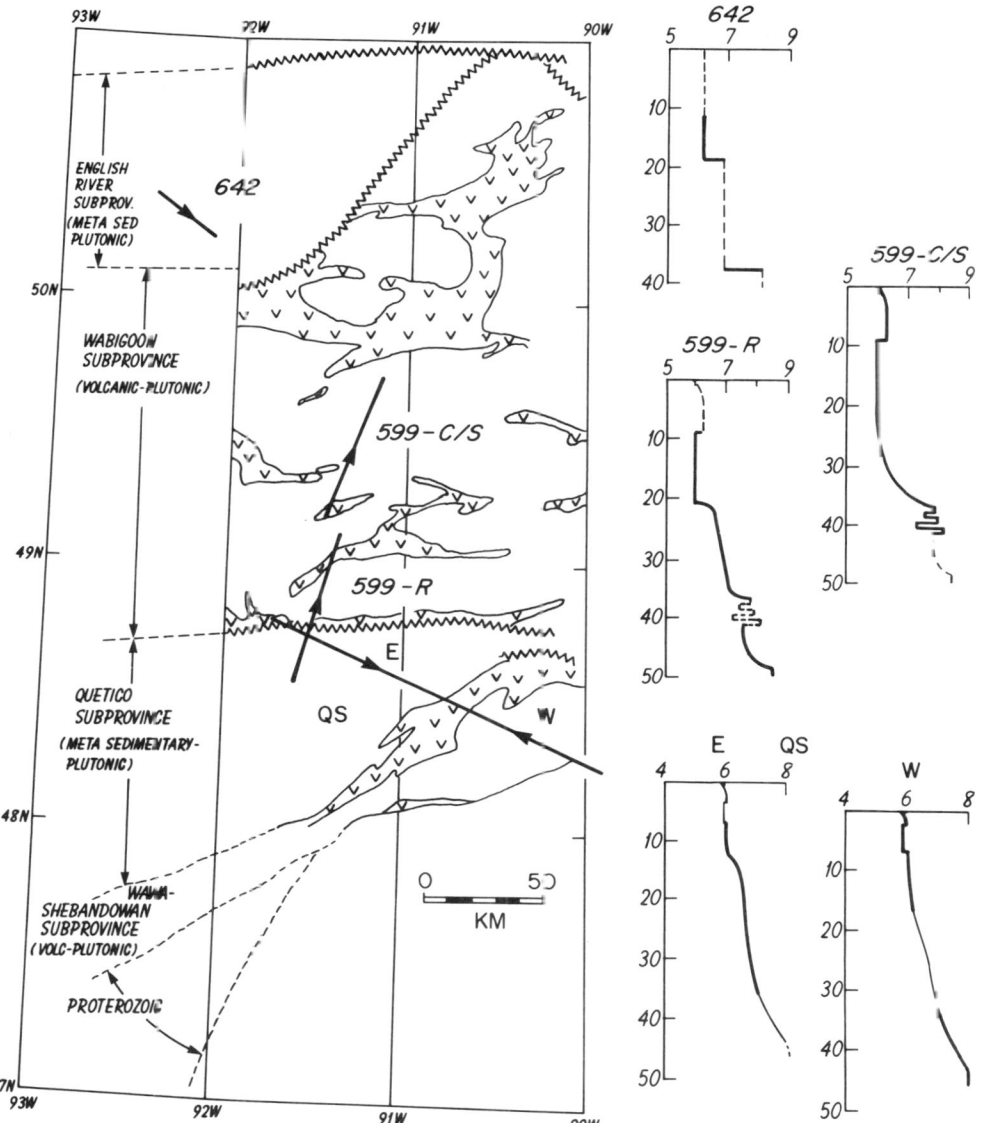

Fig. 9. Summary of the velocity models grouped according to approximate sub-surface coverage and terrain type. Dashed segments are poorly constrained by the present seismic data. Heavy segments highlight the best constrained features.

granitoids. The presented laboratory velocity measurements, as well as many others in the literature, indicate a P wave velocity of about 6.0 km/s for felsic gneiss, whereas mafic metavolcanics above lower greenschist facies exhibit velocities of about 6.6 km/s. In our short range profiles which sample metavolcanics, an average velocity of about 6.3 km/s is almost always attained. Interpretation of the refracted-reflected events indicates a somewhat lower velocity of about 6.0 km/s (models 599-C/S and QS) in the top several kilometers of the upper crust. This is explicable as a return at depth from volcanics to a granitoid rock type, so the seismic evidence supports the view that greenstone belts are relatively thin remnants of formerly thicker volcanic piles. A limited depth extent has been suggested by gravity surveys in the same area (Dusanowskyj, 1976). Thus what we have called low velocity zones in the models are not zones of anomalously low velocity; they result from higher velocity metavolcanics dispersed in and capping the more universal upper crustal granitoids.

The velocity structure of the crust in this part of the shield is not simply layered; significant variation over lateral distances of 100-200 km is definitely present. The occurrence of considerable reverberant energy with a short correlation distance indicates scattering, presumably from local velocity structure. Yet, only the QS profile might possibly be described as the stepchild of a three-dimensional random variation superimposed on a systematic increase of velocity with depth. On the other hand, well-organized, persistent events on the 599 and 642 profiles certainly conform to the patterns expected from layered structures.

The main difference between the several interpreted velocity structures is in the transition from upper to lower crust. In all cases, the velocity rises from an upper crustal value of less than 6.3 km/s to at least 7 km/s before the M discontinuity is reached. However, the depth and form of the increase are highly variable. In the 599-C/S section, the velocity only rises above granitoid values below 30 km depth, where it begins a rapid but featureless increase to mafic values. In the QS profile shot west, there is a steady velocity increase from about 18 km depth. In the remaining cases, a distinct jump in velocity occurs at depths of 21, 19, and 13 km respectively in the 599-R, 642, and QS --> E profiles.

It is not immediately clear how to correlate the differences in lower crustal structure with the geology of the area. When beginning the seismic program, we thought that there might be distinct variations in crustal structure in crossing from one type of subprovince to another. However, the observed variations seem to be on a smaller scale, and may be related to the distribution of the major batholithic and supracrustal units.

The lamellar structure of the M discontinuity and the sub-M rise in velocity to 8.3 km/s were interpreted only for the 599 profile. Unfortunately, the limited scope of the surveys did not make it possible to determine if these features are present in the other sections. The absence of a clear, late, wide-angle M reflection in the 100-150 km range on the QS sections may indicate either that lamination is absent, or that the discontinuity is too laterally irregular to produce a correlatable reflection.

The results of this program demonstrate that crustal structure surveys using open pit mine blasts are feasible. However, fewer blasts into a larger number of recorders would have aided event correlation, and reversal of the 599 line would have constrained analysis beyond simply layered structures. Simple surveys such as ours can easily detect lateral inhomogeneity, but substantially more data is needed to resolve the lateral structure in an unambiguous manner.

Acknowledgments. Professor C.H. Chapman was of great help in computing, and he and Dr. Ralphe Wiggins loaned us synthetic seismogram programs. Discussions with Drs. Wooil Moon and Rick Allis often sparked an exchange of views with R.Y. The Ontario Geological Survey kindly shared their field camp with us. Mr. Khader Khan and Mr. Dale Parsons drafted the diagrams, and Mrs. Frances Young lent able editorial assistance. The cooperation of the following mining companies is gratefully acknowledged: Butler Taconite, Caland Ore Company, Falconbridge Mining Company, Mattagami Lake Mines, National Steel Pellet Company, NBU Mines, Reserve Mining Company, Steel Company of Canada, and Steep Rock Iron Mines. This work was supported financially by a Research Agreement with the Department of Energy Mines and Resources and by operating grant A1187 from NSERC (Canada).

References

Bau, A., Structures in the Kashabowie-Upsala Region, Northwest Ontario, Proceedings of the 1975 Geotraverse Workshop, Dept. of Geol., Univ. of Toronto, 14-1 through 14-6, 1975.

Berry, M.J., and J.A. Mair, The nature of the earth's crust in Canada, in The Earth's Crust, Amer. Geophys. Union Mon. 20, pp. 319-348, 1977.

Berry, M.J., and J.A. Mair, Structure of the continental crust: a reconciliation of seismic reflection and refraction studies, in Geol. Assoc. of Can., Special Paper 20, pp. 195-214, 1980.

Berry, M.J., and G.F. West, A time-term interpretation of the first-arrival data of the 1963 Lake Superior experiment, in The Earth Beneath the Continents, Amer. Geophys. Union Mon. 10, pp. 166-180, 1966.

Birch, F., The velocity of compressional waves in rocks to 10 kilobars, I, J. Geophys. Res., 65, 1083-1102, 1960.

Birch, F., The velocity of compressional waves in rocks to 10 kilobars, II, J. Geophys. Res., 66, 2199-2224, 1961.

Brewer, J.A., and J.E. Oliver, Seismic reflection studies of deep crustal structure, Ann. Rev. of Earth and Planet. Sci., 8, 205-230, 1980.

Chapman, C.H., A new method for computing synthetic seismograms, Geophys. J.R. astr. Soc., 54, 481-518, 1978.

Clowes, R.M., and E.R. Kanasewich, Seismic attenuation and the nature of reflecting horizons within the crust, J. Geophys. Res., 75, 6693-6705, 1970.

Dusanowskyj, T., Gravity study of the Sturgeon Lake area, Proceedings of the 1976 Geotraverse Workshop, Dept. of Geol., Univ. of Toronto, 14-1 through 14-10, 1976.

Godlewski, M.J.C., and G.F. West, Rayleigh-wave dispersion over the Canadian Shield, Bull. Seismol. Soc. Amer., 67, 771-779, 1977.

Goodwin, A.M., and G.F. West, The Superior Geotraverse Project, Geoscience Canada, 1(3), 21-29, 1974.

Green, A.G., D.H. Hall, and O.G. Stevenson, A sub-critical seismic crustal reflection survey over the Aulneau batholith, Kenora region, Ontario, Can. J. Earth Sci., 15(2), 301-315, 1978.

Green, A.G., N.L. Anderson and C.G. Stevenson, An expanding spread seismic reflection survey across the Snake Bay-Kakagi Lake greenstone belt, northwestern Ontario, Can. J. Earth Sci., 16(8), 1599-1612, 1979.

Hajnal, Z., and M.R. Stauffer, The application of seismic reflection techniques for subsurface mapping in the Precambrian shield near Flin Flon, Manitoba, Can. J. Earth Sci., 12, 2036-2047, 1975.

Hale, L.D., and G.A. Thompson, The seismic reflection character of the continental Mohorovicic discontinuity, J. Geophys. Res., 87, 4625-4635, 1982.

Hall, D.H., and W.C. Brisbin, Overview of regional geophysical studies in Manitoba and northwestern Ontario, Can. J. Earth Sci., 19(11), 2049-2059, 1981.

Hall, D.H., and Z. Hajnal, Deep seismic crustal studies in Manitoba, Bull. Seismol. Soc. Amer., 63, 885-910, 1973.

Hall, D.H., and Z. Hajnal, Crustal structure of northwestern Ontario: refraction seismology, Can. J. Earth Sci., 6, 82-99, 1969.

Iyer, H.M., L.C. Pakiser, D.J. Stuart, and D.H. Warren, Project Early Rise: Seismic probing of upper mantle, J. Geophys. Res., 75, 4409-4441, 1969.

Karson, J., unpublished manuscript, 1979.

McConnell, R.K., and G.H. McTaggart-Cowan, Crustal seismic reflection profiles--a compilation, Institute of Earth Sciences, Univ. of Toronto, Scientific Report #8, AF Contract 19(628)-22, 1963.

Mereu, R.F., and J.A. Hunter, Crustal and upper mantle structure under the Canadian Shield from Project Early Rise, Bull. Seismol. Soc. Amer., 59, 147-165, 1969.

Schwerdtner, W.M., and G.F. West, The Superior Geotraverse Project, Can. J. Earth Sci., 16(10), 1903-1905, 1979.

Smith, T.J., J.S. Steinhart and L.T. Aldrich, Crustal structure under Lake Superior, in The Earth Beneath the Continents, Amer. Geophys. Union Mon. 10, pp. 181-197, 1966.

Sollogub, V.B., D. Prosen and co-workers, Crustal structure of central and southeastern Europe by data of explosion seismology, Tectonophysics, 20, 1-33, 1973.

Wiggins, R.A., and J.A. Madrid, Body wave amplitude calculations, Geophys. J. Roy. astr. Soc., 37, 423-433, 1974.

Wright, J., Seismic crustal studies in northwestern Ontario, Ph.D. thesis, 212 pp. Univ. of Toronto, Toronto, Ontario, September 1976.

Young, R.A., Seismic crustal structure northwest of Thunder Bay, Ontario, Ph.D. thesis, 326 pp. Univ. of Toronto, Toronto, Ontario, October 1979.

A SEISMIC CROSS SECTION OF THE NEW ENGLAND APPALACHIANS:
THE OROGEN EXPOSED

Robert A. Phinney

Department of Geological and Geophysical Sciences
Princeton University, Princeton, NJ 08544

Abstract. Several marine multichannel seismic reflection lines collected by the U.S. Geological Survey in its assessment of the Eastern U.S. continental margin for oil and gas are found to constitute a high quality deep reflection profile of the continental crust on the Long Island platform. In the time range 4-11 seconds (10-35 km), the CDP section shows nearly continuous, dense, well-correlated reflections. The systematic variation in local dip angle of these reflections defines large-scale tectonic packets which constitute the bulk of the crystalline crust in this area. Most conspicuous is a single large packet in the eastern end of the line which deps to the W at 25 degrees and extends from the sediment-basement boundary to the lower crustal boundary layer (Moho) at 25-30 km. This packet is interpreted as an accretionary structure formed and thickened to continental thickness during the late Paleozoic accretion of Avalonia and the Appalachian margin of North America. In the center and western portions of the line a low angle complex packet dipping east at depth is interpreted as the pre-Acadian margin of North America. A "keystone" packet lying between these two bodies, and forming most of the basement subcrop under the central portion of the line, appears to correlate with the high grade medial zone of south central New England, and appears to be the strongly compressed, thickened, and uplifted remains of the oceanic basin, volcanic islands, and marginal sedimentary wedges which separated Avalonia from North America before their collision. The lowest 1-2 seconds of the crust appears as a nearly horizontal layered complex, with at least two pronounced local structural breaks. I suggest that this "Moho" forms the lower carapace of a compressional orogen which was thickened to at least 45 km during Paleozoic collision(s). It is interpreted as the highly strained tectonic boundary layer established during the subduction of oceanic crust at an active continental margin.

1. Introduction

The effort to delineate the geological structure of the continental crust by reflection seismology has passed its tenth year in this country. The COCORP program of land vibrator work has revealed examples of tectonic architecture which have cast quite a different light on traditional geological debates. Perhaps the greatest frustration with these fine results has been the difficulty of finding ideal near-surface conditions, and the resultant degradation of signal quality of many deeper crustal reflections. This paper gives an example of both the quality of continental reflection data which can be obtained when surface conditions are ideal, and of the particular geological significance of high quality data. The 1978-79 United States Geological Survey (USGS) marine reflection surveys on the mid-Atlantic sector of the eastern continental margin made a number of passes over the Long Island Platform, where 12 seconds of high quality CDP data was acquired over crystalline continental crust. Hutchinson, Klitgord, and Detrick [1985a], Hutchinson et al [1985], and Hutchinson, Klitgord, and Detrick [1985b] have also discussed this area, including both the cited USGS regional lines and more recent high resolution lines.

2. Data Acquisition and Processing

From 1973-79, the USGS conducted a major program of multichannel seismic exploration of the U.S. Atlantic continental margin, to evaluate the sedimentary basins of the shelf, slope, and rise for their hydrocarbon potential. Among and over 20000 km of line collected in this pro-ram, portions of lines 9, 16, 21, 22, 23, 24, 36 and 37 were run on the Long Island Platform over a thin (<2 km) flat-lying sedimentary section [Grow et al, 1980]. The resultant CDP sections, processed to 12 seconds of two-way time, provide seismic sections through the cratonized crust of the Southern New England Appalachians and its transition from continental thickness on the Long Island Platform to oceanic crust across the passive rifted margin. In this paper I deal with Line 36, which runs along the platform, midway between the mainland coast and the shelf break (Figure 1), and with Line 23, an important cross-line.

157

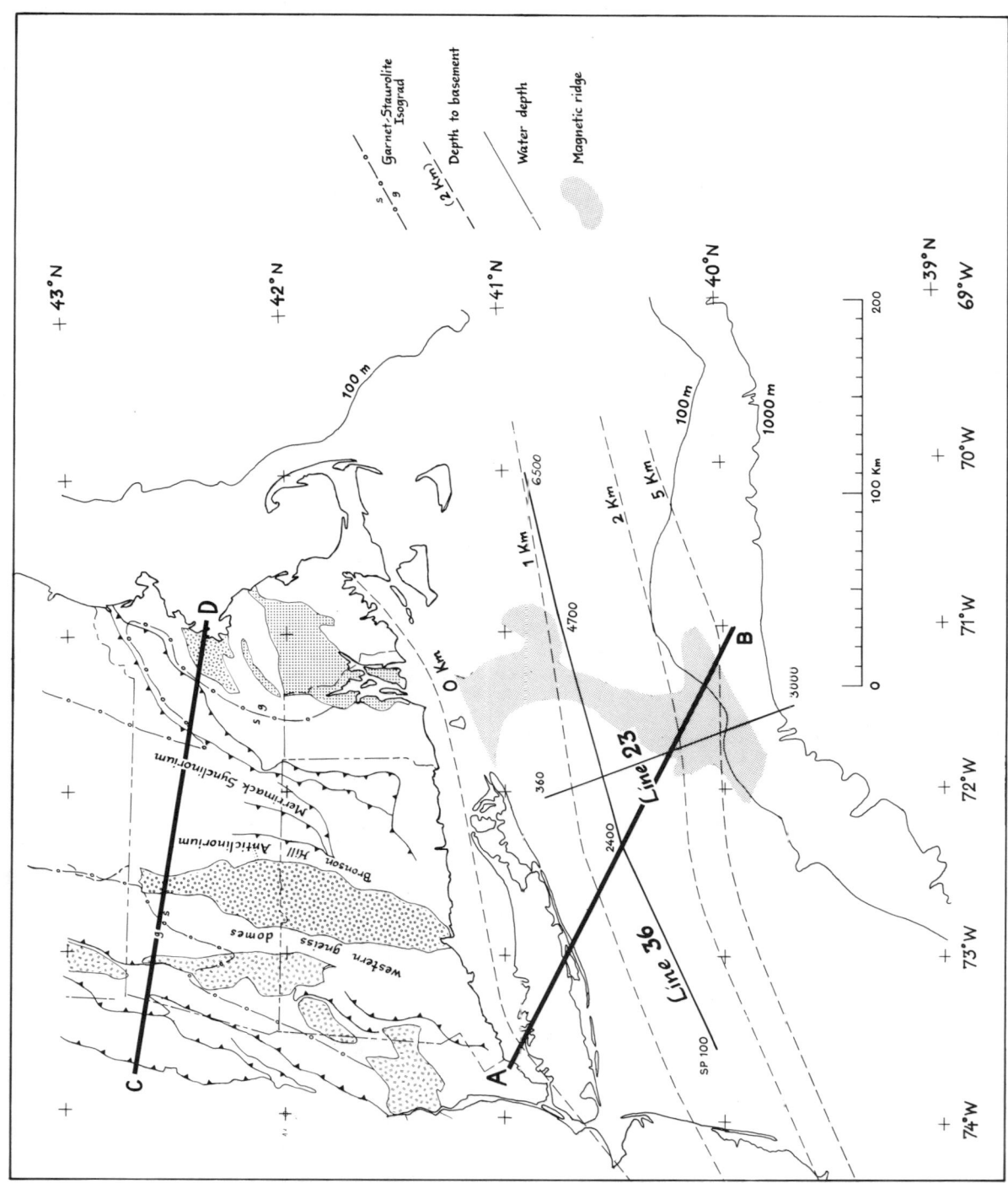

Fig. 1. Map of Southern New England and the Long Island Platform, showing tracks for USGS multichannel lines 36 and 23, with generalized geological and geophysical information. Magnetic anomaly from Klitgord and Behrendt [1977], garnet-staurolite isograd from Thompson and Norton [1968], and sediment thickness estimated from all the regional USGS seismic sections (lines 9, 16, 21, 22, 23, 24, 36, and 37).

The data were collected and processed into CDP sections by Texas Instruments, Inc. (TI), under contract to the USGS. A 2000 cu in airgun array was fired at 50 m intervals into a 48 channel marine hydrophone streamer with a maximum offset of 4 km, for nominal 48-fold coverage at 50 m CDP spacing. Processing was conventional; a time variant filter with breakpoints at 3 sec and 5 sec demands some care on the part of the interpreter. Although stacking velocity models were produced by the contractor, their usefulness in discussing the present data is limited. Not only are the velocity analysis spectra not available, the contractor's responsibility lay in the direction of optimum resolution of the sedimentary cover, and the normal moveout would not be enough with a 4 km cable to discriminate mid- to deep-crustal velocities.

3. Seismic character of the section

This paper is based on an interpretation of the stacked sections produced by the contractor. At the same time, we are also engaged in a long range effort to reprocess this line, both pre- and post-stack, to optimize signal quality and to produce an adequate quality migrated section. Most of the line shows, from 4 to 10 seconds, numerous, coherent reflection events which show systematic variation in local dip angle over the section. The high quality of the signals on the CDP section may be attributed to the ideal field conditions for shot and receiver placement afforded by the marine survey. Moreover, the sediments and the sediment-basement interface are ideally horizontal, and produce minimal distortion of the deep signals. Some confirmation of the importance of this factor may be obtained by noting that the quality of the deep signals appears to be somewhat degraded under the irregular basin at shotpoints (SP) 200-900.

The seismic character of the section is illustrated by Figures 2 and 3. A line drawing (Figure 4) of the principal coherent reflections is used as the basis of a structural interpretation. This is done without migration of the section. Some modification of the shapes of the major units may occur upon migration. As a rule, I have omitted from this line drawing those few events with apparent slowness greater than about 0.1 sec/km, which would experience significant displacement in the migration process.

A few dip angles are indicated in Figure 4. These are estimated by (1) projecting the section onto a dip line (at 45°; see below), and (2) applying the migration formula for dip in a constant velocity medium:

$$\tan \psi = \sin \phi$$

where ψ is the apparent dip on the unmigrated section and ϕ is the true dip. The effect is always to increase the dip.

Beneath the thin (~50 m) water layer lie the inner continental shelf platform sediments (about 1.0 sec = 1.25 km). The sediment-basement contact is represented by a strong horizontal reflection, at 0.8-1.2 sec. This is produced by the basal zone of the sediments, for the event continues across the tops of the two intrabasement fault basins (SP's 200-900 and 1950-2250).

From 1.0 to about 4 seconds, few good reflections can be seen, apart from horizontal multiples produced in the water-sediment column. If this zone were to be characterized by complex structures with horizontal scales less than about 5 km, it would be difficult with standard field geometry and processing to obtain more than a few weak diffractions. Moreover, sediment-produced multiples may be masking the desired signals. Both factors appear to be relevant; further understanding of the issue must await experimental reprocessing. Changes in reflection character in the window 2-6 sec (Figure 3) do not correlate with the 3 and 5 sec breakpoints in the time variant display filter.

Four groups of dipping events can be seen in this shallow domain, all at apparent dips of about 30°:
(a) at SP 150-200, dipping eastward, associated with the margin of an irregular intrabasement fault basin.
(b) at SP 1900-2200, dipping eastward, associated with the margin of an intrabasement half-graben, merging into a major dipping boundary zone which goes to at least 9 seconds.
(c) at SP 3700-4100, the surface continuation of a west-dipping intracrustal boundary which appears paired with (b);
(d) at SP 5200-5700, also west-dipping, and penetrating well into the deep crust. After allowance for the obliquity of the line to the regional strike (see below), the dips associated with these events are found to lie between 25° and 50°.

Given the clearly poorer signal to noise environment in the shallow domain, these signals represent instances of enhanced reflectivity. From their association with the boundaries of fault basins and with major through-cutting crustal boundaries, I interpret these features as major tectonic shears or dislocations, and use them to help demarcate the principal tectonic units in the section.

Coherent, flat to moderately dipping signals fill most of the record from 4 to 11 seconds. Signal to noise ratio is good enough that a local apparent dip angle can be assigned to the reflected energy at nearly every point in the section. Except for a region around SP 600, diffraction crossings of these events are uncommon. The dip angles remain nearly constant over large regions; this behavior can be used to define major seismo-structural domains, or tectonic units (see Figures 2 and 3). The transition upward to a band of weaker signals showing multiple diffraction arcs and undulations (Figure 2, 4-5 seconds) is interpreted as a significant change in structural style. In numerical models of reflection from complex folded and faulted structures, Wong et al [1982] highlighted the factors which differentiate these two contras-

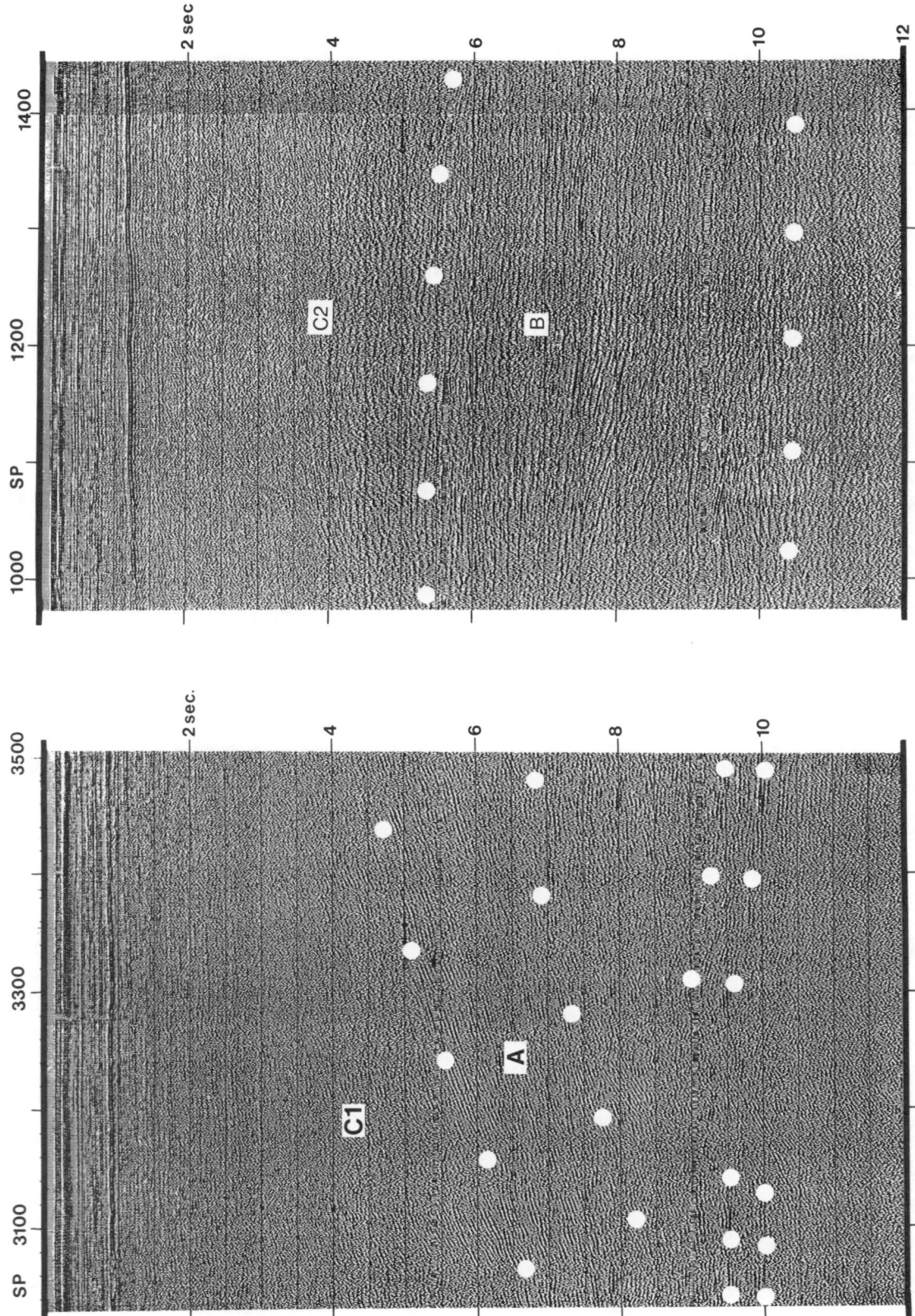

Fig. 2. View of portion of the CDP section for line 36, from SP 3030 to SP 3500. (1) Bottom (10 sec), a pair of M reflections lying conformably beneath a horizontally layered domain. A possible fault in M is indicated by the diffractions. (2) Middle, domain A, with strong west-dipping laterally coherent reflections. (3) Above, domain C1, showing (below 4 seconds) weaker reflections of a more undulatory character, which are downwardly conformable with A. While the M events marked with white dots are identifiable events, the dots used to identify the top and bottom of A are placed to identify changes in gross dip angle or character. (4) At the top, the portion from the Basal sediment reflector at 0.8 sec to about 3.0 sec is dominated by weak horizontal multiples from the sediments, and carries little information about the basement.

Fig. 3. View of portion of the CDP section for line 36, from SP 980 to SP 1430. (1) Bottom (10-10.5 sec), a consistently horizontal M reflection group, with transition upward to the nearly horizontal subdomains of (2) domain B, with strong laterally coherent reflections, ranging from horizontal to 20° dips eastward. (3) Above, domain C2, with weaker reflections, downward conformable with B, and showing (from 4 to 5.5 sec) the undulatory character I associate with fold structures.

ting seismic characteristics. It was found that tight folds with moderate to steep axial plane dips were characteristic of complex structures which produce the weak, diffraction-dominated seismic records; in contrast, consistent sub-horizontal layering is produced by layered units with low dips and broad horizontal scales.

Between 9.5 and 11 seconds, the complex of intracrustal reflections is underlain by events which are nearly horizontal, and which show pronounced continuity across the section. These events define a special tectonic domain which ranges in thickness from 0.5 to 2.0 seconds. Below these events, the number of strong, coherent signals is markedly reduced. As a matter of convention, I refer to the layered complex as the lower crustal boundary layer (LCBL), and the bottom of the LCBL as the reflection-Moho. The reflection-Moho is commonly, if casually, regarded as identical to the Moho as defined by seismic refraction, although no suitable case of correlative refraction data has permitted a demonstration that this is the case at better than 10% agreement. Taylor and Toksoz [1982] in a regional summary of existing refraction data and teleseismic P-delays obtain, from various sources, depths to the refraction-Moho ranging from 36 to 42 km in southern New England. The model of Chiburis and Ahner [1979] for southeastern New England is most relevant (Table 1). If we adopt their velocity model for layers 2 and 3 and pick reflection times from Figure 3 at SP 1200, we arrive at 30 km for the depth to the reflection-Moho. Although the discrepancy between this value and the previously reported values may be due to crustal thinning between Southern New England and the Long Island Platform (toward the rifted margin), no such indication appears on cross line 23 (Figure 4). Later in this paper I suggest a model in which the refraction-Moho may lie several km deeper than the reflection-Moho. N. Christensen [personal communication, 1985] has suggested that substantial transverse anisotropy in crustal gneisses might give rise to an apparent discrepancy of this sort.

2. Architecture of the seismic cross section

The lower portion of Figure 4 is an interpretive cross-section, in which the major seismo-structural units are distinguished. Units are characterized by consistency and continuity of dip angle, and by consistency of the character of the signals. Line 36, between SP 1000 and SP 5000 has the most characteristic architecture. Here, the crust is seen to be composed of four major domains and to have 3 significant boundaries (sutures ?) which cut the crust at angles of $36° \pm 14°$.

1: The lower crustal boundary layer, with little or no dip. From SP 3200 to SP 5000, it ranges in thickness from 1 sec (3 km) to 3 sec (9 km). West of SP 2800 it tends to be around 0.5 sec thick. It is broken by zones of discontinuity,

TABLE 1. Comparison of Refraction and Reflection Models

Layer No.	A	B	C	D
1	1.2 sec	2.5 km/s	---	1.5 km
2	5.3 sec	6.1 km/s	13 km	14.0 km
3	10.2 sec	6.6 km/s	35 km	30.1 km

(A) Reflection times (Figure 3); (B) Refraction velocity model and (C) Depth to bottom of layer [Chiburis and Ahner, 1979], except for layer 1 velocity estimated; (D) Depths computed from columns A and B.

at 2000 and 3000, which are discussed in a later paragraph.

2: A tectonic wedge, (domain A: Figure 2), with dip about 25°-42° west, which subcrops continuously to the top of the LCBL at around 9-9.5 sec (27-29 km). Its western apex forms a tongue which tapers out at SP 2800, and is bounded below by a thickened piece of the M-complex.

3: A tectonic wedge, (domain B: Figure 3), underlying the western area from SP 800 to SP 2800, and from about 4.5 seconds down to the LCBL. The apparent dip is eastward, between 0 and 15 degrees. This domain shows somewhat more internal variety of appearance than domain A. Its easternmost apex seems to touch the western apex of domain A.

4: A pronounced east-dipping narrow belt of discontinuity which separates domain C1 from the domains to the west (fault zone α). It can be traced from the east-dipping (50°) reflections near SP 1900 at the west end of the half-graben. At greater depth, these reflections merge into the east-dipping east end of domain B. This zone divides the eastern domains, A and C1 from the western domains b and C2.

5: Transparent zones C1, C2, and C3. C1 is a wedge-shaped "keystone" block which is nested above and between A and B. C2 overlies B along a horizontal to east-dipping boundary. In each case, the layering above and below, and the transition are conformable. Characteristically, the layering above the transition, in C1 or C2, begins to develop an undulatory form with diffraction edges, which suggests that these zones are characterized by folds or similar steeply dipping structures on scales of 1-4 km (Figures 2 and 3). C3 is a transparent, rhomboidal block underlying SP 5800-6500. The existence of a few weak west-dipping events under SP4400-5000 has led me to leave this zone as a part of A.

6: A pronounced west-dipping alignment of reflections which divides A from C1, across which dips are conformable, but seismic character changes. This probable shear zone (labeled β) can be followed easily from subcrop to LCBL.

7: A similar appearing alignment (labeled γ) which subcrops around SP 5200-5700, dipping to the west, which flattens somewhat as it finally terminates in the lower crustal boundary layer.

Estimation of regional strike. Other lines collected by the USGS in this area also show good deep reflections which have character comparable to line 36. Of these (Figure 1), line 23 crosses line 36 at a critical point (SP 3260), and can be used to establish the strike of the structure. The section appearing on line 23 (Figure 4) is almost identical to the segment of line 36 between SP 2600 and SP 4200, with the tie point correlating well [line 36, SP 3260; line 23, SP 1000]. The structure seen in cross section on line 36, then, has a regional strike which approximately bisects the two lines ... at N 25° E. This coincides with the regional orientation of the medium wavelength ridge shown on Figure 1, which then correlates with the subcrop of domain A.
This estimated regional strike for the structure on the Long Island Platform is concordant with the large-scale syntactical bend in the Appalachians between New England and Pennsylvania.

5. Geological Evaluation

It is helpful first to consider the broad regional geology as a basis for the working assumptions to be used in interpreting these sections. The seismic lines lie on normal thickness continental crust, as a southward extension of southern New England, and athwart the center of the Appalachian orogen. Along the center of the orogen, terranes of lower paleozoic clastic rocks and arc-related clastics and volcaniclastics have been tectonically thickened, polydeformed, metamorphosed, and wedged between the North American craton on the west and the late Precambrian Avalon terrane to the east. To a very great degree, then, the architecture of the crust in this region is the architecture of continental crust which has been assembled out of terrane fragments during several orogenic periods which range from late Precambrian to Permian (Table 2).

Combination of detailed field work with multiple age-dating methods is now yielding a fairly consistent scenario for the sequence in which the terranes were assembled [Zartman and Naylor, 1984; O'Hara and Gromet, 1985]:

1: Taconic accretion of an island arc to the North American rifted margin as the North American plate attempted to subduct eastward beneath the arc.

2: Acadian collision of a continental fragment of late Precambrian age (Avalonian) with the post-Ordovician margin, and entrainment of oceanic basin sediments, islands and Avalonian basement into the core of the orogen. Establishment of normal or extra-thickness continental crust over the affected region.

3: Alleghenian collision of a second fragment of

TABLE 2. Orogenic movements in the New England region[†]

Episodes/dates	Area of Influence	Characteristics
Alleghenian: 234-260 my	Southeastern New England	Folding, medium-grade metamorphism, granite intrusion
Acadian: 360-400 my	Central New England: from PG allochthons in west to central R.I. and E. Mass.	Medium-high grade metamorphism, ductile folding, granite intrusion
Taconic: 450-500 my	Western belt: from PC allochthons westward to Hudson Valley	Major thrusting, at least low-grade metamorphism, ultramafic emplacement
Avalonian: 600-650 my[*]	Dominates in S.E. New England, overprinted by Acadian westward	Granitic intrusion, folding, low to medium grade metamorphism, thrusting

[†]: after Rodgers [1970]
[*]: after Zartman and Naylor [1984]

Avalonian crust with the post-Acadian margin [O'Hara and Gromet, 1985].

The onshore geology is characterized by the following structural styles: (a) extensive metamorphic terranes with regionally consistent dips (e.g. Merrimack Synclinorium), and, in places with the strong mylonitization characteristic of deep-seated faulting (e.g. Nashoba belt); (b) domes and tight folds with near-vertical axial planes, and with concordant cores and layers of granitic rocks (e.g. Bronson Hill anticlinorium); (c) discordant granitic intrusive bodies (e.g. Dedham granodiorite, (d) supracrustal basins of Pennsylvanian, Triassic, and Jurassic age. Based on extensive numerical and theoretical modeling studies (e.g. Wong et al [1982]), I work with the following correspondence between seismic character and structural style:

1. Major through-cutting seismic boundaries are interpreted as fault zones of the sort which separate terranes.
2. Deep-seated seismo-structural domains which have strong, coherent reflections are interpreted as regionally coherent metamorphic terranes with dips generally less than 45°. For this region, I strongly prefer this interpretation to other possibilities which have appeared in the literature: (a) originally horizontal large-scale igneous intrusive sill cimplexes; (b) sedimentary rocks and (c) mylonites. The regional geology does not show the large batholithic complexes which would be associated with igneous underplating. Nearly all sedimentary rocks which have been entrained in the center of the orogen have been converted to gneissic or schistose metamorphic rocks which are found between central Rhode Island and the New York-Connecticut line, deformation on ductile shear zones with mylonitic fabric is common, and from the megascopic perspective of seismic waves, is an intrinsic part of the overall layered fabric.
3. Domains with weaker, more diffraction-like events may be complex, polydeformed zones with near-vertical axial planes for the dominant folding, or nearly homogeneous, but irregular intrusive bodies. In southern New England, in the central high-grade part of the orogen, complex, ductilely deformed structures are in any case closely associated with both sorts of intrusives. In other studies of crustal reflections, it has not been possible to establish criteria for identifying intrusive bodies as such, save the possibility that a large, homogeneous batholith is seismically transparent.

The post-accretional history is marked by the breakup of the continent and the opening of the Atlantic beginning in Triassic time. The Triassic-Jurassic rift basins form the onshore record of this period, while offshore seismic studies have delineated the development of the passive Atlantic margin as one side of the Atlantic megarift. Lying beneath the sediment cover of the modern margin are numerous fault basins, the offshore correlates of the onshore basins, and satellite to the main breakup. The two intrabasement basins seen on line 36 lie in this class. Inspection of the irregular basin 200-900 on line 24 shows that this basin opnes and deepens toward the continental edge a few miles to the southeast [Hutchinson et al, 1985]. The basin 1950-2250 is a half-graben about 15 km across and 4 km deep which developed by normal faulting along the western boundary, which is continuous with fault zone α.

Recent studies of extensional provinces have

yielded many examples of good laminated reflectors associated with ductile basement deformation produced in crustal extension. The extent to which extension has produced the intracrustal fabrics cannot be resolved on the ground of seismic character, but must be based on the regional geological context. In this area, the magnitude of Mesozoic or late Paleozoic crustal extension is substantially smaller than the degree of compression which characterized the Paleozoic accretionary phases; late stage extension must be regarded as a fractional reversal of the compressional tectonics which led to the building of the crust in the first instance. On the other hand, while the Acadian and Alleghenian orogenic periods, which both affected this region, were indeed major compressional episodes, the region may well have stood high, with tectonically thickened crust over most of that very long interval, and been subjected to more than one phase of extensional relaxation. My interpretation of the major fault zones which make up the architecture of this section, then, is that, while some degree of extension is required in the Mesozoic, they tell the story of how the crust was assembled in the first instance by compression due to collision of different terranes.

Relationship to the onshore geology

Recent detailed mapping, fabric studies and age-dating, in combination with the notion of suspect tectonostratigraphic terranes, have helped clarify the principal structural and age units of southern New England. Figure 5 (top), adapted from Hatcher [1981] is a representative geological cross section without vertical exaggeration. Most authors now consider the crust along this section to consist of at least three terranes, which were accreted to the eastern margin of North America during the orogenic periods of the Paleozoic [Williams and Hatcher, 1983, Zen, 1983, Zartman and Naylor, 1984]. The terranes are identified on the basis of their characteristic lithologic associations and tectonic histories, and are separated by major fault zones, across which stratigraphic correlation fails. In general, such terranes may have originated at quite distinct times and places.

The principal elements of this section, from east to west, consist of:
(a) The southeast New England platform, lying east of the Lake Char - Bloody Bluff fault, includes all of Rhode Island, Massachusetts southeast of the Worcester area, and a narrow strip of eastern Connecticut, now agreed to be a part of the Avalon terrane. The basement is a late Precambrian plutonic complex, which cores a supracrustal sequence of metasedimentary and metavolcanic rocks of Precambrian age; together these form a north-south oriented arch in Rhode Island [Barosh and Hermes, 1981]. Metamorphic grades are low, exceeding chlorite only in western Rhode Island, where folding and faulting associated with the Lake Char - Bloody Bluff fault are found. O'Hara and Gromet [1985] argue that the narrow western portion of the Avalonian block is a separately transported sub-block with distinct lithostratigraphic, metamorphic, and isotopic imprint, much like the adjacent Putnam-Nashoba zone.
(b) The Putnam-Nashoba zone, a narrow belt of high grade gneisses and schists dipping about 25° west, and lying between the Lake Char - Bloody Bluff fault and the Clinton-Newbury fault.
(c) The center of the orogen (labeled the Gander terrane by Williams and Hatcher [1983]) shows quite high metamorphic grades in southern New England, indicating substantial crustal thickening associated with accretion. Starting in the east, in contact with the Nashoba belt, the thick, west-dipping package of metaclastics of the Merrimack belt, grades westward into upright domes and folds of the Bronson Hill anticlinorium, in which volcanic and volcaniclastic rocks are associated with a thin cover sequence correlative with the Merrimack clastics. Further west, in the Connecticut Valley synclinorium, east-dipping folds and thrusts complete the fan-shaped cross-sectional geometry of this terrane. Local relations among major geological units in this terrane are complex, made obscure by numerous cross-cutting Acadian plutons. On the basis of available age dating of igneous rocks and correlation with unmetamorphosed equivalents in Maine, the terrane may be considered as a thick Siluro-Devonian sequence of often mature clastics deposited on a basement giving Ordovician and late Proterozoic ages. Although no oceanic crust appears in the basin, the setting does not suggest a cratonic basement for this deposition either.

West of these terranes, Taconic orogenic elements are thrust westward over the basement of the North American craton, with the transported thrust bodies of the Housatonic (Berkshire, Green Mountain) massifs passing westward to the thin-skinned Taconic thrust sheets.

Interpretation

In Figure 5 I have compared the interpretation of Figure 4 with the geological cross section by Hatcher [1981] as follows: At bottom, the seismic section is shortened by projection into a dip line; at the top, the Hatcher section is extrapolated down to the base of the crust using the seismic section as a model. The resemblance is striking. From the regional perspective, the architecture of the southern New England Appalachians varies sufficiently slowly along strike that it is not unreasonable to anticipate such a correlation. I believe that the agreement is good enough to give a reasonable first approximation to the structure of the orogen down to the base of the crust, and

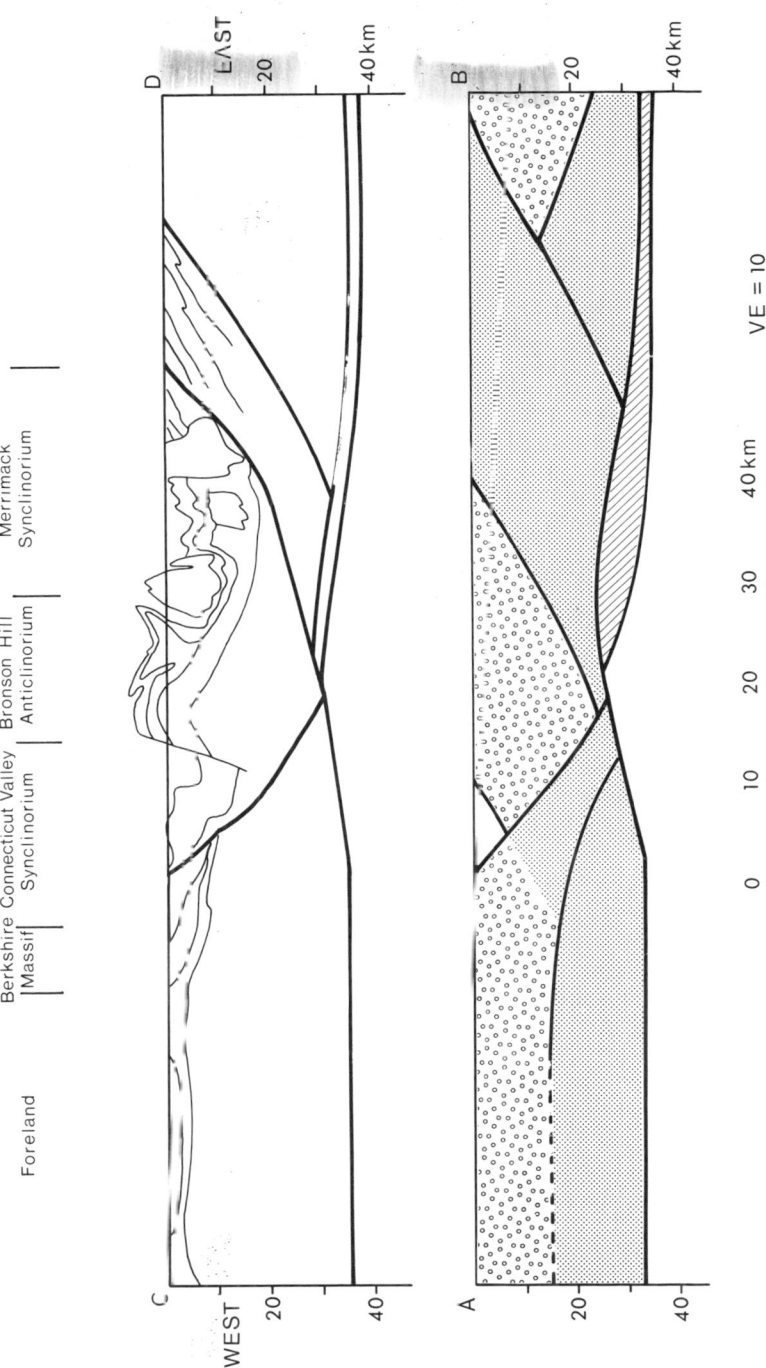

Fig. 5. Top: Geological cross-section of southern New England, from Hatcher [1981], without vertical exaggeration, along section line CD in Figure 1. An interpretive extrapolation to the base of the crust is based on the Long Island platform section (bottom), scaled to the dip direction.

should be used as a basis for tying down other issues.

A closer attempt to correlate is not irrelevant, although the possibility of along-strike changes between the geological and the seismic data must be kept in mind. The wedge-shaped seismic domain C1 is clearly associated with the core of the orogen, the Gander terrane. Moreover, both are coincident with the domain of high metamorphic grade across southern New England [Thompson and Norton, 1968] (Figure 1), which indicates a zone of particularly great crustal thickening at the time of orogenesis. The existence of weak hyperbolic arcs in C1 is interpreted as due to reflection from the axes of upright structures, like the Bronson Hill Anticlinorium. These are seen in the seismic section to flatten out with increasing depth, and to pass into a much simpler layered geometry. Perhaps the most natural correlation of the major faults would be to assign β to the Lake Char fault and γ to an intra-Avalonian fault such as the Hope Valley shear zone [O'Hara and Gromet, 1985]. However, these faults may be no better as reflectors than other layering in the west-dipping region which ranges from western Rhode Island to central Connecticut, and includes portions of three terranes. With these correlations, the shallow transparent zone C3 would lie in the center of the eastern Avalonian terrane, and would correspond to the late Precambrian plutonic basement of this region.

In southeastern Connecticut the Lake Char fault swings around to the west as the Honey Hill fault, as the Avalonian basement forms a domed promontory to the west. This feature correlates closely with a characteristic complex short wavelength magnetic pattern which runs east-west and lies athwart the coastline between Groton and the Rhode Island border. Combined with a general lack of correlation between the marine and land magnetic patterns, this has led to the suggestion that a dislocation or discontinuity along the axis of Long Island Sound disrupts the longitudinal continuity of the orogen. The regional Bouguer gravity similarly lacks characteristic features which can be easily associated with a cylindrical geometry for the orogen. A close look at the gravity and magnetic data in light of the seismic results reported here would be the subject of another study. In east-central Connecticut the Willimantic Dome is an oval tectonic window of Avalonian basement surrounded by rocks of the Merrimack synclinorium, in which both Lake Char and Clinton-Newbury fault equivalents are exposed [Wintsch and Fout, 1982]. By this model, the Honey Hill fault may be taken as a similar, slightly domed dislocation between terranes. In this case, substantial departures of the surface outcrop of the fault from the NNE regional strike line may occur. An antiformal structure in domain A, near SP 4600 (Figure 4) is an example of exactly this sort of low angle doming in section, with the right scale (∼15 km).

If the medium wavelength magnetic anomaly (Figure 1) used to establish regional strike on the Long Island Platform is also used for onshore correlation, homoclinal domain A ends up 50 km too far east, in Rhode Island. Some departures from strict two dimensional geometry are doubtless required. Uncertainty occurs in attempting to correlate fault zone α onshore. It may or may not be correlative with the western border of the Hartford Basin. There is no evidence that the basins themselves connect. Further west, there are no characteristic features on the section which can be correlated with western Connecticut geology. The line does not go far enough west (through New York City!) to tie in with the Precambrian highland massifs at the edge of the pre-Ordovician North American continent.

The COCORP lines in Vermont-New Hampshire [Ando et al, 1984] and in Georgia [Cook et al, 1981], although having lower signal/noise than the USGS marine lines, show features which appear quite similar to the deep-rooted domains on line 36, with appropriate dips of 25° to 50°. In northern New England, this is true of the entire New Hampshire transect, while in Georgia, this style appears eastward of the Kings Mountain Belt in the Inner Piedmont.

6. Remarks on the LCBL and the reflection-Moho

The LCBL shows (Figure 4) a characteristic thickness of about 0.5 sec west of SP 2800 and varies from 1.0 km to 1.5 km or more, east of SP 2800. This is strongly suggestive of distinct histories for these two sections. Between SP 2000 and SP 3000 relief in the reflection-Moho looks like an attenuated image of the fault α. The gap at SP 2000 is undoubtedly due to defocusing by the basin directly above. An estimate of the velocity pulldown caused by the basin accounts for part of the east-dip at SP 1850-1950 and hardly affects the west-dipping section at SP 2100-3000 at all.

More critical is whether this latter zone, where the reflection-Moho seems to shallow from 10.5 sec to 9 sec, is produced by velocity pullup due to high velocities in domain C1. An evaluation of this issue can be based on a joint assessment of the gravity and seismic information. Suppose that domains A and B are associated with an average sialic crustal composition with velocity 6.2 km/sec and density 2.75, and that domain C1 is substantially enriched in mafic materials due to entrainment of oceanic crust ... with a bulk velocity of 6.7 km/sec and density of 2.95. This velocity contrast would produce the observed pullup of 1.5 sec. If isostatically balanced, this load would place the M-discontinuity 1.9 sec (∼8 km) deeper than observed. It is certainly not supported in place by crustal strength however, for gravity anomalies in excess of 100 mgal would be seen, and the gravity map shows weak anomalies <10 mgal. I conclude that the average density of C1 is very close to the average elsewhere in this region; consequently, the average velocity of C1 is no different from the regional average velocity. This is more in accord with the geology, for the center of the Acadian orogen seems to have late Precambrian age sialic basement. In the next section I suggest that the thinning of the crust below the C1 wedge is connected with Mesozoic crustal extension and slip along the fault zone A.

7. Speculations on the tectonic history

The concepts of accretion of crustal material at subduction zones and the building of continental crust by lateral accretion of suspect terranes have become particularly well developed in studies of Alaska and the North American Cordillera [Coney, 1978; Coney et al, 1980; Jones et al, 1981]. Zen [1983] and Williams and Hatcher [1983] have made a strong case for the applicability of the suspect terrane idea in the Appalachians. The seismic section discussed in this paper provides clear examples of the cross-sectional structure of such accreted terranes. Probable terrane boundary faults penetrate to the LCBL (in this example, at least) at dip angles of 25°-50°. Where the bounding faults of a given terrane dip in opposite directions, wedge-shaped cross sections result. This provides an easy mechanism for significant vertical uplift in an active orogen.

The cartoons in Figure 6 illustrate a suggested scheme for the assembly and cratonization of the crust in southern New England. The seismic section is most informative as regards the mode by which terranes were accreted in the late Paleozoic and subsequently modified by extension in the Mesozoic; it contains no characteristic features which permit strong statements about the structure and evolution of the Taconic orogen or about the division of the accretionary movement into Acadian and Alleghenian phases.

The eastern margin of North America in Silurian time (a) begins with a package accreted by the Taconic collision. West-directed subduction of oceanic crust is causing the accretion of additional sediments from the possibly thick clastics in the basin. The Avalon terrane, with its Precambrian plutonic core, and mantled by Precambrian to Silurian shelf and marginal sediments, approaches collision with North America. The accretional margin builds up to an intermediate crustal thickness, perhaps 15 km, prior to the actual continent-continent collision.

Seismic multichannel reflection studies of active margins, where the wedge is less than 10 km thick, have revealed the characteristic geometry of an accretional margin. A complexly deformed, thickened wedge of accreted sediment overlies a detachment zone lying within sediments directly above the subducting oceanic crust. Such intermediate thickness crust thus acquires a basal boundary layer of strongly sheared sediments which shows up well on reflection data. On conversion to a full thickness crust, this basal layer must be raised to amphibolite or granulite grade metamorphism, and its reflectivity may be enhanced by further layer parallel ductile shear. The original oceanic M-discontinuity would then lie perhaps 5 km below the depth of the strong layered reflections and would be even more difficult to image than it is on normal oceanic crust. It is doubtful whether the characteristic diffraction patterns seen as the top of oceanic basement would be preserved in the accretion process.

By Middle Devonian (b), the traveling terrane has collided with the North American margin. The leading edge of the terrane is accreted to the orogen by underthrusting, and the marginal sediment package is entrained into a huge compressed wedge of material derived from the sedimentary and volcanic rocks of the basin. This low strength wedge is shortened, thickened, uplifted, and obducted westward over the continental margin along the fault A. The traveling terrane serves as a buttress which supports the thickening of the accreting wedge to about 50 km. A fan-shaped cross section is produced in the wedge, with isoclinal folds verging outward from the center. Near the margins of the wedge, recumbent isoclinal folds pass over into a strongly sheared homocline of high grade rocks. During this stage, dewatering of the lower crust driven by differential thermal expansion and high pore pressure may be responsible for a number of familiar phenomena: the production of mid- to upper-level granites, the strong relative enrichment in large ion lithophile elements toward the surface, the induction of retrograde metamorphism, and the emplacement of ore deposits. As collision proceeds, the LCBL continues to be derived from the detachment zone atop subducting oceanic crust, and its shape is an isostatic mirror of the orogenic topography.

Following the end of subduction and compression (c), the newly accreted crust is eroded to elevations closer to 1 km, as isostatic rebound brings the crustal thickness back to around 30 km. Uplift of the core of the orogen brings amphibolite and granulite grade rocks to the surface. Onshore age data require at least two phases of late Paleozoic accretion. According to O'Hara and Gromet [1985], Acadian accretion of a western Avalonian terrane (Hope Valley terrane) was followed by Alleghenian accretion of the larger eastern Avalonian terrane. Consequently, multiple occurrences of an accretion-relaxation cycle must have occurred, and the geometric details for the area of known geology will differ somewhat from Figure 6.

Finally (d), the extensional stresses associated with the Mesozoic rifting of Pangaea thin the crust about 10%. Some of the strain is taken up by normal faulting on reactivated faults (such as α, β, and γ), while the remainder occurs by ductile deformation. The 1.5 sec shallowing of the reflection-Moho between SP 2000 and SP 4000 is commensurate with the amount of horizontal extension indicated by the geometry of the basin at SP 2000, although details would be model-dependent.

8. Discussion

Only at SP 600 does the seismic section show the disrupted appearance with multiple crossing

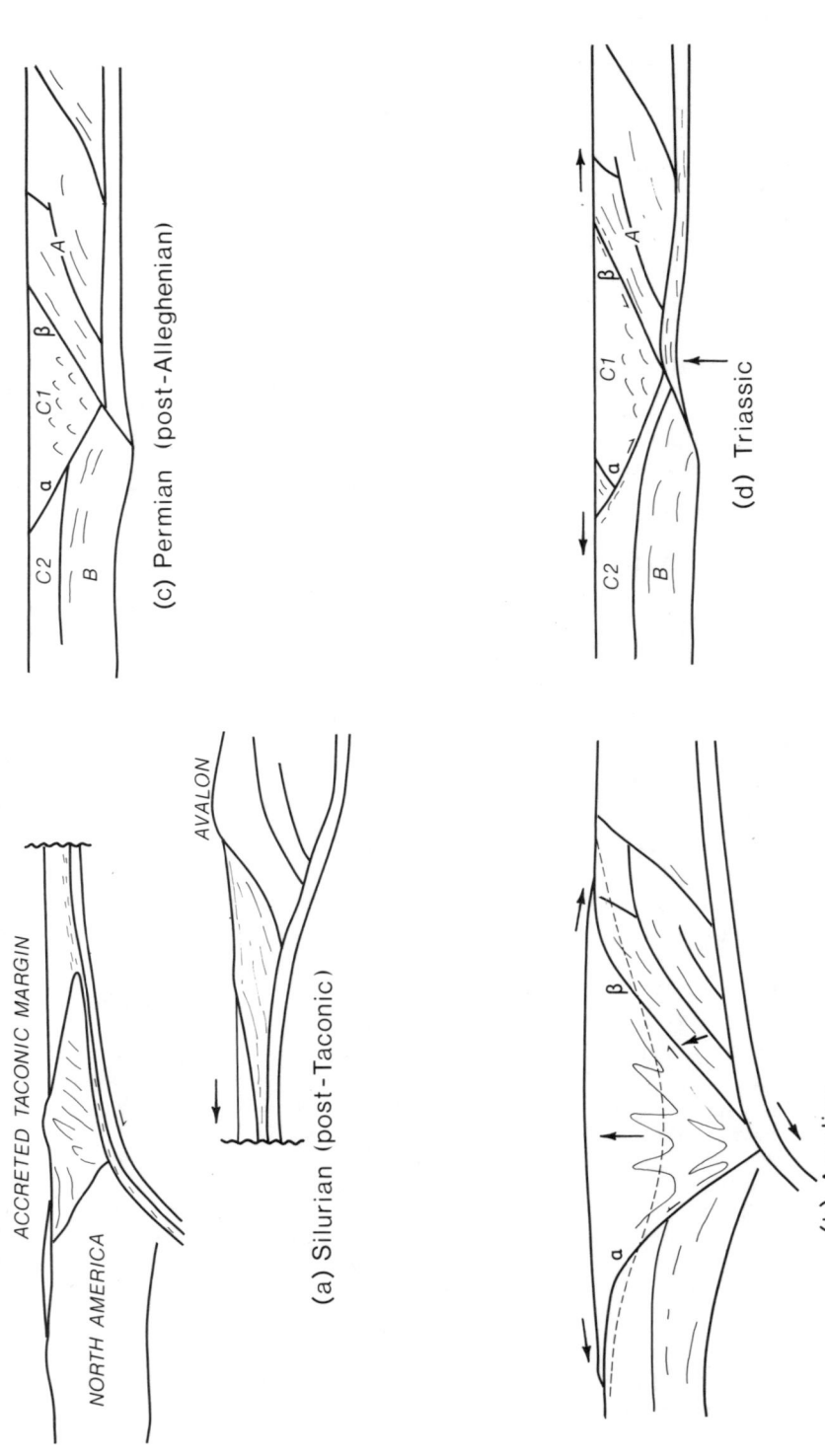

Fig. 6. Simplified model for stages in the development of the continental crust in southern New England: (a) Post-Tectonic: Subduction of oceanic crust toward the west under the margin of North America closes the basin between North America and Avalonia. (b) Acadian to Alleghenian: Collision of the traveling terrane with the North American margin in several episodes gives rise to detachment of the bulk of continental basement of Avalonia from the mantle and accretion along west-dipping sutures. Clastic, volcanic, and arc rocks from the basin are obducted into a wedge which is drastically shortened and thickened. Dashed line is the trace of the modern land surface, extending down to 20 km beneath the paleosurface. The keystone wedge is a mixture of both Avalonian and North American materials. (c) Post-Alleghenian: Erosion reduces land elevation to near sea level. Isostatic rebound causes the loss of 20 km of crustal thickness and the uplift of the center of the obducted wedge. (d) Triassic: Extensional stresses associated with the formation of the Atlantic by rifting extend the crust by about 10%, reactivating the inter-terrane boundaries as normal faults, producing rift basins at the surface and thinning of the crust at depth.

diffractions which would characterize a strike-slip zone across the section. Thus, the undisrupted texture elsewhere is more in accord with the recent demonstration by Irving and Strong [1984] that major strike-slip motion between the Avalon and Gander terranes as proposed by Kent [1982] is not required.

Ubiquitous smaller post-tectonic Acadian plutons throughout the region are principally alkaline or peralkaline [Zartman and Naylor, 1984], and probably derived from anatexis of crustal material in the center of the thickened crust, assisted by dewatering deeper in the pile. Their existence illustrates the need for some time delay between the principal isoclinal folding and the intrusion of the magmas. The lack of a classical large-scale calc-alkaline intrusive belt suggests that collision was probably turned off or minimized following the main Acadian deformation.

Rocks of the Merrimack synclinorium have a sedimentary protolith; it would be useful to estimate the volume of such accreted sediments from the seismic section (Figure 4). If the western half of domain A is identified with these sediments, its volume comes to about 800 cu km (per km along strike). The modern continental rise of the U.S. east coast has been accumulating sediment for 150 my, and tapers seaward for more than 300 km from a maximum thickness of about 15 km at the slope-rise break; its cross-sectional volume is thus about 2000 cu km. Given that the Avalonian microplate is not expected to be as prolific a sediment source as the Appalachian orogen, the quantities involved are not unreasonable. Particular interest inheres in having domain A derived from sediments, for this represents a particularly effective way to inject large quantities of water into the mid- and lower crust. Crawford and Mark [1982] discussed the problem of regional retrograde metamorphism for southeastern Pennsylvania in terms of finding a source for the water. The model introduced here provides a mechanism for the necessary water input from below and provides temporal and spatial significance to the occurrence of a particular retrograde event.

The keystone block of the collisional orogen constitutes a "root zone" which has been uplifted by compression between an active subducting wedge and a passive wedge. It is perhaps ironic that "the root zone has no root." This idea is similar to a proposal made by Hudson et al [1979] for an early Tertiary belt along the Gulf of Alaska margin.

9. Conclusions

The details of crustal structure revealed in this section provide a close look at the characteristic geometry assumed by accreted terranes in the building of continental crust in this region. The data yield of a specific model for the origin and composition of the continental Moho although it is quite unlikely that this model applies everywhere under continents. Moreover, this example can be extremely helpful in making sense of deep crustal reflections observed by land surveys, where the signal quality is not nearly as good. In this regard, a goal of this kind of work, "wet continental reflection seismology," would be to obtain enough examples of different crustal architectures and building blocks to facilitate interpretation of the land data.

Acknowledgements. This research was supported by NSF grant EAR 80-24108. I am grateful to John Suppe and Cliff Ando for their very helpful comments and suggestions, and to the reviewers for a very careful job. The USGS seismic sections were purchased from the National Geophysical and Solar-Terrestrial Data Center, Boulder, Colorado.

References

Ando, C.J., B.L. Czuchra, S.L. Klemperer, L.D. Brown, M.J. Cheadle, F.A. Cook, J.E. Oliver, S. Kaufman, T. Walsh, J.B. Thompson, Jr., J.B. Lyons, and J.L. Rosenfeld, Crustal profile of Mountain Belt: COCORP deep seismic reflection profiling in New England Appalachians and implications for architecture of convergent mountain chains, Bull. Am. Assn. Petrol. Geologists, 68, No. 7, 819-837, 1984.

Barosh, Patrick J. and O. Don Hermes, General structural setting of Rhode Island and tectonic history of southeastern New England, in Guidebook to Geological Field Studies in Rhode Island and Adjacent Areas; New England Intercollegiate Geologic Conference, edited by Jon C. Boothroyd and O Don Hermes, Department of Geology, University of Rhode Island, Kingston, Rhode Island 02881, p. 1-16, 1981.

Coney, P.J., Mesozoic-Cenozoic Cordilleran plate tectonics, Geological Society of America, Memoir No. 152, 38 pp., 1978.

Coney, P.J., D.L. Jones, and J.W.H. Monger, Cordilleran suspect terranes, Nature, 288, 329-333, 1980.

Cook, Frederick A., Larry D. Brown, Sidney Kaufman, Jack E. Oliver, and Todd A. Peterson, COCORP seismic profiling of the Appalachian orogen beneath the coastal plain of Georgia, Geol. Soc. Am. Bull., 92, No. 10, 738-748, 1981.

Crawford, Maria Luisa and Lawrence E. Mark, Evidence from metamorphic rocks for overthrusting, Pennsylvania piedmont (U.S.A.), Canadian Mineralogist, 20, 333-347, 1982.

Grow, J.A., J.S. Schlee, and W.P. Dillon, Multichannel seismic reflection profiles collected along the U.S. continental margin in 1978, U.S. Geological Survey Open File Report, No. OF 80-834, 1980.

Hatcher, Robert D., Jr., Thrusts and nappes in the North American Appalachian orogen, in Thrust and Nappe Tectonics, edited by C.R. McClay and N.J. Price, Geological Society of London Special Pub. 9, 491-499, 1981.

Hudson, Travis, George Plafker, and Zell E. Peterman, Paleogene anatexis along the Gulf of Alaska margin, Geology, 7, 573-577, 1979.

Hutchinson, Deborah R., John A. Grow, Kim D. Klitgord, and Robert S. Detrick, Moho reflections beneath the Long Island platform, eastern United States, in Results from Reflection Seismology, edited by Muawia Barazangi and Larry Brown, American Geophysical Union, Geodynamics Series, 1985 (this volume).

Hutchinson, Deborah R., Kim D. Klitgord, and Robert S. Detrick, The Block Island Fault, a Paleozoic crustal boundary on the Long Island platform, (submitted), Geology, 1985.

Hutchinson, Deborah R., Kim D. Klitgord, and Robert S. Detrick, Rift basins of the Long Island platform, (submitted), Geol. Soc. Am. Bull., 1985.

Jones, D.L., N.J. Silberling, H.C. Berg, and George Plafker, Tectonostratigraphic terrane map of Alaska, U.S. Geological Survey, Open File Report No. 81-792, 1981.

Klitgord, K.D. and J.C. Behrendt, Aeromagnetic anomaly map of the United States Atlantic continental margin, U.S. Geological Survey Miscellaneous Field Studies Map, No. MF-913, 1977.

Rodgers, John, The tectonics of the Appalachians, Wiley-Interscience, 270 pp., New York, 1970.

Taylor, Steven R. and M. Nafi Toksoz, Crust and upper mantle velocity structure in the Appalachian orogenic belt: Implications for tectonic evolution, Geol. Soc. Am. Bull., 93, No. 4, 314-329, 1982.

Thompson, J.B. and S.A. Norton, Paleozoic regional metamorphism in New England and adjacent areas, in Studies of Appalachian geology: northern and maritime, edited by E-an Zen, W.S. White, J.B. Hadley, J.B. Thompson, Jr., Interscience Publishers, 319-327, 1968.

Williams, Harold and Robert D. Hatcher, Jr., Appalachian suspect terranes, in Contributions to the tectonics and geophysics of mountain chains, edited by R.D. Hatcher, Jr., Harold Williams, and Isidore Zietz, Geological Society of America Memoir 158, 33-53, 1983.

Wintsch, Robert P. and James S. Fout, Structure and petrology of the Willimantic dome and the Willimantic fault, eastern Connecticut, in Guidebook for fieldtrips in Connecticut and South Central Massachusetts, New England Intercollegiate Geological Conference, edited by R. Joesten and S. Quarrier, State Geological and Natural History Survey of Connecticut, Guidebook No. 5, 465-482, 1982.

Wong, Yun K., Scott B. Smithson, and Ronald L. Zawislak, The role of seismic modeling in deep crustal reflection interpretation, Part I., Contributions to Geology, 20, 91-109, 1982.

Zartman, Robert E. and Richard S. Naylor, Structural implications of some radiometric ages of igneous rocks in southeastern New England, Geol. Soc. Am. Bull., 95, 522-539, 1984.

Zen, E-an, Exotic terranes in the New England Appalachians - limits, condidates, and age: a speculative essay, in Contributions to the tectonics and geophysics of mountain chains, edited by R.D. Hatcher, Jr., Harold Williams, and Isidore Zietz, Geological Society of America Memoir 158, 55-82, 1983.

MOHO REFLECTIONS FROM THE LONG ISLAND PLATFORM, EASTERN UNITED STATES

D.R. Hutchinson[1,2], J.A. Grow[3], K.D. Klitgord[1], R.S. Detrick[2]

[1] U.S. Geological Survey, Woods Hole, Mass. 02543,
[2] Graduate School of Oceanography, U.R.I., Kingston, R.I., 02882
[3] U.S. Geological Survey, Denver, Colorado 80225

Abstract. Strong reflections from 9.5-12 s depth (two-way travel time), which were recorded on a grid of seismic-reflection profiles on the Long Island platform of the U.S. Atlantic continental margin, are interpreted as reflections from the Mohorovicic discontinuity. The character of the reflection is generally sharp although some laminations occur to the east. The southerly dip of the Moho surface is explained by the effects of thickening low-velocity sediment and water layers. A region of lower travel times east of the Long Island rift basin may be crust that was thinned during Mesozoic rifting. A region of apparent crustal thickening in the central platform may have been relatively unaffected by Mesozoic extension. Travel times through the crust (excluding sediments and water) decrease towards the basement hinge zone, in agreement with theories of crustal thinning across passive continental margins. The velocity of the crust beneath the platform is not well enough known to permit accurate conversion of the time-sections to depth or migration of the deep events.

Introduction

The Mohorovicic discontinuity (Moho), which is best known from seismic refraction experiments, has been frequently observed in deep seismic-reflection profiles from continents during the last decade. The seismic character of Moho beneath the continents has been observed as laminated and discontinuous, leading to models of a layered transition zone between the crust and mantle [Meissner, 1973; Hale and Thompson, 1982]. Recent deep reflection profiles from the continental shelf around Scotland have shown a strong reflector at Moho depths and indicate that continental Moho can be characterized as a sharp interface [Matthews, 1982; Brewer and Smythe, 1984]. To date, our knowledge of reflections from the Moho beneath continents is restricted to widely spaced, isolated profiles such as those collected by the Consortium for Continental Reflection Profiling (COCORP) and the British Institutions Reflection Profiling Syndicate (BIRPS) [Matthews, 1982; Oliver et al., 1983].

In this paper, we present examples of reflections from the Moho observed on a grid of seismic profiles located on the Long Island platform segment of the U. S. Atlantic continental margin (fig. 1). High-amplitude reflectors can be identified on many of the profiles at times ranging from 9.5-12.0 seconds (s), in general agreement with inferred Moho times of 11-12 s obtained in earthquake and reflection studies to the north [Taylor et al., 1980; Taylor and Toksoz, 1982; Brown et al., 1983]. Despite its location in a marine environment, the Long Island platform is considered to be continental crust [Klitgord and Behrendt, 1979]. The basement hinge zone, which marks the zone of rapid increase of subsidence, sediment thickening, and major crustal extension within the Baltimore Canyon trough to the south and the Georges Bank basin to the east marks the seaward edge of the platform (fig. 1). Our data set provides details on the lateral variation of Moho reflections. We compile a traveltime map of the Moho surface, compare it to published crustal maps for New England, and speculate on its implications for crustal structure.

Seismic-reflection data

The seismic-reflection data from the Long Island platform consist of about 1300 km of profiles contracted by the U.S. Geological Survey in the 1970's (fig. 1). About 430 km of these data (USGS lines 2, 5, 9, and 12) do not record reflections from the Moho because of short record lengths. Moho reflections can be identified on 710 km of the remaining 870 km of data (USGS lines 16, 21-24, 36 and 37), which were recorded to 12 s. None of these profiles have been migrated. Representative profiles showing reflections from the Moho from beneath the Long Island platform are shown in Figures 2-9. The profiles will be discussed in a west-to-east progression across the platform.

USGS line 37/line 36: This profile (fig. 2)

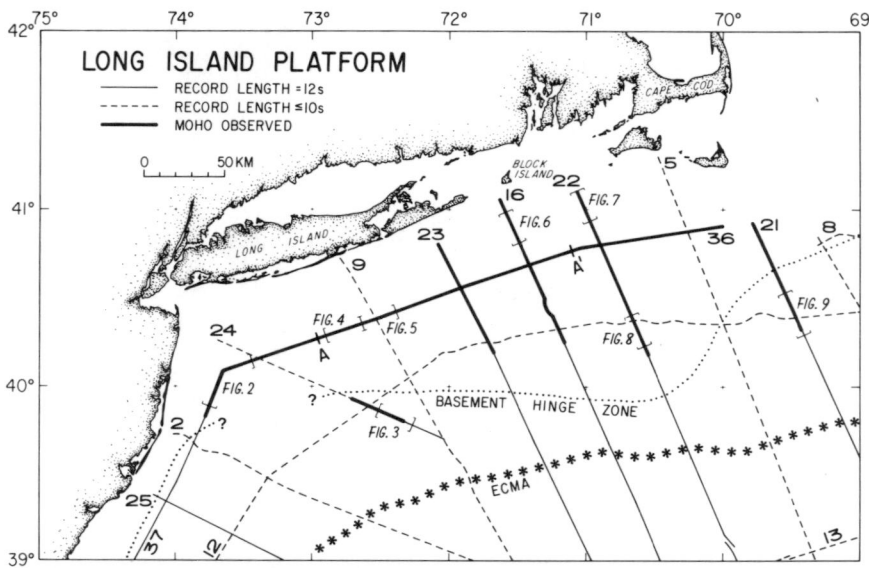

Figure 1: Location of deep seismic-reflection profiles on the Long Island platform. Profiles containing Moho reflections are shown by heavy lines. Line numbers are shown next to each line. The basement hinge zone (dotted line) is from Klitgord and Schouten (in press). The axis of the East Coast Magnetic anomaly (ECMA) is shown in stars [after Klitgord and Behrendt, 1979]. Locations of figures 2-9 are shown by brackets. Except for Lines 21 and 24, only data landward of the basement hinge zone were used in this study.

is characterized by a single strong reflector at about 10.5 s that can be traced continuously over 35 km from SP 4500 (line 37) to SP 400 (line 36). This Moho event marks an abrupt change in reflection character from lower amplitude, less continuous and more dipping events in the lower crust to infrequent, discontinuous and generally subhorizontal events in the upper mantle. This profile illustrates the consistency in reflection style and depth on lines shot in different directions, since line 37 strikes north and line 36 strikes east. It also contains the best signal-to-noise ratio of most of the data on the platform. Postrift sediments on this profile are generally less than 1.5 s thick.

USGS line 24: Line 24 (fig. 3) illustrates deep reflections that continue seaward of the basement hinge zone. The Moho reflection is interpreted to dip slightly from 10.8 s at SP 2650 to 11 s at SP 2800 and more steeply from just over 11 s at SP 2900 to just over 11.5 s at SP 3000. The generally lower signal-to-noise ratios on this deep event may be due to the thicker section of postrift sediments (3 s) and therefore greater attenuation of the seismic signal relative to profiles landward of the hinge zone. The steep dip of the event between SP 2900 and SP 3000 is probably a velocity pull-down caused primarily by rapid thickening of low velocity postrift sediments across the hinge zone.

USGS line 36: Figures 4 and 5 show a continuous record from SP 1500 to SP 2400 and illustrate some of the relief on the Moho surface. The inferred Moho event is subhorizontal at 10.5 s from SP 1500 to SP 1800, then dips eastward to 11 s at SP 1950. It cannot be easily identified between SP 1950 and SP 2200, but reappears at 10.2 s beneath SP 2200 and shallows to 9.5 s near SP 2350. The Moho reflection distinctly separates dipping reflectors in the lower crust from an essentially transparent upper mantle. The change in travel time for the Moho event on Figure 4 is about 0.5 s, and on Figure 5 is just less than 1.0 s. This occurs beneath a localized basin in the upper crust that is inferred to be a rift graben containing up to 2 s of sediments [Hutchinson, 1984]. As on line 24, we interpret most of this time relief to a velocity pull-down effect caused by low velocity sedimentary rock within the rift basin.

USGS line 16: Beneath the central portion of the Long Island platform, the Moho event continues to be a boundary between a reflective lower crust and transparent upper mantle. In Figure 6, the Moho reflection dips gently seaward from 10 s at SP 200 to 10.4 s at SP 500. Despite the poor quality of line 16, the Moho reflection can be identified as a distinct and relatively continuous event.

USGS line 22: Clear reflections from Moho depths are evident just east of line 16 on

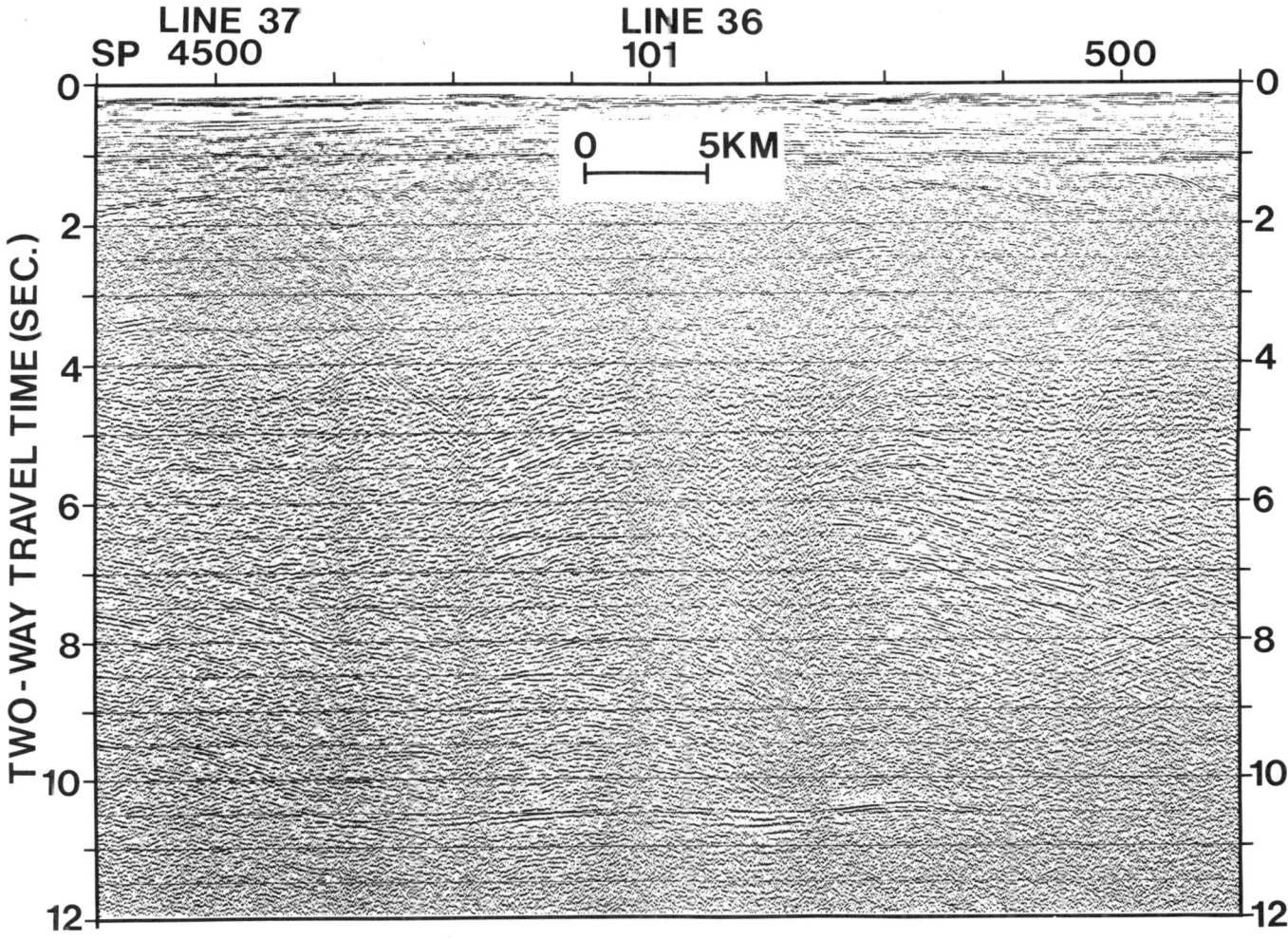

Figure 2: Multichannel seismic profile along the intersection of Lines 37 and 36. The strong event at 10.5 s, which is interpreted as Moho, separates a highly reflective lower crust from a relatively transparent upper mantle. Postrift sediments are at 0-1 s depth. The irregular basement surface at 1-2 s is the inferred bottom of a synrift basin that underlies the postrift sediments.

Line 22 (figs. 7 and 8). The reflections are more banded than on other lines, as illustrated by the subhorizontal reflections between 9.2 and 10.0 s near SP 1000. Within this zone, the reflection at 9.8 s is the most continuous and may connect with the high-amplitude reflector at 9.7 s near SP 1300. The banded zone near SP 1000 may consist of lower crustal reflections which are subparallel to the Moho reflector, making identification of the more continuous Moho reflection difficult. Beneath the outer shelf, the inferred Moho reflection is again a sharp reflector (fig. 8). It dips from about 10.3 s at SP 2600 to about 10.6 s at SP 2800 and can be seen to truncate lower crustal reflections at SP 2700.

USGS line 21: At the eastern end of the Long Island platform, on line 21 (fig. 9), weak subhorizontal reflections can be seen between 10.0 and 10.5 s and are inferred to be from the Moho. They are similar to the Moho reflections on the landward end of line 22 (fig. 7) in that they define a zone of reflectors about 0.3-0.4-s thick. Two features distinguish this line: First, the Moho does not dip seaward despite the seaward thickening of low-velocity sediments in the upper crust, which suggests that higher velocity material must compensate for the effect of low-velocity sedimentary rock or that the crust must thin. We lack velocity information to investigate this effect, since these reflection data do not contain reliable velocity measurements for the deeper crustal section. Second, the subhorizontal events at 10.0-10.5 s

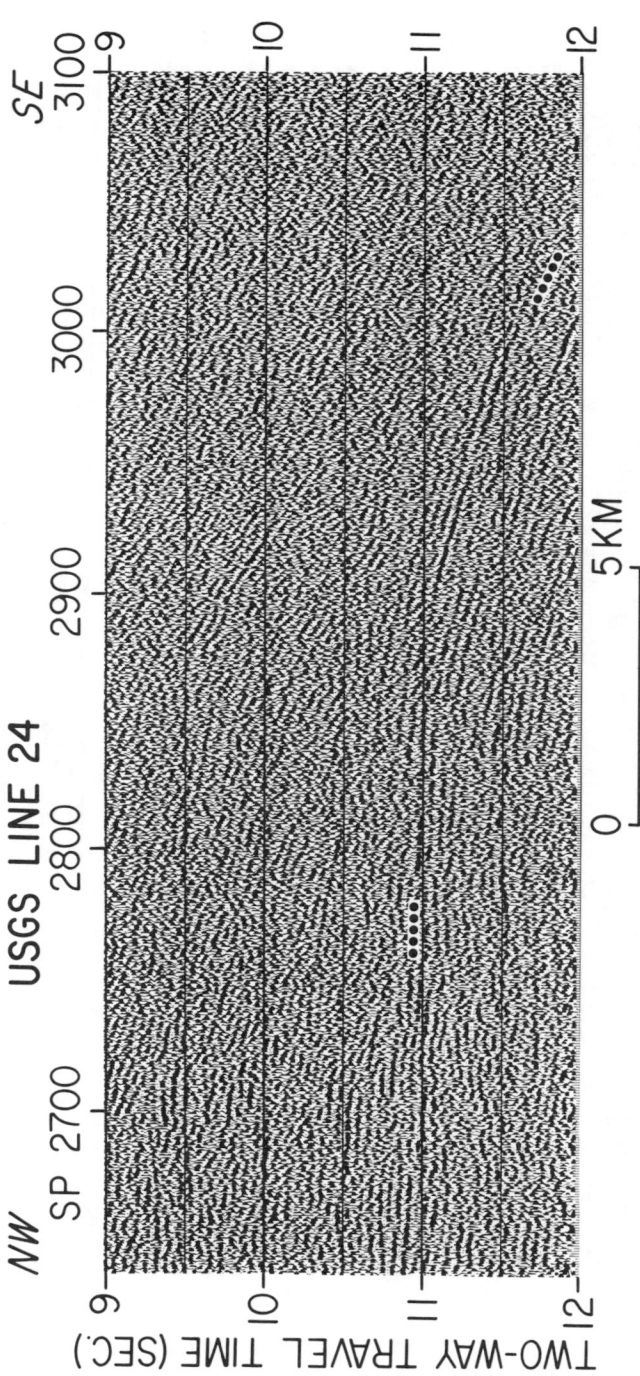

Figure 3: Multichannel seismic-reflection profile along Line 24. Moho, marked by dots, occurs near 11 s at SP 2700 and dips to 11.5 s at SP 3000.

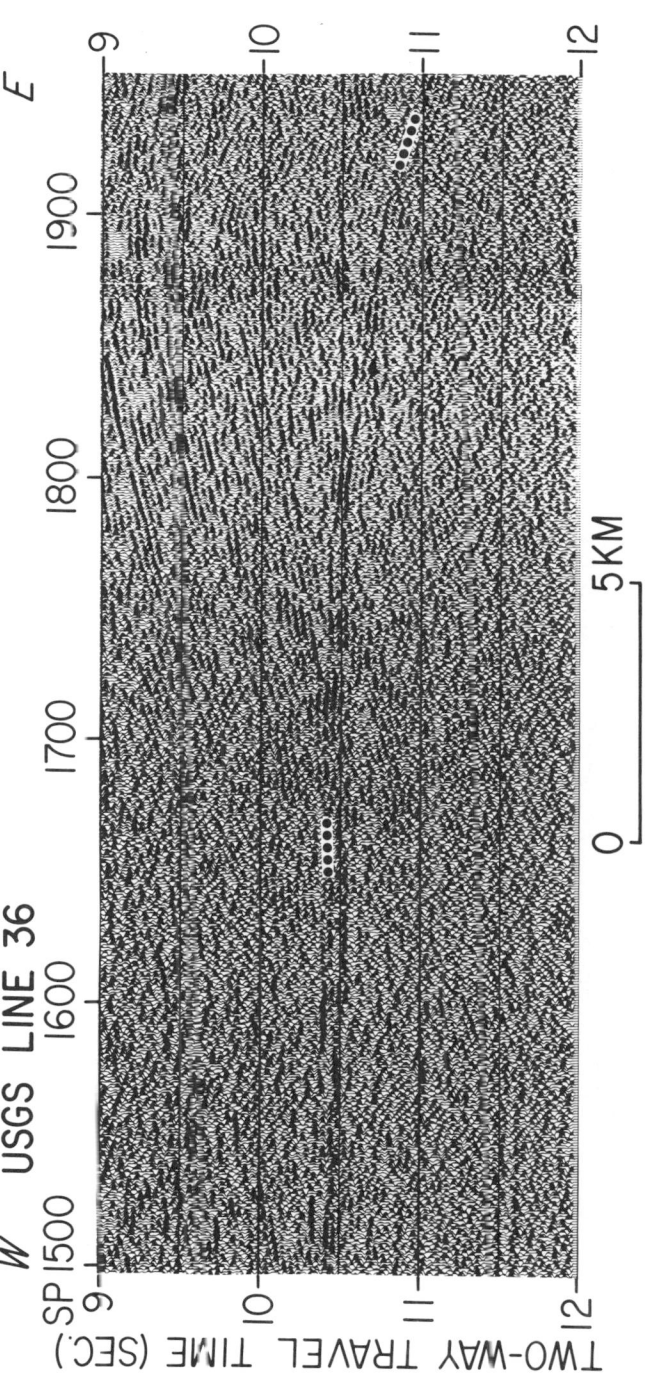

Figure 4: Multichannel seismic-reflection profile along Line 36 showing Moho (dots) at about 10.3 s (SP 1500), at 10.5 s (SP 1800) and near 11 s (SP 1950). This dip is interpreted as a velocity pull-down related to a buried synrift basin in the upper crust. This profile connects to Figure 5.

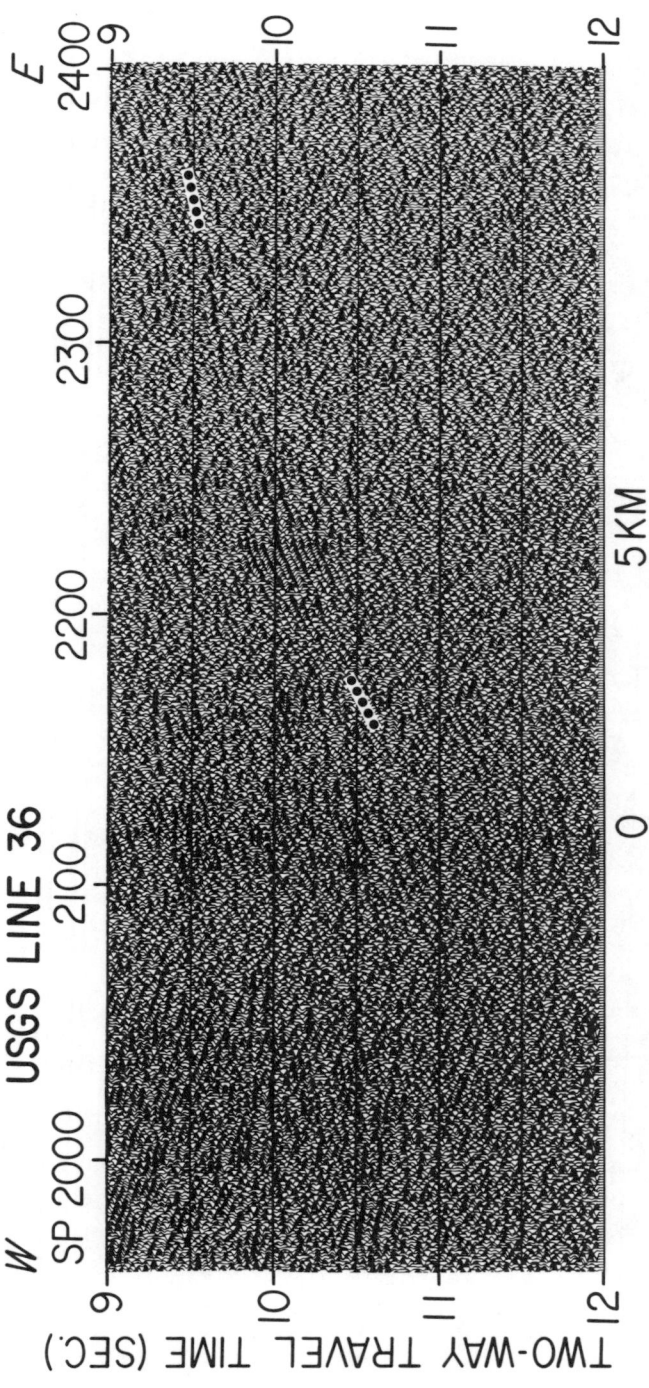

Figure 5: Multichannel seismic-reflection profile along Line 36. Moho, marked by dots, is evident near 10.0 s (SP 2200) and 9.7 s (SP 2300). See Figure 4 caption.

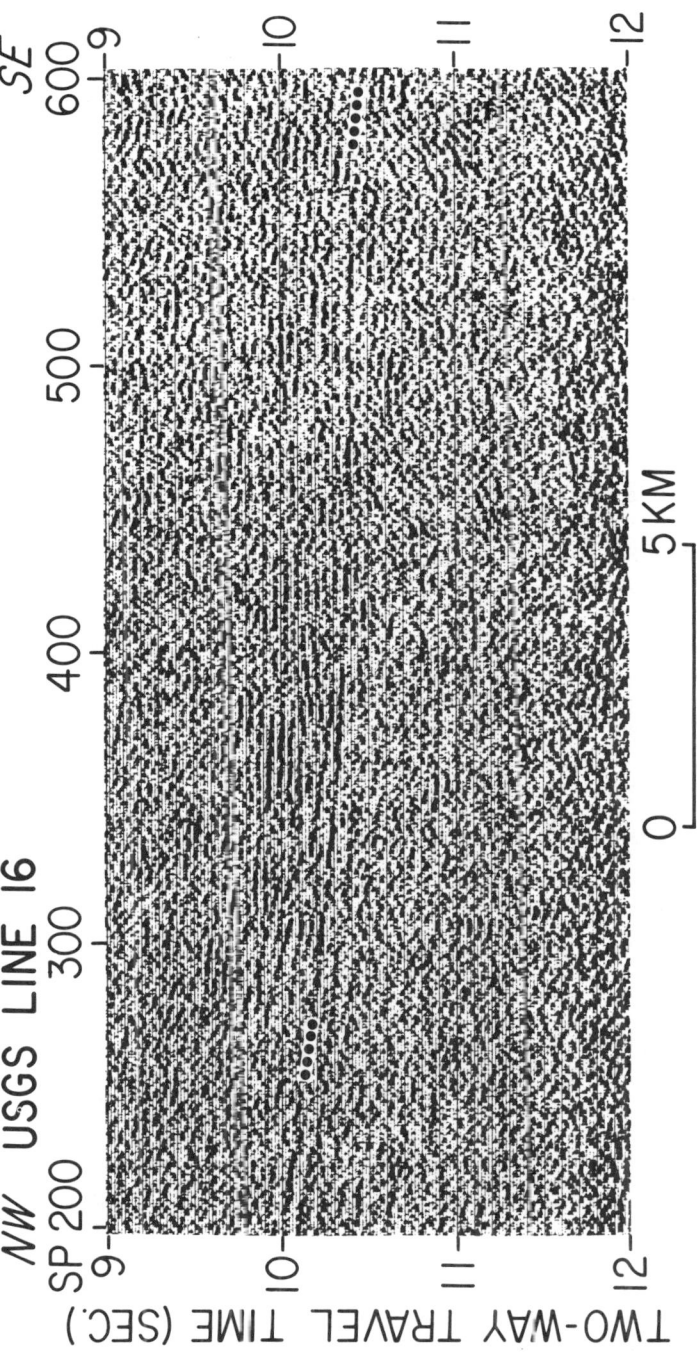

Figure 6: Multichannel seismic-reflection profile along the north end of Line 16. Moho, marked by dots, is at 10.0 s (SP 200), at 10.3 s (SP 400), and at 10.4 (SP 600) and appears to truncate dipping reflectors near SP 300.

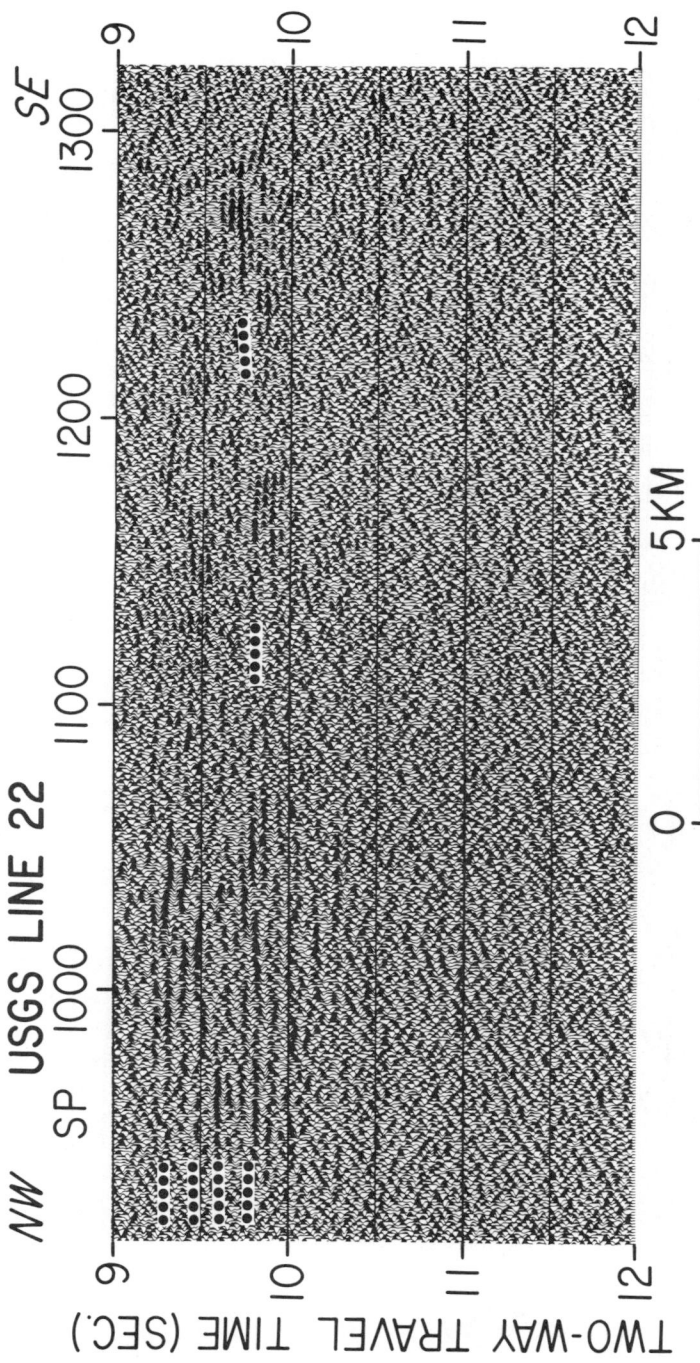

Figure 7: Multichannel seismic-reflection profile along Line 22 on the inner shelf. The Moho reflection, which is marked by dots, consists of parallel reflections from 9.2-10.0 s. The reflector at 9.8 s (SP 1000) is more continuous than the others and may connect with the strong event at 9.7 s (SP 1280).

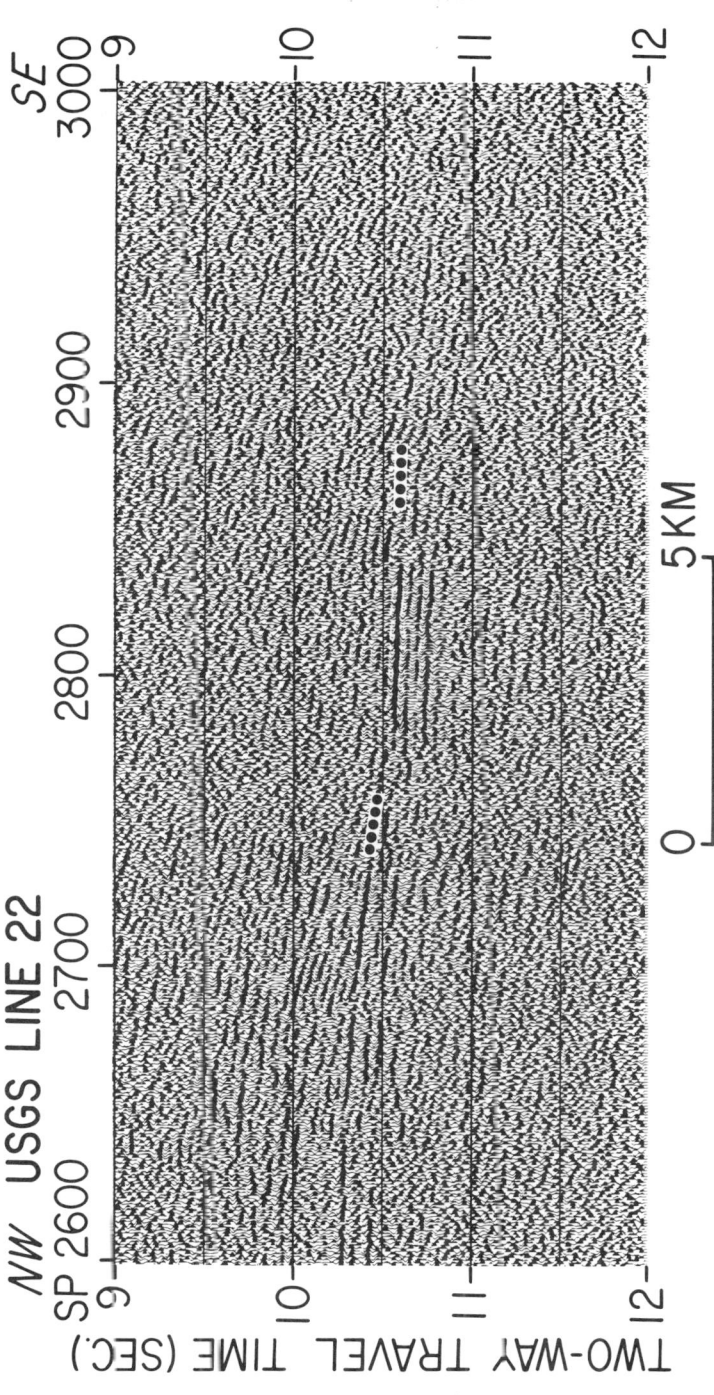

Figure 8: Multichannel seismic-reflection profile along Line 22 on the outer shelf. Moho (dots) is the strong reflector at 10.3 s (SP 2600) that dips to 10.6 s (SP 2800). Several subparallel wavelets occur in the zone 0.1-0.2 s beneath the reflector. Dipping reflectors in the lower crust are truncated near SP 2700.

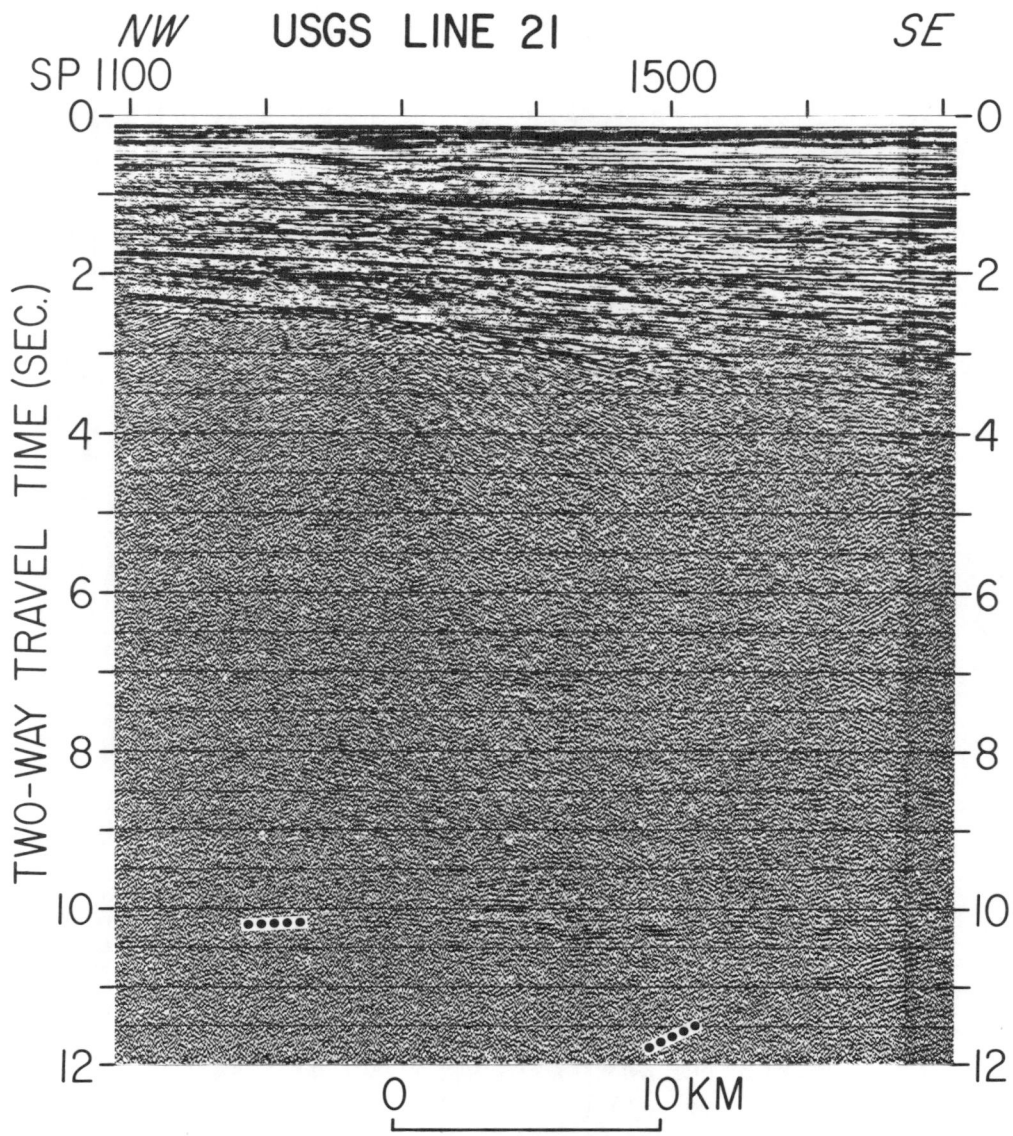

Figure 9: Multichannel seismic-reflection profile along Line 21. Moho, which is marked by dots, consists of discontinuous subhorizontal reflectors from 10.0-10.5 s across the profile. A band of dipping events that shallows from 12 s (SP 1450) to 10.2 s (SP 1700) truncates the subhorizontal event near SP 1700. The basement hinge zone occurs near SP 1300, where the strength and dip of the postrift unconformity (2.6 s) changes. See text for discussion.

are truncated by a dipping event that shallows from 12.0 s near SP 1500 to 10.2 s near SP 1700. This event is not continuous beyond SP 1700 and cannot be easily detected beyond SP 1800.

Regional variations in the reflections from Moho

The sharpest and most distinct reflections from Moho occur on the western and central parts of the platform where Moho is represented by a single, double or triple wavelet that is generally 0.1-0.2 s wide (figs. 2-4, 6). The same wavelets can be traced continuously for distances up to 35 km (fig. 2) and almost continuously for much greater distances. The style of these Moho reflections resembles parts of the BIRPS profiles from the continental shelf in the northern British Isles, [Smythe et al., 1982; Brewer et al., 1983; Brewer and Smythe, 1984].

The Moho is characterized by banded reflections in the eastern part of the platform

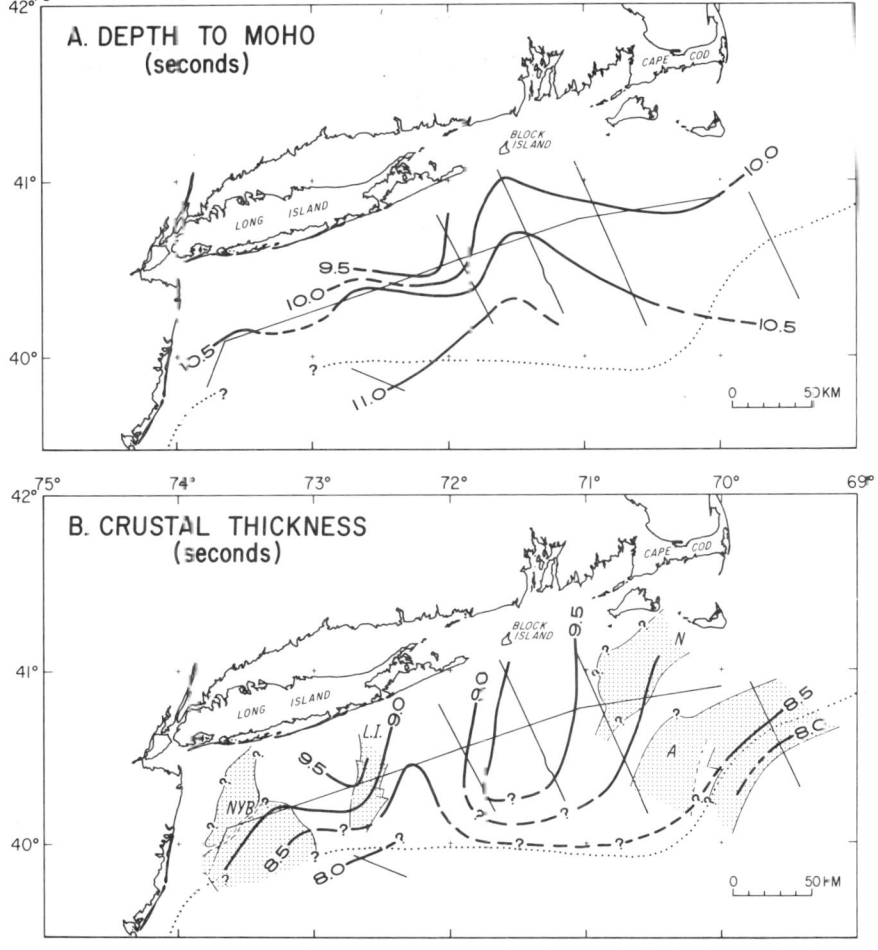

Figure 10: A. Depth to Moho in two-way travel time, dashed where inferred. B. Crustal thickness in two-way travel time. This represents the removal of water and postrift sediments from the depth-to-Moho measurement. The locations of buried Mesozoic rift basins are shown in stippled pattern: NYB-New York bight basin; L.I.-Long Island basin; N-Nantucket basin; A-Atlantis basin. Buried basins are from Hutchinson [1984]. Dotted line marks the basement hinge zone.

(fig. 9) and in localized regions in the central portion (figs. 7-8). Whereas the bands may be up to 0.9-s thick, they are generally only 0.3-0.5-s thick, and sometimes one wavelet within the zone of laminae dominates in amplitude and continuity (fig. 8). Most reflections from the Moho beneath continents are laminated in zones 1-2 s thick [Meissner, 1973; Hale and Thompson, 1982; Oliver et al., 1983], which is somewhat thicker than the banded packets interpreted on the Long Island platform.

Part of the reason that Moho is such a regionally identifiable reflector on the Long Island platform may relate to the nature of the surrounding crustal and mantle rocks. In general, the lower crust of the platform is highly reflective with dipping sequences, whereas the upper mantle is relatively transparent (figs. 2, 4, 6, 8). In the central and eastern portions of the platform, where Moho appears laminated, the lower crust is sometimes less reflective or is characterized by subhorizontal reflections. Faint subhorizontal reflectors also persist in the inferred upper mantle. Because the lower crust and mantle are similar in seismic character, identification of the Moho event is considerably more difficult in these areas.

A second reason that Moho is so readily observed is that the thin postrift sediment cover (≤ 1 km) minimizes attenuation and loss of seismic energy shallow in the section. The quality of the reflection from Moho deteriorates beneath inferred rift basins, where the sedimentary fill may reach 2-4 km (figs. 4, 5), and towards the

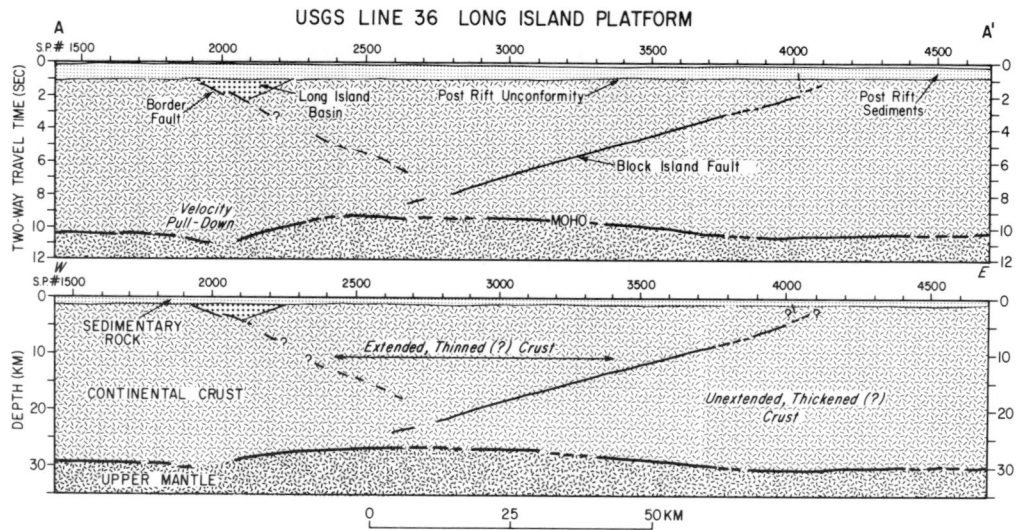

Figure 11: A cartoon cross-section of line 36 across the central Long Island platform showing interpretations in two-way travel time (upper half) and depth (lower half). The interpretations of the Long Island basin and Block Island fault are from Hutchinson [1984] and Hutchinson et al. [in press]. Location of A-A' is shown on figure 1. See text for an explanation.

basement hinge zone, where the sediments thicken to 4-5 km (fig. 9).

Relief on the Moho surface

From the grid of reflection profiles, we have constructed a traveltime map representing the depth to the base of the crust (fig. 10A). The Moho reflector was picked at the top of the strongest and most continuous reflection from the western and central parts of the platform. In the eastern region, it was picked in the middle of the laminated zone if no reflection dominated. The depth-to-Moho map shows an undulating surface that dips southerly from 9.5 s to 11.0 s. The regional southerly dip can be explained by the effect of thickening low-velocity sediments and increasing water depth across the continental shelf. No abrupt changes in Moho depth occur across the hinge zone.

The regional effects of increasing water and sediment thickness have been removed in the crustal-thickness map (fig. 10B) by subtracting the thickness of sediments and water from the depth-to-Moho measurement. The base of sediments was picked at the postrift unconformity, which is a conspicuous acoustic reflector found everywhere on the platform (and on most of the U.S. continental margin) that separates prerift or synrift deformed rocks from undeformed postrift continental margin sediments [Schlee, 1981]. Rift-related grabens or basins were included in the crustal-thickness measurement since the bottom of these structures is not always acoustically well defined. Outlines of the known locations of these basins beneath the platform [Hutchinson, 1984] are shown in Figure 10B for reference. The most-pronounced and best-constrained zone of crustal thickening is that which is south of Block Island, where the crust thickens from 8.5 s on the west and 9.0 s on the east to slightly more than 9.5 s. The slight increase in traveltime near the Long Island rift basin (fig. 10B) is probably caused mostly by velocity pull-down related to sedimentary rocks within the basin, as discussed earlier. This velocity pull-down is asymmetrical, with traveltimes to Moho decreasing from 9.0 s west of the basin to 8.5 s east of the basin. This asymmetry is difficult to explain as a pull-down effect, and may indicate that the crust east of the Long Island basin is either thinner or of higher-velocity than that to the west. No similar pull-down effect is observed beneath the Nantucket, New York Bight, or Atlantis basins.

The crust thins across the platform towards the hinge zone. Whereas the total travel time to Moho near the hinge zone appears to increase to 10-11s (fig. 10A), crustal thickness excluding sediments and water actually decreases across the platform from 9.0-9.5s to about 8.0s (fig. 10B).

Implications for crustal structure

A comparison of the traveltime through the crust of the Long Island platform with that measured on the New England Appalachian COCORP profile shows that the platform has lower traveltimes. Moho reflections were observed on Line NY 7 in the Adirondacks at about 11 s, and Line NY 1 in the Taconics at 10-12 s dipping east [Brown et al., 1983]. These measurements should be compared to the traveltimes through the platform omitting sediments and water which are 8.5-9.5 s (fig. 10B). Whether the lower traveltimes of the platform represent a higher velocity crust or a truly thinner crust cannot be resolved without velocity information. Possible reasons that the crust might be thinner include its position within a tectonically different part of the Appalachian orogen, or its location on crust that was thinned during the Mesozoic rifting event.

Mesozoic extension causing crustal thinning probably explains most of the regional decrease in travel times near the hinge zone (fig. 10B). Our data are insufficient to identify or map the nature of changes in Moho reflections seaward of the hinge.

We have drawn a schematic cross section of line 36 to illustrate the relationship between the Moho surface and other crustal features within the central part of the platform (fig. 11). The depth section was calculated from the time section by applying interval velocities to the postrift sedimentary rock and the Long Island rift basin in the upper crust, and assuming a constant velocity of 6.0 km/s for the rest of the crust. No velocity changes were assumed across the Block Island fault. The western border fault of the Long Island basin projects into a zone of east-dipping reflections that are truncated by the Block Island fault [Hutchinson, 1984]. The Block Island fault is interpreted as a low-angle thrust(?) fault that formed in the late Paleozoic and has been multiply reactivated with normal and/or reverse motion in late Paleozoic, early Mesozoic and late Mesozoic-Cenozoic times [Hutchinson and others, in press]. An alternative interpretation for these data is given by Phinney [this volume].

In both the time and depth sections, the crust at SP 2400-3200 east of the Long Island basin is thinner than that at SP 1400-1800 to the west by about 0.5-1.0 s (1.5-3.0 km). The zone of crustal thinning at SP 2400-3200 is probably the region that underwent crustal extension resulting in formation of the Long Island basin. Recent models of extensional tectonics allow rift basins to be located at large distances from zones of crustal extension because of low-angle normal faults [Wernicke, 1981; Anderson, et al., 1983]. Here, the Long Island basin is adjacent to, but not directly over, the zone of apparent thinning. The Block Island fault appears to have formed a barrier which restricted extension from continuing across the fault into the lower plate. We have no evidence to evaluate extension along the Block Island fault because no synrift sedimentary basin exists along its surface.

The zone of crustal thickening south of Block Island (fig. 10B) is spatially associated with the lower plate of the Block Island fault (fig. 11). The crust thickens by about 1.0 s (3 km) relative to other parts of the platform. This zone of thickening may connect with a similar zone of thickening (or low-velocity crust) identified in central New England by Taylor and Toksoz [1979]. They calculated that the crust beneath central Vermont and New Hampshire is 3-5 km thicker than crust in northern New York or Maine and that slight thickening continues southward into eastern Connecticut and Rhode Island. A negative Bouguer gravity anomaly of -50 mgal correlates with the crustal thickening beneath Vermont/New Hampshire and a -25 mgal anomaly correlates with that beneath Connecticut/Rhode Island [Taylor and Toksoz, 1979]. These negative gravity values extend offshore beneath the Long Island platform with values between 0 and -20 mgal [Hutchinson, 1984].

An alternative and more probable interpretation is that the region of crustal thickening represents a part of the platform unaffected (or minimally affected) by Mesozoic extensional tectonics. It only appears as a zone of thickening relative to the surrounding extended or thinned crust. The distribution of rift basins both west and east of the region (fig. 10B) supports this interpretation.

The change in reflection character of Moho from sharp in the west to laminated on the east does not seem to correlate with crustal boundaries or thickness. Both sharp and laminated reflections occur at depths of 10.0 to 10.5 s (compare figures 2 and 7). Laminated reflections at the eastern end of the platform are structurally located at the base of the lower plate of the Block Island Fault, but sharp reflections also occur on parts of line 23 (not illustrated) and line 16 (fig. 6) at the base of the lower plate. The change in the continuity and thickness of the reflection Moho is interpreted to be most closely related to the thickness and structure of sedimentary rock in the uppermost crust and secondarily to the strength and dip of reflectors in the lower crust, as discussed earlier.

The only apparent truncation of the Moho reflector occurs on line 21 where a reflector shallows from a depth of 12 s (fig. 9). On most continental profiles, reflections from the Moho are too discontinuous to reliably identify offsets [Meissner, 1973; Hale and Thompson, 1982, Oliver et al., 1983]. If this feature is real, it may be similar to the Flannan Thrust identified on BIRPS profiles off Scotland [Smythe et al., 1982; Brewer et al., 1983; Brewer and Smythe, 1984], and provides another example of such a feature. Unfortunately, this example is much less convincing and cannot be traced for any distance into the crust, presumably because of attenuation of the sound source by the overlying, thickening sediments. Possibly, the event might be a low-velocity diffraction originating out of the plane of the section.

The lack of a velocity model is a serious deficiency in our understanding of the crustal structure of the Long Island platform. Short-range refraction experiments on the platform showed that the basement surface beneath the sediments had velocities of 5.2-6.0 km/s [Ewing et al., 1950; Oliver and Drake, 1951]. A refraction line in New York and another across New Jersey showed crustal thicknesses of 36.0 and 32.7 km, respectively, and crustal velocities of 6.28 ± 0.02 km/s and 6.01 ± 0.3 km/s, respectively [Katz, 1955]. The refraction Moho from these analyses need not correspond to the reflection Moho observed offshore. Regional earthquakes in the northeast have been used to construct an average crustal model for the New England Appalachians, which consists of a two-layered crust 40 km thick. Its 15-km-thick upper layer has a of velocity 6.1 km/s and its 25-km-thick lower layer, a velocity 7.0 km/s [Taylor et al., 1980]. Three-dimensional inversion of teleseismic P-wave arrivals has been used to define variations in crustal velocity and crustal thickness for New England [Taylor and Toksoz, 1979]. The average model of velocity 6.5 km/s and thickness of 38 km (which was consistent with the model of Taylor et al., [1980]) showed velocity variations from 6.4 to 6.7 km/s and thickness variations of 35 to 41 km. These studies suggest that significant lateral variations in velocity characterize the crust of the Appalachian orogen north of the Long Island platform, and imply that a single velocity function cannot be realistically applied to the entire region. Our traveltime data have shown that differences in crustal thickness and/or velocity exist, but until we can get reliable velocity data, we cannot translate our time measurements to meaningful depths. Our lack of velocity functions also precludes satisfactory migration of the deep seismic data.

Conclusions

We conclude by summarizing the major points of our study of the Mohorovicic discontinuity beneath the Long Island platform.

1) Moho can be identified over most of the platform where reflection profiles were recorded to 12 s. Acoustically, it is a laterally variable boundary: a narrow, sharp reflector occurs on the central and western portions of the platform, and a more banded package of reflectors occurs to the east.

2) The acoustic properties of the lower crust and upper mantle contribute to the clarity of the Moho reflection. Moho is most distinct where it separates a reflective lower crust containing dipping events from a relatively transparent upper mantle. Regions where Moho is less identifiable tend to be characterized by a less reflective lower crust or one in which reflections are subhorizontal.

3) The upper crustal structure also effects the quality of Moho reflections. Rift grabens and thickening postrift sediments, for instance, are two areas beneath which reflections from the Moho become difficult to identify.

4) The two-way travel time through the crust varies. A region of lower traveltimes east of and adjacent to the Long Island rift basin may be caused by crust that was extended or thinned during Mesozoic rifting. A region of apparent crustal thickening in the central part of the platform may be the offshore extension of a similar zone mapped in central and southern New England or may be a zone that was relatively unaffected by Mesozoic extension. The traveltimes through the crust decrease towards the basement hinge zone in accordance with ideas of crustal thinning and extension on passive continental margins.

5) The evidence for velocity and/or thickness variations of the crust across the platform points to the need for determining good velocity functions. Until these are available, depth conversions and migrations of the deep seismic data cannot be accurately done.

Acknowledgements. We thank A. Trehu, J. Schlee and two anonymous reviewers for critically reviewing the manuscript; C. Daly, M. C. Mons-Wengler, K. DeMello, P. Forrestel, J. Zwinakis, and D. Blackwood assisted in preparation of the manuscript.

References

Anderson, R.E., M.L. Zoback, and G.A. Thompson, Implications of selected subsurface data on the structural form and evolution of some basins in the northern Basin and Range province, Nevada and Utah, Bull. Geol. Soc. Amer., 94, 1055-1072, 1983.

Brewer, J.A., D.H. Matthews, M.R. Warner, J. Hall, D.K. Smythe, and R.J. Whittington, BIRPS deep seismic reflection studies of the British Caledonides, Nature, 305, 206-210, 1983.

Brewer, J.A., and D.K. Smythe, MOIST and the continuity of crustal reflector geometry along the Caledonian-Appalachian orogen, J. Geol. Soc. London, 141. 105-120, 1984.

Brown, L., C. Ando, S. Klemperer, J. Oliver, S. Kaufman, B. Czuchra, T. Walsh, and Y.W. Isachsen, Adirondack-Appalachian crustal structure: the COCORP northeast traverse, Bull. Geol. Soc. Am., 94, 1173-1184, 1983.

Ewing, M., J.L. Worzel, N.C. Steenland, and F. Press, Geophysical investigations in the emerged and submerged Atlantic coastal plain, Bull. Geol. Soc. Amer., 61, 877-892, 1950.

Hale, L. D., and G.A. Thompson, The seismic reflection character of the continental Mohorovicic discontinuity, J. of Geophys. Res., 87, 4625-4635, 1982.

Hutchinson, D. R., Structure and Tectonics of the Long Island platform, unpubl. PhD thesis, Kingston, R. I., Univ. of Rhode Island, 289 p, 1984.

Hutchinson, D.R., K.D. Klitgord, and R.S. Detrick, The Block Island fault: a Paleozoic crustal boundary on the Long Island platform, Geology, in press.

Katz, S., Seismic study of crustal structure in Pennsylvania and New York, Bull. Seism. Soc. Am., 45, 303-325, 1955.

Klitgord, K.D., J.C. and Behrendt, Basin structure of the U.S. Atlantic margin, in Geological and geophysical investigations of continental margins, edited by J.S. Watkins, L. Montadert, and P.W. Dickerson, pp. 85-112, Am. Assoc. Pet. Geol. Memoir 29, 1979.

Klitgord, K.D., and H. Schouten, Tectonic and magnetic structure: Baltimore Canyon Trough and adjacent magnetic quiet zone, Scale 1:1,000,000, Misc. Field Studies Map, U.S. Geol. Surv., Reston, Va., in press.

Matthews, D., BIRPS: deep seismic reflection profiling around the British Isles, Nature, 298, 709-710, 1982.

Meissner, R., The Moho as a seismic transition zone, Geophys. Surv., 1, 195-216, 1973.

Oliver, J., F. Cook, and L. Brown, COCORP and the continental crust, J. Geophys. Res. 88, 3329-3347, 1983.

Oliver, J.E., and C.L. Drake, Geophysical investigations in the emerged and submerged Atlantic Coastal Plain, Part IV: the Long Island area, Bull. Geol. Soc. Am., 62, 1287-1296, 1951.

Phinney, R. A., A seismic cross section of the New England Appalachians: the orogen exposed, this volume.

Schlee, J.S., Seismic stratigraphy of the Baltimore Canyon Trough, Bull. Am. Assoc. Pet. Geol., 65, 26-53, 1981.

Smythe, D.K., A. Dobinson, R. McQuillan, J.A. Brewer, D.H. Matthews, D.J. Blundell, and B. Kelk, Deep structure of the Scottish Caledonides revealed by the MOIST reflection profile, Nature, 299, 338-340, 1982.

Taylor, S.R., and M.N. Toksoz, Three-dimensional crust and upper mantle structure of the northeastern United States, J. Geophys. Res., 84, 7627-7644, 1979.

Taylor, S.R., and M.N. Toksoz, Crust and upper-mantle velocity structure in the Appalachian orogenic belt: implications for tectonic evolution, Bull. Geol. Soc. Am., 93, 315-329, 1982.

Taylor, S.R., M.N. Toksoz, and M.P. Chaplin, Crustal structure of the northeastern United States: contrasts between Grenville and Appalachian provinces, Science, 208, 595-597, 1980.

Wernicke, B., Low-angle normal faults in the Basin and Range province: Nappe tectonics in an extending orogen, Nature, 291, 645-648, 1981.

THE QUEBEC-WESTERN MAINE SEISMIC REFLECTION PROFILE: SETTING AND FIRST YEAR RESULTS

D. B. Stewart, J. D. Unger, J. D. Phillips, and R. Goldsmith

U.S. Geological Survey, Reston, Virginia 22092

W. H. Poole

Geological Survey of Canada, Ottawa, Ontario K1A 0E8

C. P. Spencer and A. G. Green

Earth Physics Branch, Ottawa, Ontario K1A 0Y3

M. C. Loiselle

Maine Geological Survey, Augusta, Maine 04333

P. St-Julien

Department of Geology, Laval University, Quebec, Quebec G1K 7P4

Abstract. The Quebec-Western Maine seismic reflection profile is part of a nearly continuous profile about 1000 km long across the Northern Appalachian orogen from the craton to the ocean basin. During 1983, 219 km of 800-channel sign-bit data were collected for 15 seconds two-way travel time using VIBROSEIS sources. Variable upsweeps from 7 to 45 Hz with a 12 km -0- 12 km spread, 30 m group intervals, and 90 m vibration points were used. The profiles obtained have nominal 133 fold, and numerous reflectors can be seen at depths corresponding to two-way travel times of 1 to 12 seconds (about 3-42 km). This paper summarizes the regional geology, gravity and magnetic fields, and velocity structure from seismic refraction.

Preliminary interpretations are given for parts of three seismic reflection profiles. Rocks of the Connecticut Valley-Gaspe synclinorium and Chain Lakes massif are allochthonous above a major regional decollement that dips south from a two-way travel time of 3.5 seconds in Quebec to 7.8 seconds beneath the southern part of the massif. Profiles in central and coastal Maine image the shapes of plutons to below 3 seconds, offsets on steep faults to over 7 seconds, and many subhorizontal reflectors at various depths. The 800-channel sign-bit method gave high quality shallow and deep data that correlate well with geologic, gravity, and magnetic data.

Introduction

The Quebec-Western Maine seismic reflection profile extends about 330 km from northwest of Lac Megantic, Quebec, to the Maine seacoast at Penobscot Bay. This profile (fig. 1) is part of a nearly continuous deep seismic reflection profile normal to the axis of the Appalachian orogen beginning close to the exposed craton 50 km southwest of Quebec City. A deep seismic marine profile for the U.S. Geological Survey (USGS) will link the Quebec-Maine profile to USGS line 19, which extends across the Gulf of Maine to the continent-ocean margin. These profiles comprise a craton-to-ocean-basin traverse approximately 1000 km long that originated in proposals by P. H. Osberg et al. to COCORP [Oliver, 1982, fig. 2], and R. T. Haworth et al., to LITHOPROBE [CANDEL, 1981, profile 1D].

The northwestern 150 km of the craton-to-ocean profile was obtained for the Ministère de l'Energie et des Ressources, Québec as profile 2001 (MERQ in fig. 1). These 12-fold data were processed to 4 seconds two-way travel time in the northwestern part of the profile, and to 6 seconds for the southeastern 107 km. Field tapes for the southeastern 45 km have generously been made available to the USGS by Ministère de l'Energie et des Ressources, Québec and are being reprocessed to yield data for 10 seconds of two-way travel time. The MERQ profile has been in-

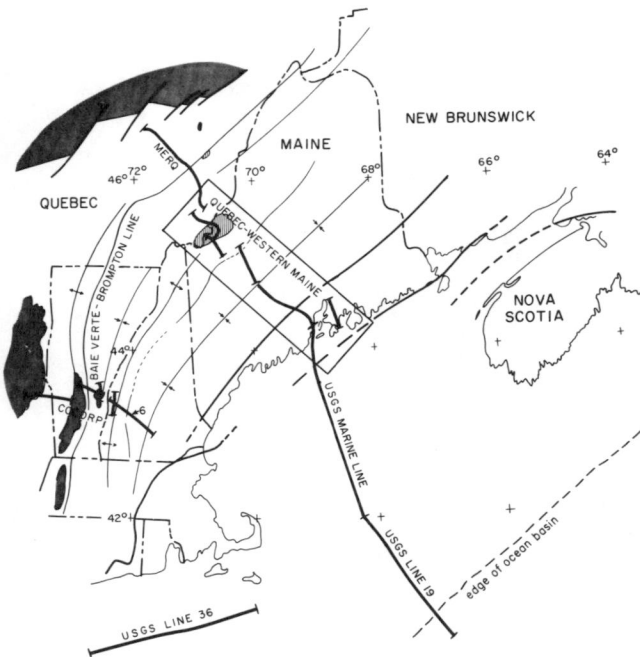

Fig. 1. Deep seismic reflection profiles in New England and adjacent Canada. The Chain Lakes massif is shown by a ruled pattern and the Precambrian Grenville rocks of the craton are shown by the gray shading. The area of Figure 2 is shown also.

terpreted by St-Julien et al. [1983], Ando et al. [1983], and LaRoche [1983] to indicate extensive northwestward thrusting with progressively deeper levels of detachment being exposed toward the southeast through a stratigraphic section that thickens southeasterly.

The northwesternmost 12 km of line 1 of the Quebec-Western Maine profile parallel the southeastern portion of the MERQ profile and lie 11 km southwest of it (fig. 1). Other deep reflection profiles in the region shown in Figure 1 are the New England COCORP lines [Ando et al., 1984], USGS line 19 [Klitgord et al., 1982], and USGS line 36 [Phinney, 1982; and Hutchinson et al., this volume].

The four seismic reflection profiles discussed here (1, 2, 3A, 3B) are part of the Quebec-Western Maine profile. They total 219 km and were obtained between October and December, 1983, through a cooperative project organized with the extensive participation of the organizations and individuals listed on Table 1. Collection of field data was funded jointly by the U.S. Geological Survey and the Geological Survey of Canada.

The principal goals of the project were to obtain geologic interpretations of the highest possible quality, using a combination of state-of-the-art field technology and processing methodology together with geological field data, gravity and magnetic data, and other relevant information. A secondary goal was to help the development of seismic reflection methodology, a principal reason for gathering field data with a 800-channel sign-bit data system.

Regional Data

Geologic Map and Cross Section

The geologic map of an area approximately 100 km wide and 300 km long containing the seismic reflection profiles, the southeastern end of the MERQ line, and the northern end of the marine profile is shown in Figure 2a. The schematic geologic cross section accompanying the map is based on cross section E-E' in Osberg et al. [1984] as modified where possible by our preliminary interpretations of the seismic reflection data. Figure 2a also shows schematically how the offset segments of our seismic profile overlap each other when projected along regional structures.

The Quebec-Western Maine seismic reflection profile lies entirely southeast of exposures of Precambrian basement of Grenville gneisses that form the adjacent craton (fig. 1). These rocks were metamorphosed to granulite facies about 1.1 b.y. ago. The basement beneath the folded and faulted rocks of southeasternmost Quebec and northwestern Maine is thought to be either Grenville gneisses or the Precambrian rock of the Chain Lakes massif. The massif is a terrane of formerly high-grade metasedimentary and metaigneous gneisses that were retrograded during the early Paleozoic. A thick diamictite containing quartz and mafic-rock clasts in a feldspar-rich matrix characterizes the massif and has not been recognized in the Grenville terrane. The nature of the contact between basements of Grenville and Chain Lakes types is not known and was an objective of our project.

Silurian and Devonian metasedimentary rocks of the Connecticut Valley-Gaspe synclinorium are the cover rocks at the northwest end of the profile. In interpretations of the MERQ profile by St-Julien et al. [1983] and Ando et al. [1983], these rocks are considered to have been thrust northwestward in the Acadian orogeny. Faulted belts of lower Paleozoic rocks border the Chain Lakes massif on the northwest, and the southeastern flank of the Chain Lakes is overlain along thrust contacts by a Cambrian-Ordovician ophiolite and olistostrome assemblage [Boudette, 1982] and by Silurian-Devonian turbidite metasedimentary rocks. The Merrimack synclinorium is a thick (>10 km?) sequence of turbidites with minor marble and quartzite. Regional metamorphic grades of the lower Paleozoic rocks along the profiles are mostly greenschist to lower amphibolite facies, but reach second sillimanite grade just west of Penobscot Bay.

Another objective of our project is to attempt

TABLE 1. Organizations and Individuals Participating in
Quebec-Western Maine Seismic Reflection Project

Organization	Participating Individuals	Technical Speciality
U.S. Department of the Interior		
Geological Survey	D. B. Stewart	Project coordination, regional geology
	J. D. Unger, J. Luetgert	Seismic reflection and refraction
	J. D. Phillips	Regional gravity and magnetics
	R. Goldsmith, N. L. Hatch, Jr., E. L. Boudette, E. Zen, D. W. Rankin	Regional geology
	F. C. Frischknecht	Regional aeromagnetic data
	Carl Koteff	Surficial deposits of New Hampshire
	R. A. Ayuso	Regional geochemistry of plutonic rocks
Canada Department of Energy, Mines and Resources		
Geological Survey	W. H. Poole	Project coordination, regional geology
	R. M. Gagne	Shallow seismic survey of surficial deposits
	P. J. Hood	Regional aeromagnetic data
Earth Physics Branch	A. G. Green, C. Spencer	Seismic reflection and refraction
	M. D. Thomas	Regional gravity data
Maine Department of Conservation		
Geological Survey	W. A. Anderson	Coordination of Maine state agencies, media coverage
	M. C. Loiselle	Bedrock geological map of Maine
	W. B. Thompson	Surficial deposits of Maine
	O. C. Gates, A. M. Hussey, II	Regional geology
Laval University	P. St-Julien	Regional geology, logistics in Quebec
U. of Maine at Orono	P. H. Osberg, C. V. Guidotti, J. Biederman	Regional geology and metamorphism
U. of New Hampshire	W. A. Bothner	Regional gravity, magnetics, and geology
	C. E. Jahrling, II	Gravity measurements along profiles
Syracuse University	G. M. Boone	Regional geology
Weston Observatory, Boston College	J. E. Ebel, A. Kafka	Regional seismic velocity structure
Boston University	D. W. Caldwell	Surficial deposits of Maine
Cornell University	S. Kaufman, L. D. Brown, J. Oliver	Collection and processing of VIBROSEIS data
	B. Thompson	Electro magnetotelluric data
Princeton University	R. A. Phinney	Wide angle seismic reflection data
Societe Québécoise d'Initiatives Petrolières	A. Trépanier	Seismic reflection noise tests
Ministère de l'Energie et des Ressources, Québec	Y. Tessier	Tapes of MERQ line 2001
Geophysical Systems Corporation	G. W. Fercho, R. Kolb, J. Hood	Acquisition of seismic reflection data
Interseis	F. Mixon	Processing of seismic reflection data

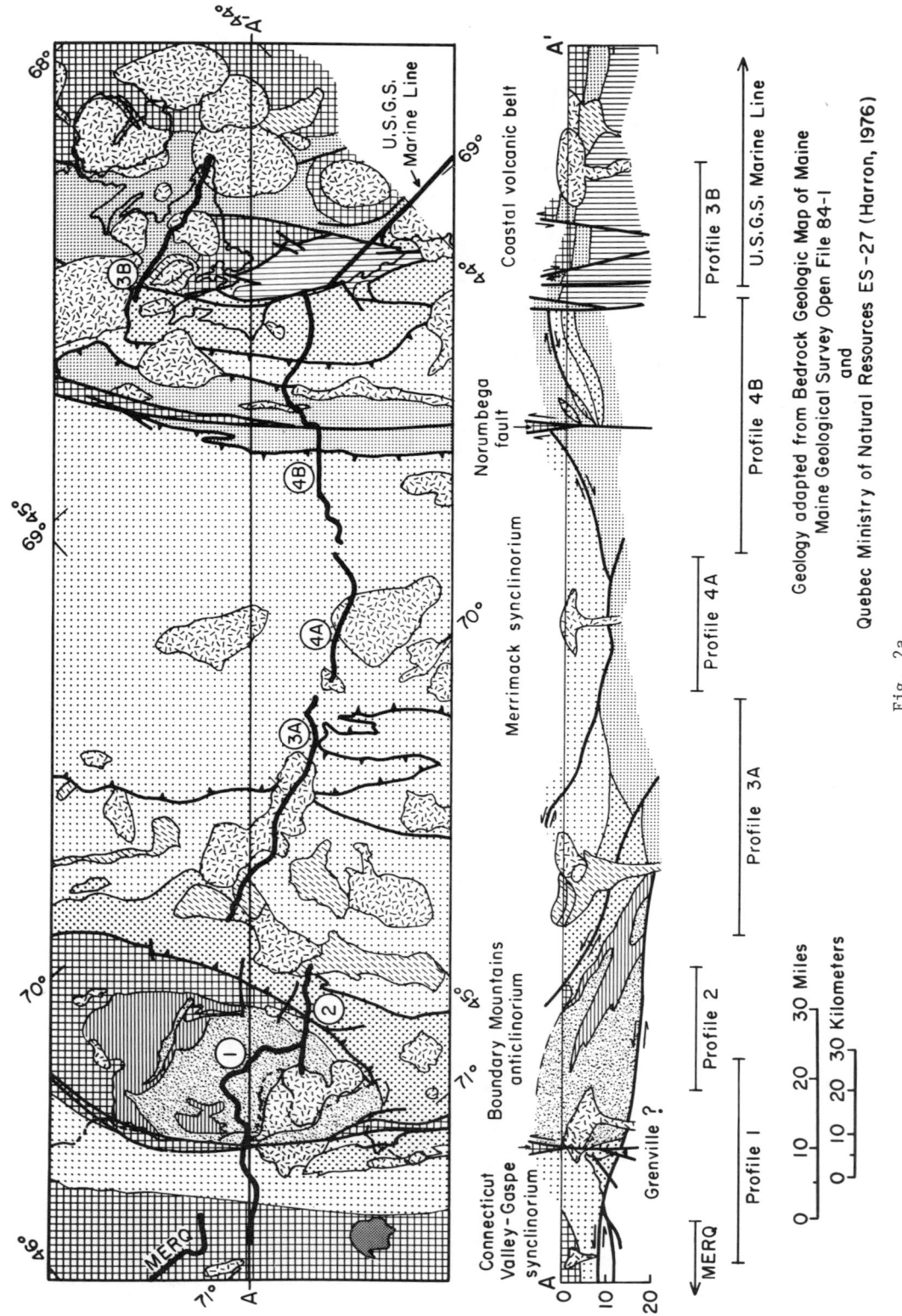

Fig. 2a.

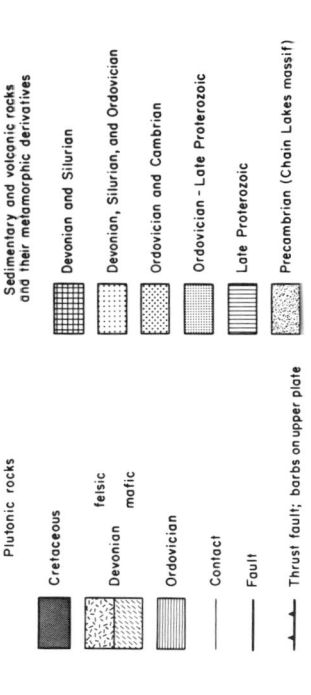

Fig. 2a. Geologic map and schematic cross section AA' for the region of the Quebec-Western Maine seismic reflection profile. The overlap of the profiles when projected along regional structures is shown also. Lines 4A and 4B were shot in 1984.

Fig. 2b. Regional gravity map and cross section AA' for the same region as figure 2a.

Fig. 2c. Regional aeromagnetic map and cross section AA' for the same region as figure 2a.

(Figures 2b and 2c are in pocket inside back cover)

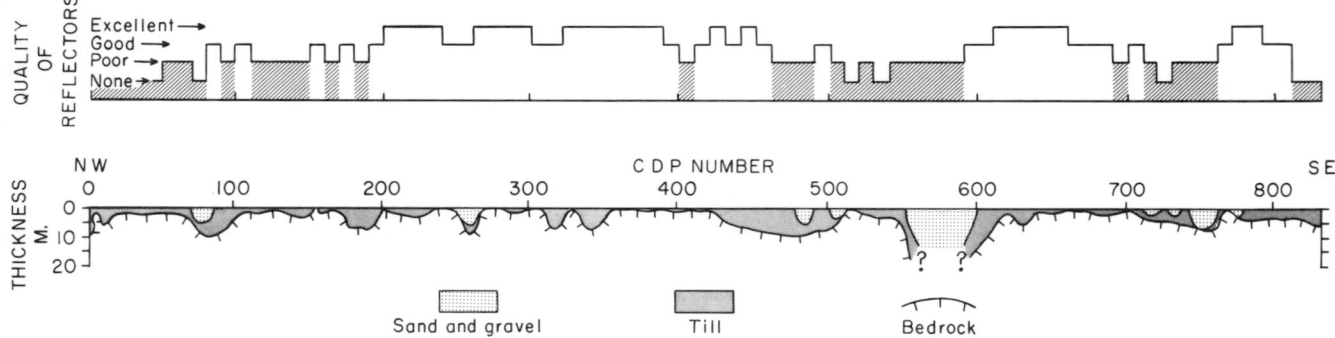

Fig. 3. Thickness of surficial deposits and visually estimated quality of reflectors along COCORP New England transect profile 6 in southern New Hampshire. Excellent = strong continuous reflectors, good = medium or interrupted reflectors, poor = weak discontinuous reflectors. Poor or no reflectors were observed under thick surficial deposits.

to determine the nature of the basement in central Maine and its depth. In southeastern Maine a Late Proterozoic(?) sequence of high-grade gneisses and schists of sialic composition is complexly folded and faulted with early Paleozoic metasedimentary rocks in structures with southeast vergence. Late Proterozoic metasedimentary rocks underlie the Coastal Volcanic Belt.

Gravity

The Bouguer gravity anomaly map and profile for the region (fig. 2b) displays three major regions with characteristic fabrics. The northwestern half of the study area is dominated by low gravity values, indicating thick, low density crust. Closed highs and lows correspond to exposed mafic and felsic plutons respectively, interpreted to extend to depths from 1 to 6 km [Koller, 1979; Carnese, 1981]. Because plutons of similar composition tend to lie within northeast-trending belts, some northeast elongation of the anomalies is seen. This fabric terminates against a strong positive gravity gradient in the center of the study area. Southeast of the gradient, the level of the gravity field is higher and relatively smooth. Toward Penobscot Bay the fabric returns to one of closed highs and lows produced by mafic and felsic plutons extending to depths of up to 7.5 km [Hodge et al., 1982]. The overall level of the gravity field remains high in this region and increases offshore, indicating that the crust is relatively more mafic in the southeastern part of the study area.

Line 1 crosses the relatively low gravity field over the Connecticut Valley-Gaspe synclinorium and over the Chain Lakes massif. It is noteworthy that the gravity field is only moderately affected by the presence of the massif. Line 2 starts in a gravity low over the Seven Ponds pluton, crosses a low density portion of the Chain Lakes massif, and terminates on a gravity high over the mafic Flagstaff complex. Beginning on the same gravity high, line 3A crosses the gravity low produced by the Lexington batholith and terminates over low to moderate gravity values of the northwest portion of the Merrimack synclinorium. Lines 4A and 4B cross the central Maine gravity gradient and the high smooth gravity field of the southwest portion of the synclinorium. Line 3B crosses the heavily intruded portion of the Coastal Volcanic Belt, but it does not cross the centers of the major intrusions so that the gravity field varies by less than 20 mgal along this line.

Aeromagnetic Data

The aeromagnetic anomalies (fig. 2c) also define several regional fabrics. Along the international boundary, positive magnetic anomalies delineate belts of mafic volcanic rocks and mafic plutons. The intensity of the magnetic anomalies decrease southeastward across the northwestern half of the study area. The central third of the study area is relatively quiet magnetically. No magnetic expression corresponding to the central gravity gradient is seen. A band of positive magnetic anomalies corresponds to the outcrop belt of Late Proterozoic(?) gneisses in southeastern Maine. An area of low magnetic relief separates this structure from a belt of intense positive magnetic anomalies just offshore.

Seismic Refraction Data

The seismic velocity structure of the region is crudely known from studies of regional seismic data from earthquakes and large explosions and was summarized by Taylor and Toksoz [1982]. The crust in northwestern Maine appears to be approximately 44 km thick, thinning southeastward to about 36 km near the seacoast. Velocity models have near-surface P-wave velocities about

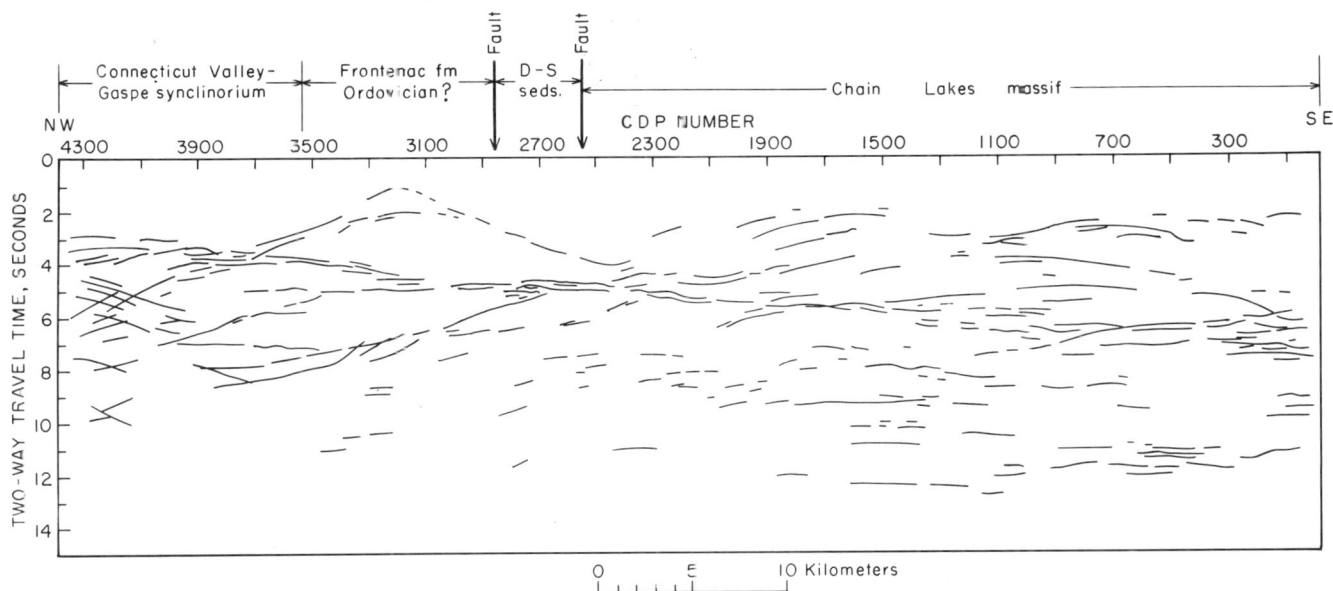

Fig. 4. Line drawing of unmigrated data for Profile 1, Quebec-Western Maine seismic reflection profile. The strong reflector between 3.5 sec two-way travel time at the left (NW) side of the figure and 7.8 sec at the right (SE) marks a major regional decollement. Horizontal exaggeration about 2:1.

6 km/sec, increasing to near 6.6-7.0 km/sec at 12-15 km, and about 8.1 km/sec at the base of the crust and represent average velocities for several terranes.

A major program of seismic refraction studies using large explosive sources was conducted in September-October 1984, by the U.S. Geological Survey, Earth Physics Branch of Energy, Mines and Resources Canada, and collaborators from several universities to determine crustal structures and to obtain velocity data for use in migrating the seismic reflection data.

Surficial Geology

We were advised by colleagues in COCORP to give special attention in route selection to surficial geology and cultural development to ensure field data of the highest quality.

The COCORP profiles in New Hampshire [Ando et al., 1984] crossed surficial deposits and bedrock similar to those we know exist in Maine. At our request, Carl Koteff mapped surficial deposits along COCORP lines 6 and 9. In Figure 3 the thickness of surficial deposits and a visual four-part estimate of the quality of reflectors are plotted on the same horizontal scale. A strong decrease in reflector quality occurs when the thickness of surficial deposits exceeds 10 m. Accordingly, before selecting routes in Maine, D. W. Caldwell of Boston University and the Maine Geological Survey prepared maps and approximate cross sections of the thickness of surficial deposits along all candidate routes.

Data from the Maine Department of Transportation, well logs from the files of the Maine Geological Survey and field mapping were used. In Quebec, the excellent geological and shallow seismic data of Shilts [1981] were supplemented by a shallow seismic survey by R. M. Gagne of the Geological Survey of Canada along portions of the preferred route. Careful route selection allowed us to minimize the lengths of profile that crossed surficial deposits (till, clay, sand, gravel) more than 25 m thick. This effort undoubtedly enhanced the quality of data collected and minimized statics corrections. There is no correlation between known unavoidable thick surficial deposits and reflector quality in our data. The high fold from our long spread of geophones significantly reduces this problem.

Seismic Reflection Data

Field Operating Parameters

The field data were collected using sign-bit recording methods and VIBROSEIS energy sources. Four vibrators (each with a maximum base plate force of 21,200 lbs) were in operation at all times, and a fifth vibrator was used for field testing and then kept in reserve as a backup unit. A symmetrical, split-spread pattern of 800 geophone groups of six geophones equally spaced over a 30 m group interval yielded a 12,000 m - 0 - 12,000 m geometry. The vibration interval was 90 m, giving a nominal CDP fold of 133; all channels were operational during

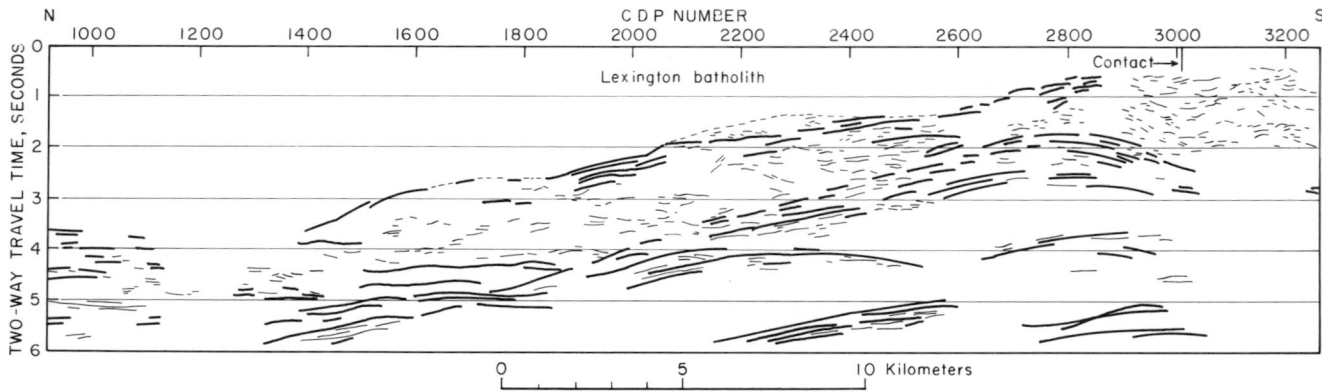

Fig. 5. Line drawing of unmigrated data for a portion of Profile 3A processed from CDP gathers range-limited to stations with less than 5 km offset and less than 4 sec two-way travel time, combined with normally-processed data for greater depths. Reflectors are absent in the batholith and a consistent reflector marks the bottom of the pluton. Arrow indicates mapped contact of batholith with country rock. Horizontal exaggeration about 2:1.

vibrating so there was effectively no offset at near distances. The vibrators advanced 3 m after each upsweep. Six different frequency bands of upsweep were used (7-27 Hz, 9-30 Hz, 11-34 Hz, 13-39 Hz, 15-42 Hz, and 17-45 Hz), each three times, resulting in symmetrical tapered input spectrum with a relatively flat maximum between 17 and 25 Hz. This spectrum reduced the energy in the side lobes of the input signal after correlation. Individual sweeps were 15 seconds long, and a total recording time of 30 seconds with a 4-millisecond sample rate was used that gave a record with a 15 second two-way travel time. The data were recorded on a GEOCOR-IV sign-bit system, which cross-correlates, sums a total of 18 sweeps, and records the sum.

The advantages of the sign-bit method for deep crustal experiments are higher quality estimates of the stacking velocities at mid-crustal depths (due to the long offsets possible with a 800 channel system) and the capability of recording high resolution shallow data. When not being used for data acquisition, the GEOCOR-IV system was used for data processing, and within five days of data collection on any part of the line we obtained brute stacks of selected portions of the data. Finally, the large number of data channels allowed us to collect CDP data with high average fold, although the long offsets along the crooked lines caused large variations in the number of traces used in the stack at each CDP point.

Data Processing

The data were processed using a VAX 11/780 minicomputer running Digicon Geophysical Corporations DISCO processing software. Early analysis of the shot gathers and brute stacks showed that the data would not be degraded by resampling from 4 m/sec to 8 m/sec; after filtering to eliminate the possibility of aliasing, this step was carried out. Velocity analyses were carried out each ~10 km using constant velocity stacks of 50 adjacent CDPs.

We were able to improve the quality of the shallow data by employing special processing techniques: 1) traces to be included in the CDP gathers were range-limited to less than 5 km offset; 2) the minimum possible mute and narrow window AGC were applied both before and after stacking; and 3) additional velocity analyses were done for the top four seconds of the records. These techniques resulted in marked improvements in certain sections of the profiles.

After stacking the data, the deconvolution, time-variant band pass filtering and wide window AGC operations were applied. In the final stacks adjacent traces were summed to increase the signal to noise ratio. Data were plotted with approximately 2:1 horizontal exaggeration.

Preliminary Interpretations of Profiles

Introduction

All processing has not been completed so that only preliminary interpretations are possible. Migrated sections have not yet been prepared. It is apparent that high quality data with numerous reflectors at 1 to 12 seconds two-way travel time have been obtained for most of the profiles, and that considerable detail has been obtained about the shapes of plutons crossed on profiles 3A and 3B. More complete interpretations will appear as separate publications on profiles 1 and 2, 3A, and 3B.

Most of the steep (>65°) faults and folds with steep limbs known to exist in the region were not imaged, and thus critical elements of the regional geology remain unresolved.

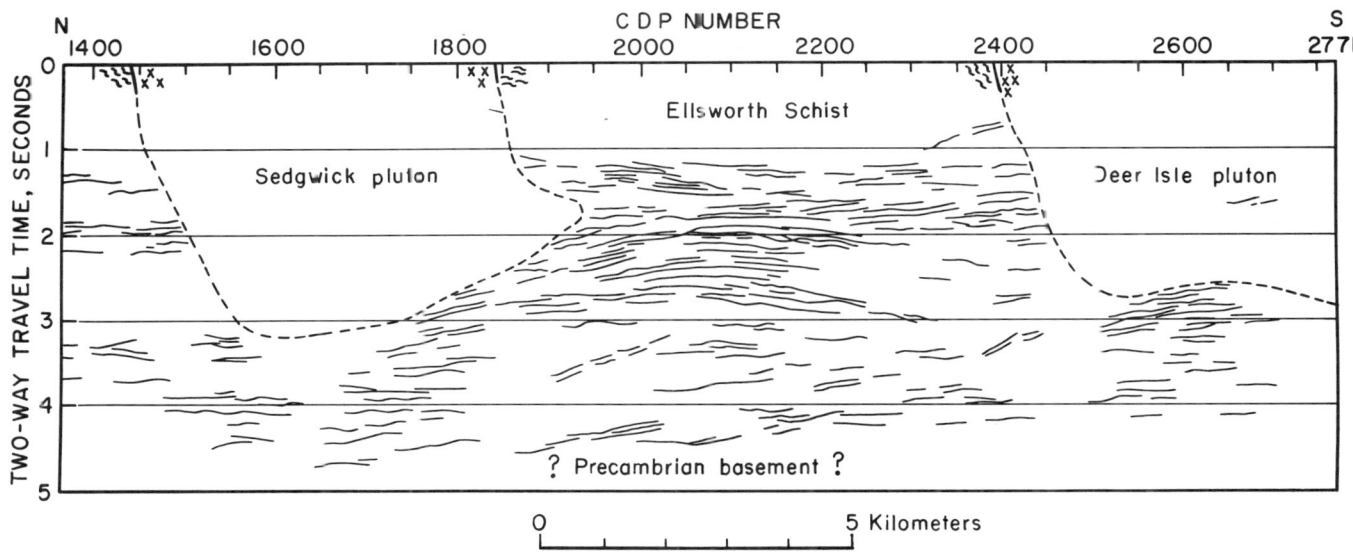

Fig. 6. Line drawing of unmigrated data for a portion of Profile 3B. The shapes of two granite plutons are shown by the contrast between reflector-free plutons and reflector-rich country rock. The mapped contacts are shown by patterns. The contact of the Ellsworth Schist and Precambrian basement is inferred to be at the base of the abundant reflectors at about 4 sec. Horizontal exaggeration about 2.5:1.

Profiles 1 and 2

A line drawing for the 75-km long profile 1 (for location see figs. 1 and 2A) is given in Figure 4. A portion of the seismic section is shown as Figure 3 in Green et al., this volume. The most prominent feature on this profile is a strong reflection that has an apparent dip to the south from a depth of 3.5 seconds at the northwestern end of the profile to 7.8 seconds at the southern end of the profile. This reflection is offset little, if it all, where it comes under the most prominent geologic contact along the profile, the fault between Chain Lakes massif and Devonian sediments on the northwest side of the massif. It continues uninterrupted beneath the sediments of the Silurian and Devonian Connecticut Valley-Gaspe synclinorium (fig. 2A), but it may be offset under the Ordovician(?) Frontenac Formation. This reflector, which correlates in strength and depth with a reflector seen at about 3.0 seconds on the southern end of the MERQ line, is interpreted to be a complex decollement with southeast over northwest thrusting of either Taconian or Acadian age. Where this thrust reaches the surface is problematical in the MERQ data. One possibility is the Guadeloupe fault at the northwestern boundary of the Connecticut Valley-Gaspe synclinorium; another is near the Baie Vert-Brompton line (fig. 1) at the eastern contact of the St. Daniel Formation. If the thrust reaches the surface along the Guadeloupe fault, an interpretation of the Connecticut Valley-Gaspe synclinorium as an Acadian allochthon is favored. However, a large sliver of rocks similar to those found in the Chain Lakes massif, 10 km long and 0.5 km thick, occurs to the north in the St. Daniel Formation at Beauceville, Quebec, [Williams and St-Julien, 1982] suggesting transport by Taconian faulting for tens of kilometers. Thus, it is probable that the decollement is a Taconic structure. Complex interpretations such as Acadian reactivation of a Taconic structure are also possible at this stage of understanding. A more extensive discussion of the geologic significance of this major feature is given by Green et al., this volume.

Numerous other reflectors at depths up to 12 seconds two-way travel time are present along the profile. The deepest of these is interpreted to be the Moho, estimated to be at a depth of 44 km in this region by Taylor and Toksoz [1982]. The 30-km long profile 2 has not yet yielded sufficiently strong reflectors for interpretation.

Profile 3A

This profile begins over a section of Cambrian and Ordovician sediments 19 km eastward along regional strike from the southeastern end of line 2. It then passes southeastward over several Devonian gabbroic and granitic plutons for 50 km before crossing medium-grade metasedimentary rocks on the northwestern flank of the Silurian-Devonian Merrimack synclinorium. The line drawing in Figure 5 shows the central portion of the profile where it passes over the granitic Lexington batholith. The bottom of the batholith appears as a contrast between reflector-free batholith and reflector-rich country rock. The batholith is seen to be shallow (2-3 seconds = about 6-9 km) and its base dips gently northward in agreement with gravity models [Koller, 1979].

Profile 3B

This 47 km profile begins 7 km to the north of a major regional fault and terrane boundary, inferred to be near vertical, and passes over metavolcanic rocks of the Ellsworth Schist in the Coastal Volcanic Belt and crosses the margins of several plutons before reaching the seacoast. The line drawing in Figure 6 shows the southern portion of the profile where the shapes of the bottoms of two equigranular granite plutons and the base of the metavolcanic section can be inferred from contrasts of the reflection-rich metavolcanic section and the reflection-free plutons and basement. Shallow (7 km) pluton bottoms also are inferred from gravity data [Hodge et al., 1982]. It appears that the plutons open out only after magma has risen into the metavolcanic section above the basement. This conclusion is in agreement with the shallow depth of emplacement of these plutons inferred from the fact that they are of Devonian age and intrude volcanic rocks as young as Devonian and from the low pressure metamorphic assemblages of their contact zones. Numerous other reflectors at various depths are discernable.

Conclusions

The 800-channel sign-bit seismic reflection method gave high quality shallow and deep data that correlate well with geologic, gravity, and magnetic data. We used this method to complete the Quebec-Western Maine profile and will continue to develop our interpretations by integrating the new geologic, gravity, magnetic, and seismic refraction data sets.

References

Ando, C. J., Cook, F. A., Oliver, J. E., Brown, L. D., and Kaufman, S., Crustal geometry of the Appalachian orogen from seismic reflection studies, in Contributions to the Tectonics and Geophysics of Mountain Chains, edited by R. D. Hatcher, Jr., H. Williams, and I. Zietz, Geological Society of America, Memoir 158, pp 83-101, 1983.

Ando, C. J., Czuchra, B., Klemperer, S., Brown, L. D., Cheadle, M., Cook, F. A., Oliver, J. E., Kaufman, S., Walsh, T., Thompson, J. B., Jr., Lyons, J. B., and Rosenfeld, J. L., A crustal profile of a mountain belt: COCORP deep seismic reflection profiling in the New England Appalachians, American Association of Petroleum Geology, Bull., 68, pp. 819-837, 1984.

Boudette, E. L., Ophiolite assemblage of early Paleozoic age in central western Maine, in Major Structural Zones of the Northern Appalachians, edited by P. St-Julien and J. Beland, Special Paper 24, Geological Association of Canada, pp. 209-230, 1982.

CANDEL (Canadian Committee on the Dynamics and Evolution of the Lithosphere), LITHOPROBE: Geoscience studies of the third dimension - a coordinated natural geoscience project for the 1980s, Geoscience Canada, 8, p. 117-125, 1981.

Carnese, M. J., Gravity study of intrusive rocks in west-central Maine, Masters thesis, 97 p., University of New Hampshire, 1981.

Green, A. G., Berry, M. J., Spencer, C. P., Kanasewich, E. R., Chiu, S., Clowes, R. M., Yorath, C. J., Stewart, D. B., Unger, J. D., and Poole, W. H., Recent seismic reflection studies in Canada, Am. Geophys. Union, Geophysical Monograph Series, this volume.

Harron, G. A., Metallogeny of sulfide deposits in the eastern townships, Quebec, Publication ES-27, 42 pp., Quebec Ministry of Natural Resources, 1976.

Hodge, D. S., Abbey, D. A., Harbin, M. A., Patterson, J. L., Ring, M. J., and Sweeney, J. F., Gravity studies of subsurface mass distributions of granitic rocks in Maine and New Hampshire, American Journal of Science, 282, pp. 1289-1324, 1982.

Hutchinson, D. R., Grow, J. A., Klitgord, K. D., and Dietrick, R. S., and Moho reflection from the Long Island platform, eastern United States, Am. Geophys. Union, Geophysical Monograph Series, this volume.

Klitgord, K. D., Schlee, J. S., and Hinz, K., Basement structure, sedimentation, and tectonic history of the Georges Bank Basin, in Geological Studies of the COST Nos. G-1 and G-2 wells, United States North Atlantic Outer Continental Shelf, edited by P. A. Scholle and C. R., Wenkam, pp. 160-186, Circular 861, U.S. Geological Survey, 1982

Koller, G. R., Geophysical and petrologic study of the Lexington batholith, west-central Maine, Ph.D. thesis, 167 p., Syracuse University, 1979.

LaRoche, P. J., Appalachians of Southern Quebec seen through seismic line 2001, in Seismic Expression of Structural Styles, A. W. Bally, ed., Studies in Geology Series #5, v. 3, pp. 3.2.1 to 3.2.1-22, American Association of Petroleum Geologists, 1983.

Oliver, J. E., Probing the structure of the deep continental crust, Science, 216, pp. 689-695, 1982.

Osberg, P. H., Hussey, A. M., II, and Boone, G. M., Bedrock geologic map of Maine, 1:500,000, Open-File 84-1, 13 pp and map, Maine Geological Survey, Augusta, Maine, 1984.

Phinney, R. A., Deep structure of the Appalachian

orogen on the Long Island platform [abstract], EOS Trans. AGU, 63, 1112, 1982.

Shilts, W. W., Surficial geology of the Lac Megantic area, Quebec, Geological Survey of Canada, Memoir 397, p. 102, 1981.

St-Julien, P., Slivitsky, A., and Feininger, T., A deep structural profile across the Appalachians of southern Quebec, in Contributions to the Tectonics and Geophysics of Mountain Chains, edited by R. D. Hatcher, Jr., H. Williams, and I. Zietz, Geological Society of America, Memoir 158, pp. 103-111, 1983.

Taylor, S. R., and Toksoz, M. N., Crust and upper-mantle velocity structure in the Appalachian orogenic belt: Implications for tectonic evolution: Geological Society of America Bulletin, 93, pp. 315-329, 1982.

Williams, H., and St-Julien, P., The Baie Verte-Brompton Line: Early Paleozoic continent-ocean interface in the Canadian Appalachians, in Major structural zones and faults of the northern Appalachians, edited by P. St-Julien and J. Beland, eds., Special Paper 24, pp. 177-207, Geological Association of Canada, 1982.

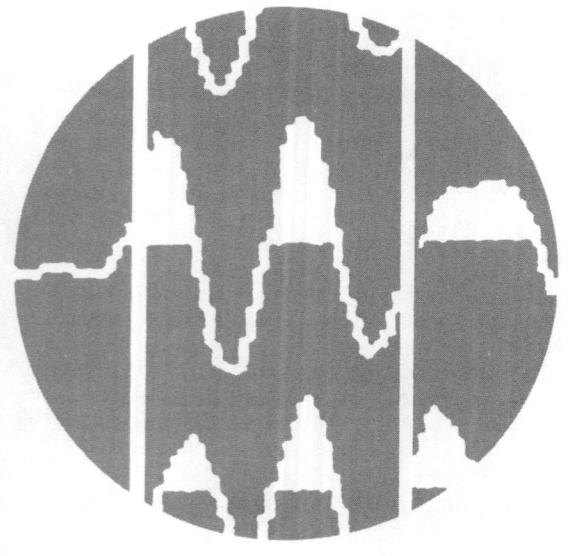

STRUCTURAL INTERPRETATION OF MULTICHANNEL SEISMIC REFLECTION PROFILES
CROSSING THE SOUTHEASTERN UNITED STATES AND THE ADJACENT CONTINENTAL MARGIN--
DECOLLEMENTS, FAULTS, TRIASSIC(?) BASINS AND MOHO REFLECTIONS

John C. Behrendt

U.S. Geological Survey, Denver, Colorado 80225

Abstract. In 1981 the U.S. Geological Survey (USGS) acquired 1350 km of 96- channel, 24-fold, multichannel seismic-reflection data along three profiles (S4, S6, and S8), recorded to 6 s and 8 s, extending across South Carolina and Georgia from the Appalachians to the Atlantic coast. Previously, in 1979, a 6-line grid (CH 1 - CH 6) comprising 650 km of 64-channel, 32-fold data recorded to 12 s was surveyed over the continental shelf near Charleston, S. C. That offshore grid is tied to line S4 onshore and to the regional survey of the Atlantic continental margin. The result is a transect of four lines (including published COCORP data for Tennessee-Georgia) across the southeastern United States, extending, on a number of offshore deep-reflection lines, to oceanic crust.

The Appalachian decollement can be seen discontinuously on S6 and S8 from the Appalachian Mountains southeastward as far as the Carolina Slate Belt; it is not apparently continuous to the surface interpreted as the Charleston decollement offshore. A series of reflections on lines S4, S6, and S8 and on the COCORP line is interpreted as evidence of southeastward-dipping imbricate faults, from the Brevard fault on the northwest to beyond the Augusta fault, which marks the southeastern extent of the Eastern Piedmont fault zone. The Carolina Slate Belt is characterized on the four seismic profiles by a complex series of diffractions and reflections extending from less than 1 s to 8 s. A number of Triassic(?) basins are apparent in the reflection data for the rifted Charleston terrane identified from low-gradient magnetic anomalies. These basins are bounded by normal faults reactivated in the meizoseismal area of the Charleston earthquake of 1886 and elsewhere, in a compressional reverse or strike-slip sense during Late Cretaceous and Cenozoic time. It appears probable that the seismicity in the Charleston terrane is related to movement on these fault zones bounding the basins; movement on the faults identified at depth in the eastern Piedmont fault zone may be related to seismicity there. Good reflections from the Moho are observed in the 6 CH lines offshore of Charleston in the range of 8-11 s, which is consistent with COCORP reflection data for land surveys.

Introduction

Over the past decade, the U.S. Geological Survey (USGS) has been investigating the cause of the Charleston, S.C., earthquake of 1886 and the likelihood of future earthquakes of similar size. As part of that work, multichannel reflection surveys were started in 1979 on land and offshore in the Charleston area [Behrendt et al., 1981; Hamilton et al., 1983; Behrendt et al., 1983]. The data for lines over the continental margin were tied into the USGS offshore seismic regional survey in the area discussed by Dillon et al., [1979]. At about the same time (1978-79) Consortium for Continental Reflection Profiling (COCORP) lines in Georgia and in the Charleston, S. C., area were recorded [Cook et al., 1979; Cook et al., 1981; and Schilt et al., 1983]. The COCORP data for Georgia [Cook et al., 1979] and other reflection data to the northeast, as discussed by Harris and Bayer [1979], indicated the presence of the Appalachian decollement or detachment, extending seaward from the Appalachian Mountains. The authors of these papers inferred that the Appalachian detachment might extend across the Piedmont and Coastal Plain to the continental shelf. Subsequently, Iverson and Smithson [1982] suggested, on the basis of their reprocessing of the COCORP line in Georgia, that the decollement was rooted in the Kings Mountain-Carolina Slate Belt area (Figure 1).

The multichannel seismic-reflection data for the Charleston, S. C., area [Behrendt et al., 1981, 1983; and Schilt et al., 1983] provided evidence, particularly offshore, of the existence of a reflecting surface at a depth of 11.4±1.5 km that was suggested as a decollement. Behrendt et al. [1981, 1983] suggested that the Charleston earthquake of 1886 might have been caused by movement on the decollement or on associated reactivated (listric?) faults bounding a Triassic(?) basin. Seeber and Armbruster [1981] suggested that movement on the Appalachian

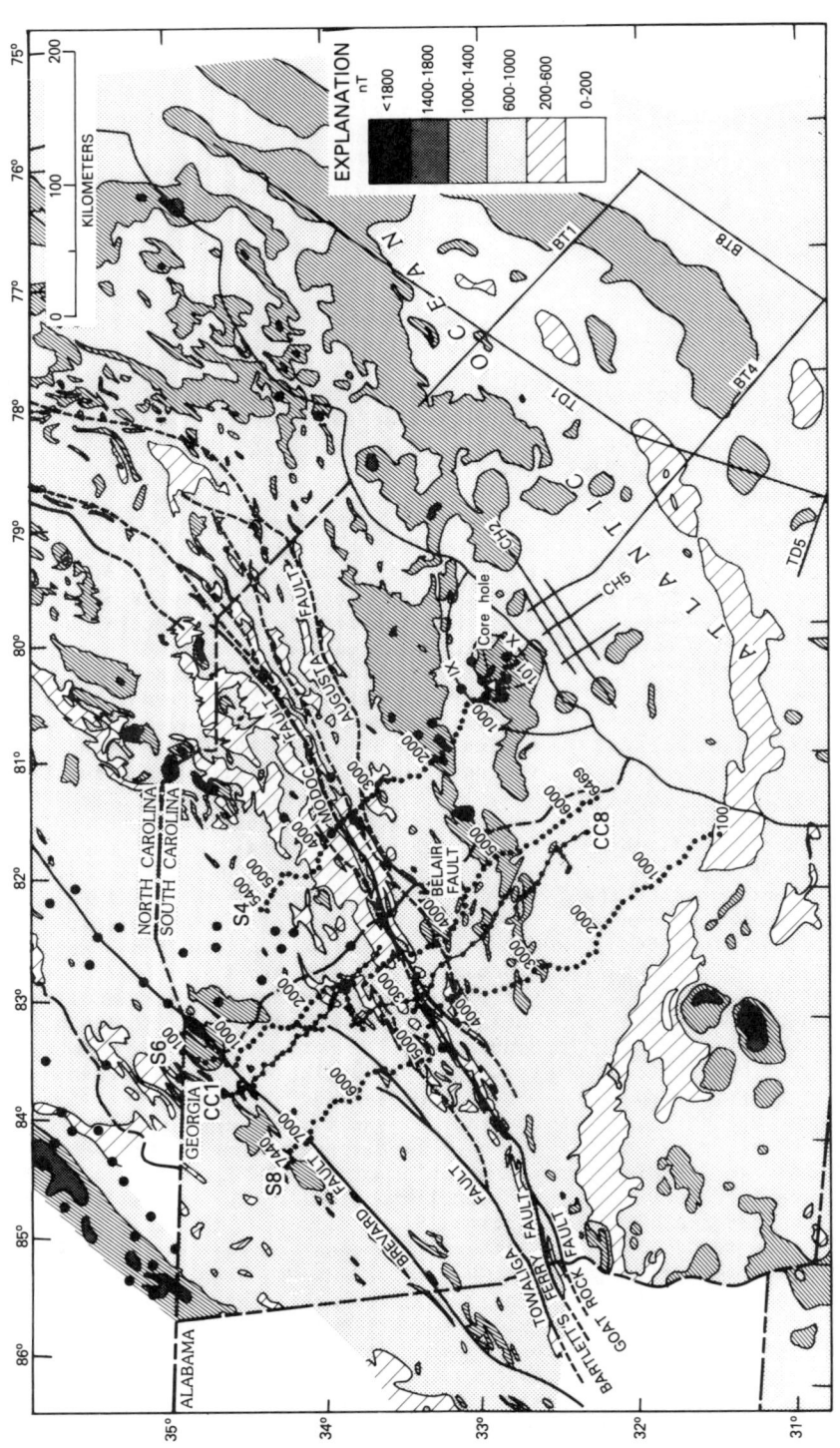

Fig. 1. Aeromagnetic map modified from Zietz and Gilbert [1980] covering South Carolina, parts of Georgia, adjacent states and the continental margin. Contour interval 400 nT. Faults are indicated from the eastern Piedmont fault system [Hatcher and others, 1977]; the Augusta fault approximately marks the Piedmont-Coastal Plain boundary. All multichannel seismic lines 5 s or greater in time are shown, including deep crustal multichannel seismic reflection lines S4, S6, S8, with shotpoints indicated, and COCORP lines CC1 - CC8 [Cook et al., 1981] on land. Marine lines CH2, CH5 and the adjacent CH grid [Behrendt et al., 1983] and BT1, BT4, BT8, TD1, TD5 [Dillon, et al., 1979] are indicated. Intensity IX and X isoseismal lines in the meizoseismal area of the 1886 Charleston earthquake are shown from Bollinger [1977]. Epicenters shown are from Bollinger [1975] and [Dewey, 1983]. Jedburg and Branchville basins are crossed by S4 between shotpoints 900 and 1800. Zone of low magnetic gradient used to define Charleston terrane can be seen southeast of S4, shotpoint 1700, and north of Brunswick anomaly, which crosses coast with an arcuate east-west trend about 75 km north of Georgia-Florida border.

decollement, if it continued coastward to Charleston, might have caused the Charleston earthquake of 1886. The most precisely determined focal depths for recent seismicity are shallower than 13 ± 2 km [Tarr et al., 1981], or above the suggested Charleston decollement. Three long deep crustal multichannel seismic reflection profiles were acquired by the USGS to address these problems.

Deep Crustal Reflection Profiles

Description of Data

Figure 1 shows the location of the three profiles crossing South Carolina and Georgia obtained by the USGS, the COCORP reflection profile from Cook et al. [1981] discussed above, and other data offshore. The 96-channel, 24-fold COCORP data have been described and discussed in the papers referenced previously. Lines S4, S6, and S8 (Figure 1) were contracted on a non-exclusive basis, with the USGS as an original participant, and collected in 1981. The spread length was 6.7 km, group interval 67 m; there were 24 geophones per group, and 96 channels. The shotpoint (vibration point) was at the center of the spread, 200 m from the groups on either side. Four vibrators were used for the data collection, and shotpoints were spaced at 134 m intervals; the sweep length of 24 s was down from 48-12 hz. The sample rate was 4 ms, and the record lengths were 8 s for line S4 and 6 s for lines S6 and S8. The data, as discussed and illustrated in this report, were processed 24 fold by the contractor and have not been migrated. Only the record sections obtained from the contractor are available, and the interpretations presented here were made using these. The S4, S6, and S8 data are generally superior in quality in the upper 2 s to the COCORP line but below that depth the COCORP data have clearly much greater penetration. The reason is indicated by a comparison of the field parameters. The contractor for the S lines used a 48-12 Hz sweep, 24 long, with 4 vibrators. Hence the field effort was:

$$\frac{24 \times 4}{48 - 12} = 2.7 \text{ s/Hz.}$$

In contrast, the COCORP survey used 8-32 Hz sweeps, 30 s long, with 5 vibrators for a field effort of:

$$\frac{30 \times 5}{32 - 8} = 6.5 \text{ s/Hz, more than two times greater.}$$

The grid of data offshore (CH lines in Figure 2) were collected, under contract, using a 33 1 (2000 in ^{3}) airgun array. The data were recorded to 12 s using a 64-channel 3200 m streamer and were processed 32 fold. The record sections and line drawing interpretations were presented on large plots by Behrendt et al. [1983] and, with the exception of two examples, will not be reported here.

Geologic Setting of Profiles

The profiles (Figures 2-4) discussed here extend from the Appalachian mountains to the continental shelf, crossing the Piedmont province of Paleozoic crystalline rocks and the Coastal Plain province of Late Cretaceous and Cenozoic sedimentary rocks. Williams and Hatcher [1982] have described the various northeast trending accreted terranes that they interpret as making up the pre-Cretaceous rocks of the area. Also, Hatcher et al. [1977] defined the extent of the northeast trending Eastern Piedmont fault system that has probably been active in various senses of movement from Paleozoic to possibly present time, as suggested by the recent seismicity (Figure 1). Geologic features indicated along the interpreted seismic profiles discussed below are taken from these papers and from Williams [1978].

Interpretations of Seismic Record Sections

In making the interpretations presented in this report, all of the lines in Figures 2-5 were produced by visually correlating a large number of adjacent seismic traces using characteristic wave forms. Of course, some multiple reflections may have inadvertently been identified and certainly there are many diffractions shown. I have labeled certain reflections along the northwest ends of profiles from S6 and S8 (Figures 3 and 4) as "D" to indicate my inference that they are from a decollement. In like manner, I used "D" at the southeast end of S8 (Figure 4) and offshore for the surface mapped by diffractions (many of which migrated to flat reflections), also inferred to be from a decollement [Behrendt et al., 1983] without regard to the geologic continuity, or lack thereof, of the geologic structure responsible for these arrivals. If a mean crustal velocity of 6.0 km/s for the crystalline terrane is assumed, a horizontal stretch of about 1.5 results in these figures. Similarly, I labeled as "M" reflections offshore in the range of 8-11s (lines CH 2 and CH 5, Figure 5) to indicate my interpretation that these are from the Moho.

Because of space limitations, the line drawings of the seismic record sections from lines S4, S6, and S8 are shown at a greatly reduced scale in Figures 2, 3, and 4. Larger scale versions of these profiles are available in Behrendt [1985]. The gravity and magnetic profiles shown along the tops of the figures were compiled from USGS open file maps of this area.

The offshore grid of data (CH lines) is tied approximately to the end of S4 (to CH 5, Figure 1). Line drawings for CH 2 and the crossing line CH 5 are shown in Figure 5. Airguns fired in the sea are much more efficient than vibrators coupled to the ground, so record quality is higher

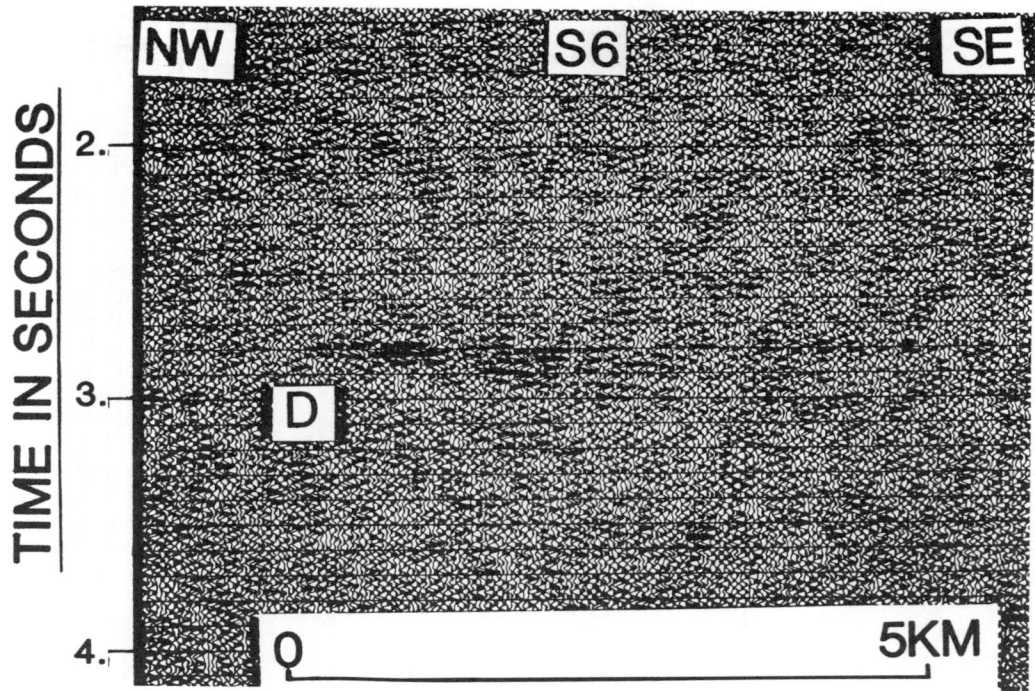

Fig. 6a. Line S6 near SP 300. D indicates arrival from decollement.

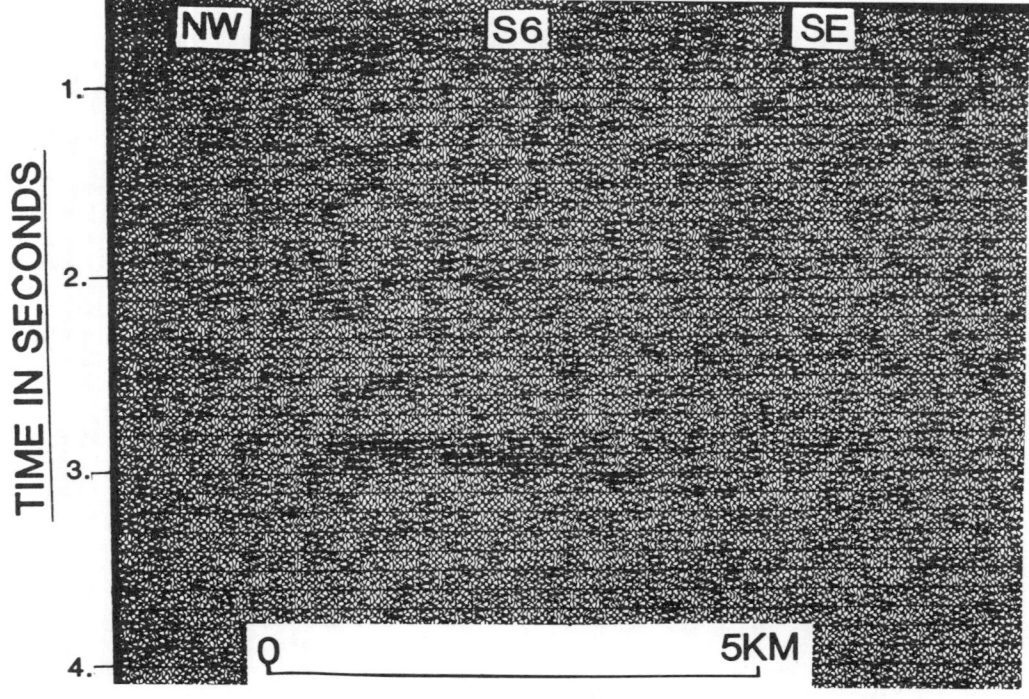

Fig. 6b. Line S6 near SP 1500.

Fig. 6. Examples of seismic record sections from S6 and S8 (Figures 1, 3, and 4) showing reflections from Appalachian decollement.

Fig. 6c. Line S3 near SP 7200. Note earlier arrival time for D reflection, compared with that shown in 6a-d.

Fig. 6d. Line S8 near SP 5900.

offshore over the same Coastal Plain sedimentary rock section.

Discussion

Appalachian Decollement

Reflections from the Appalachian decollement can be seen at the northwest end of lines S6 and S8 (Figures 1, 3, and 4) and on the COCORP line [Cook et al., 1981] shown in Figure 1. Examples from record sections for S6 and S8 are shown in Figure 6. The inferred reflections from the detachment are labeled D in Figures 3 and 4. On S4 (Figure 2), I was not able to identify any reflections that I could interpret as the Appalachian decollement (Figures 1 and 2), but this could be due to record quality rather than to the absence of the decollement.

On S6, the D reflection is about 2.9 s deep where shown in Figures 6a (near SP 300) and 6b (near SP 1500). The line drawing (Figure 3) shows that it varies from earlier to later about this time and suggests deepening of D slightly towards the southeast to about 3.4 s near the Modoc fault. Line S6 crosses the Brevard fault about 20 km southwest of the COCORP Georgia line (Figure 1). The average time to the D reflection on both S6 and the COCORP Georgia line [Cook et al., 1979] is about 2.6 s.

By contrast to that shown in S6, the reflection from the detachment is significantly earlier on line S8, to the southwest. There are strong arrivals at about 1.6 s near SP 7200 (Figures 4, 6c, 6d). Assuming a mean velocity above D of 6.0 km/s, the depth difference between S6 and S8 would be some 3.9 km shallower to the southwest. There is not a clear reflection on S8 where it crosses the Brevard fault, but interpolating between picked reflections (e.g., SP 7200 and SP 6700, Figure 4) suggests 2.0 s for the D reflection, or apparently 1.8 km shallower than on line S6. By comparison, on COCORP Tennessee line 1, the D reflection is at about 2.0 s at the Tennessee-North Carolina border and shallows to the northwest to about 1.5 s [Cook et al., 1979]. Of course, the general deepening of the Appalachian detachment to the southeast has been known for some time [Cook et al., 1979; Harris and Bayer 1979] but possible variation in depth along strike, although not surprising, has not been reported before. On the short COCORP Georgia line 2 near, and parallel to, the Brevard fault [Cook et al., 1979], D is at 2.5 s and does not show any dip.

There has been a great deal of discussion in the literature recently about the southeastward extent of the Appalachian detachment. Cook et al. [1979 and 1981] and Harris and Bayer [1979] first suggested its existence beneath the Coastal Plain and even beneath the continental shelf [Harris and Bayer, [1979]. Harris and Bayer [1979] interpreted USGS offshore profiles (lines 31 and 32, not shown here) to show reflections at 11-s depth (or deeper than the Moho in the area), which they interpreted as a decollement continuous with the Appalachian decollement. Hutchinson et al. [1983] subsequently presented a more reasonable interpretation of these lines and did not report any reflections from the same time. Iverson and Smithson [1982] reprocessed and reinterpreted the COCORP Georgia profile and concluded that the Appalachian decollement was rooted in the northwestern part of the Carolina Slate Belt, near the Kings Mountain Belt. Interpretation by Cook et al. [1983] also shows the decollement rooted in the Charlotte and Slate Belts, in about the same area as that shown by Iverson and Smithson, as a suggested alternative to their earlier [Cook et al., 1979, 1981] interpretation, which was also shown.

Nevertheless, reflections and diffractions off a mappable surface, referred to previously, are shown in the grid of data collected offshore of Charleston (the CH lines Figure 1). These arrivals define a surface which was interpreted as evidence of a decollement by Behrendt and others [1983], here called the Charleston decollement. This surface at 3.7±0.5 s is quite different in appearance [e.g. Figure 5] from that shown by the reflections from the Appalachian decollement. The depth of 11.7±1.4 kms corresponding to this time, based on the velocity function used to determine hyocenters [Tarr et al., 1981], defines the base of the upper crust and is well above the Moho reflections (Figure 5) at 8-11 s that are discussed in a later section. Behrendt et al., [1983] discussed the probable continuation of this surface beneath the meizoseismal area of the Charleston 1886 earthquake of 1886 (Figure 1) suggested by weak reflections on S4 (Figure 2) and in the COCORP Charleston data [Schilt et al., 1983]. Other reflections labeled D, along the southeast part of S8 (Figure 4) may be continuous with the Charleston decollement.

Careful study of the seismic record sections for S6 and S8 shows reflection D interpreted from the Appalachian decollement extending (discontinuously) southeast, beyond the Brevard fault, and the Kings Mountain Belt into the Carolina Slate Belt area (Figures 2, 3 and 4). However, because of the low energy penetration at depth on S4, S6 and S8 (discussed previously) compared with the COCORP data, I cannot conclude anything about the continuity of the Appalachian decollement further southeast across the Piedmont and Coastal Plain on S4, S6 and S8.

Coastward Dipping Faults

One of the most interesting results of the study is the interpretation of a series of southeastward dipping reflections (compare Figures 1-4), extending from as far west as the Brevard zone to southeast of the Augusta fault, beneath the Coastal Plain (Figures 2-4). An example of reflections from one of these inferred faults is shown in Figure 7. These data have not been

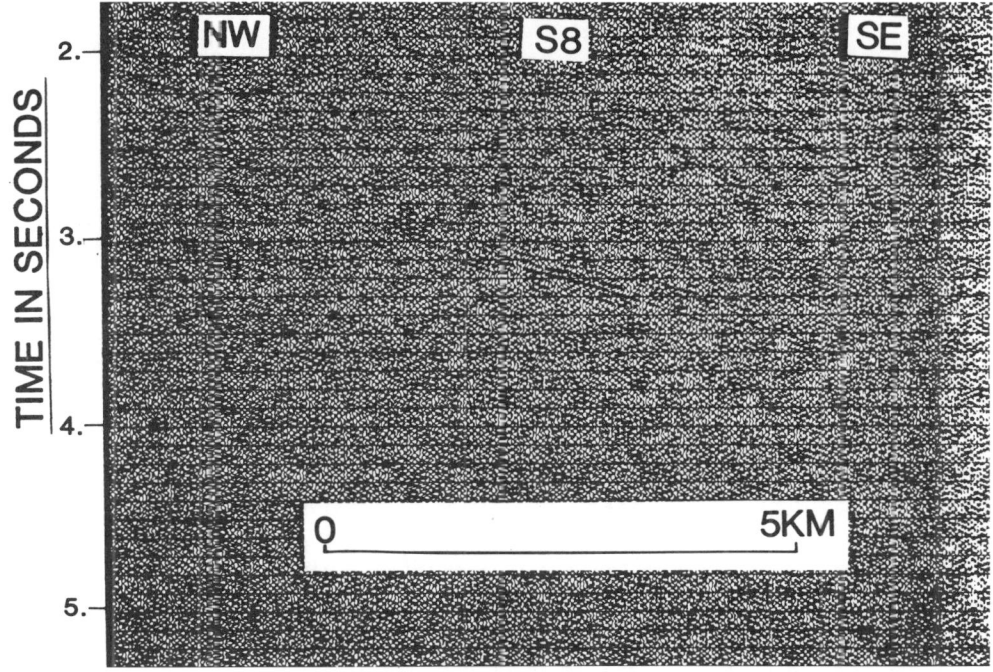

Fig. 7. Example of reflection from line S8 about SP 5100, interpreted as having as its source a southeast-dipping fault. The weak arrival would migrate up dip, to left in figure.

migrated. Comparison with migrated COCORP data [Cook et al., 1981; Petersen et al., 1984] indicates that the dipping reflections would migrate up dip significantly so that the apparent crossing of the Appalachian decollement by the Brevard or Towaliga faults, for example on S6 and S8 (Figures 3 and 4), is not real. Migration would not change the dip, however, on these straight line segments. Correcting for a horizontal stretch of 1.5, the apparent angles of dip (9°-15°) on the line sections would be 14°-22°.

These coastward-dipping faults in the Eastern Piedmont fault zone [Hatcher et al., 1977] have been identified in the surface exposures as large zones of cataclastic rocks, interpreted as associated with mylonite zones. Their locations as shown in Figures 1-4, taken from Hatcher et al. [1977] were inferred by those authors on the basis of aeromagnetic data. The Brevard, Towaliga, Modoc and Augusta faults appear to correlate with reflections in Figures 2-4. Others in the Piedmont do not correlate with known surface features. The northeast trending linear magnetic anomalies (Figure 1) are shown in the profiles of Figures 2-4. The suggestion on S4 and S6 of the continuation of these faults southeast of the Augusta fault (Figures 2 and 3) beneath the Coastal Plain is not inconsistent with the magnetic data (Figure 1), although, of course, no faults have been mapped geologically in the buried crystalline terrane. Other unanswered questions stimulated by these seismic interpretations are concerned with the relation of the faults to dated plutons and terrane boundaries.

Carolina Slate Belt

Conspicuous groups of curved arrivals, interpreted to be diffractions, characterize the record sections within the Carolina Slate Belt on S4, S6, S8 (Figures 2-4), and on the Georgia COCORP line [Cook et al., 1983].

These curved arrivals (convex upward) are similar in appearance to the diffractions reported offshore of Charleston, S. C., in the CH lines (Figures 1 and 5), which were inferred to mark the Charleston decollement [Behrendt et al., 1983]. The significant difference between the arrivals in the CH lines and the arrivals observed in the Slate Belt is that the latter appear within 1 s of the surface and do not define a particular depth, whereas those observed in the CH lines (e.g., Figure 5) underlie an acoustically transparent zone at a particular depth. Lines S4, S6 and S8 (Figures 2-4) as well as in the Slate Belt COCORP data [Cook et al., 1983] display subhorizontal reflections at 4-6 s beneath the complex reflections.

The Coastal Plain sedimentary rock provided poor coupling of energy from vibrators, as compared with airguns in water offshore, but the diffractions associated with the Slate Belt continue across the Piedmont-Coastal Plain

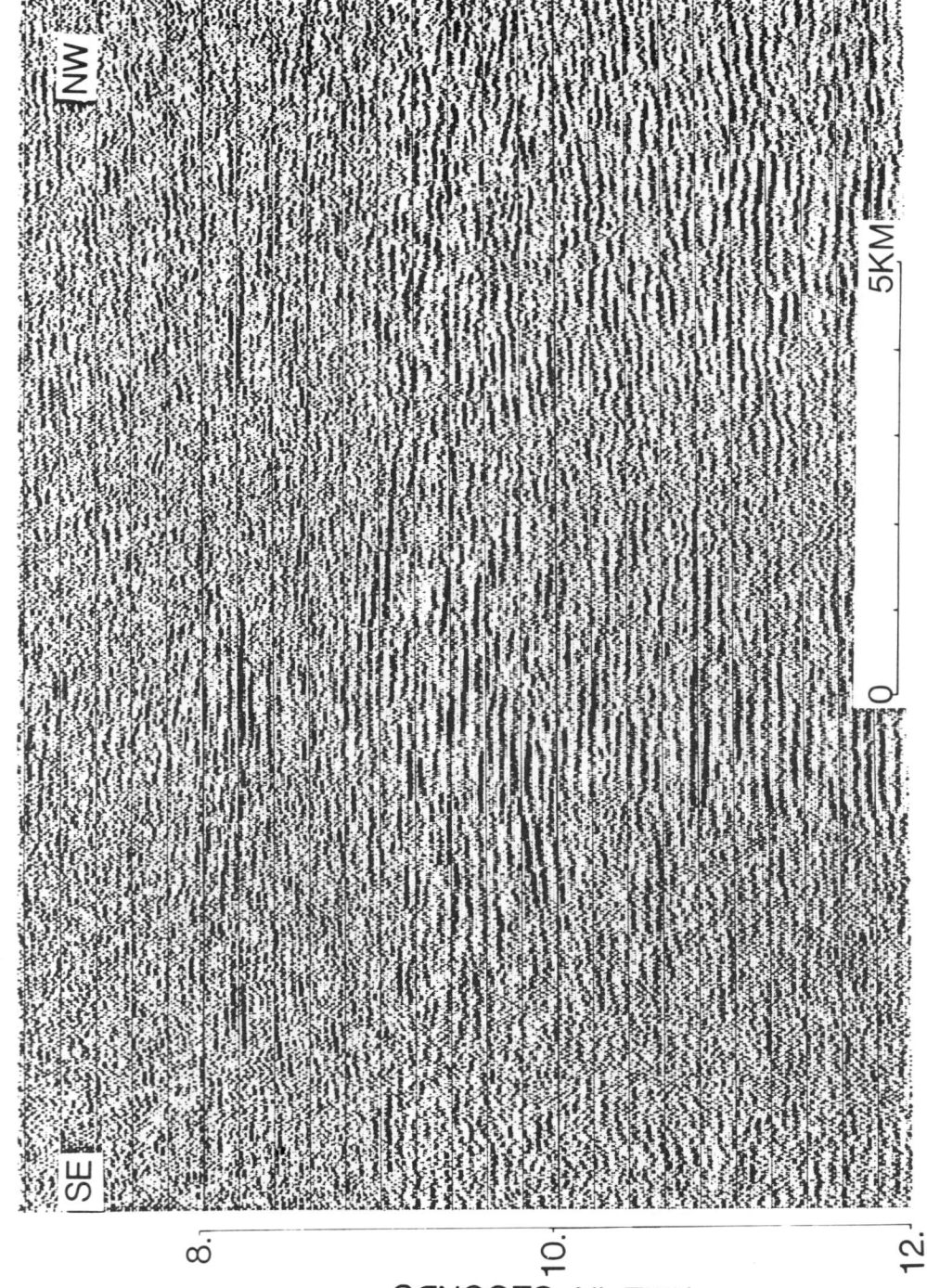

Fig. 8. Reflections from Moho, from line CH5 (Figures 1 and 5). Note laminated, discontinuous character of arrivals from 8 s to bottom of figure.

boundary (e.g., see Figure 3, from S6, where the diffractions are apparent later than 3 s, as far southeast as SP 5100). On the S4 interpretation (Figure 2) which extends to 8 s, the pattern of diffractions and complex reflections extends to the base of the record section. The source of these diffractions and reflections is not understood, but they are probably related to layering in the metasedimentary rocks of the Slate Belt, complexly disrupted by the faults discussed previously. Of course, the top of the diffractions may only appear to be as deep as shown, and many are likely to be from sources out of the plane of the section. Migration would tend to collapse the diffractions, as discussed by Behrendt et al., [1983] for the CH lines offshore, but would not make the tops of the diffractions earlier.

Pre-Cretaceous Unconformity

A prominent reflection, labeled J in Figures 2-5, marks the very strong reflecting surface (reflection coefficient of 0.5 [Behrendt et al., 1983]) associated with the pre-Cretaceous unconformity at the base of the Coastal Plain section. This reflection extends southeast to the Charleston, S. C. area, where it has been identified where drilled through at the core hole indicated in Figures 1 and 2 as caused by the contrast between sedimentary rock (V_p=2.0 km/s) and a Jurassic age basalt layer (V_p<5.8 km/s) [Hamilton et al., 1983]. The strong reflection extends over a great area offshore [Behrendt et al., 1983; Dillon et al., 1979]. Probably the basalt does not extend inland beyond the limit of the Charleston terrane [Popenoe and Zietz, 1977] marked by the change in magnetic gradient near SP 2000 on S4 (Figures 1 and 2).

Triassic(?) Basins

The edge of the Charleston terrane [Popenoe and Zietz, 1977; Behrendt, 1983], which was also called the Brunswick terrane [Williams and Hatcher, 1982], is marked by a change in the magnetic gradient from short wavelength to the broader, smoother anomalies characterizing the terrane. Klitgord et al. [1983] have discussed the regional tectonic setting in relation to the geophysical anomalies. Within the Charleston terrane are a number of Triassic(?) basins observed in magnetic data and six in the seismic reflection data presented here (four are shown in Figures 2-4) and elsewhere [e.g. Hamilton et al., 1983; Petersen et al., 1984; Behrendt, 1983]. The Jedburg basin (Figure 2), in the meizoseismal area of the Charleston earthquake of 1886, possibly was penetrated by a core hole (Figures 1 and 2) where Triassic(?) red bed sedimentary rocks were encountered [Gohn et al., 1983]. It is also possible, as inferred on the basis of reflection data from S4 and other profiles, that the unfossiliferous redbeds post-date rifting and that the actual basin underlies these rocks at a greater depth. The Jedburg basin is not particularly well defined by S4 but can be seen more easily in other nearby reflection profiles [Hamilton et al., 1983; Behrendt, 1983].

The Branchville basin, northwest of the Jedburg basin on S4 (Figure 2), is particularly well shown by these seismic data. The northwest boundary of this basin, if it has a northeast trend as suggested by the magnetic data (Figure 1), projects into the Bowman epicenters identified by Dewey [1983] and by Tarr et al. [1981].

On line S8 (Figure 4), a basin (here called the Kibbee basin) can be seen extending about 0.5 s beneath the J reflection just southeast, between SP 2200-2450. Nothing is known about this structure, although it is probably Triassic and/or Jurassic in age.

Petersen et al., [1984] discussed the Riddleville basin in the COCORP Georgia line; the basin(?) centered about SP 5600 on S6 (Figure 3) is probably the Riddleville basin.

Moho Reflections

Lines S4, S6, and S8 were not recorded over a time long enough to observe reflections from the Moho. The CH lines (Figures 1 and 5) were however (12 s), and bands of prominent arrivals can be seen in all six lines in the range of 8-11 s. Their appearance is laminated and discontinuous, Figure 8 from CH 5 shows an example of these bands; see the large-scale record sections in Behrendt et al., [1983]. The discontinuous nature of the reflections from the Moho in the CH lines is probably the result of geology rather than of poor signal-to-noise ratio of the data, as inferred on the basis of comparisons between the lines at the nine crossing points (Figure 1). This is the only data set I know of on a closely spaced grid that recorded Moho reflections on all lines. These reflections from the Moho are similar in their laminated, discontinuous character to those reported in the same time interval near Charleston, S. C., by Schilt et al. [1983] from COCORP data and from other areas [e.g., Hale and Thompson, 1982].

Relation to Seismicity

The meizoseismal area of the Charleston earthquake of 1886 (Figure 1) overlies the normal boundary fault zone of the Jedburg basin, that was reactivated in a compressional reverse [Behrendt, 1983], or strike-slip (based on unpublished reflection data with which I am working) sense during Late Cretaceous and Cenozoic time. The meizoseismal area of the 1886 earthquake also overlies the hypothesized Charleston decollement, which probably is not continuous with the Appalachian decollement., Possibly the earthquake and other seismicity in the meizoseismal area and along the boundary of the Branchville basin (Figure 2) was caused by reactivated (listric ?) faults as suggested by Behrendt et al. [1983], and Behrendt [1983]. Possibly, movement on a series of imbricate seaward dipping faults (Figures 2-4) crossing the Piedmont from the Brevard fault to at least as far as the Augusta fault in the Eastern Piedmont fault zone is related to the seismicity there as noted earlier [Hatcher et al., 1977 and Talwani, 1975].

Acknowledgments. I thank my colleagues in the USGS and elsewhere for helpful discussions in the course of this work, particularly M. D. Zoback, R. M. Hamilton, J. A. Grow, G. A. Bollinger and F. A. Cook. The S4, S6, and S8 lines were measured by Seisdata Services, Inc., and the CH lines by Whitehall Corp. The U.S. Nuclear Regulatory Commission provided partial financial support for this work.

References

Behrendt, J. C., Hamilton, R. M., Ackermann, H. D., and Henry, J. J., Cenozoic faulting in the vicinity of the Charleston, S.C. 1886 earthquake: Geology, 9, 3, 117-122, 1981.

Behrendt, J. C., Did movement on a northeast trending listric fault near the southeast edge of the Judburg Triassic-Jurassic(?) basin cause the Charleston, South Carolina, 1886 earthquake?, in Hays, W. W., and Gori, P. L., eds., Proceedings of Conference XX, a workshop on "The 1886 Charleston earthquake and its implications for today": U.S. Geol. Surv. Open-File Rep. 83-843, 127-131, 1983.

Behrendt, J. C., Hamilton, R. M., Ackermann, H. D., Henry, V. J., and Bayer, K. D., Marine multichannel seismic reflection evidence for Cenozoic faulting and deep crustal structure near Charleston, South Carolina, in Gohn, G. S., ed., Studies related to the Charleston, South Carolina earthquake of 1886--Tectonics and seismicity: U.S. Geol. Surv. Prof. Pap. 1313, J1-J28, 1983.

Behrendt, J. C., Multichannel seismic reflection profiles crossing South Carolina and Georgia from the Appalachian Mountains to the Atlantic coast--deep crustal interpretations, U.S. Geol. Surv. Map MF-1956, 1985.

Bollinger, G. A., A catalog of southeastern United States earthquakes--1754 through 1974, Vir. Poly. Tech. Inst. and State Univ. Res. Div., Bull. 101, 68 p., 1975.

Bollinger, G. A., Reinterpretation of the intensity data for the 1886 Charleston, South Carolina, earthquake in Rankin, D. W., ed., Studies related to the Charleston, South Carolina, earthquake of 1886--A preliminary report: U.S. Geol. Surv. Prof. Pap. 1028, 17-32, 1977.

Cook, F. A., Albaugh, D. S., Brown, L. D., Kaufman, S., Oliver, J. E., and Hatcher, R. D., Jr., Thin-skinned tectonics in the crystalline southern Appalachians, COCORP seismic-reflection profiling of the Blue Ridge and Piedmont: Geology, 7, 563-567, 1979.

Cook, F. A., Brown, L. D., Kaufman, S., Oliver, J. E., and Peterson, T. A., COCORP seismic profiling of the Appalachian orogen beneath the Coastal Plain of Georgia: Geol. Soc. of Amer. Bull., Part 1, 92, 738-748, 1981.

Cook, F. A., Brown, L. D., Kaufman, S., and Oliver, J. E., The COCORP southern Appalachian traverse, in Bally, A. W., ed., Seismic expression of structural styles - a picture and work atlas, AAPG Studies in Geology Ser. 15, 3, Amer. Assoc. of Pet. Geol., Tulsa, 3.2.1-1-6, 1983.

Dewey, J. W., Relocation of instrumentally recorded pre-1974 earthquakes in the South Carolina region: in Gohn, G. S., ed., Studies related to the Charleston, South Carolina earthquake of 1886--Tectonics and seismicity: U.S. Geol. Surv. Prof. Pap. 1313, 01-09, 1983.

Dillon, W. P., Paull, C. K., Buffler, R. T., and Fail, J. P., Structure and development of the southeast Georgia embayment and northern Blake Plateau, in Watkins, J. S., Montadert, L., and Dickerson, P. W., eds., Geological and geophysical investigations of continental margins: Amer. Assoc. of Pet. Geol. Mem. 29, 27-43, 1979.

Gohn, G. S., Houser, Brenda B., and Schneider, R. R., Geology of the lower Mesozoic(?) sedimentary rocks in Clubhouse Crossroads test hole #3, near Charleston, South Carolina, in Gohn, G. S., ed, Studies related to the Charleston, South Carolina, earthquake of 1886, U.S. Geol. Surv. Prof. Pap. 1313, C1 - C18, 1983.

Hale, L. D., Thompson, G. A., The seismic reflection character of the continental Mohorovicic discontinuity, J. of Geophys. Res., 87, B6, 4625-4635, 1982.

Hamilton, R. M., Behrendt, J. C., and Ackermann, H. D., Land multichannel seismic-reflection evidence for tectonic features near Charleston, South Carolina, in Gohn, G. S., ed., Studies related to the Charleston, South Carolina, earthquake of 1886--Tectonics and seismicity: U.S. Geol. Surv. Prof. Pap. 1313, I1-I17, 1983.

Harris, L. D., and Bayer, K. C., Sequential development of the Appalachian orogen above a master decollement--a hypothesis: Geology, 7, 568-572, 1979.

Hatcher, R. D., Jr., Howell, D., and Talwani, P., Eastern Piedmont fault system: speculations on its extent: Geology, 5, 636-640, 1977.

Hutchinson, D. R., Grow, J. A., Klitgord, K. D., Crustal reflections from the Long Island platform of the U.S. Atlantic continental margin, this volume, 1985.

Hutchinson, D. R., Grow, J. A., Klitgord, K. D., and Swift, B. A., Deep structure and evolution of the Carolina trough, in Watkins, J. S., and Drake, C. L., eds., Studies in continental margin geology: Amer. Assoc. of Pet. Geol. Mem. 34, 129-152, 1983.

Iverson, W. P., and Smithson, S. B., Master

decollement root zone beneath the Southern Appalachians and crustal balance: Geology, 10, 241-245, 1982.

Klitgord, K. D., Dillon, W. P., and Popenoe, P., Mesozoic tectonics of the southeastern United States Coastal Plain and Continental Margin, in Gohn, G. S., ed., Studies related to the Charleston, South Carolina, earthquake of 1886--Tectonics and seismicity: U.S. Geol. Surv. Prof. Pap. 1313, P1-P15, 1983.

Petersen, T. A., Brown, L. D., Cook, F. A., Kaufman, S., Oliver, J. E., Structure of the Riddleville Basin from COCORP sesimic data and implications for reactivation tectonics: J. of Geol., 92, 3, 261-271, 1984.

Popenoe, P., and Zietz, I., The nature of the geophysical basement beneath the Coastal Plain of South Carolina and northeastern Georgia, in Rankin, D. W., ed., Studies related to the Charleston, South Carolina, earthquake of 1886-- A preliminary report: U.S. Geol. Surv. Prof. Pap. 1028, I119-I137, 1977.

Schilt, F. S., Brown, L. D., Oliver, J. E., and Kaufman, S., Subsurface structure near Charleston, South Carolina--Results of COCORP reflection profiling in the Atlantic Coastal Plain, in Gohn, G. S., ed., Studies related to the Charleston, South Carolina, earthquake of 1886--Tectonics and seismicity: U.S. Geol. Surv. Prof. Pap. 1313, H1-H19, 1983.

Seeber, L., and Armbruster, J. G., The 1886 Charleston, South Carolina, earthquake and the Appalachian detachment: Geophys. Res., 86, 9, 7874-7894, 1981.

Talwani, P., Crustal structure of South Carolina: U.S. Geol. Surv. Second Tech. Rept., contract no. 14-03-0001-14553, 70 p, 1975.

Tarr, A. C., Talwani, P., Rhea, S., Carver, D., and Amick, D., Results of recent South Carolina seismological studies: Bull. of the Seis. Soc. of Amer., 71, 6, 1883-1902, 1981.

Williams, H., Tectonic lithofacies map of the Appalachian orogen: Mem. Univ. of Newfoundland, St. Johns, Newfoundland, Map 1, 1:1,000,000, 1978.

Williams, H., and Hatcher, R. S., Jr., Suspect terranes and accretionary history of the Appalachian orogen: Geology, 10, 530-536, 1982.

Zietz, I., and Gilbert, F. P., Aeromagnetic map of part of the southeastern United States in color: U.S. Geol. Surv. Map GP 936, scale 1:2,000,000, 1980.

CRUSTAL THICKNESS, VELOCITY STRUCTURE, AND THE ISOSTATIC RESPONSE FUNCTION IN THE SOUTHERN APPALACHIANS

Leland T. Long and Jeih-San Liow

School of Geophysical Sciences, Georgia Institute of Technology, Atlanta

Abstract. The COCORP southern Appalachian traverse in eastern Tennessee shows relatively little evidence of differential vertical movement in the seismic reflectors associated with the lower portion of the column of Paleozoic sediments under the Valley and Ridge and Blue Ridge Provinces. On the other hand, estimates of crustal thickness from seismic refraction studies and gravity data analysis imply thickening of the crust in areas of significant topography. The crustal thickness from a time term analysis of Pn arrivals and an analysis of gravity anomalies varies from 33 km in the Georgia Piedmont and 35 km in central Alabama to greater than 50 km under the mountainous areas of eastern Tennessee and northern Georgia. The apparent association of the topographic load on the overthrust crust with isostatic compensation at the base of the lower crust is difficult to explain without evidence of differential vertical movement. We suggest that, in a thrust regime, the existing relief of the underlying plate was compensated at depth and the topography of the upper plate exists because the thrust sheets are draped over the lower plate bulge.

Introduction

The southern Appalachian COCORP line in southeastern Tennessee and eastern Georgia (Figure 1) has confirmed early speculations [Hatcher, 1978; Clark et al., 1978] of large scale thrusting of crust underlying the Blue Ridge and Piedmont Province. The seismic reflection data in the northwestern portion of the southern Appalachian COCORP line [Cook et al., 1979; Cook et al., 1983] clearly show seismic reflections associated with Paleozoic sediments which served as the basal detachment fault and some major splays such as the Great Smoky thrust and the Brevard shear zone. However, reflections from below the decollement surface are sparse and their interpretation in terms of coherent structures in the crust is difficult. Reflections from the Mohorovicic discontinuity are observed only under the Charlotte and Carolina Slate belts in the southeastern portion of this line.

Studies of the southern Appalachian COCORP line have established that thin-skin tectonics is a major contribution to mountain building in the southern Appalachians. Indeed, the preservation and flatness of the decollement interpreted from seismic reflection data would seem to preclude differential vertical movement or crustal block tectonics in the last 200 to 400 ma along the northwestern end of the southern Appalachian COCORP line. In contrast with the long-term apparent stability of the southern Appalachians, releveling data [Brown and Oliver, 1976; Jurkowski and Reilinger, 1981] suggest rapid contemporary vertical movement, but the releveling data have been questioned because the uplift rates are so high that they cannot be sustained over long periods and because the correlation of uplift with topography is suspicious. Also, in contrast with the long-term apparent stability, the isostatic response function from gravity data covering the continental United States [Lewis and Dorman, 1970], suggests a linear relation between topography and isostatic compensation and, hence, that the crust adjusts to topographic loads. Studies of Bouguer gravity anomalies and topographic relief in the southeastern United States confirm that the crust of this region is in approximate isostatic equilibrium [Long, 1974]. Hence, the preservation of the decollement as a relatively undeformed feature and evidence for normal isostatic compensation in the lower plate of topography on the overthrust plate presents a paradox.

A distinctive gravitational signature of mountain chains derived from continental plate collision following convergent plate movements is a positive-negative paired gravity anomaly [Thomas, 1983; Karner and Watts, 1983]. In the southern Appalachians, the Piedmont Gravity Gradient separates the negative anomalies to the northwest from the positive anomalies in the Charlotte and Carolina Slate Belt of the Piedmont Province (Figure 1). The positive-negative paired gravity anomaly, which is related to the leading edge of continental crust of the lower plate and the suture with the upper plate, contributes to the paradox because it is not everywhere related to the surface topography in a manor expected of

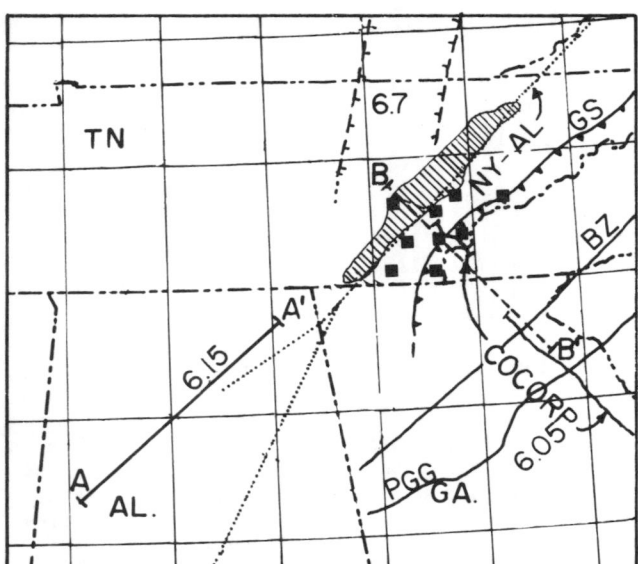

Fig. 1. Location map for profile lines and crustal structures in the southern Appalachians. Shaded area denotes area of shallow crustal negative anomaly. Heavy lines AA' and BB' are the central Alabama seismic refraction line and the cross-section in Figure 5. The New York-Alabama lineament is denoted by the dotted lines. Two interpretations of its location in Alabama are shown. The interpreted rift of the East Continent gravity anomaly is indicated north of the shaded negative gravity anomaly in eastern Tennessee. Seismic monitoring stations used in this study are indicated by squares. P-wave velocities are in km/s.

isostatic equilibrium. The association of the negative anomalies with increased crustal thickness is discussed by Hutchinson et al. [1983] and in this study. Karner and Watts [1983] were able to explain the general properties of the positive-negative paired gravity anomaly and regional isostatic equilibrium by flexure of a rigid lithosphere and a density contrast model of the suture zone.

In this study, the crustal thicknesses and gravity anomalies under consideration occur largely west of the axis of the positive-negative paired gravity anomaly. We examine evidence for variations in crustal thickness and seismic velocity anomalies that complement the COCORP reflection data by providing information on the lower crust and upper mantle. We then use gravity data to delineate major units within the crust and propose a structural model for the southern Appalachians. Then, this framework and related data are used to support the hypothesis that, in a thrust regime, the underlying plate has relief that is compensated at depth and that the upper plate topography is draped over the lower plate bulge.

Mantle Velocity Anomalies

A velocity structure for the upper mantle along the COCORP line was derived from teleseismic travel time delays for earthquakes recorded in the region. A modified version of Aki's [1977] three-dimensional P-wave travel-time inversion was applied by Volz [1979] to the seismic stations then existing in Georgia and South Carolina. Variations in arrival times from sources at various azimuths were used to determine zones of anomalous velocity. Modified formulations of the inversion equations were necessary to accommodate an irregularly spaced seismic array. First, instead of using relative arrival times in order to eliminate origin-time uncertainties, a source correction was found for each event. Second, flexible block size and geometry was introduced to help equalize the sampling of the blocks used in the reduction. Finally, instead of damped least squares, the mean velocity anomalies for each layer were constrained to zero in the simultaneous least squares solutions for velocity perturbations and source corrections.

The percent perturbations in velocity are averaged and projected onto a cross section perpendicular to the Appalachian structure (Figure 2). The percent perturbations and standard errors are comparable to typical results for such inversion techniques [Aki et al., 1977]. In a zone dipping at approximately 30 degrees southeast from the surface near the Brevard zone the velocities are consistently lower than average. This dipping low-velocity zone resembles the geometry of subducted lithosphere, but contemporary subduction zones show positive velocity anomalies, attributed to the lower temperature of the subducted lithosphere. Numerical modeling of the subduction process, however, indicates that thermal equilibrium is achieved within 50 ma [Toksöz et al., 1973]. Since the subduction process ceased nearly 400 ma ago, thermal equilibrium should now be achieved, and it is possible that an initial high velocity zone evidently matures into a low-velocity zone. The negative anomalies observed under the southern Appalachians may be explained by compositional variations of the subducted lithosphere or, perhaps, increased temperatures from radioactive decay, if significant continental crustal materials were included [Toksöz et al., 1973]. The velocity perturbations of the upper mantle in Figure 2 exclude the velocity perturbations of the crust. Also, east of the Piedmont Gravity gradient, the lower (subducted) lithosphere may have been capped primarily by oceanic crust. The interpretation by Cook and Oliver [1981] of dipping reflections below the Charlotte and Carolina Slate belt as shelf edge sediments supports the transition to oceanic crust. Hence, the component of continental type crustal materials included in the subducted lithosphere has not been resolved.

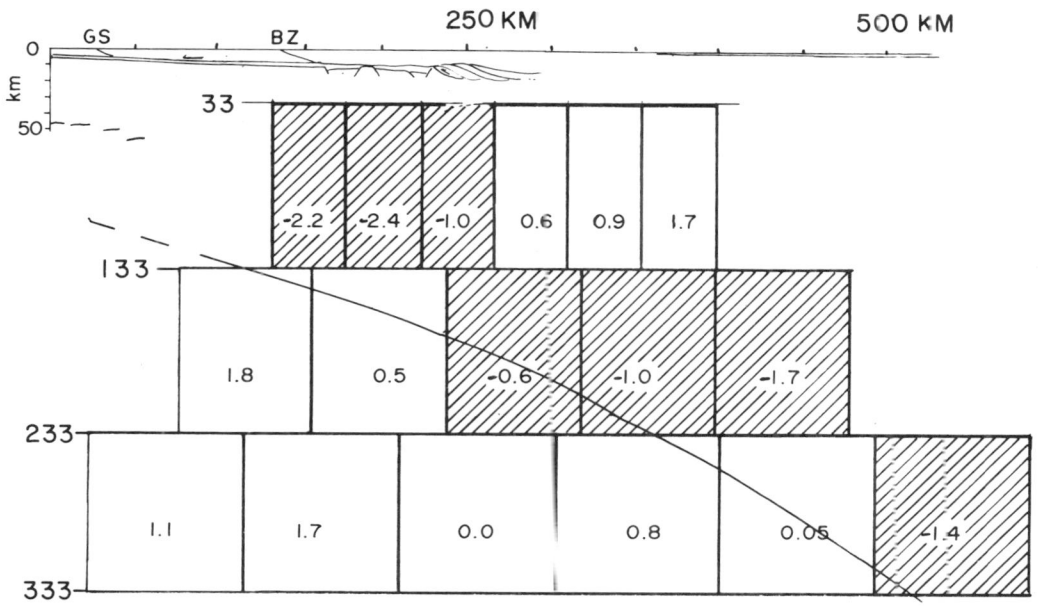

Fig. 2. Generalized interpretation of the southern Appalachian COCORP line [after Ando et al., 1984] compared to percent perturbations in mantle velocity, after Volz [1979]. One standard deviation on the velocity values is approximately 0.8. The negative values indicate lower velocities. Heavy line indicates base of the inferred lithosphere. [PGG = Piedmont Gravity Gradient, BZ = Brevard Zone, and GS = Great Smoky fault]

If the interpretation of the velocity perturbations as subducted lithosphere is correct, then the subducted lithosphere has remained an identifiable undetached unit under and to the southeast of the southern Appalachians. The curvature preserved in the velocity perturbations suggests a lithospheric rigidity at the time of emplacement comparable to subduction zones of today associated with oceanic crust. The interpretation of a rigid lithosphere at the time of emplacement and the analysis of Karner and Watts [1983] support continuity through time of a rigid lithosphere, capable of supporting the existing topographic load.

Seismic Refraction Data

Earthquake and explosion data from seismic stations in two zones parallel to the southern Appalachians have been assembled for interpretation as refraction lines (see Figure 1 for station locations). These refraction lines were used to provide a base for time-term analysis of crustal thicknesses near the southern Appalachian COCORP line in southeastern Tennessee. In the refraction line along the Charlotte and Carolina Slate belt which is southeast of the Piedmont Gravity Gradient (Figure 1) [Kean and Long, 1980] a crustal thickness of 33 km (velocity of 6.05 km/s) was observed. Reflections occur at a depth of 33 km (11 s) in the COCORP southern Appalachian line over the Charlotte and Carolina Slate belt. No direct evidence for a crustal layer with velocity of 6.7 km/s was found. However, a model study of reflection amplitudes in this area [Lee and Dainty, 1982] indicates a 3 to 6 km thick layer with velocity of 6.7 km/s at the base of the crust. The 6.7 km/s is suggestive of relic oceanic plate or rift components to the lower crust. In the Charlotte and Carolina Slate belt, the lack of significant material of velocity 6.7 km/s suggests that the positive-negative paired gravity anomaly is related to crustal thickness variations as discussed by Hutchinson et al. [1983] and not to a lateral change in density as modeled by Thomas [1983].

Arrival time data from large explosions in Alabama, which were recorded on seismic stations in Alabama, indicate a 6.15 km/s velocity for the crust (Figure 3). The velocity model includes a

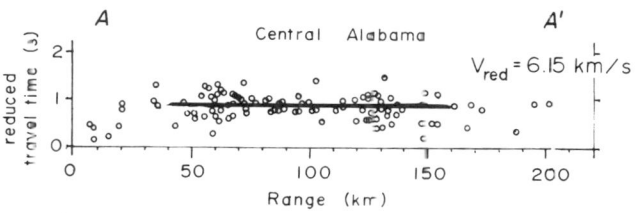

Fig. 3. Reduced travel time for the central Alabama refraction line. The circles are arrival times. A least squares error estimate of the velocity gives 6.15 ± 0.02 km/s.

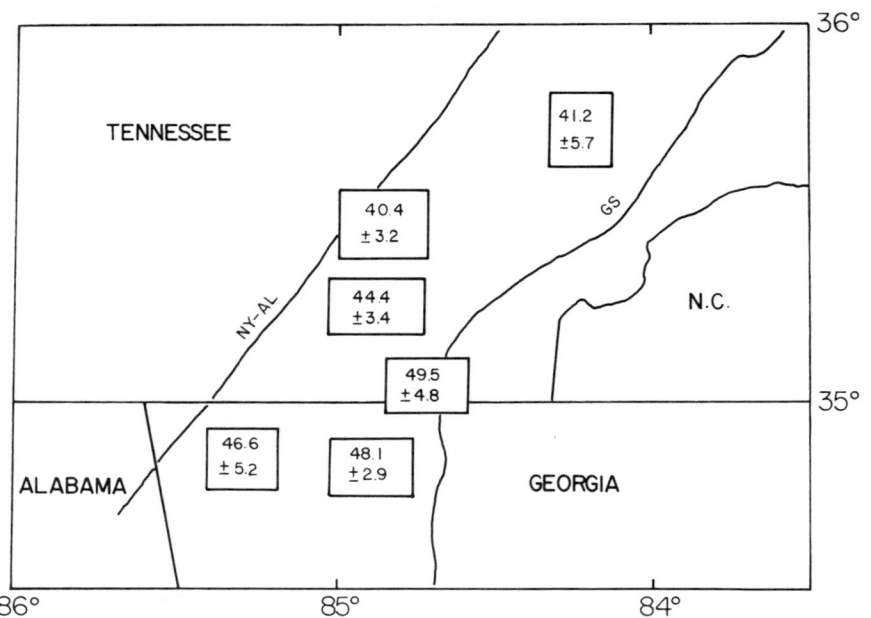

Fig. 4. Crustal thickness values in km derived from modified time-term analysis of Pn arrivals. The mantle velocity was fixed at 8.2 km/s. The time-term method was modified to collect on delay times for rays that intercept the mantle in the areas indicated in the figure. Only areas with three or more rays are included. The error estimates are for one standard deviation of the observed estimates of crustal thickness.

4.5 to 5.5 km/s, 2.4 km thick sediment layer and a 35 km thickness for the crust. No significant evidence was found for a 6.7 km/s layer at the base of the crust in the southern Appalachians of Alabama. Earlier velocity models for the northwestern flank of the southern Appalachians [Tatel et al., 1953; Steinhart and Meyer, 1961; Borcherdt and Roller, 1966] show evidence for a 6.7 km/s layer which constitutes over half of the crustal thickness. The 6.7 km/s layer appears with decreased thickness in interpretations of refraction lines to the southeast [Hutchinson et al., 1983]. The refraction lines with the thicker 6.7 km/s layer all pertain to eastern Tennessee. Recently, Owens et al. [1984], in an analysis of teleseismic P waves recorded at the Cumberland Plateau Observatory, Tennessee, provided evidence for a Precambrian rift in northeastern Tennessee. Teleseismic P waves from events to the northeast indicate crustal velocities of the order of 6.7 km/s in the vicinity of the east continent gravity anomaly. Teleseismic P waves from events to the southwest do not indicate the presence of a rift structure to the southwest, in the direction of the Alabama refraction data. Hence, the distribution of the rift structure with a dominant 6.7 km/s layer may be limited to eastern Tennessee, and the velocity structure observed in Alabama may be more appropriate for the crust throughout the remainder of the COCORP line (Figure 1).

Crustal Thickness Variations From Pn

A modified version of the time term method, similar to Barry's method [Sheriff and Geldart, 1982, p. 220] was used to compute the average thickness of the crust in limited areas (see Figure 4) of the mountains in southeastern Tennessee. We fixed the mantle velocity at 8.2 km/s and formulated the time term equations by collecting delay times for rays striking the mantle in the areas defined in Figure 4. The crustal thickness varied from 40 km to 50 km from west to east toward the greater topography of the Smoky Mountains. The computed depths are projected onto the COCORP southern Appalachian profile in Figure 5. The crustal thickening from 40 km in the Valley and Ridge Province to as deep as 50 km under parts of the mountains exceeds the compensation necessary (about 9 km/km) to satisfy local Airy isostatic compensation.

Potential Data

Gravity data are appropriate for modeling the large structural elements of the crust. One of the more prominent features in the gravity and magnetic data is the New York-Alabama lineament (NY-AL) [King and Zietz, 1978]. The NY-AL extends for over 1600 km and is interpreted as a major near-vertical discontinuity of the crust. In the regional interpretation (Figure 1) the

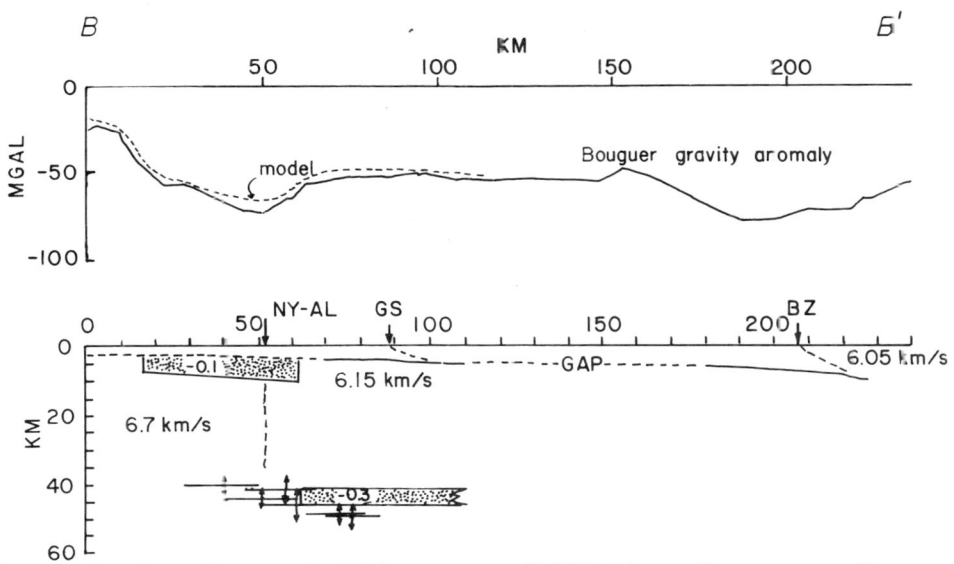

Fig. 5. Projection of crustal thickness onto COCORP line. Bouguer gravity anomalies are modeled by two zones of lower density. The density anomalies are in units of $kgm^3 \times 1000$. [NY-AL = New York-Alabama Lineament, GS = Great Smoky fault, and BZ = Brevard Zone]

east continent gravity anomaly is a relic rift structure in which 6.7 km/s crustal velocities are found. The rift structure terminates against the NY-AL.

In southeastern Tennessee, the NY-AL is defined by the trend of a linear negative gravity anomaly. This anomaly must be derived from low-density materials below the thrust plane since the gravity anomaly contours do not exactly follow the grain of the near-surface geologic structures. The COCORP southern Appalachian profile ends short of the NY-AL, but in Figure 5 we have extended the profile to include gravity data from the Brevard to 50 km northwest of the NY-AL. The NY-AL marks a gradient in Bouguer gravity anomaly which is more negative to the northwest. However, the source of this negative anomaly is in the 5 to 10 km depth range (see Figure 5). The gradient of the Bouguer anomaly with the shallow crustal structure removed is toward more positive anomalies consistent with a decrease in crustal thickness to the northwest. Gravity models for the southern end of the COCORP profile have been presented in some detail by Long [1979] and by Dainty and Frazier [1984].

Isostatic Response Function

Dorman and Lewis [1970] derived an expression for direct computation of an isostatic response function for topographic loads. The isostatic response function is the gravity anomaly generated by the compensating mass for an impulse in topography. In the Lewis and Dorman [1970] data analysis, the response was assumed to be radially symmetric. However, for the southern Appalachians the dominant linear structures preclude an assumption of radial symmetry. Instead, the data require a two-dimensional analysis perpendicular to structure and nearly parallel to the COCORP southern Appalachian line. Figure 6 shows the Bouguer gravity anomalies and topography averaged along strike of the major trend of the southern Appalachians. The isostatic response function derived from these data for the southern Appalachians (Figure 6) is sharper than the Lewis and Dorman [1970] response function for the continental United States. The half width is on the order of 50 km for the continental United States and significantly less for the southern Appalachians. The half width of 50 km or less in the isostatic response function apparently implies that lithospheric flexure in response to a topographic load is limited to distances of 50 to 100 km. Hence, for the southern Appalachians, the paradox of a rigid crust coexisting with Airy style isostatic equilibrium exists.

Discussion and Conclusions

As an alternative to lithospheric flexure, the isostatic response function may be interpreted as the differential vertical movement of crustal blocks with horizontal dimensions on the order of 75 km, in effect simulating local Airy style isostatic equilibrium. The P_n delay times confirm that areas of high topography are associated with increased crustal thicknesses, in agreement with local compensation, either by short-wavelength flexure or crustal block vertical

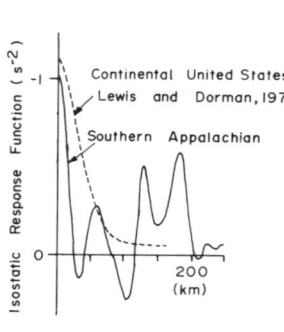
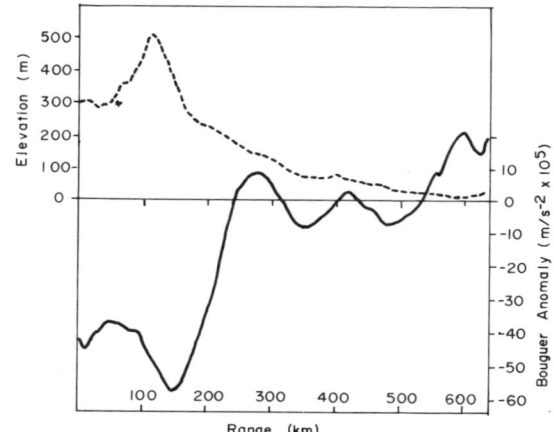

Fig. 6. Elevations and Bouguer anomalies averaged for Georgia and western South Carolina. The isostatic response function was obtained from these profiles in the Fourier Transform domain and then transformed back into the space domain.

displacement. On the other hand, the interpretation of the subducted lithospheric plate (Figure 2) and the analysis of Karner and Watts [1983] implies a rigidity at the time of implacement that is comparable to the rigidity of the crust observed today.

The long wavelengths of flexure in subduction and the short wavelengths implied by the isostatic response function and crustal thicknesses illustrate the apparent paradox in flexural response of the lithosphere in the southern Appalachians. Karner and Watts [1983] similarly noted that the short wavelengths in the isostatic response function seriously underestimate the effective thickness of the lithosphere and reconciled the apparent contradiction by suggesting that subsurface loads are a major contributor to the deflection of the Mohorovicic discontinuity. McNutt and Parker [1978] reconciled the contradiction by proposing that the crust adjusts toward isostatic equilibrium with time through a viscoelastic response.

Since the assumptions normally used to derive the isostatic response function assume that the influence of the geologic structures average to zero, the coherent geologic belts in the southern Appalachians could easily contribute to the character of the isostatic response function. For example, the near-surface mafic units of the Charlotte and Carolina Slate belt cause the positive Bouguer anomalies [Long, 1979] of the positive-negative paired gravity anomalies that do not correspond directly to a change in elevation. The elevation gradient in the Piedmont is more gradual, and consequently, the coherent geologic signature of the Charlotte and Carolina Slate belt tends to destabilize the isostatic response function. However, even if we accept the contamination of the isostatic response function by coherent geologic structures in the Piedmont, the correlation of crustal thickness with topography from seismic data in the Blue Ridge remains in apparent contradiction with the flexural rigidity of the lithosphere.

We offer a possible solution to the apparent contradiction. The topography of the overthrust plate may exist in response to the topography and crustal thickness of the underlying plate. During collision the thin overthrust sheets flex easily and conform to the topography of the thrust plane surface. Hence, topographic relief is preserved with no significant change in isostatic equilibrium, since the load is equivalent to a plate of uniform thickness. No new need exists for differential vertical movement subsequent to the collision episodes because the thrusting of sheets does not create isostatic imbalance. The question of the origin of the topography and crustal thickness of the underlying crust remains unanswered. In some areas the topography may reflect regional variations in crustal composition. Other possible explanations include a thrust as proposed for the Wind River Mountains [Smithson et al., 1978] or periods of tectonic activity characterized by reduced lithospheric rigidity (for example, rifting). In the latter case, the isostatic response function derived from gravity and elevation data would be frozen in from times of more mobile crustal dynamics.

Acknowledgments. This work was supported by the Alabama Geological Survey through a contract with the U.S. Nuclear Regulatory Commission, Office of Nuclear Regulatory Research, Earth Sciences Branch. The work on the isostatic response function was originally supported by the Army Research Office--Durham (grant DAHC04-74-G-0003). Appreciation is extended to the Tennessee Earthquake Information Center and the Tennessee Valley Authority for cooperation in seismic data acquisition.

References

Aki, K., A. Christoffersson, and E. S. Husebye, Determination of the three dimensional seismic structure of the lithosphere, J. Geophys. Res., 82, 277-296, 1977.

Ando, C. J., B. L. Czuchra, S. L. Klemperer, L. D. Brown, M. J. Cheadle, F. A. Cook, J. E. Oliver, S. Kaufman, T. Walsh, J. B. Thompson, Jr., J. B. Lyons, and J. L. Rosenfeld, Crustal profile of mountain belt: COCORP deep seismic reflection profiling in New England Appalachians and implications for architecture of convergent mountain chains, The American Association of Petroleum Geologists Bulletin, 68, 819-837, 1984.

Borcherdt, R. D., and J. C. Roller, A preliminary summary of a seismic-refraction survey in the vicinity of the Cumberland Plateau Observatory, Tennessee, U.S. Geol. Survey Technical Letter, Crustal Studies - 43, 1966.

Brown, L. D., and J. E. Oliver, Vertical crustal movements from leveling data and their relation to geologic structure in the eastern United States, Rev. Geophys. and Space Physics, 14, 13-35, 1976.

Clark, H., J. Costain, and L. Glover, Structure and seismic reflection studies on the Brevard Zone ductile deformation zone near Rosman, North Carolina, Am. J. Sci., 278, 419-441, 1978.

Cook, F. A., D. Albaugh, L. Brown, S. Kaufman, J. Oliver, and R. Hatcher, Thin-skinned tectonics in the crystalline southern Appalachians: COCORP seismic-reflection profiling of the Blue Ridge and Piedmont, Geology, 7, 503-567, 1979.

Cook, F. A., and J. E. Oliver, The late Precambrian-early Paleozoic continental edge in the Appalachian orogen, Am. J. Sci., 281, 993-1008, 1981.

Cook, F. A., L. D. Brown, S. Kaufman, and J. E. Oliver, The COCORP seismic reflection traverse across the southern Appalachians, AAPG Studies in Geology, No. 14, 1983.

Dainty, A. M., and J. E. Frazier, Bouguer gravity in northeastern Georgia: A buried suture, a surface suture, and granites, Geol. Soc. Am. Bull., 95, 1168-1175, 1984.

Dorman, L. M., and B. T. R. Lewis, Experimental isostasy--1. Theory of the determination of the Earth's isostatic response to a concentrated load, J. Geophys. Res., 75, 1970.

Hutchinson, D. R., J. A. Grow, and K. D. Klitgord, Crustal structure beneath the southern Appalachians: Nonuniqueness of gravity modeling, Geology, 11, 611-615, 1983.

Hatcher, R. D. Jr., Tectonics of the western Piedmont and Blue Ridge, southern Appalachians: Reviews and speculation, Am. J. Sci., 278, 276-304, 1978.

Jurkowski, G., and R. Reilinger, Recent vertical crustal movements: The eastern United States, NUREG/CR-2290, 74 pp., 1981.

Karner, G. D., and A. B. Watts, Gravity anomalies and flexure of the lithosphere at mountain ranges, J. Geophys. Res., 83, No. B12, 10449-10477, 1983.

Kean, A. E., and L. T. Long, A seismic refraction line along the axis of the southern Piedmont and crustal thicknesses in the southeastern United States, Earthquake Notes, 51, No. 4, 3-14, 1980.

King, E. R., and I. Zietz, The New York-Alabama lineament: Geophysical evidence for a major crustal break in the basement beneath the Appalachian basin, Geology, 6, 312-318, 1978.

Lee, C. K., and A. M. Dainty, Seismic structure of the Charlotte and Carolina Slate belts of Georgia and South Carolina, Earthquake Notes, 53, No. 2, 23-38, 1982.

Lewis, B. T. R., and L. M. Dorman, Experimental isostasy--2. Isostatic model for the U.S.A. derived from gravity and topography data, J. Geophys. Res., 75, 3367-3386, 1970.

Long, L. T., Bouguer gravity anomalies of Georgia, in Symposium on the Petroleum Geology of the Georgia Coastal Plain, Bulletin 87, Ga. Geol. Surv., Atlanta, GA, 141-166, 1974.

Long, L. T., The Carolina Slate belt--evidence of a continental rift zone, Geology, 7, 180-184, 1979.

McNutt, M. K., and R. L. Parker, Isostasy in Australia and the evolution of the compensation mechanism, Science, 199, 773-775, 1978.

Owens, T. J., G. Zandt, and S. R. Taylor, Seismic evidence for an ancient rift beneath the Cumberland Plateau, Tennessee: A detailed analysis of broadband teleseismic P waveforms, J. Geophys. Res., 89, 7783-7795, 1984.

Sheriff, R. E., and L. P. Geldart, Exploration Seismology, Volume 1, History, Theory, and Data Acquisition, 253 pp., Cambridge University Press, 1982.

Smithson, S. B., J. A. Brewer, S. Kaufman, J. E. Oliver, and C. Hurich, Nature of the Wind River thrust, Wyoming (from COCORP deep reflection data and from gravity data), Geology, 5, 648-652, 1978.

Steinhart, J. S., and R. P. Meyer, Explosion Studies of Continental Structure, 409 pp, Carnegie Inst. of Washington Publication 622, Washington, DC, 1961.

Tatel, H. E., L. H. Adams, and M. A. Tuve, Studies of the earth's crust using waves from explosions, Proc. Amer. Phil. Soc., 97, 658-669, 1953.

Thomas, M. D., Tectonic significance of paired gravity anomalies in the southern and central Appalachians, Geol. Soc. Am., Memoir 158, 113-124, 1983.

Toksöz, M. N., N. H. Sleep, and A. T. Smith, Evolution of the downgoing lithosphere and the mechanisms of deep focus earthquakes, Geophys. J. R. Astr. Soc., 35, 285-310, 1973.

Volz, W., Travel time perturbations in the crust and upper mantle in the southwest, Masters Thesis, 198 pp., Georgia Institute of Technology, Atlanta, GA, 1979.

NATURE OF THE LOWER CONTINENTAL CRUST: EVIDENCE FROM BIRPS
WORK ON THE CALEDONIDES

Jeremy Hall

Department of Geology, University of Glasgow, Glasgow G12 8QQ, Scotland

Abstract. The Western Isles and North Channel ('WINCH') seismic reflection profile traverses the metamorphic Caledonides of northern Britain. The lower crust appears to be more reflective than the crust above or the mantle below. The base of the reflective sequence, assumed to be the base of the crust, can be drawn with confidence over most of the profile. The top of the reflective sequence can also be picked, but with greater uncertainty. Across the Caledonide Dalradian metasediments the profile indicates an antiformal Moho and a synformal top of the reflective sequence. There is little variation in the gravity field over this feature. In effect the Moho relief is compensated isostatically by necking of the lower crustal reflective layer. Modelling this using velocity-density systematics indicates that the lower crust here must have a density of around 3100 kg m^{-3}, and a mean P-wave velocity of about 7.3 km s^{-1}. Along strike the lower crust has high electrical conductivity. It is suggested that the lower crust is of basic composition, of variable metamorphic grade, and containing free water trapped by contraction during cooling.

Introduction

The aims of this paper are to provide estimates of, and then explain, the physical properties of deep, layered continental crust. The deep crust often shows strong reflectivity on seismic sections [Mueller, 1977; Smithson, 1978], well shown on much of the data collected by the British Institutions Reflection Profiling Syndicate (BIRPS) [Brewer et al., 1983; Brewer and Smythe, 1984]. A section of layered crust from the BIRPS Western Isles - North Channel (WINCH) profile is used as an example.

The Western Isles - North Channel Profile: 'WINCH'

Initial British deep reflection work was directed to looking at the NW margin of the Caledonides on marine profiles (Figure 1) just offshore from known basement geology. The Moine and Outer Isles Thrusts profile (MOIST) [Smythe et al., 1982; Brewer and Smythe, 1984] crossed the seaward extensions of the Moine and Outer Isles Thrusts which carry basement rocks westwards over the undeformed Archaean basement rocks of the Caledonian foreland. These thrusts were clearly seen on the MOIST section together with a deeper third - the Flannan Thrust - which appears to cut the crust-mantle boundary. The thrusts may have long histories and were certainly active in post-Caledonian sedimentary basin development, when they were reactivated as soles to listric normal faults in the extensional phase. The WINCH line was intended to follow up MOIST by adding the beginnings of a three-dimensional view of the thrusts and then proceeding southwestwards across the foreland before turning to the southeast to cross all the major units of the British Caledonides except the Welsh shelf edge. Thus the profile would examine the marginal thrust belt to the northwest, the metamorphic Caledonides of the American (Laurentian) plate margin and the suture(s) of American and European continents, with whatever arc terranes may be trapped in between. Boundaries between the major tectonic units are often steep fault zones, which may have moved different ways at different times and include strike-slip movement (e.g. the Great Glen Fault).

A preliminary account of WINCH [Brewer et al., 1983] explains that several thrusts were seen in addition to confirmation of those seen in MOIST; very little evidence of steep tectonic boundaries is observed; sedimentary basins are often found in the hanging walls of thrusts; and the lower crust is variably reflective right across the section, while the upper crustal basement is invariably unreflective. The seismic reflection sections are available from the offices of the British Geological Survey in Edinburgh. A detailed interpretation of the middle section of the WINCH line [Hall et al., 1984] includes the major features shown in Figure 2. This shows the possible nature of the obduction of the

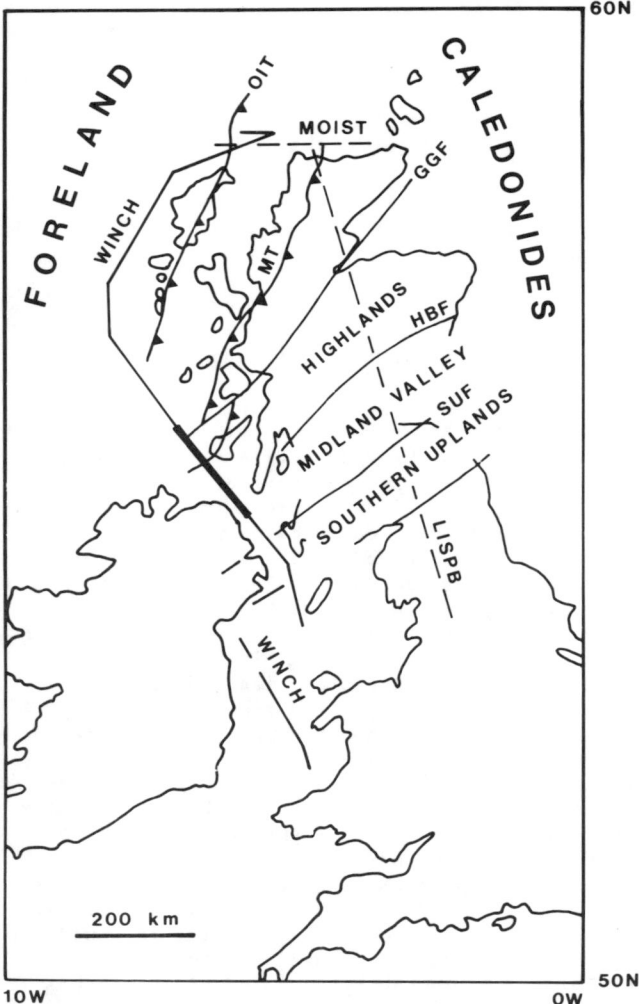

Fig. 1. Map showing location of MOIST and WINCH deep seismic reflection lines, which run from the foreland of NW Britain across various tectonic units of the Caledonides. OIT and MT are the Outer Isles and Moine Thrusts, respectively. HBF and SUF are the Highland Boundary and Southern Uplands Faults. GGF = Great Glen Fault. Thickened part of WINCH line is that examined in this paper. Location of the LISPB refraction profile is also shown.

European continental margin against the Archaean foreland of the American continent with an intervening island arc (below the Midland Valley) shunted below the trench and ophiolite remnants exposed at the surface of the Southern Uplands. The wedging mechanism inferred to have juxtaposed the Midland Valley basement against the European continent includes a deep crustal northward dipping thrust which is, in effect, the Iapetus Suture. A similar wedging mechanism is believed to have operated to bring the shelf-edge sediments of the Caledonide Dalradian basin against the foreland. Record quality varies along the line but reflective lower crust (Figure 3) is discernible along most of the line and is terminated suddenly downwards at what is assumed to be the crust-mantle boundary. This correlation has been established in the Scottish region in two places: on MOIST at its crossing with the LISPB refraction profile [Bamford et al., 1978], and in the North Sea [Barton et al., 1984]; moreover the Moho delay time concurs roughly with those from crustal refraction surveys conducted nearby, the Hebridean Margin Seismic Experiment [Bott et al., 1979], for example.

It is surprising that after such a complex history the variation in depth to the Moho should appear so modest. It has been suggested [Hall et al., 1984] that the density contrast across the Moho is so large that significant relief on the Moho surface implies gravitational stress which is relaxed during later basin formation.

A Moho 'High'

There is one place along the WINCH line where quite substantial relief of the Moho surface appears to be present. This is between shot-points 12000 and 14000. In two-way time there is a Moho 'high' here of over 1 s amplitude. This is about an order of magnitude greater than any pull-up effects between later basins. Assuming a mean crustal velocity of 6 km s^{-1}, a first order estimate of the depth relief would be 3 km. If such relief is associated with a density contrast of say 500 kg m^{-3} between crust and mantle, then a gravity anomaly of 600 gu (60 mgal) would result. No such anomaly exists (Figure 4): the Bouguer gravity rises steadily to the north-west with lows of the order of 100 gu (10 mgal) over known sedimentary basins.

Just above the Moho at its high, the lowermost crust is quite reflective, with the reflectivity diminishing rapidly both downwards across the Moho and upwards across a boundary picked as the top of the lower crustal reflective layer. Figure 3 illustrates the record section at shot point 13600, on the high. Moving both ways from shot-point 13600, the Moho remains distinct but the top of the lower crustal layer less so. The latter rises to the south and may be picked with moderate certainty. To the north the reflective lower crust tapers (see Figure 2) but becomes overlain by another (but less) reflective lower crust within the wedge of Archaean foreland. In Figure 4 the best estimates of reflection times from the top of the lower crustal layer (or layers, where composite) and the Moho show an obvious inverse correlation, and because of that, a possible explanation of the lack of an

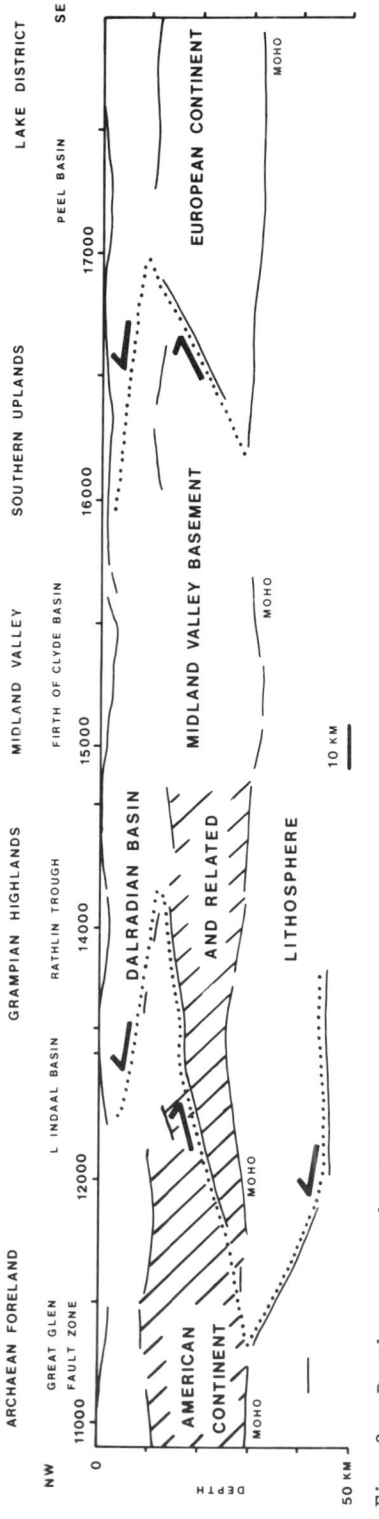

Fig. 2. Depth-converted, migrated section of part of the WINCH line showing principal deep reflectors. Dotted lines indicate thrust boundaries along which wedge obduction is believed to have occurred during the closure of the Iapetus Ocean. The shot-point numbers at the top of the section show a jump of 1000 between 12000 and 14000, signifying a ship turn round between 12500 and 13600 (both marked). 100 shot points = 5 km. Note the Moho high at shot-point 13600. Lower crustal 'neck' indicated by diagonal ruling.

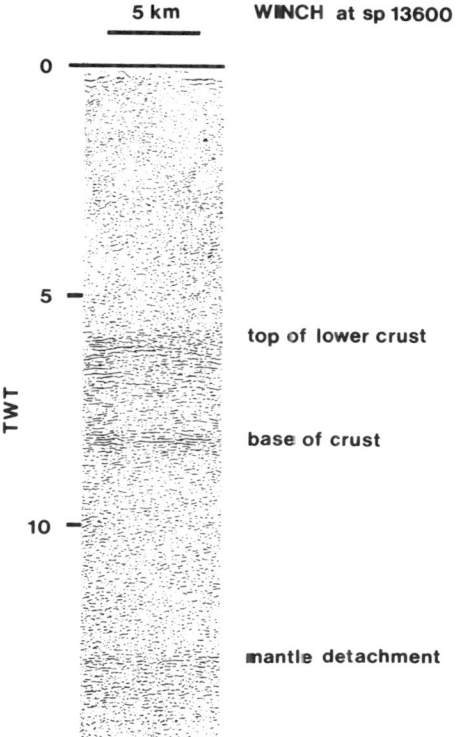

Fig. 3. A strip of the WINCH section at shot point 13600 on the Moho high. Note the reflective lower crust with well-defined upper and lower boundaries. Mantle event forms part of the lower obduction wedge shown at the left of Fig. 2.

observed gravity anomaly from the Moho high. If there is an appropriate density contrast across the top of the lower crustal reflective layer the depression in it may result in a gravity low which negates the high from the Moho.

In the following sections, the argument is quantified. In essence it is assumed that the cancellation of the two gravity effects is due to perfect local isostatic compensation. This is unlikely in practice though if the elastic lid were very thin during some late Caledonide or post-Caledonide event, compensation might be achieved locally to within the noise level (a few mgals) of the Bouguer gravity signal. It is also assumed that there are no lateral variations in density. This is manifestly invalid for surface rocks, but here the variations are negligible to first order especially as sedimentary basin effects (velocity pull-down) have been removed in the calculations and the Dalradian succession here lacks the large thicknesses of low density metaquartzites present further to the NE. The problem then reduces to that of calculating the density of the lower crustal layer, given the

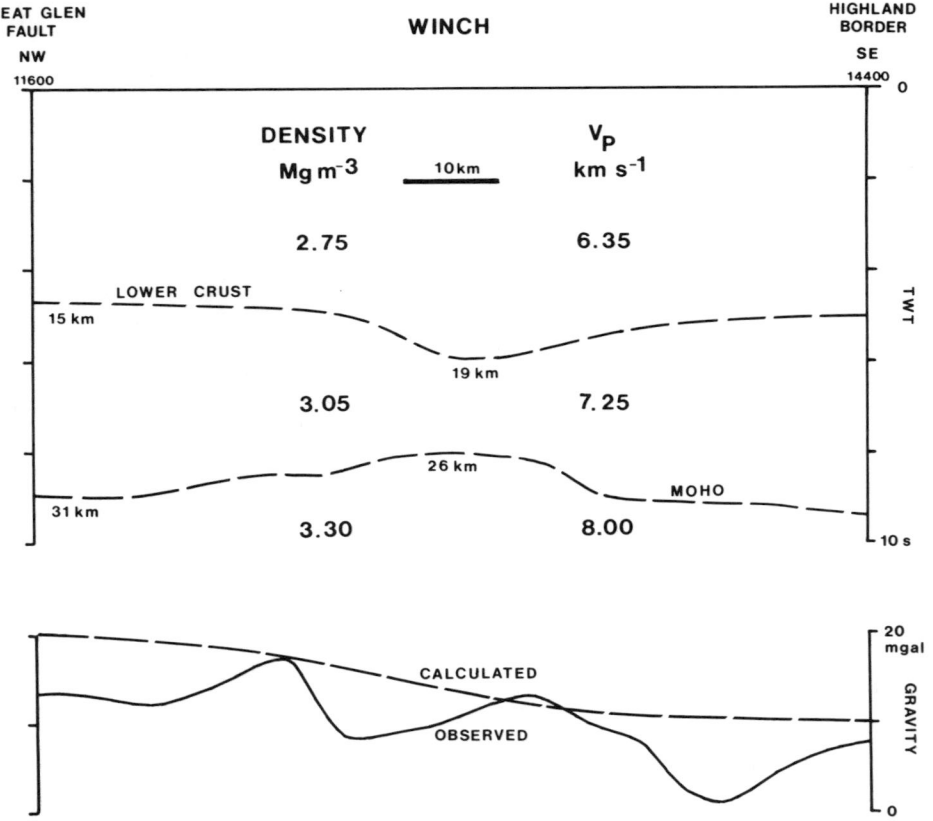

Fig. 4. Two way travel time (TWT) section of part of the WINCH line around shot-point 13600, showing inverse relationship of reflecting boundaries at top and base of the lower crust. Resulting velocity-density model shown gives depths to reflectors as indicated. Two-dimensional gravity model gives good fit of calculated to observed gravity, with remnant lows in observed gravity related to known sedimentary basins.

reflection time variation across the apparent necking of that layer (Figure 4), together with assumed densities of upper crust and mantle, and a suitable relationship of density with P-wave velocity.

Velocity-Density Systematics and Isostasy

In Figure 5, P-wave velocity is plotted against density for isotropic aggregates of the mineral groups likely to be predominant in rocks of the deeper crust and mantle. The data are estimates for pressures (6 kb) and temperatures (300°C) of deep cratonic crust (i.e. about 20-30 km and 250-400°C) taken from a variety of sources summarised in Anderson and Liebermann [1968], Birch [1960, 1961], Christensen [1982] and Simmons and Wang [1971]. Values plotted are for surface temperature and pressure since the effects of increases of pressure and temperature balance to within 0.05 km s^{-1}. The figure includes data from single crystals and from near monomineralic rocks (for which the high pressure asymptote is extrapolated back to zero pressure to remove effects of open cracks). Certain determinations lying well outside the ranges shown in Figure 5 have been omitted either because they are suspiciously low (extrapolations from low-pressure measurements on possibly cracked specimens) or are of members of the mineral group not commonly encountered in deep crustal rocks. The variations in V_p and density on crossing the upper and lower boundaries of the lower crustal reflective layer are to be determined. The lower crust is likely to gain in amphibole or pyroxene (and possibly garnet) at the expense of quartz and feldspar relative to the upper crust. The mantle is believed to be olivine rich [Ringwood, 1975], with no quartz and little feldspar. Given the uncertainty of the 'rules' and the spread of (V_p, density) pairs for each mineral group, a

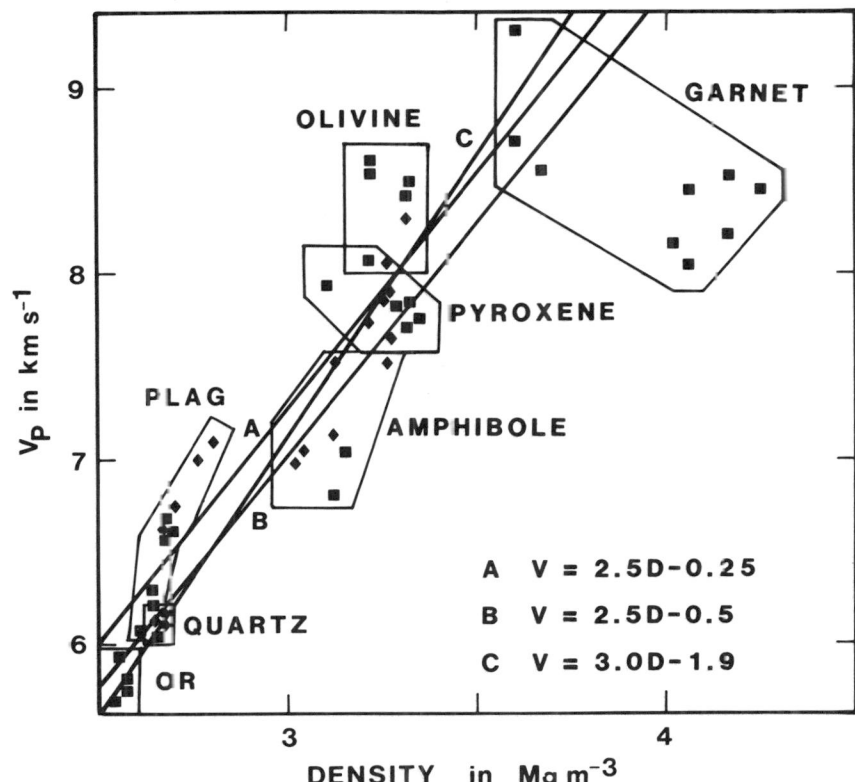

Fig. 5. Plot of P-wave velocity (V_p) against density (D) for common mineral groups at lower crustal conditions. Sources listed in text. Variation generalized by a choice of three linear velocity-density equations used in estimating lower crustal density. Square symbols indicate values from single crystal data, diamonds from near-monomineralic aggregates. OR = orthoclase, PLAG = plagioclase.

variety of linear relationships of V_p with density are used to assess their effect on estimates of lower crustal density; though the same linear rule is used for both transitions in any one application. As a guideline, Birch's [1961] law has been used, with a mean atomic weight of 21 thought appropriate to these rocks (line C on Figure 5). However Birch's law, even allowing variation in mean atomic weight, is too restricted to represent all the velocity, density variations in crossing major lithological boundaries. Consequently two other linear relationships (A and B on Figure 5) with different constants (constant and gradient) have also been used.

Lower crustal density is then estimated as follows. Suppose the top of the lower crustal layer increases laterally in reflection time by dT_1 and the Moho increases correspondingly in reflection time by dT_m.
The inverse correlation of the two reflection times may be expressed as

$$dT_m = -k \cdot dT_1.$$

Assuming a linear relationship of velocity (V) with density (D) implies that

$$V = a + bD,$$

where a, b are constants.
Given that there is perfect isostatic balance, it can be shown that the density of the lower crust is given by

$$2bD_1 = -(a-bD_m)$$
$$\pm [(a-bD_m)^2$$
$$+ 4b[aD_m - (a+bD_u)(D_m-D_u)/(k+1)]]^{1/2} \quad (1)$$

where D_u is the density of the upper crust and D_m the density of the mantle.

Thus given estimates of k, the regression coefficient of the reflection time correlation,

TABLE 1. Estimates of Lower Crustal Density for Example of WINCH Data

Velocity (V, in km s^{-1}) Density (D, in Mg m^{-3}) Relationship	Mantle Density in kg m^{-3}	Lower Crustal Density in kg m^{-3} for	
		k = 1.4	k = 1.0
V = -0.25 + 2.5D	3300	3110	3060
	3400	3170	3120
V = -0.5 + 2.5D	3300	3100	3050
	3400	3170	3120
V = -1.9 + 3.0D	3300	3100	3060
	3400	3170	3110

Calculations based on equation (1). Upper crustal density assumed 2750 kg m^{-3}. Uncertainties of ± 50 kg m^{-3} in mantle and upper crustal densities and ± 0.3 in k, yield uncertainties in estimates of lower crustal density of ± 70 kg m^{-3}. k is the ratio of time relief on the Moho reflection to the opposed relief on the reflection from the top of the lower crust.

and the densities of the upper crust and mantle, the density of the lower crust may be estimated using equation (1).

Application to Example from WINCH

Three different velocity-density relationships are used, as indicated on Figure 5 and discussed above. Surface basement rocks have densities averaging around 2700 kg m^{-3}, but given evidence (e.g. from LISPB) that the middle crust has rather higher seismic velocities than the upper basement, the upper crustal density is assumed to be 2750 ± 50 kg m^{-3}. Mantle density is estimable from refraction data; several refraction experiments in northern Britain [Bamford et al., 1978; Bott et al., 1979; Smith and Bott, 1975] and adjacent seas yield velocities of about 8.0 km s^{-1} for the topmost mantle. The corresponding density is likely to be in the region of 3300-3400 kg m^{-3}. The inverse correlation of the two reflectors is less easily estimated given the uncertainties in the pick of the top of the lower crustal reflective layer. From the data on Figure 4 a value for k of 1.2 may be obtained allowing for the 'regional' rise to the NW in both reflectors; but a weighting of observations according to reflection quality criteria could change the value by at least 0.2. In estimates here, both 1.0 and 1.4 are used for k to indicate the effect of this uncertainty on estimates of lower crustal density.

In Table 1, the effects of all these variables on D_l may be discerned. Estimates of D_l vary from 3050 to 3170 kg m^{-3} and the conclusion to be drawn is that the lower crustal density is estimated to be about 3110 ± 60 kg m^{-3}. The corresponding P-wave velocity is about 7.2 - 7.4 km s^{-1}. It is noticeable that the choice of velocity distribution has rather little effect on D_l, whereas D_m and k affect the estimate more substantially. Given that the error estimates of each determination of D_l in Table 1 are rather small (± 70 kg m^{-3}), it seems inescapable that the data suggest that the lower crust is more like the mantle than it is like the upper crust. It could be argued that the uncertainties used in the calculations are too small. For example, the mantle density could be as high as 3500 kg m^{-3} for V_P = 8.0 km s^{-1} for an eclogitic composition. The implication from Table 1 is that such a high density mantle would make the lower crust even more dense, by about 60 kg m^{-3}, than it would be for a mantle density of 3400 kg m^{-3}. It is unlikely that the upper crustal density has been overestimated, indeed data from Hipkin and Hussain [1983] suggest that surface basement densities are more likely to be underestimated here, in which case the estimates of lower crustal density would again be marginally too low. Regarding the range of velocity distributions used in Table 1 to be a little restricted and doubling the uncertainty of upper crustal and mantle densities, the estimates of lower crustal density estimates would still lie in the range 3000-3250 kg m^{-3}.

Using a density estimate and corresponding velocity, toward the lower end of the range of possibility, the depths to the top of the lower crust and the Moho, and the gravity effect of the structure are shown on Figure 4. The Moho

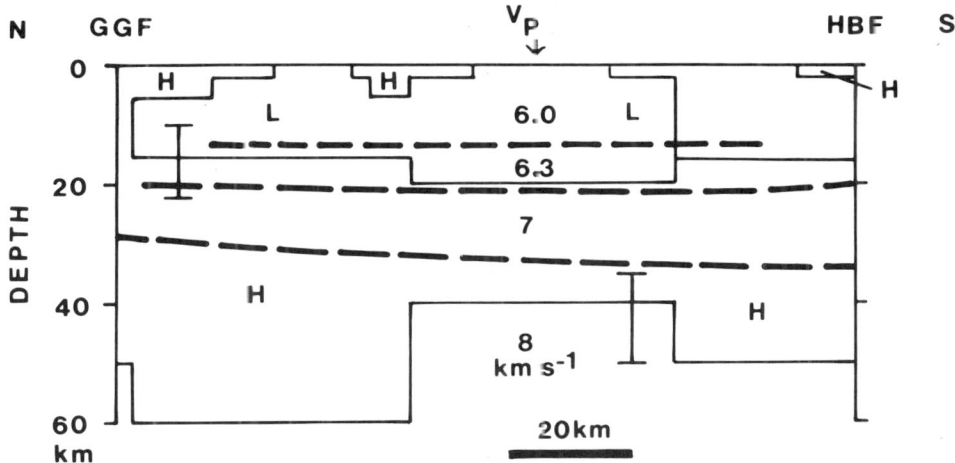

Fig. 6. Section of the LISPB refraction profile across the Scottish Highlands from the Great Glen Fault (GGF) to the Highland Boundary Fault (HBF) based on Bamford et al. [1978]. H and L denote blocks of high (greater than 0.003 S m^{-1}) and low (less than 0.001 S m^{-1}) electrical conductivity as determined by Hutton et al. [1980]. Error bars from Mbipom and Hutton [1983].

relief is 5 km and the lower crust is necked down to only 7 km at the Moho high. The calculated gravity matches the northwestward regional rise of the observed gravity [Hipkin and Hussain, 1983], which differs only in also showing the effects of known sedimentary basins.

It is hoped that a future wide-angle refraction-reflection traverse may cross this feature, perhaps along strike, and so determine the velocities deep in the crust, as a test of this thesis.

Physical Properties of the Lower Crust below the Caledonides

Estimation of the density of the lower crust from reflection and gravity data requires multiple assumptions as detailed above. The conclusion that the lower crust below the part of the WINCH line considered has a P-wave velocity of 7.0 - 7.5 km s^{-1} and a density of 3000 - 3250 kg m^{-3} must be regarded as speculative but the gravity data are well matched by such a model. To explain the lack of gravity anomaly over the Moho 'high' any other way requires more arbitrary assumptions.

It is significant that 200 km along strike, on the LISPB refraction profile [Bamford et al., 1978] the lowest crustal layer also has V_p about 7 km s^{-1} (Figure 6). It could well be the same kind of crust. If so the lower crust is also characterised here as having a Poisson's Ratio of 0.249 $\pm$ 0.017 [Assumpcao and Bamford, 1978] and electrical conductivity of about 0.01 S m^{-1} [Hutton et al., 1980]. In the interpretation which follows, a lower crustal model is constructed to match these properties.

Interpretation

From the above estimates, a number of conclusions may be drawn about the nature of the lower crust in this area. The P-wave velocity and density fall over the amphibole field on Figure 5. The calculation of the velocity of an isotropic aggregate of different minerals may be sufficiently well approximated, and the density is given, by arithmetic means of mineral properties weighted in proportion to their fractional volume. Thus possible compositions may be estimated from Figure 5 and include (a) 100% amphibole; (b) 65% pyroxene + 35% quartz; (c) 50% pyroxene + 50% feldspar; mixtures of (a), (b) and (c). Presence of pyroxene, garnet and olivine require balancing proportions of quartz and/or feldspar.

Most of the possible combinations include between 50% and 100% of mafic minerals, indicating that the lower crust is likely to be at least as basic as basic igneous rocks as suggested by Mueller [1977]. More acid compositions [Smithson, 1978] are not ruled out but are relatively unlikely since they require rather large amounts of garnet (less than 30%) rarely found in xenolith suites and Poisson's Ratios lower than found because of the high quartz content. Ultramafic compositions are unlikely since olivine would need to be serpentinised to provide low enough velocity and density, and this would be manifest in a much higher Poisson's Ratio than that observed.

If the composition of the lower crust is basic, how are the Poisson's Ratio and the electrical conductivity explained? The high conductivity cannot be explained in terms of the major mineral phases [Parkhomenko, 1982] other than by particular amphiboles such as riebeckite, not found in high grade metamorphic rocks. Appeal has been made to conducting accessory minerals such as graphite [Garland, 1981; Fyfe, this volume]. Few exhumed granulite terrains show evidence of widespread interconnected graphite, compared with the commonplace occurrence of high electrical conductivity in the lower crust [Hutton et al., 1980]. The more likely explanation of the high conductivity is that the lower crust contains free aqueous solution in interconnected pores [Shankland and Ander, 1983]. That such aqueous solutions are present in high grade metamorphic rocks is witnessed by their occurring as fluid inclusions, often with such low freezing points as to indicate salinities up to several times that of seawater [Touret and Dietvorst, 1983]. Strong brines at 300°C occupying less than 1% by volume of the rock can produce the required conductivity, since at this temperature molar NaCl has conductivity of 50 S m^{-1} [Quist and Marshall, 1968]. A mechanism for trapping and retaining water in the deep crust is to be presented elsewhere [Hall, 1985], but the evidence from fluid inclusions [Touret and Dietvorst, 1983] and from retrogressive metamorphism of granulite terrains (those of NW Scotland, for example [Sutton and Watson, 1951]) offer ample testimony to the flushing of water through deep crust during late metamorphic stages.

Small amounts of water in the lower crust would affect the minerals present and the seismic velocities. Unless the water present was inhibited by very low permeability or chemical activity, it would tend to cause at least local retrogression of 'dry' granulite crust to amphibolite facies. Basic amphibolites tend to have Poisson's Ratio rather higher than that observed in the lower crust in NW Britain, unless they have a very strong regional fabric [Hall and Simmons, 1979]. Free water itself would also tend to increase Poisson's Ratio [O'Connell and Budiansky, 1974]. Thus it is likely that much of the lower crust is basic granulite -- which could have appropriate Poisson's Ratio [Hall and Simmons, 1979] -- but with local patches of free water among amphibolite facies.

This model is particularly appealing. Firstly, it explains more than just the V_P, V_S, electrical conductivity and gravity data: it can explain the lower crustal reflectivity. The Dalradian rocks at the surface contain many metamorphosed basic sills intruded shortly after sedimentation in a rapidly extending basin [Graham, 1976; Anderton, 1982]. Similar sill intrusion in the deep crust could bring in water so that a sub-horizontal layering with associated free water would be established. 1% porosity in cracks of aspect ratio about 10^{-3} gives crack densities of order unity. Water saturated cracks of this density would have a P-wave velocity of 20% less than uncracked rock [O'Connell and Budiansky, 1974], so that there would be ample velocity contrast between watered and 'dry' layers to give deep crustal reflectors, provided the layering has a thickness component approaching the wavelength of the reflected waves (about 300 m).

Conclusions

The deep continental crust below part of the Caledonides may give some clues to the nature of the reflective lower crust encountered widely in deep reflection studies.

In this case, the reflective crust may have a P-wave velocity of over 7 km s^{-1} and a density of 3100 kg m^{-3}. It is likely to be associated with a Poisson's Ratio of about 0.25 and electrical conductivity greater than 0.003 S m^{-1}. From this it is inferred that the deep crust has an average composition like that of very basic igneous rocks, may have been in granulite facies but is likely to have been at least partly retrogressed to amphibolite facies with free water possibly trapped, initially, at low pressure.

Reflectivity is probably produced by sill intrusion during underplating enhanced by variations in water content across horizons.

Acknowledgements. Shell Expro UK supported my work on the BIRPS data.

References

Anderson, O.L., and Liebermann, R.C., Sound velocities in rocks and minerals: experimental methods, extrapolations to very high pressures, and results, in Physical Acoustics IV B, edited by W.P. Mason, pp.329-472, Academic Press, New York, 1968.

Anderton, R., Dalradian deposition and the late Precambrian-Cambrian history of the N. Atlantic region: a review of the early evolution of the Iapetus Ocean, J. Geol. Soc. Lond., 139, 423-431, 1982.

Assumpcao, M., and Bamford, D., LISPB-V. Studies of crustal shear waves, Geophys. J. R. Astron. Soc., 54, 61-74, 1978.

Bamford, D., Nunn, K., Prodehl, C., and Jacob, B., LISPB-IV. Crustal structure of Northern Britain, Geophys. J. R. Astron. Soc., 54, 43-60, 1978.

Barton, P., Matthews, D., Hall, J., and Warner, M., The Mohorovicic Discontinuity seen on normal incidence and wide-angle seismic records, Nature, 308, 55-56, 1984.

Birch, F., The velocity of compressional waves in rocks to 10 kbar, 1, J. Geophys. Res., 65, 1083-1102, 1960.

Birch, F., The velocity of compressional waves in rocks to 10 kbar, 2, J. Geophys. Res., 66, 2199-2224, 1961.

Bott, M.H.P., Armour, A.R., Himsworth, E.M., and Murphy, T., An explosion seismology investigation of the continental margin west of the Hebrides, Scotland, Tectonophysics, 59, 217-231, 1979.

Brewer, J.A., Matthews, D.H., Warner, M.R., Hall, J., Smythe, D.K., and Whittington, R.J., BIRPS deep seismic reflection studies of the British Caledonides - the WINCH profile, Nature, 305, 206-210, 1983.

Brewer, J.A., and Smythe, D.K., MOIST and the continuity of crustal reflector geometry along the Caledonian - Appalachian orogen, J. Geol. Soc. Lond., 141, 105-120, 1984.

Christensen, N.I., Seismic velocities, in Handbook of Physical Properties of Rocks - Vol. II, edited by R.S. Carmichael, pp.1-228, CRC Press, Boca Raton, Florida, 1982.

Fyfe, W.S., Price, N.J., and Thompson, A.B., Fluids in the Earth's crust, 383 pp., Elsevier, Amsterdam, 1978.

Garland, G.D., Correlation between electrical conductivity and other geophysical parameters, Phys. Earth Planet. Inter., 10, 220-230, 1975.

Graham, C.M., Petrochemistry and tectonic significance of Dalradian metabasaltic rocks of the SW Scottish Highlands, J. Geol. Soc. Lond. 132, 61-84, 1976.

Hall, J., Physical properties of the lower continental crust, in The Nature of the Lower Continental Crust, edited by D.A. Carswell et al., Spec. Vol. Geol. Soc. Lond., (in press).

Hall, J., and Simmons, G., Seismic velocities of Lewisian metamorphic rocks at pressures to 8 kbar: relationship to crustal layering in north Britain, Geophys. J. R. Astron. Soc., 58, 337-347, 1979.

Hall, J., Brewer, J.A., Matthews, D.H. and Warner, M.R., Crustal structure across the Caledonides from the WINCH seismic reflection profile: influences on the evolution of the Midland Valley of Scotland, Trans. R. Soc. Edinburgh: Earth Sci., 75, 97-109, 1984.

Hutton, V.R.S., Ingham, M.R., and Mbipom, E.W., An electrical model of the crust and upper mantle in Scotland, Nature, 287, 30-33, 1980.

Mbipom, E.W, and Hutton, V.R.S., Geoelectromagnetic measurements across the Moine Thrust and the Great Glen in northern Scotland, Geophys, J. R. Astron. Soc., 74, 507-524, 1983.

Mueller, S., A new model of the continental crust, in The Earth's Crust, edited by J.G. Heacock, Geophys. Monog. Am. Geophys. Un. No. 20, 289-317, 1977.

O'Connell, R.J., and Budiansky, B., Seismic velocities in dry and saturated cracked solids, J. Geophys. Res., 79, 5412-5426, 1974.

Parkhomenko, E.I., Electrical resistivity of minerals and rocks at high temperature and pressure, Rev. Geophys. Space Phys., 20, 193-218, 1982.

Quist, A.S., and Marshall, W.L., Electrical conductances of aqueous sodium chloride solutions from 0 to 800° and at pressures to 4000 bars, J. Phys. Chem., 72, 684-703, 1968.

Ringwood, A.E., Composition and petrology of the Earth's mantle, 618pp, McGraw-Hill, New York, 1975.

Shankland, T.J., and Ander, M.E., Electrical conductivity, temperatures and fluids in the Earth's crust, J. Geophys. Res., 88, 9475-9484, 1983.

Simmons, G., and Wang, H., Single crystal elastic constants and calculated aggregate properties: a handbook, 370pp. MIT Press, Cambridge, Mass., 1971.

Smith, P.J., and Bott, M.H.P., Structure of the crust beneath the Caledonian foreland and Caledonian belt of the north Scottish shelf region, Geophys. J. R. Astron. Soc., 40, 187-205, 1975.

Smithson, S.B., Modelling continental crust: structural and chemical constraints, Geophys. Res. Lett., 5, 749-752, 1978.

Smythe, D.K., Dobinson, A., McQuillin, R., Brewer, J.A., Matthews, D.H., Blundell, D.J., and Kelk, B., Deep structure of the Scottish Caledonides revealed by the MOIST reflection profile, Nature, 299, 338-340, 1982.

Sutton, J., and Watson, J.V., The pre-Torridonian metamorphic history of the Loch Torridon and Scourie areas in the northwest Highlands, and its bearing on the chronological classification of the Lewisian, J. Geol. Soc. Lond., 106, 241-308, 1951.

Touret, J., and Dietvorst, P., Fluid inclusions in high-grade anatectic metamorphites, J. Geol. Soc. Lond., 140, 635-649, 1983.

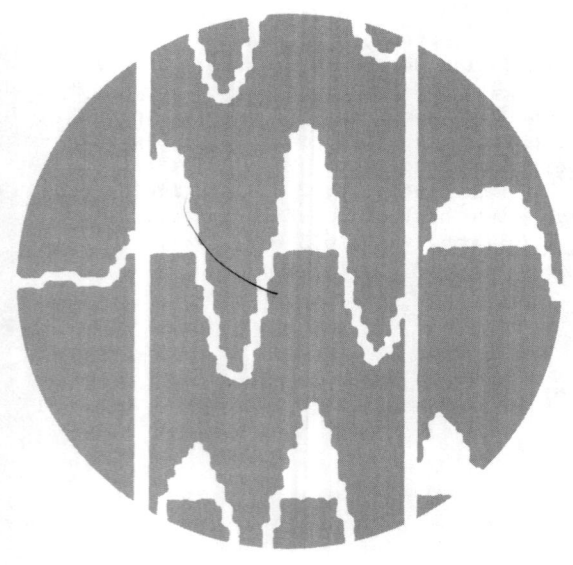

THE HERCYNIAN EVOLUTION OF THE SOUTH WEST BRITISH CONTINENTAL MARGIN

G.A. Day

British Geological Survey

Abstract. Motions of platelets in W Europe during the Devonian are not well defined, but the evidence supports northward movement with the closure of oceanic basins. Geophysical data demonstrate that the major features of the crust in the western English Channel and the area to the west out to the shelf edge, all trend WSW-ENE, and in addition that the whole southern flank of the Cornubian High is deformed by north-directed thrusting active in Devonian times. The character of the crust changes to the east and it is proposed that a transform fault, along the line of the Bray Fault in northern France, offsets the zone of Hercynian deformation in SW Britain from a similar zone in N Germany. Post-Hercynian adjustment produced further movement of the European blocks, firstly in a zone of dextral shear and finally during a period of extension, when Permian grabens appeared along the lines of the suture transform complex.

Introduction

Seismic reflection profiles provide the most powerful single tool in understanding the late Palaeozoic and Mesozoic evolution of SW Britain and the adjacent sea area. (The area of the present study is shown in figure 1.) Exposed rocks on land are predominantly of Carboniferous or older age which give way offshore to Permo-Triassic and younger rocks. Few commercial seismic lines have been shot at sea adjacent to the Cornubian Peninsula and the only land data are from farther east where some petroleum exploration lines on the western flank of the more prospective Wessex Basin have been supplemented by British Geological Survey (BGS) lines (Chadwick et al, 1983). Farther offshore, in the western part of the English Channel and its SW approaches, considerably more seismic data are available mostly shot in the early or mid-seventies but only a few wells have been drilled. In 1983 the British Institutions' Reflection Profiling Syndicate (BIRPS) commissioned four lines crossing the Channel and a fifth crossing its SW approaches as part of its SWAT programme. At the time of writing only preliminary stacks of the SWAT data are available but some BGS and commercial data have been released and interpretations based on seismic data have appeared in the literature (Day and Edwards 1983), so a reasonably comprehensive regional picture can be drawn.

Structural History of Western Europe

To understand the structural evolution of the Western Channel it is necessary to see it within the framework of Western Europe. By Permian times Western Europe had been assembled in a configuration very similar to that seen today, the main difference between the reconstructions of different authors being in the orientation and position of Iberia and the off-lying banks. There is much less agreement on pre-Permian plate tectonic evolution. Several platelets may have been involved: the Brabant Massif which includes Southern England and Belgium and probably extends into Wales, Cornubia, Armorica, Iberia, the Massif Central, the Vosges/Black Forest and the Rhenish Massif. Each of these blocks may have existed as a separate micro-continent during the Palaeozoic.

With the closure of the Iapetus suture in the Silurian, Northern Europe became part of North America and during the Devonian a marine environment existed to the south of Britain. The extent of the Devonian basin or basins to the south and their relationship to a Rheic Ocean is debatable. As Gondwana moved northwards a zone of subduction must have existed to the north. Several authors formerly placed this subduction in Southern Europe (Ziegler, 1984) with limited back-arc extension farther north, but there is evidence of subduction in several regions: Weber (1981) suggests the subduction of lithospheric mantle beneath the continental crust on the southern margin of the Rheinische Schiefergebirge and the Harz, and Lefort and Peucat (1974) describe an ophiolite complex to the west of Brittany, although this was thought to be Silurian in age. Styles and Rundle (1984) date the obduction of the Lizard ophiolite at around 370Ma.

There is palaeontological evidence for a Rheic Ocean widening during the Silurian and narrowing from some time in the Devonian until final closure with the Hercynian suture in the Carboniferous (Cocks and Fortey, 1982) and late or

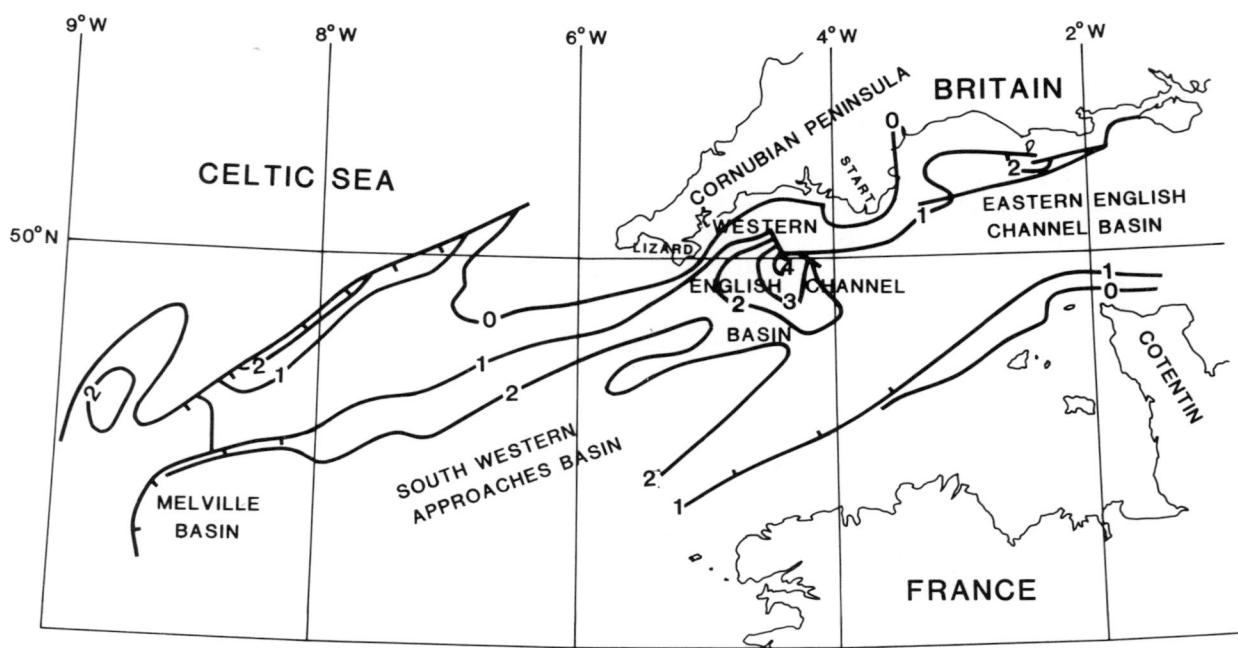

Fig. 1. Simplified isochron map of two-way travel time in seconds to the pre-Permo-Triassic basement in the western English Channel and adjacent continental shelf from seismic reflection data. Based principally on a compilation by E J Armstrong and H A Auld of BGS Hydrocarbons Research Programme. Datum is sea level.

post-Hercynian major shear movements (Arthaud and Matte, 1977; Van der Voo and Scotese, 1981;) further complicate the picture. Badham (1982) presents an argument for oblique strike-slip motion with an Ibero-Armorican arc, having been riven from SE Europe in the Lower mid-Palaeozoic, being translated westwards along the southern edge of Europe until impact at a major bend in the Europe-N America continental margin. He rejects the hypothesis of a major Rheic Ocean but there seems to be no compelling evidence for close juxtaposition of the various elements of SW Europe in their present configuration before the late Carboniferous by which time there is ample evidence for a major strike-slip orogen in central Europe. Lefort and Van der Voo (1981) present a scheme which combines a late Carboniferous central European collision with a 2000km sinistral displacement along the earlier North American margin, arguing a mechanism similar to the Molnar and Tapponier (1978) model for the India-Asia collision. But the timing of the sinistral shear is not well controlled and the directions of relative motion between the European platelets and North America are imprecise: while the 18° average divergence of Devonian pole positions for Europe and North America (Van der Voo and Scotese,1981)undoubtedly indicates large relative motions, the scatter in the data could well indicate different motions for different platelets and in particular a more N-S motion for some poles appears to be consistent with the data. In northern Britain, there is good evidence that sinistral movement occurred along the Great Glen Fault line in the Upper Carboniferous but with regard to the magnitude of this event, there is considerable debate (see also Briden et al, 1984).

Out of a great deal of somewhat conflicting evidence a pattern emerges of general northward movement of Gondwana relative to Northern Europe and North America in Devonian times during which the various blocks that now make up Western Europe were brought together. Recently several authors have argued for the main suture lying along the southern margin of Armorica and there may have been several subduction zones which have not yet been identified. Leveridge et al (1984) argue for the commencement in early or middle Devonian of a sequence of thrusting which includes obduction of the Lizard ophiolite nappe and closure of an oceanic basin. In the late Carboniferous this N-S convergence had given way to a massive dextral shear regime extending across central Europe: within this shear zone the European platelets were further jostled until finally, as the two major components of Pangaea adjusted to each other a tensional regime developed in northern Europe and Permian grabens appeared.

Distribution of Permian and Younger Sediments

Figure 1 shows the present disposition of a complex basin containing considerable thicknesses of Permian and younger sediments in the western

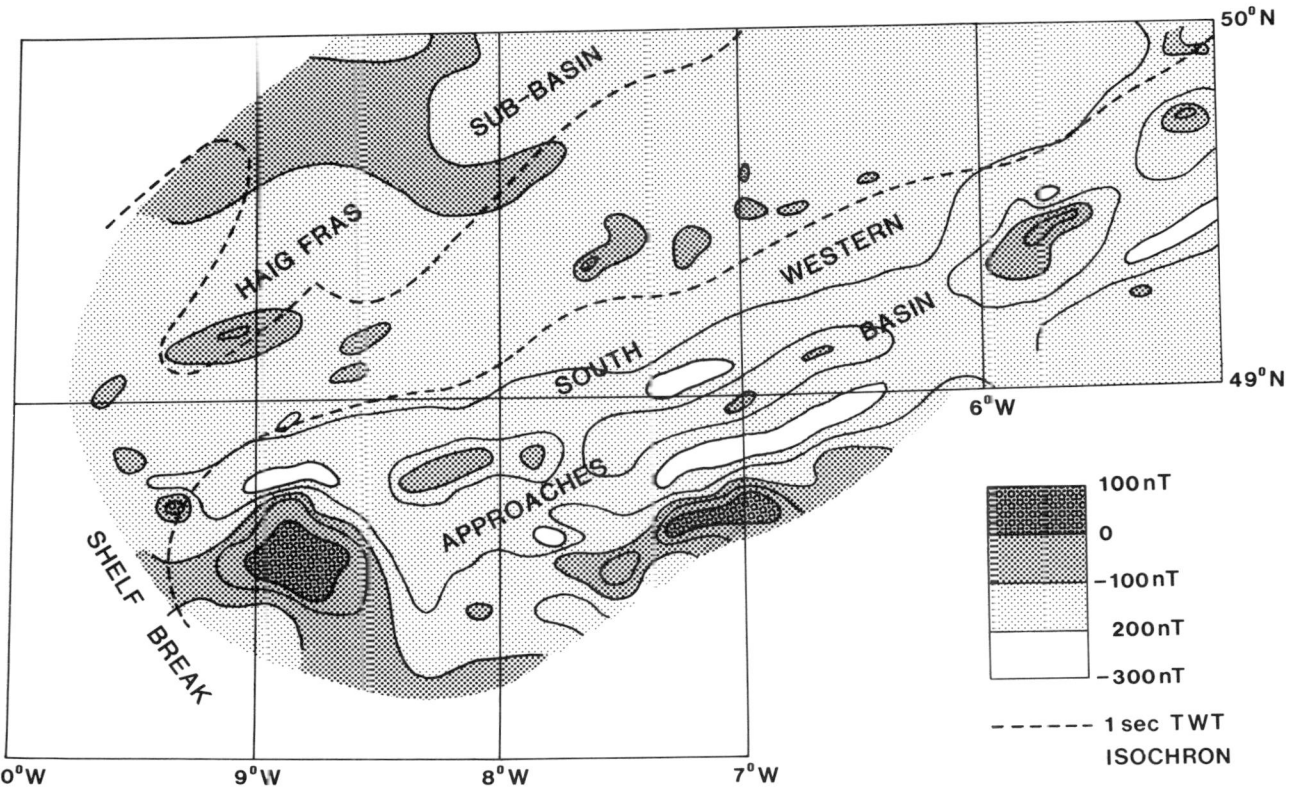

Fig. 2. Magnetic anomaly map of the south western approaches to the English Channel with the location of the basins flanking the Cornubian Ridge.

English Channel area. Figure 2 is a simplified version of a previously unpublished magnetic map of the SW Approaches which shows pronounced lineations along the axis of the SW Approaches Basin. Using earlier marine magnetic data, observed with much more widely spaced survey lines, Hill and Vine (1965) recognised this linear anomaly pattern. On the evidence of the then existing seismic refraction data, they attributed the anomalies to ridges in shallow metamorphic basement, but with the benefit of seismic reflection data it can be seen now that the group of velocities between 3.65 and 4.85km/sec (Class 3 of Day et al, 1956), attributed by them to Palaeozoic rocks, relates to rocks within the Permo-Triassic succession. In some cases velocities in excess of 5km/sec, which they attributed to the metamorphic basement, are observed from rocks within the sedimentary basins. Brooks et al (1983) reported velocities exceeding 5km/sec for Devonian and Lower Carboniferous strata from the Bristol Channel and suggested that many of the layers in the western Channel attributed by Day et al and by Avedik (1975) to the Palaeozoic are likely to be of Mesozoic age. They quite reasonably reassigned (Brooks et al, 1983, figure 10.1(c)) velocities over 5km/sec to the top of the Palaeozoic at depths between 2 and 2.2km, but interpretation of reflection seismic data (Day and Edwards, 1983), confirmed by the BIRPS line SWAT 8, demonstrates that over 3 secs TWT of sediments exists at the location of two of the refraction stations, so it must be deduced that the high velocities relate to units within the Permo-Trias. These units could be evaporites, volcanics or limestones.

The deep sedimentary basin in the Western Channel south of Plymouth, (Day and Edwards, 1983) has greater than 4 secs TWT of basin fill and a reported average velocity near the surface of over 4km/sec. A conservative estimate of the depth therefore is at least 9km. Since Permo-Triassic rocks overlie Devonian along its northern flank and crop out in the seabed at the centre it seems likely that the whole succession is Permo-Triassic, although there is obviously a possibility that there are Carboniferous rocks at its base. As the isochrons in figure 1 imply, thick sediments are present throughout the Western Channel and its south western approaches, extending to near the shelf edge, with basement shallow or cropping out on either side. The greater part of the succession is interpreted to be Permo-Triassic (with possibly latest Carboniferous); see Ziegler (1982). From at least 9km in the Plymouth Bay Basin this succession

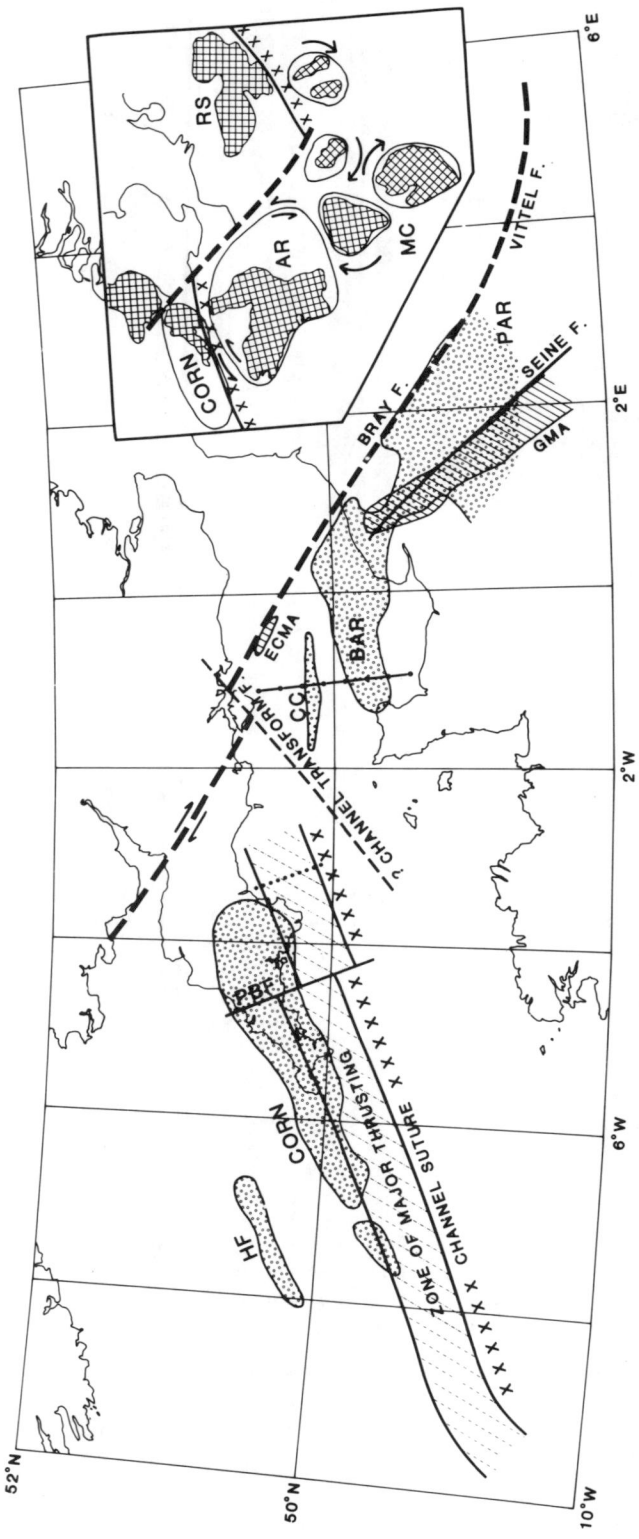

Fig. 3. Structural features of the English Channel area. The line defined by the Bray Fault and the easternmost Channel magnetic anomaly is interpreted as a major crustal boundary and hence the trace of a transform fault offsetting the Hercynian deformation zone of SW Britain from that of N Germany. Its path to the NW and SE is conjectural and a sinistral offset is inferred along a supposed transform fault active in the Triassic and Jurassic during the major opening phases of the North Sea. For clarity the large magnetic anomalies which extend westwards from the central Channel have been omitted. Stipple; gravity low anomalies mentioned in text: diag. ruling; magnetic anomalies mentioned in text: BAR; Barfleur? batholith: CC; Central Channel anomaly: CORN; Cornubian batholith anomaly: ECMA; easternmost Channel magnetic anomaly: GMA; Great Magnetic Anomaly of the Paris Basin: HF; Haig Fras batholith anomaly: PAR; Paris Basin gravity anomaly: PBF; Plymouth Bay Fault.

Fig. 3 inset. Proposed motion of platelets in late Carboniferous time after dextral translation of W. Europe along Bray Transform Fault. AR; Armorica: CORN; Cornubia: MC; Massif Central: RS; Rheinische Shiefergebirge.

shallows westwards to between 3 and 4km, then becomes deeper attaining a depth of over 6km in places.

It can be seen from figure 2 that most of the magnetic anomalies in the south-western approaches are spatially associated with the Permo-Triassic basin, which is of the order of 5km deep. Together with the refraction evidence referred to above this suggests that the principal source of the anomalies is volcanics within the Permo-Triassic succession, as proposed by Smith and Curry (1975).

Pre-Permian Basement

The pre-Permian basement south of the SW Approaches Basin is made up of pre-Cambrian and Lower Palaeozoic rocks injected by late Variscan granites. The basin is bounded on its southern flank by a major fault zone, south of which the basement is covered locally by only thin sediments. Within the basin, pre-Permian basement is at a considerable depth, as described in the previous section, and has not been penetrated by any publicly available wells. On the north side of the Channel, where rocks crop out on land, this basement consists of metamorphosed Devonian sediments intruded by Late Variscan granites, except for the Lizard and possibly Start rocks. These latter occur at the southern tips of the peninsulas of the same names and at isolated locations in the seabed between. They range in type from low grade schists at Start to rocks which are generally recognised as ophiolite at the Lizard.

Barnes and Andrews (1984) argue that the Lizard ophiolite complex was placed in its present environment by thrusting as a cold slab having been obducted (from a marginal oceanic basin) earlier. Day and Edwards (1983) interpreted south-dipping reflections from within the pre-Permian basement, seen in seismic lines from the western Channel, as thrust planes. These events have been observed in seismic reflection profiles along the whole northern flank of the South Western Approaches Basin and one event has been correlated (Leveridge et al, 1984) with a major thrust inferred on land north of the Lizard Thrust. Thrusting in this area is dated lower or middle Devonian and the thrusting becomes younger northwards across the Cornubian Peninsula (Shackleton, Ries and Coward, 1982; Coward and McClay, 1983), Westphalian rocks having been deformed in north Devon. Thus the deformation spans 80Ma or so and was preceded by an earlier tectonic episode.

Geophysical Evidence

The salient features of the area shown in figure 3 are derived principally from geophysical observations. Gravity data are from the British Geological Survey 1:250 000 Bouguer gravity anomaly map and, in the south west, the gravity map of Lalaut, Sibuet and Williams (1981). Recent data acquired by BGS in the eastern part of the English Channel have improved the cover of gravity data (Caen, Rouen and Dungeness-Boulogne sheets, in preparation). One significant feature of the Cornubian Peninsula is the linear gravity low associated with the granite batholith that crops out on the peninsula and at the Scilly Isles to the west (see figure 3). Recent gravity modelling (Edwards, 1984) suggests that a second batholith exists beneath Haig Fras where granite has been sampled from the seabed. The anomaly is parallel to the Cornubian low but offset to the NW. From the new gravity data we can infer a similarly trending granite batholith in the central English Channel (see figure 3) extending offshore from the Cotentin Peninsula where granite crops out at Barfleur near Cherbourg (BGS work in progress, see also figure 4). This anomaly is parallel to the two farther west and ends at $\frac{1}{2}°E$ longitude. East of this longitude the gravity in the Channel is uneventful.

The highly anomalous magnetic field in the Channel also dies out in about the same place, the easternmost large anomaly being a NW-SE trending linear anomaly. This anomaly lies along the extension of the Bray Fault which is a straight feature passing through the Paris Basin, seen at the surface where Cretaceous rocks to the NE are downthrown against Jurassic to the SW. Along the continuation of the Bray Fault line to the SE is a narrow Permian basin which forms the deepest part of the Paris Basin (Brunet and Le Pichon, 1982) and along the axis of this basin an elongate gravity low (Carte Gravimetrique de la France) (see figure 3) extends north-westwards so that the steep gradient along its NE flank lines up with the Bray Fault. This zone of low gravity merges with the mid-Channel anomaly interpreted as a granite batholith.

Along the western edge of the Paris Basin is a strong linear magnetic anomaly trending NNW-SSE known as the Great Magnetic Anomaly of the Paris Basin, which merges with the main field of Channel magnetic anomalies where the two gravity lows meet. Lefort and Weber (1977) associate this with a line of basic intrusions that do not perforate the basement except in the south, close to the Massif Central, and they maintain no granite occurs east of the anomaly. They report 60km of sinistral strike-slip motion along the associated fracture zone in the Westphalian, with further movement in the Stephanian. The anomaly curves south and extends into the Massif Central and the fracture zone continues into the Sillon Houiller Fault which suffered similar strike-slip movement in the Westphalian. According to Arthaud and Matte (1977) the motion was 70km.

The Massif Central, Armorica and southern Cornubia were all deformed by north-directed thrusting during the Devonian when a major foredeep basin was being formed in Belgium. The existence of a compressive thrust regime in northern Cornubia during Upper Carboniferous times

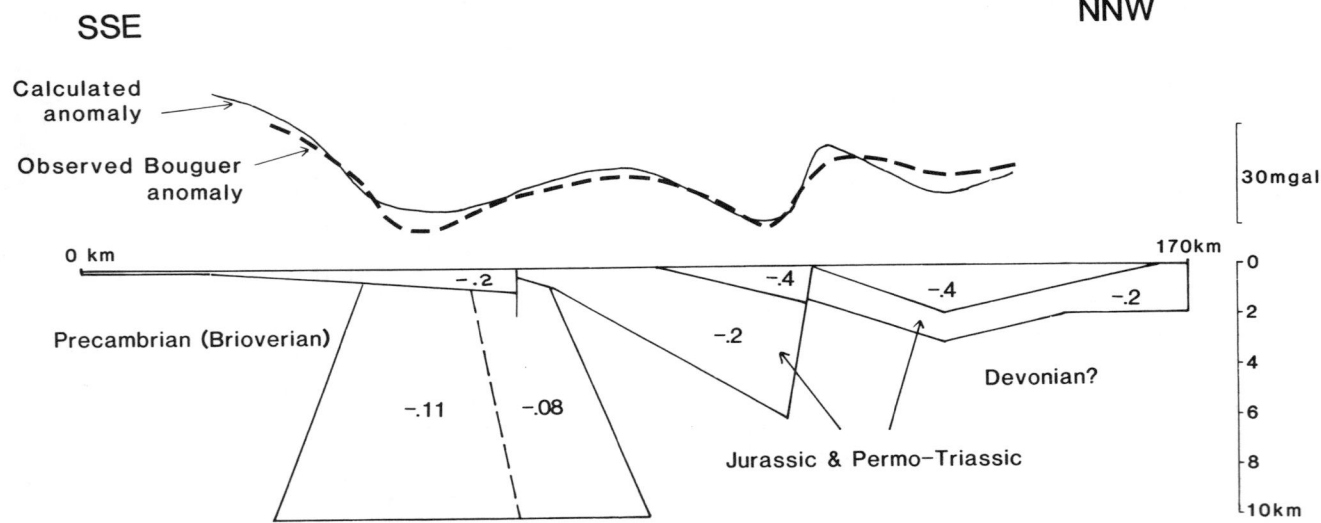

Fig. 4. Model along an observed gravity profile crossing the two gravity anomalies in the central English Channel. Numbers on bodies are density contrasts (Mg/m^3) with the basement. The large body modelled to account for the southernmost low is assumed to be a granite batholith. Uppermost sediments in the northern part of the Channel are Cretaceous and younger. The deep sedimentary basin modelled to account for the large central anomaly is not seen on the SWAT 11 seismic profile that crosses it.

is difficult to reconcile with contemporaneous southward motion of Armorica and the western half of the Massif Central relative to central Europe. However, within the context of a massive post-Hercynian dextral shear zone across central Europe at the end of the Carboniferous, a plausible model can be proposed.

Hercynian Evolution Model

Although the Great Magnetic Anomaly probably marks a line of major crustal post-Hercynian shear, gravity and seismic evidence suggests that the major boundary between dissimilar types of crust occurs along the line of the Bray Fault. It is proposed that this was the position of a transform fault against which Western Europe moved in a northerly direction having previously closed up segments of ocean with possibly several subduction zones, including one in the western part of the English Channel along which the Lizard ophiolite was finally obducted (Bard et al, 1980).

At the end of the Carboniferous these platelets were all docked onto the North American continent and as Gondwana continued in their wake a change of relative motion from N to NW or W occurred (Zeigler, 1982), causing these blocks to rotate clockwise producing the sinistral shear along the Great Magnetic Anomaly, and further dextral shearing along NW trending faults such as the Bray Fault and the South Armorican Shear Zone (see figure 3 inset).

Clockwise rotation of Armorica against the Cornubian block produced further crustal shortening in Cornubia and a minor tensional environment in the central Channel. A prominent linear E-W low gravity anomaly (see figure 3) runs through the central Channel north of the supposed granite batholith. This might be caused by a sedimentary trough (see figure 4) but there are no obvious sediments deep enough on the SWAT line that crosses these features to account for the anomaly. There is a possibility that the gravity low is associated with a buried granite body which is not imaged in the preliminary SWAT data. This problem is being addressed by current BGS studies.

With the final collision of Gondwana with Laurasia a tensional regime developed in early Permian times (Russell and Smythe, 1978). This produced grabens over much of northern Europe, among the more prominent being the Rockall Trough, and the Oslo and Horn Grabens. The South Western Approaches Basin is a graben of similar trend to the Rockall Trough. The Permian graben below the Paris Basin however parallels the supposed NW-SE transform fault along the Bray Fault line, whereas in the Rhine and Jura regions on the other side of the transform fault the Permian basins again follow a NE-SW trend (Chauve et al, 1980). In Germany, as in SW Britain, Permian basins lie

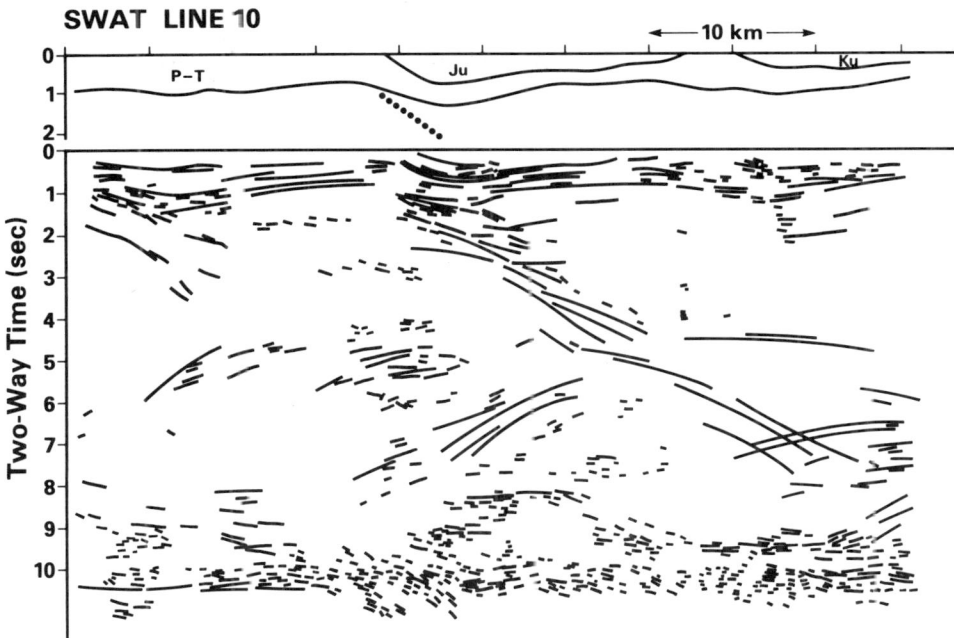

Fig. 5. Line drawing of the upper part of the northern portion of SWAT 10 seismic profile and interpretation of the top 2 seconds. P-T; Permo-Triassic: Ju; Upper Jurassic: Ku; Upper Cretaceous: dots; reactivated thrust fault.

normal to the shortening direction, in this case south of and parallel to the supposed subduction zone beneath the Mid German Crystalline Rise, along the northern boundary of the Saxothuringian zone. On geological evidence, Holder and Leveridge (in press) directly equate the Rhenohercynian zone of the German Hercynides with SW Britain, invoking dextral movement along a major transcurrent fault in latest Carboniferous. The similarity between the two areas is striking enough to suggest that the suture at which the Lizard ophiolite suite was obducted was on the north side of what is now the English Channel and that both sutures terminate at the Bray Fault transform.

The easternmost Channel magnetic anomaly lines up very strikingly with the Bray Fault but projection of the fault across the Channel into England is problematical. Allowance must be made for sinistral movement along a transform fault zone inferred to have been active in the vicinity of the Channel when the North Sea opened during the Triassic and Jurassic so the actual position of the supposed Bray transform beneath the Mesozoic rocks of Southern Britain is uncertain. An estimate of the offset could be made if the amount of North Sea opening were well defined and the later opening of the Rhine Graben taken into account. The Bray transform fault may correspond to a zone of poor reflectors on the seismic profile of Chadwick et al (1983).

Further evidence associating the Bray Fault with a major crustal shear comes from the French ECORS programme data. On a seismic reflection profile crossing the Paris Basin a step in the lower crust, with apparently thicker crust to the east, occurs beneath the position of the Bray Fault suggesting a vertical transform fault in this position (Ecors report, 1984).

The BIRPS SWAT lines (Matthews and Cheadle, this volume) confirm the existence of a major graben in the western Channel and suggest that west of Start it opened by normal movement along former thrust faults in the Devonian basement. One major thrust fault north of the Plymouth Bay Basin extends deep into the crust (Day and Edwards, 1983). East of Start the Permian basin is shallower but the same mechanism of relaxation seems to have operated (see figure 5) except that here the extension associated with the main crustal fault appears to be post Upper Jurassic.

Conclusions

Hercynian deformation in Britain and Ireland has traditionally been assumed to parallel the northern limit of foreland thrusting (the Hercynian Front) which, where observed, has an ESE-WNW direction. Consequently attempts to trace the Rhenohercynian zone across Europe produce an unconvincing sinuate pattern with Iberia, Armorica and Cornubia occupying a salient bulge. The geophysical evidence presented here, along with the geological interpretations of Leveridge et al

(1984) and Holder and Leveridge (in press) suggest a more convincing hypothesis. Post-Hercynian dextral wrench faulting occurred along several parallel faults (Arthaud and Matte, 1977) but has not formerly been recognised along the Bray Fault line because of its Mesozoic cover. Gardiner and Sheridan (1981) argued that the main Hercynian deformation in the north Celtic Sea (and subsequent reactivation by stretching) occurred along an ENE trending zone rather than the traditional ESE. This is compatible with the hypothesis developed here which requires crustal shortening during late Carboniferous either in the north Celtic Sea or farther to the NW, however further work, particularly interpretation of deep seismic profiles, is required to substantiate this.

Acknowledgements. Many of the hypotheses advanced her are the result of discussions with, or are based on the work of colleagues in the British Geological Survey. I am particularly indebted to Jed Armstrong, John Edwards and John Molloy in this respect. Deep reflection seismic data were made available through the Core Group of the British Institutions' Reflection Profiling Syndicate (BIRPS). This paper is published by permission of Director BGS (NERC).

References

Arthaud, F. and P. Matte. Late Palaeozoic strike-slip faulting in Southern Europe and North Africa: results of a right-lateral shear zone between the Appalachians and the Urals. Geol. Soc. Am. Bull, 88, 1305-1320, 1977.

Avedik, F. The seismic structure of the Western Approaches and the Armorican Continental Shelf and its geological interpretation. In: Woodland, A.W. (ed) Petroleum and the Continental Shelf of North West Europe. Vol. 1, 29-43, 1975.

Badham, J.P.N. Strike-slip orogens - an explanation for the Hercynides. J. geol. Soc. London, 139, 493-504, 1982.

Bard, J.P., J.P. Burg, Ph. Matte and A. Ribeiro. La chaine Hercynienne d'Europe occidentale en termes de tectonique des plaques. In: Cogne J. and Slansky, M. (eds) Geologie de l'Europe - Mem B.R.G.M. 108, 90-111.

Barnes, R.P. and J.R. Andrews. Hot or cold emplacement of the Lizard Complex. J. geol. Soc. London, 141, part 1, 37-40, 1984.

Briden, J.C., H.B. Turnell and D.R. Watts. British palaeomagnetism, Iapetus Ocean, and the Great Glen Fault. Geology, 12, 428-431, 1984.

British Geological Survey Bouguer gravity anomaly map 1:250,000 series.

Brooks, M., J. Mechie and D.J. Llewellyn. Geophysical investigations in the Variscides of southwest Britain. In: Hancock, P.L. (ed) The Variscan Fold Belt in the British Isles. Hilger, 186-97, 1983.

Brunet, M-F. and X. Le Pichon. Subsidence of the Paris Basin. J.Geophys. Res., 87, No. B10, 8547-8560, 1982.

Carte Gravimétrique de la France, B.R.G.M.

Chadwick, R.A., N. Kenolty and A. Whittaker. Crustal structure beneath southern England from deep seismic reflection profiles. J. Geol. Soc. London, 140, part 6, 893-911, 1983.

Chauve, P., R. Enay, P. Fluck and C. Sittler, in C. Lorenz (ed) Geolgie des pays Europeens C.N.F.G. 357-430, 1980.

Coward, M.P. and K.R. McClay. Thrust tectonics in South Devon. J. geol. Soc. London, 140, 215-28, 1983.

Cocks, L.R.M. and R.A. Fortey. Faunal evidence for oceanic separations in the Palaeozoic of Britain. Geol. Soc. London J. 139, 465-478, 1982.

Day, G.A. and J.W.F. Edwards. Variscan thrusting in the basement of the English Channel and SW Approaches. Proc. Ussher Soc., 5, 432-436, 1983.

Day, A.A., M.N. Hill, A.S. Laughton and J.C. Swallow. Seismic prospecting in the Western Approaches of the English Channel. Q.J. geol. Soc. London, 112, 15-44, 1956.

Ecors report. Deep seismic profiling of the crust in northern Frnce: the Ecors project 1984.

Edwards, J.W.F. Interpretation of seismic and gravity surveys over the eastern part of the Cornubian Platform. In: Hutton, D.W. and Sanderson, D.J. (eds) Variscan Tectonics of the North Atlantic Region. Geol. Soc. London Spec. Publ., 119-124, 1984.

Gardiner, P.R.R. and D.J.R. Sheridan. Tectonic framework of the Celtic Sea and adjacent areas with special reference to the location of the Variscan Front. J. Struc. Geol. 3, 317-331, 1981.

Hill, M.N. and F.J. Vine. A preliminary magnetic survey of the Western Approaches to the English Channel. Q.J. geol. Soc. London, 121, 463-75, 1965.

Holder, M.T. and B.E. Leveridge. Correlation of the Rhenohercynian Variscides. J. Geol. Soc. Lond. (in press).

Lalaut, P., J-C. Sibuet and C.A. Williams. Carte gravimétrique de l'Atlantique Nord-Est, CNEXO, 1981.

Lefort, J.P. and J.J. Peucat. Le socle antémésozoique submergé à l'ouest de la baie d'Audierne (Finistère) C.R. Acad. Sci. Paris, 279, D: 635-637, 1974.

Lefort, J.P. and R. Van der Voo. A kinematic model for the collision and complete suturing between Gondwanaland and Laurussia in the Carboniferous. J. of Geol., 89, no. 5, 537-550, 1981.

Lefort, J.P. and C. Weber. Le Socle ante-Permien sous le bassin Anglo-Franco-Belge, d'apres les donnes geophysiques. Essai de correlation entre les massifs Hercyniens peripheriques. In: La chaine varisque d'Europe moyenne et occidentale. Coll. intern. CNRS, Rennes, 243, 415-422, 1977.

Leveridge, B.E., M.T. Holder and G.A. Day. Thrust nappe tectonics in the Devonian of south Cornwall and the western English Channel. In: Hutton, D.W. and Sanderson, D.J. (eds) Variscan Tectonics of the North Atlantic Region. Geol. Soc. London Spec. Publ., 103-112, 1984.

Matthews, D.H. and M.J. Cheadle. Deep reflection from the Caledonides and Variscides west of Britain and comparison with the Himalayas. This volume.

Molnar, P. and P. Tapponier. Active tectonics of Tibet. J. Geophys. Res. 83, 5361-75, 1978.

Russell, M.J. and D.K. Smythe. Evidence for an early Permian oceanic rift in the northern North Atlantic. In E.R. Newmann and I.B. Ramberg (eds) Petrology and geochemistry of continental rifts, 173-179, 1978. D Reidel Publishing Company (Dordrecht).

Shackleton, R.M., A.C. Ries and M.P. Coward. An interpretation of the Variscan structures in SW England. J. Geol. Soc. London, 139, 533-541, 1982.

Smith, A.J. and D. Curry. The structure and geological evolution of the English Channel. Phil. Trans. R. Soc. Lond. A.279, 3-20, 1975.

Styles, M.T. and C.C. Rundle. The Rb-Sr isochron age of the Kennack Gneiss and its bearing on the age of the Lizard Complex, Cornwall. J. geol. Soc. London, 141, 15-19, 1984.

Van der Voo, R. and C.H. Scotese. Palaeomagnetic evidence for a large (c. 2000km) sinistral offset along the Great Glen Fault during Carboniferous times. Geology, 9, 583-589, 1981.

Weber, K. The structural development of the Rheinische Schiefergebirge. Geologie en Mijnbouw, 149-160, 1981.

Ziegler, P.A. Geological Atlas of Western and Central Europe. Elsevier (Amsterdam), 130pp, 1982.

Ziegler, P.A. Caledonian and Hercynian crustal consolidation of western and central Europe - a working hypothesis. Geologie en Mijnbouw, 93-108, 1984.

THE DEEP CRUST IN CONVERGENT AND DIVERGENT TERRANES: LARAMIDE UPLIFTS AND BASIN-RANGE RIFTS

George A. Thompson and Janice L. Hill

Department of Geophysics, Stanford University, Stanford, CA 94305
and Chevron USA, 700 S. Colorado Blvd., Denver, CO 80222

Abstract. Seismic profiles across the Pacific Creek anticline, the Wind River Range, and the Casper arch in western Wyoming provide a coherent, nearly continuous section deep into the crystalline crust through these Laramide basement uplifts. Anticlines or monoclines in the sedimentary section overlie reflective thrust faults in the Precambrian crystalline basement. Structural relief on the folds and displacements on the faults are quantitatively coupled. The thrust faults tend to splay and/or decrease in dip with depth until they disappear as distinct reflectors at mid-crustal levels. The reflection data clearly reveal transitions downward from folding to brittle thrust faulting to distributed or ductile compression in the deep crust.

Seismic sections in the rifted Basin and Range province (including the Rio Grande rift) also reveal striking transitions with depth. The near-surface, high-angle normal faults commonly merge with, and do not displace, subhorizontal detachment faults at depths of only a few kilometers. The abruptness and shallowness of the change suggest to us that it is not governed solely by rock softening due to increased temperature with depth. Instead we suggest that conditions of open hydrothermal circulation in the broken upper crust change abruptly at depth to conditions of high pore-water pressure in rocks self-sealed by mineral deposition or metamorphic processes.

Zones of subhorizontal reflectors at mid-crustal depths are much better developed in extensional regions such as the Basin and Range provinces than in the compressional Wyoming province. On reflection sections from both regions, however, the basal crust appears laminated. A tectonic explanation is suggested for the subhorizontal reflectors by the flattening of both thrusts and normal faults in the deep crust, but an origin coupled also to metamorphic and magmatic events is likely.

Introduction

Exploration by deep seismic reflections is not only shedding light on deformational processes in the crust but also raising intriguing new questions. Our purpose here is to synthesize selected data from our own experience bearing, first, on the origin of the Laramide uplifts of Late Cretaceous to early Tertiary age in the western United States and, second, on Cenozoic rift structures of the Basin and Range province (Figs. 1 and 2). On reflection sections from these areas the basal continental crust commonly appears layered or laminated through a transitional region bounded below by the comparatively transparent upper mantle (Hale and Thompson, 1982). A zone of discontinuous subhorizontal reflectors which is one-half to several kilometers thick thus marks the Mohorovicic discontinuity. At shallower depth, thick zones of subhorizontal reflections are especially prominent in rifted terranes such as the Basin and Range province; they may be characteristic of much, but not all, continental crust (e.g. Mueller, 1977; Lynn et al., 1981; Allmendinger et al., 1983; Mathur, 1983). The challenging problem is to understand the thicknesses and acoustic impedances of these reflectors within the crystalline crust in order to identify rock types and the tectonic, metamorphic and perhaps igneous processes by which they originated. Our ultimate objective is to gain an understanding of processes in the deeper continental crust.

Laramide Uplifts

A long-standing debate about the origin of Laramide basement uplifts--whether or not they are produced by horizontal compression--was resolved in most people's minds by seismic reflections from the Wind River thrust fault extending to great depth in the basement (Smithson et al., 1979; Brewer et al., 1980; Zawislak and Smithson, 1981). The difficult task of depth migration of the seismic section, necessary to properly image the fault zone, revealed that the zone splays and flattens into the ductile lower crust (Lynn, 1979; Lynn et al., 1983). The Wind River Range derives its characteristic eastward tilt from compressional slip on this curved, upward-steepening fault zone.

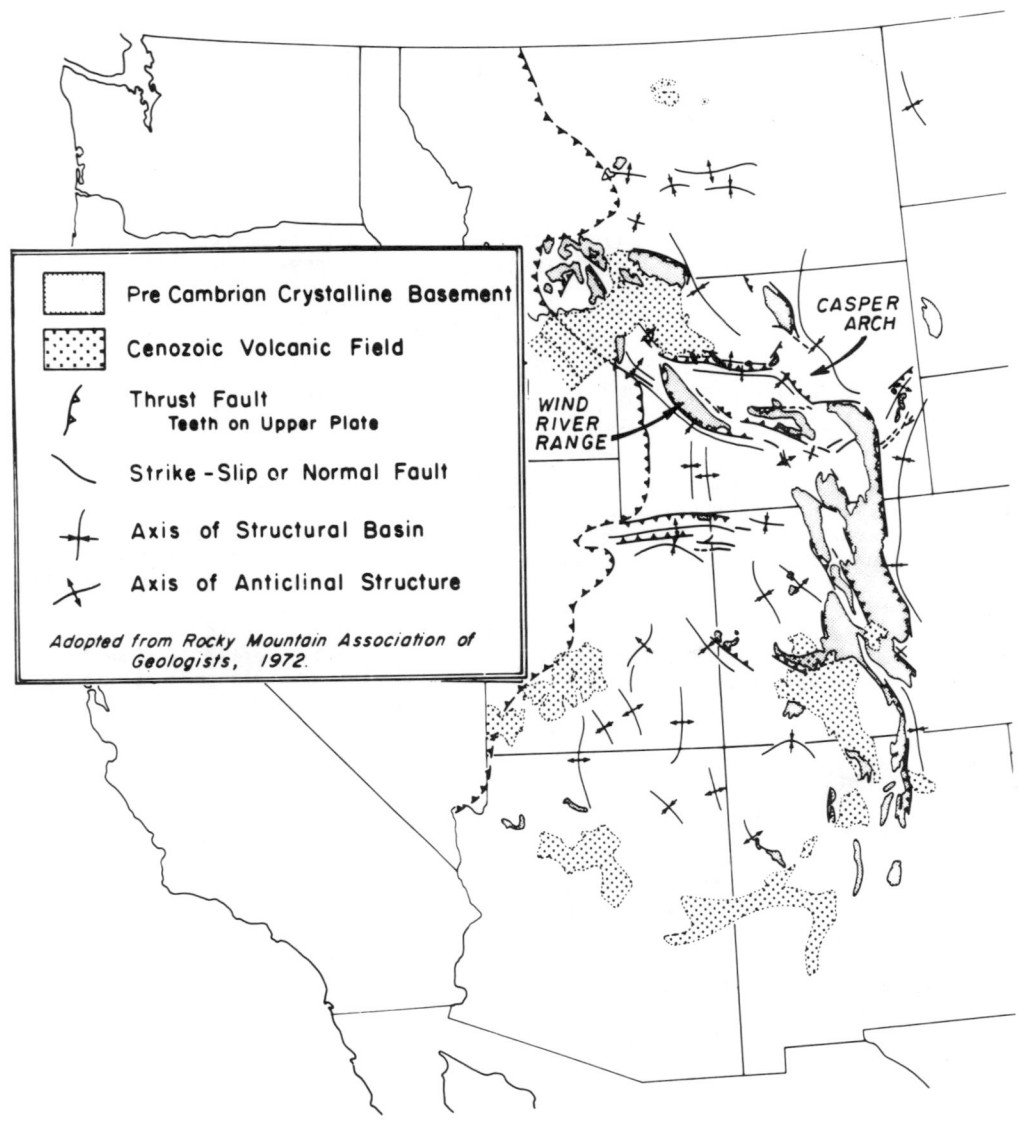

Figure 1. Regional setting of the Late Cretaceous-Early Cenozoic Laramide uplifts and their relationship to the overthrust belt.

Intensive study of seismic sections made for petroleum exploration in the Pacific Creek anticline, southwest of the Wind River Range (MacLeod, 1981) and the Casper arch, northeast of the Wind River Range and basin (Hill et al., 1981) provide coherent overlapping sections deep into the crystalline crust (Figure 3). The sections reveal interrelations among folding in the Mesozoic and Paleozoic sedimentary rocks, thrust faulting in the brittle upper part of the crystalline basement, and distributed faulting or ductile deformation in the deep crust.

Pacific Creek Anticline

The Pacific Creek anticline, which lies in the Green River basin southwest of the Wind River uplift, is broad and asymmetrical, steeper on the west side. Beneath the anticline, there is evidence on the COCORP seismic line of a thrust fault in the Precambrian basement similar in character to the Wind River thrust but of much smaller displacement. MacLeod (1981) studied detailed seismic lines from industry sources (Fig. 4) in addition to the COCORP line. The Rainbow Resources deep exploratory well, which bottomed in Mississippian Madison Formation at a depth of 7.85 km, is at the center of the detailed study area.

Unmigrated and migrated sections (Figs. 5,6) clearly reveal the Pacific Creek thrust in its genetic relationship with the asymmetrical anticline. Reflectivity of the fault zone within

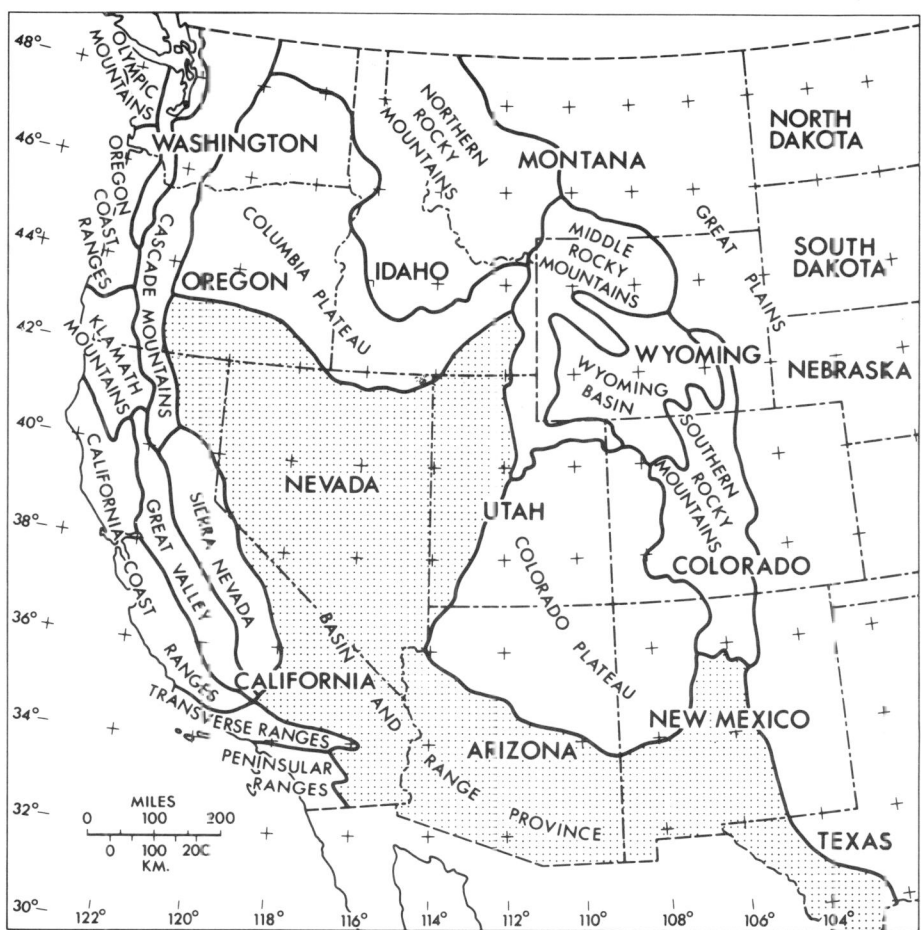

Figure 2. Location of the Basin and Range province, a region of Cenozoic rifting.

basement is probably the result of fabric anisotropy and mineralogic changes in the fault zone constituents (Jones and Nur, 1983). In the same depth section, details of the transition from faulting in the basement to anticlinal folding in the sedimentary section were worked out with the aid of the industry lines (1, 1A, and 2, Fig. 4; Fig. 7).

The Pacific Creek thrust is at the base of the section (Fig. 7) and it displaces the Madison Formation approximately 200 meters. The other faults are all small reverse faults and were evidently formed as part of the adjustment from basement faulting to anticlinal folding. MacLeod's isochronous maps detail thinning between the Latest Cretaceous Lance and Lewis Formations that can be seen also in Figure 7. The axis of thinning is parallel to the northwest trend of faults. The seismic evidence thus indicates that Laramide uplift in this region was coincident with faulting, that it began in Latest Cretaceous and that it continued to deform the Paleocene Fort Union Formation. The basement Pa-

cific Creek thrust fault, which has a dip of about $30°$, clearly buckled the overlying 8-km sedimentary section to form the Pacific Creek anticline. As seen in the depth-migrated reflection section, the fault is a concave-upward zone at least 1 km thick traceable to a depth of 15-16 km.

Wind River Uplift

The largest structure in the region, the Wind River uplift, lies to the northeast of the Pacific Creek anticline. In its central part the uplift is essentially an east-tilted, upthrust block of Precambrian rocks with the sedimentary cover stripped away. To the north, where the structural relief decreases, the cover is preserved, and the uplift has more of the character of an asymmetrical anticline. The thrust fault underlying the uplift, strikingly shown as a zone of dipping reflections in the COCORP profile, was originally interpreted as a nearly planar fault to a depth of about 25 km, where the

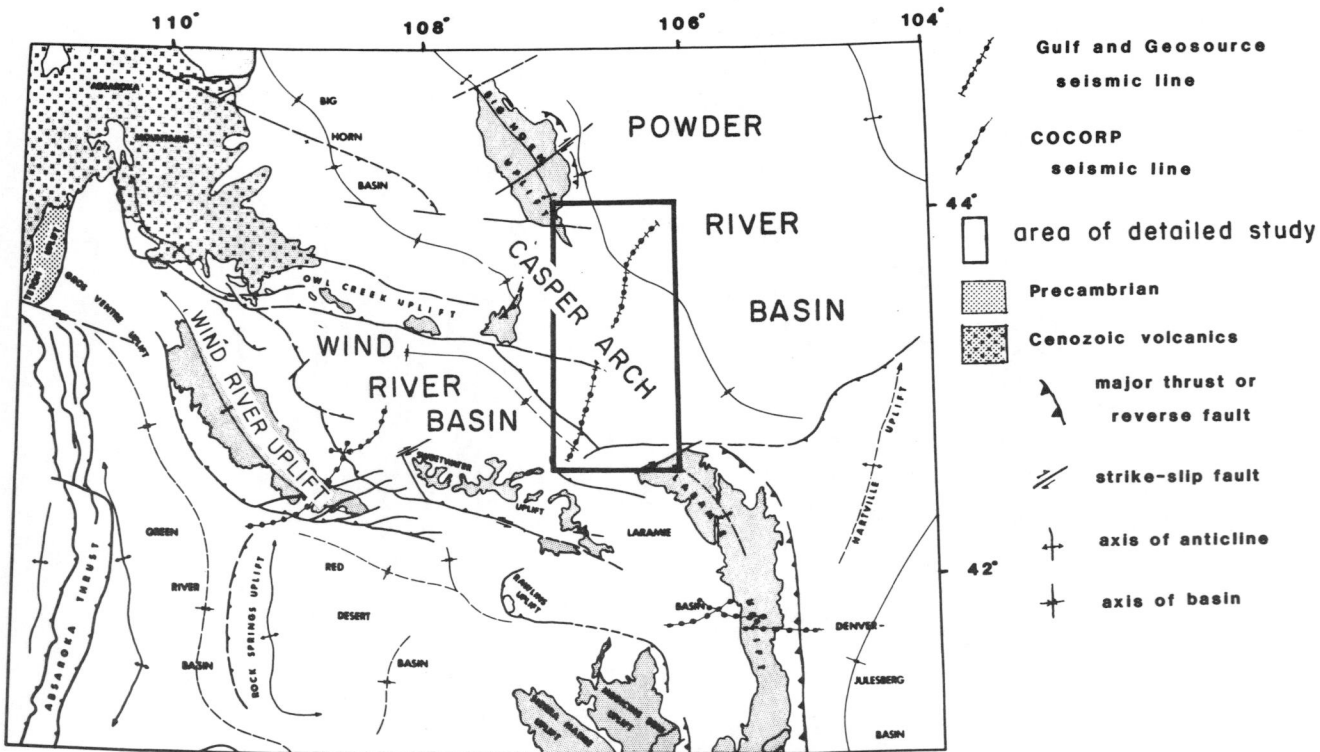

Figure 3. Generalized tectonic map of the Rocky Mountain foreland of Wyoming, showing location of the COCORP seismic line across the Wind River uplift and the Gulf-Geosource line across the Casper arch.

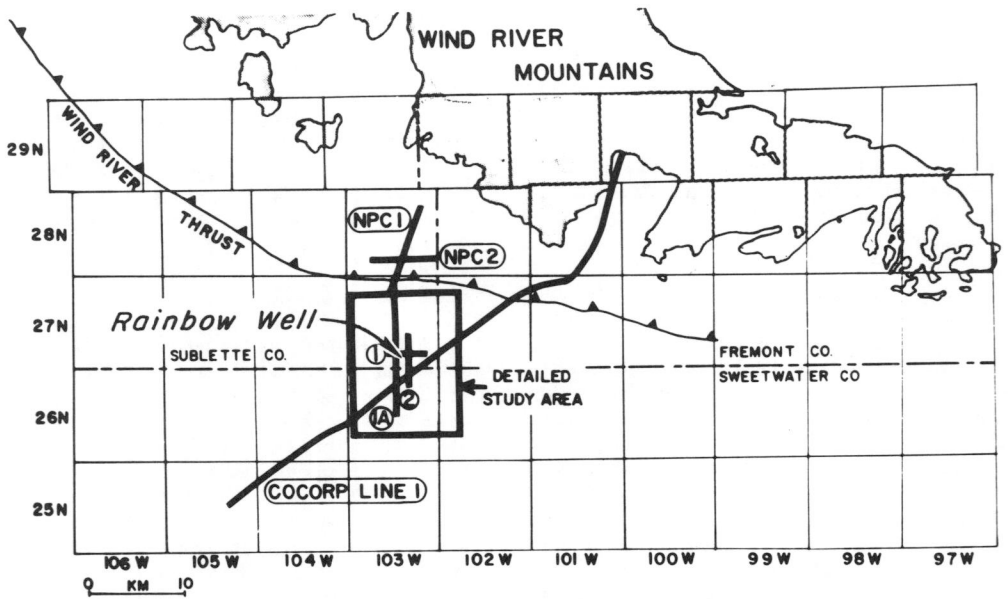

Figure 4. Location of seismic lines in the Pacific Creek area. From MacLeod, 1981.

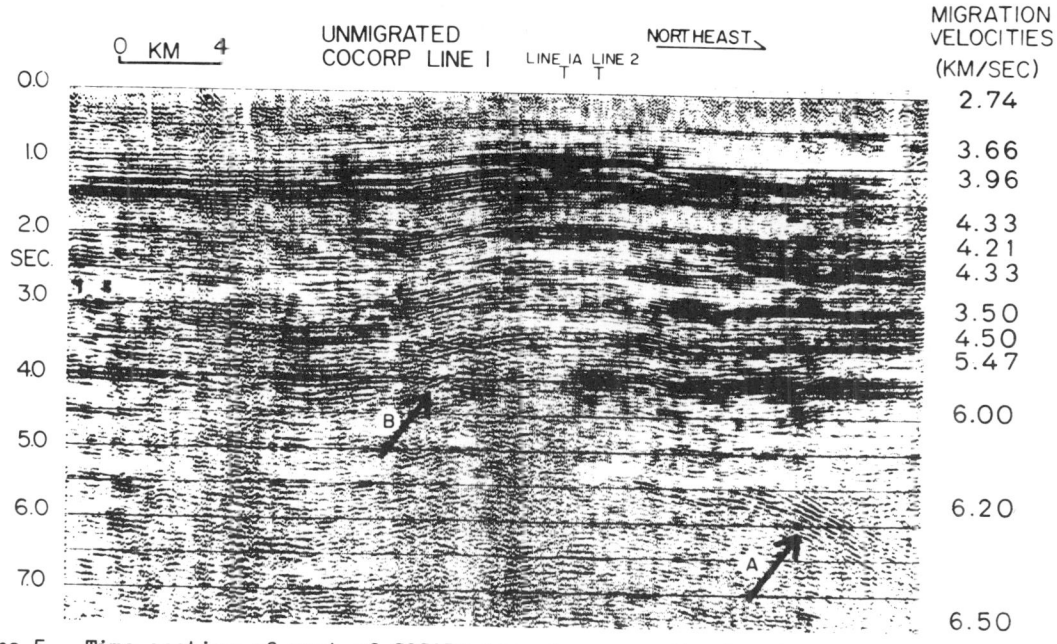

Figure 5. Time section of part of COCORP Line 1 across the Pacific Creek anticline; A = Pacific Creek thrust reflections; B = buried focus effect. Velocities are those used for the migration in Figure 6. From MacLeod, 1981.

reflections become less distinct (Smithson et al., 1979; Brewer et al., 1980; Zawislak and Smithson, 1981). True dip of the upper part of the fault is about 33° as determined by MacLeod (1981) from the two northern seismic lines shown on Figure 4. With a vertical component of displacement of about 13 km, the Wind River fault cuts through the entire Paleozoic and Mesozoic section; it thus represents a similar but much more advanced stage of faulting and uplift than the Pacific Creek anticline.

Depth migrations (Lynn, 1979; Lynn et al.,

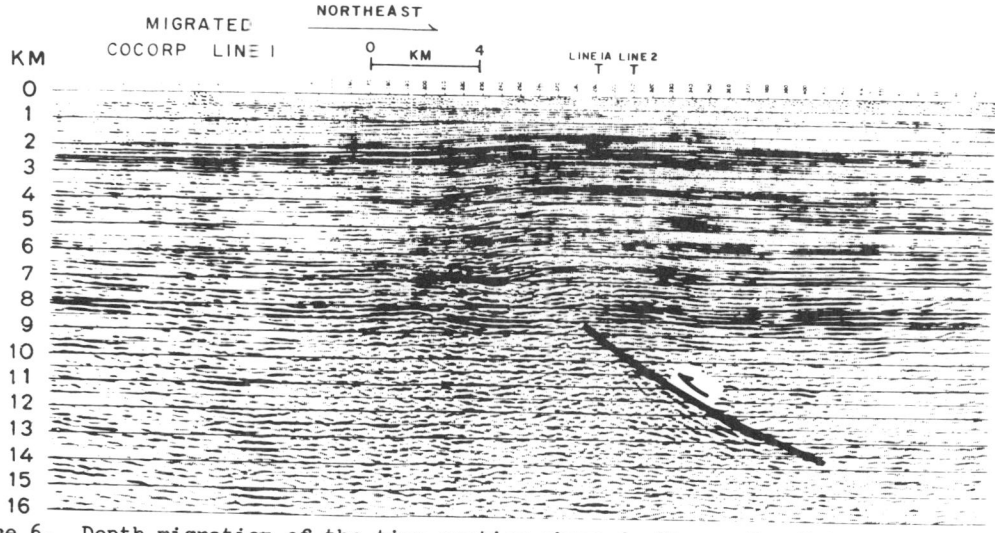

Figure 6. Depth migration of the time section shown in Figure 5. Note position of Pacific Creek thrust beneath the anticline. No vertical exaggeration. From MacLeod, 1981.

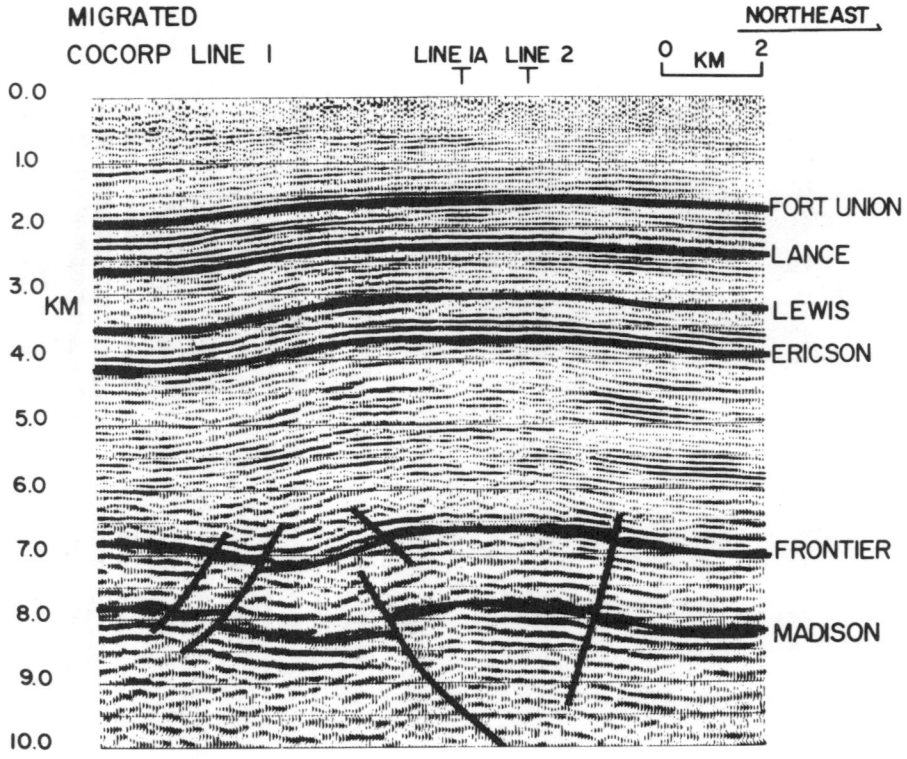

Figure 7. Detail over Pacific Creek anticline, showing transition from faulting to folding; vertical exaggeration 1.3X. The formations are identified from drill logs. From MacLeod, 1981.

1983) of the Wind River seismic sections are essential to understanding the origin and development of the uplift. The key finding is the splaying and flattening of the Wind River thrust zone along an over-all arc, the curvature of which nicely explains the tilting of the range as it was thrust up (Fig. 8). Below a depth of about 30 km, deformation presumably was more nearly uniform and ductile.

Sharry et al. (1984), of the Gulf Research and Development Co., thoroughly reprocessed and reinterpreted the COCORP seismic data, and their results, with many added details, are in fundamental agreement with Lynn's.

Together the Pacific Creek anticline and Wind River uplift record the progression of basement thrusting and the response of the overlying sedimentary section. Properties of the continental crust have been aptly compared to those of a jelly sandwich, the upper crust and upper mantle being relatively strong and brittle whereas the hot lower crust, closer to its melting temperature, is softer (e.g. Matthews, 1984; Sibson, 1982). To this picturesque analogue one needs to add a deformable topping of sedimentary rocks, which had a maximum thickness of more than eight kilometers during Laramide deformation. As shown in the Pacific Creek anticline, the sedimentary rocks rode upon the basement thrust block and were folded. In advanced stages, as shown by the Wind River uplift, the basement thrust cut upward through the entire sedimentary section. In both cases the thrust faults flatten and disappear downward.

On the Casper arch, sedimentary cover is still preserved and it neatly records the response to small and large thrusts in the basement.

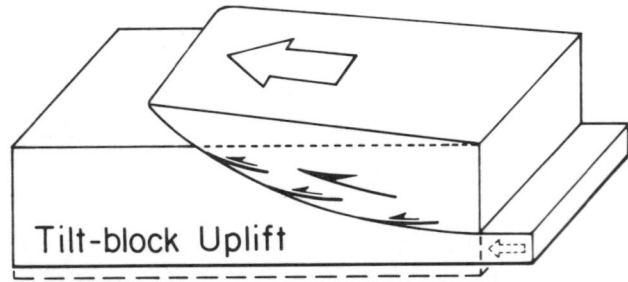

Figure 8. Diagrammatic representation of Wind River block uplifted and tilted along curved, splayed Wind River thrust zone. The tilt angle = arctan D/R, where D is the fault displacement, about 26 km, and R is the radius of curvature of the fault zone, about 120 km; the resulting tilt is about 12°.

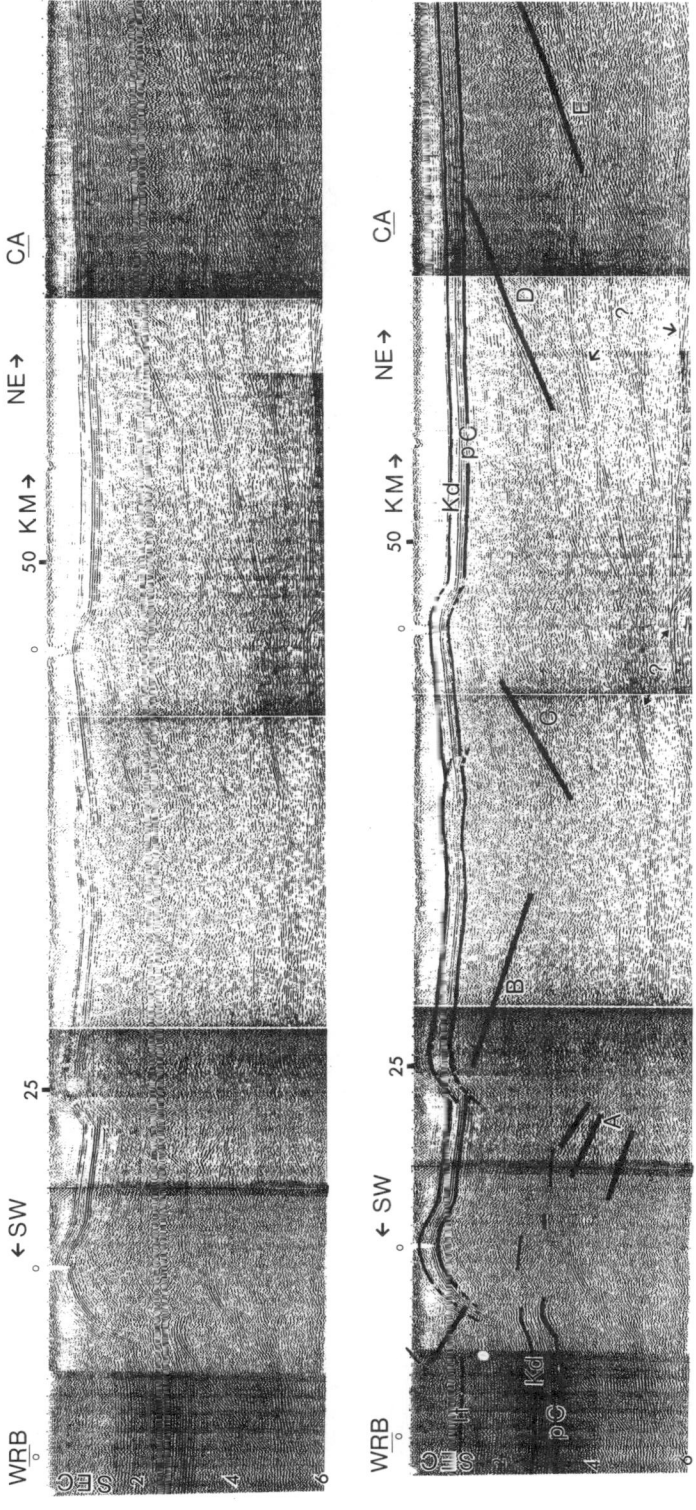

Figure 9a. Southwestern half of the Casper arch seismic section; WRB = Wind River Basin; CA = Casper arch. Top: the stacked section; Bottom: superimposed line drawing, showing faults A-E, the top of Precambrian basement (pC), Dakota (Kd) and Fort Union (Tf).

CONVERGENT AND DIVERGENT TERRANES   249

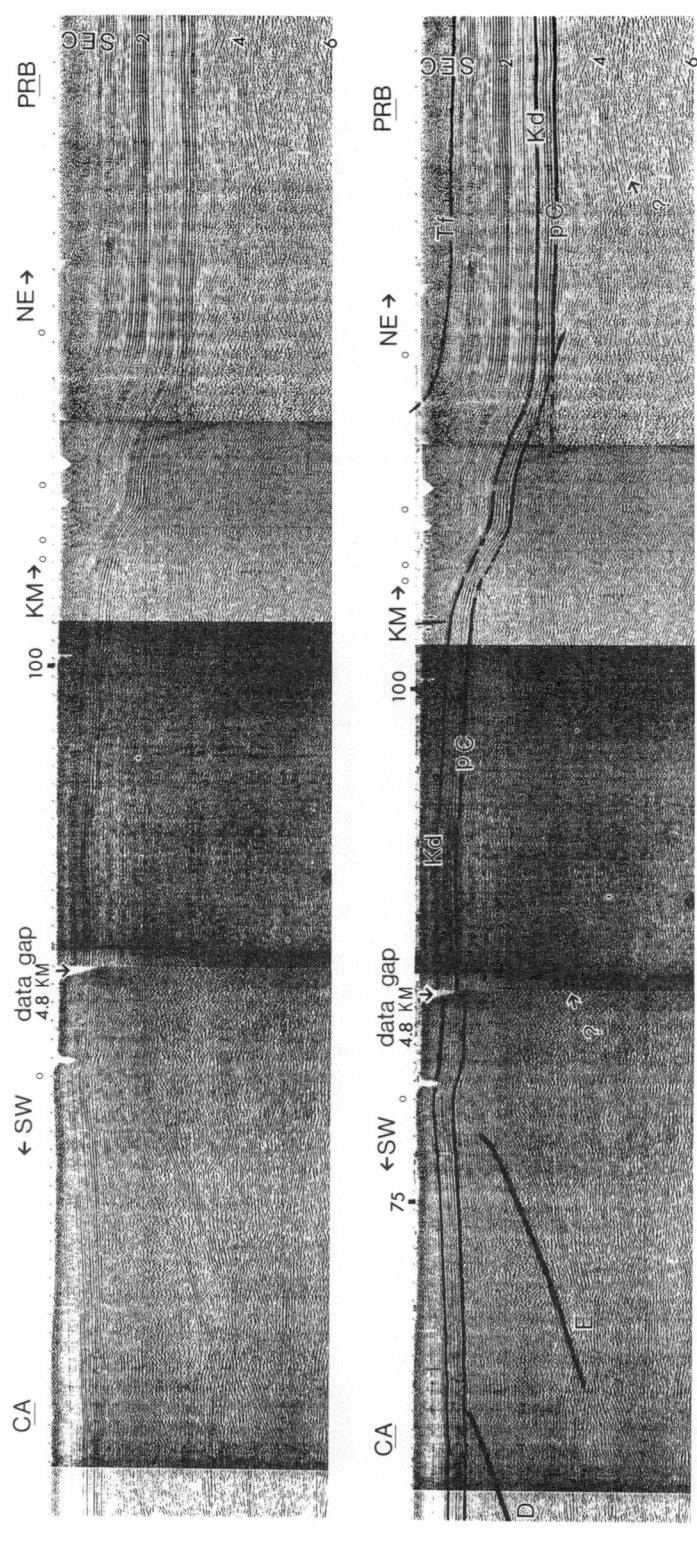

Figure 9b. Northeastern half of the Casper arch seismic section. Top: the stacked section; Bottom: superimposed line drawing; symbols same as in Figure 9a.

Figure 10a. Detail of seismic section, from Wind River basin (WRB) to Casper arch (CA).

The Casper Arch

The Casper Arch, the next uplift northeast of the Wind River Range, trends northwest and separates the Wind River basin from the Powder River basin (Fig. 3). A 114-km reflection line contributed by Gulf and Geosource has allowed us to integrate reflection data, subsurface well data, and surface geology to form a structural interpretation of the Casper Arch (Hill et al., 1981). Because this work has not been published before, we will present it in somewhat greater detail. Keen and Ray (1983) have made a detailed study of the southwest side of the Casper arch and have named the boundary structure the Casper arch thrust fault. Gries (1983) has also published seismic and geologic sections across the southwest boundary.

Seismic data and stratigraphic correlation.

As shown on Figure 3, the seismic line crosses the arch approximately perpendicular to the general structural trend. These eight-fold dynamite data were shot in 1973 with a 48-geophone group in a split-spread configuration and maximum geophone offset of 3.2 km. The data were sampled at a 4-msec interval and frequency filtered with a 12-45 Hz bandpass. The resulting 6-sec seismic section (Fig. 9a,b) has a datum at 1677 m. The great structural relief (about 4.6 km) and the disruption of seismic reflectors across the margin of the Wind River basin are striking (Figs. 9 and 10). On the opposite side of the arch, reflectors can be traced continuously from the arch into the Powder River basin. The line drawings emphasize the major events. Some of the events are inexplicable, as noted by question marks. Note especially, however, the five dipping reflectors (A-E, Fig. 9a,b) within basement that tie into deformation of the overlying sedimentary rocks. "A" is the Casper arch thrust system of Skeen and Ray (1983).

Correlation of the seismic section with the stratigraphic column is based upon published lithologic descriptions. Four synthetic seismograms were derived from sonic logs, using a 10-25 Hz zero-phase wavelet convolved with the reflection coefficients, and these synthetic seismograms were used to identify formations and carry

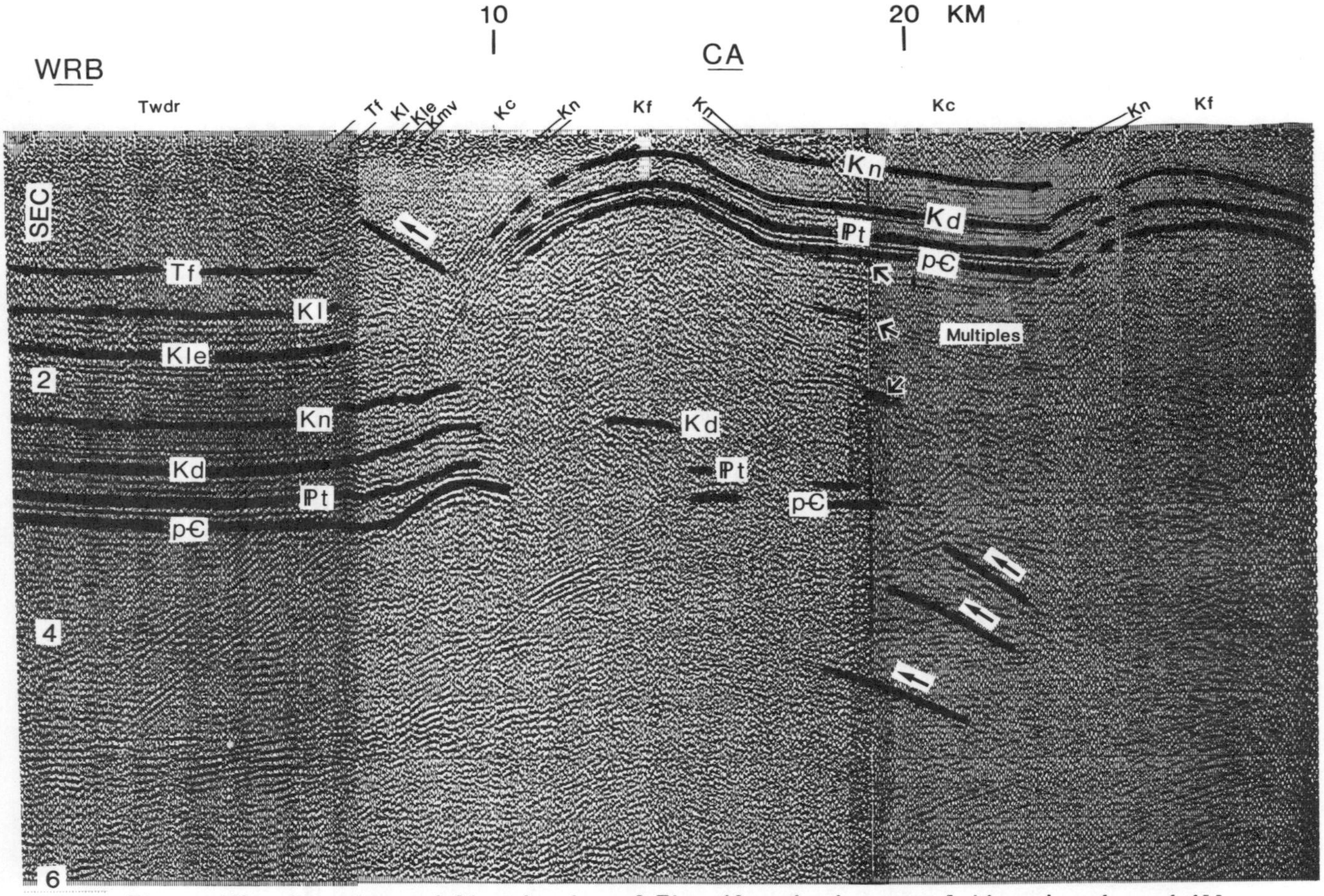

Figure 10b. Superimposed line drawing of Fig. 10a, showing correlation, based on drill logs and surface geology. Top of Precambrian basement (pC), Tensleep (Pt), Dakota (Kd), Frontier (Kf), Niobrara (Kn), Cody (Kc), Mesa Verde (Kmv), Lewis (Kle), Lance (Kl) and Fort Union (Tf).

them across the section (Fig. 9a,b). The basement reflector was tied to wells in the arch but its position in the Wind River basin and Powder River basin is extrapolated from formation thicknesses, because wells there do not penetrate to basement.

Structural interpretation. Because digital data tapes were not available, the basement reflector and dipping basement events A-E were two-dimensionally migrated using a variation of the procedure described in Dobrin (1976). The migration was simplified to a two-layer problem

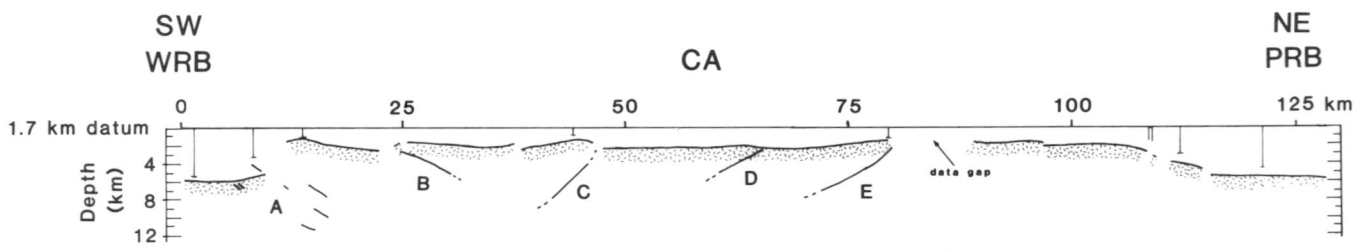

Figure 11. Depth-migrated section across the Casper arch, showing the top of Precambrian basement and the faults A-E. WRB = Wind River basin; CA = Casper arch; PRB = Powder River basin.

TABLE 1. Plane-of-Section Observations of Thrust Faults A-E

Event (Fault)	Vertical Offset (meters)	Horizontal Offset (meters)	Reflective to: Time (sec)	Depth (km)	Dip Range and Direction (degrees)
A	4600	5183	4.7	11.5	48-26° NE
B	800	1021	2.2	4.8	30-37° NE
C	800	200	3.6	8.0	49-42° SW
D	100	~0	2.8	5.1	38-29° SW
E	500	200	3.5	7.0	52-30° SW

with a sedimentary velocity of 4.0 km/s and a basement velocity of 6.0 km/s. The average sedimentary velocity is derived from sonic logs and the basement velocity is taken from Lynn (1979).

In the resulting migrated depth section (Fig. 11) basement events A-E migrate updip to effectively tie deformation of the basement into that of the overlying sedimentary section, and these events are therefore interpreted as thrust faults. Faults such as E are concave upward, like the Pacific Creek and Wind River thrusts; others appear planar but extrapolate downward toward flatter reflectors (Fig. 9a). Monoclines in the overlying sedimentary strata are characterized by one steep limb and a gently tilted back slope (Fig. 9a,b). The top of the basement is clearly involved in the folding (e.g. C, Fig. 9a) and we think this apparent bending of the crystalline rocks may be the result of discrete slip and minor adjustment along many small faults and joints that are not resolvable in the seismic section.

Table 1 summarizes data on the basement thrusts along this line. Vertical and horizontal offsets of the top of basement are measured as shown in Figure 12. The reflections weaken or disappear downward at the time and depth shown, and dip tends to decrease downward.

Because the seismic line is not exactly perpendicular to the strikes of faults, Table 1 necessarily records apparent dips that are less than the true dips. The three-dimensional geometry was analyzed by Dianna Shelander to determine the possible range of true strike and dip that could produce each fault reflection. For modeling purposes the deeper part of each fault was approximated by a plane and the rock velocities by the same two layers used in the migration. Surface and subsurface geology was then used to choose the most probable fault trends, and the dips shown in Table 2 were calculated. The several segments of fault A have a range of dips. Moreover, Skeen and Ray (1983), in a series of seismic sections across this fault zone, found dips increasing from about 20° in the north to nearly 40° in the south.

Summary. In summary, the Casper arch reflection profile reveals a linkage between reflective Laramide thrust faults in the Precambrian crystalline basement and folds in the overlying sedimentary rocks. The vertical components of displacement range from only 100 m in fault D (less than that of the Pacific Creek thrust) to 4.6 km in fault system A (about one-third that of the Wind River thrust). Displacement on fault system A increases markedly to the northwest along strike (Gries, 1983). The faults tend to flatten downward in the basement, as do the Pacific Creek and Wind River thrusts. Only fault system A, with the largest displacement, cuts upward through the entire sedimentary section. The others, with vertical displacements of 800 m or less, generate unbroken folds in the overlying sedimentary section. In the absence of subsurface data, one could confidently predict the presence and estimate the dip and displacement of an underlying thrust from the structural relief and amount of shortening across a fold. Consequently, it is reasonable to infer that the northeastern boundary of the Casper arch, which consists of a double monocline, may be underlain by zones of thrust faulting wholly within the basement. Why these particular fault zones are not visible seismically is unknown; perhaps they are too diffuse and gradational.

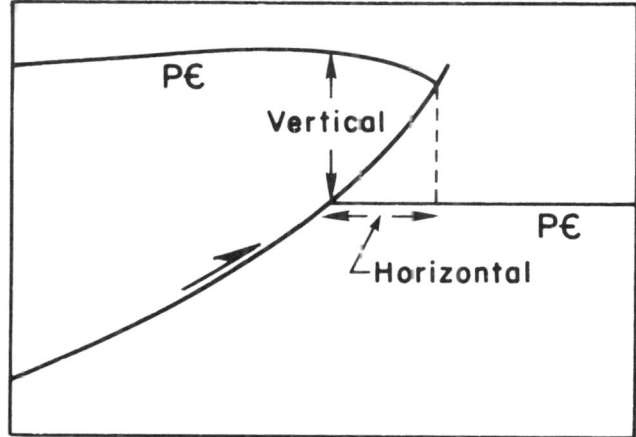

Figure 12. Measurement of vertical and horizontal offsets compiled in Table 1.

TABLE 2. Estimated Strike and Calculated Dip of Thrust Faults A-E

Event (Fault)	Angle between Strike and Seismic Line	Strike of Fault	Dip
A	65°	N47° W	29-55° NE
B	55°	N42° W	34° NE
C	58°	N40° W	63° SW
D	46°	N32° W	42° SW
E	55°	N24° W	37° SW

Divergent Terranes

The Cenozoic rifted region of western North America includes the Basin and Range province and the Rio Grande rift subprovince (Fig. 2). A growing body of deep reflection data in this region shows some similarities and some major differences in reflection character from the convergent Laramide region (e.g. Allmendinger et al., 1983; McCarthy, 1984; Okaya, 1984). Both regions display a tendency for moderately to steeply dipping fault reflectors in the shallow crust to give way to flatter structures at large depths. But whereas the Laramide thrusts splay and flatten with depth, the Basin and Range normal faults tend to be planar in their upper parts and to merge abruptly into subhorizontal detachment faults. Furthermore the intermediate and deep crust is more often characterized by abundant subhorizontal reflectors that give it a laminated appearance. The transition to subhorizontal detachment faults as seen in reflection sections has been well documented in the Basin and Range province by McDonald (1976) and by Allmendinger et al. (1983), and a deeply penetrating planar fault has been studied by Okaya and Thompson (1984). Detachments are also well mapped geologically (e.g. Miller et al., 1983).

The data base of deep reflection sections in the Basin and Range province is expanding so rapidly (e.g. Allmendinger et al., 1983; McCarthy, 1984; Okaya, 1984) that a comprehensive discussion is not practical. However, one example, in the Rio Grande rift, illustrates general problems; we summarize briefly here.

Rio Grande Rift

Focusing on the shallow structure of the rift, Cape et al. (1983) used migrated COCORP sections along with drill-hole information and gravity data. Although the normal faults had originally been interpreted as high-angle planar faults to great depth (Brown et al., 1980), the migrated sections are convincing in showing that the normal faults merge into subhorizontal reflectors at a depth of only about 5 km. The crust below this depth contains many subhorizontal reflectors, including a well-documented tabular magma body (Brown et al., 1980). The structure is shown diagrammatically in Figure 13. The detachment fault is at a depth of about 5 km, and all of the high-angle structures in Figure 13 below that depth are hypothetical. Our reasons for hypothesizing the deeper structures are: (1) Earthquakes occur deeper than the detachment and have focal mechanisms appropriate for normal faults (Sanford et al., 1973; Sanford et al., 1977). (2) Basaltic volcanism in the rift requires feeder dikes from the mantle; and (3) Other rift anomalies, such as the heat flow, require deep sources.

The Rio Grande rift thus exemplifies important general problems of the Basin and Range province: Why do normal faults flatten abruptly at shallow to moderate depths? And what is the nature of the abundant subhorizontal reflectors in the crust?

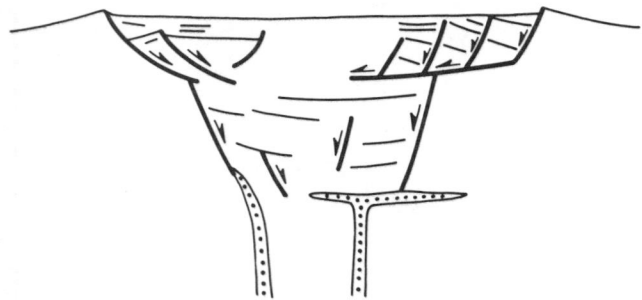

Figure 13. Diagrammatic section across the Rio Grande rift, showing how the high-angle faults near the surface are decoupled along subhorizontal detachments from the deeper structure. Structure below the subhorizontal detachment faults is mostly hypothetical in this sketch.

Conclusions

In the convergent Laramide terrane, shortening and structural relief of folds is closely linked to the dip and displacement of underlying thrusts. In the crystalline basement the thrusts are concave upward, and large thrusts splay subhorizontally as well as curve, thus producing tilt in the overlying block. Thrusts disappear at depths where the crust would be expected to be hotter and weaker (paleodepths of 10 to 30 km).

Possibly subhorizontal shearing adjustments within a soft zone of deep crust between the brittle blocks of the upper crystalline crust and a strong mantle are responsible for some of the deep subhorizontal reflectors. Alternatively, igneous and metamorphic processes may be partly responsible (Lynn et al., 1981; Hale and Thompson, 1982).

In the rifted Basin and Range region many of the high-angle normal faults terminate on detachment faults at surprisingly shallow depths (less than 5 km). The shallow crustal blocks are like books spreading and lying over on a slippery bookshelf. On the other hand, evidence from deeper earthquakes (deeper than the bookshelf) and from basaltic magmatism indicate that there are important high-angle structures deeper in the crust. The abruptness of transitions from shallow high-angle faults to detachment faults requires more explanation than a simple thermal weakening of rocks due to the geothermal gradient. We suggest that the cooling effect of deep water circulation in the broken upper plates, combined with the high heat flow in the Basin and Range province, could lead to a very abrupt transition to high pore water pressure at depth. A change from open circulation to tight rock self-sealed by mineral deposition and by ductility could be abrupt and could explain the sharpness of the structural transition.

Reflective laminations in the deep crust appear to be a characteristic of continental crust in many places and are not an artifact of seismic profiling. One likely cause is shearing adjustments beneath coherent structural blocks of the upper crust. This mechanism would be effective in strike-slip regimes as well as compressional and extensional terranes.

Acknowledgments. We owe special thanks to John Garing and Dianna Shelander for their participation in the analysis of the Casper arch section and to Peter Montecchi of Gulf for his generous help in obtaining the Casper arch data. Rick Allmendinger's critical review improved the manuscript.

References

Allmendinger, R.W., J.W. Sharp, D. Von Tish, L. Serpa, L. Brown, S. Kaufman, and J. Oliver, Cenozoic and Mesozoic structure of the eastern Basin and Range province, Utah, from COCORP seismic-reflection data, Geology, 11, 532-536, 1983.

Brewer, J.A., S.B. Smithson, J.E. Oliver, S. Kaufman, and L.D. Brown, The Laramide orogeny: Evidence from COCORP deep crustal seismic profiles in the Wind River Mountains, Wyoming, Tectonopysics, 62, 165-189, 1980.

Brown, L.D., C.E. Chapin, A.R. Sanford, S. Kaufman, and J. Oliver, Deep structure of the Rio Grande rift from seismic reflection profiling, Journ. Geophys. Res., 85, 4773-4800, 1980.

Cape, C.D., S. McGeary, and G.A. Thompson, Cenozoic normal faulting and the shallow structure of the Rio Grande rift near Socorro, New Mexico, Geol. Soc. Amer. Bull., 94, 3-14, 1983.

Dobrin, M.B., Introduction to Geophysical Prospecting, 630 pp., McGraw-Hill, New York, 1976.

Gries, R., Oil and gas prospecting beneath Precambrian foreland thrust plates in Rocky Mountains, Amer. Assoc. Petr. Geol. Bull., 67, 1-28, 1983.

Hale, L.D., and G.A. Thompson, The seismic reflection character of the continental Mohorovicic discontinuity, Journ. Geophys. Res., 87, 4625-4635, 1982.

Hill, J.L., D.L. Garing, and N. Josten, Seismic structural interpretation of the Casper Arch, Wyoming (abs.), EOS, Amer. Geophys. Un. Trans., 62, 1040, 1981.

Jones, T., and A. Nur, Seismic velocity and anisotropy in mylonites and the reflectivity of deep crustal fault zones, Geology, 10, 160-163, 1982.

Lynn, H.B., Migration and Interpretation of Deep Crustal Seismic Reflection Data, 170 pp., Ph.D. Thesis, Stanford University, Stanford, CA., 1979.

Lynn, H.B., L.D. Hale, and G.A. Thompson, Seismic reflections from the basal contacts of batholiths, Journ. Geophys. Res., 85, 10,633-10,638, 1981.

Lynn, H.B., S. Quam, and G.A. Thompson, Depth migration and interpretation of the COCORP Wind River, Wyoming, seismic reflection data, Geology, 11, 462-469, 1983.

MacLeod, M.K., The Pacific Creek anticline: Buckling above a basement thrust fault, Univ. of Wyoming Contr. to Geology, 19, 143-160, 1981.

Mathur, S.P., Deep crustal reflection results from the central Eromanga basin, Australia, Tectonophysics, 100, 163-173, 1983.

Matthews, D., Deep reflection from the Caledonides and Variscides west of Britain and comparison with the Himalayas, This Symposium, 1984.

McCarthy, J., Reflection profiles from the Snake Range metamorphic core complex: windows into the deep crust, This Symposium, 1984.

McDonald, R.E., Tertiary tectonics and sedimentary rocks along the transition: Basin and Range province to plateau and thrust belt province, Utah, Rocky Mt. Assoc. Geologists Symp., 281-327, 1976.

Miller, E.L., P.B. Gans, and J. Garing, The Snake Range decollement: an exhumed mid-Tertiary ductile-brittle transition, Tectonics, 2, 239-263, 1983.

Mueller, S., A new model of the continental crust, in The Earth's Crust, edited by J.G. Heacock, pp. 289-317, Amer. Geophys. Union Mon. 20, 1977.

Okaya, D., Seismic profiling of the lower crust: Dixie Valley, Nevada, This Symposium, 1984.

Okaya, D., and G.A. Thompson, Geometry of Cenozoic extensional faulting, Dixie Valley, Nevada, Tectonics, 4, 107-125, 1985.

Sanford, A.R., O. Alptekin, and T.R. Toppozada, Use of reflection phases on microearthquake seismograms to map an unusual discontinuity beneath the Rio Grande Rift, Bull. Seism. Soc. Amer., 63, 2021-2034, 1973.

Sanford, A.R., R.P. Mott, P.J. Shuleski, E.J. Rinehart, F.J. Caravella, R.M. Ward, and T.C. Wallace, Geophysical evidence for a magma body in the crust in the vicinity of Socorro, New Mexico, in Heacock, J.G., ed., The Earth's Crust: Its Nature and Physical Properties, Washington, D.C., Amer. Geophys. Un. Geophysical Monograph 20, 385-403, 1977.

Sharry, J., R.T. Langan, D.B. Jovanovich, G.M. Jones, N.R. Hill, and T.M. Guidish, Enhanced imaging of the COCORP Wind River line, This Symposium, 1984.

Sibson, R.H., Fault zone models, heat flow, and the depth distribution of earthquakes in the continental crust of the United States, Bull. Seism. Soc. Amer., 72, 151-163, 1982.

Skeen, R.C., and R.R. Ray, Seismic models and interpretation of the Casper arch thrust: application to Rocky Mountain foreland structure, in Rocky Mt. Assoc. of Geologists, edited by J.D. Lowell and R. Gries, pp. 99-124, 1983.

Smithson, S.B., J.A. Brewer, S. Kaufman, J.E. Oliver, and C.A. Hurich, Structure of the Laramide Wind River uplift, Wyoming, from COCORP deep reflection data and gravity data, Journ. Geophys. Res., 84, 5955-5972, 1979.

Zawislak, R.L., and S.B. Smithson, Problems and interpretation of COCORP deep sesmic reflection data, Wind River Range, Wyoming, Geophysics, 46, 1684-1701, 1981.

PHANEROZOIC TECTONICS OF THE BASIN AND RANGE - COLORADO PLATEAU TRANSITION FROM COCORP DATA AND GEOLOGIC DATA: A REVIEW

Richard W. Allmendinger, Harlow Farmer[1], Ernest Hauser, James Sharp[2], Douglas Von Tish[3], Jack Oliver, and Sidney Kaufman

Institute for the Study of the Continents and Department of Geological Sciences
Cornell University, Ithaca, New York 14853

Abstract. The COCORP 40°N transect in Utah and easternmost Nevada crosses the eastern Basin and Range Province and the northwest Colorado Plateau. This part of the Cordillera of western North America has been the site of late Precambrian rifting, Mesozoic and early Cenozoic thrust faulting, and middle and late Cenozoic extension. The COCORP data, in combination with drilling and surface geologic data, suggest that a major change in seismic character and crustal structure occurs at the location of the autochthonous hingeline, which was formed during the Precambrian rifting and subsequent passive margin stage. East of the hingeline, there is little evidence that neither thrust faults nor low-angle normal faults cut deeply into the continental basement. West of the hingeline, a prominent, west-dipping seismic fabric may correspond to Cenozoic and perhaps Mesozoic low-angle structures that cut down to middle and lower crustal levels. The deepest reflections in the Basin and Range (28-30 km) correspond in depth to a prominent mid-crustal horizon (~27 km) in the Colorado Plateau; their relation is uncertain. The Colorado Plateau crust appears to be more than 15 km thicker than the Basin and Range crust. Also, the Colorado Plateau data is dominated by diffractions, where as the Basin and Range has a more persistent west-dipping or layered fabric. These contrasting seismic characters may be typical of cratonic versus orogenic crust, respectively.

Introduction

The eastern half of the COCORP 40°N transect of the western U.S. Cordillera is located on rocks autochthonous to the North American continent and spans the eastern Basin and Range in eastern Nevada and western Utah to the northwestern Colorado Plateau (Fig. 1). These Cenozoic morphotectonic provinces are only the youngest manifestations of a series of tectonic events that have modified the western margin of the continent. The region has also been the site of: 1) latest Precambrian rifting to form the Cordilleran miogeocline, 2) the Late Jurassic(?) through Early Tertiary Sevier foreland thrust belt, 3) the Early Tertiary Laramide basement uplifts, and 4) Middle and Late Cenozoic extensional tectonics. The COCORP data and available surface and subsurface data summarized here suggest that the structures and sedimentary basin geometries related to these events are closely interwoven to form the current fabric of the continental crust. Although it is difficult to prove that any particular fault has been reactivated, older structures and basins have clearly influenced the general position and geometry of younger features. The features of the COCORP seismic lines reviewed here are described in detail in Allmendinger et al., [1983], Von Tish et al., [1985], and Sharp [1984]. Industry seismic data bearing on these same topics have also been published by Royse et al., [1975], McDonald [1976], and Smith and Bruhn [1984].

Geology of the Cordilleran Hingezone

The Cordilleran hingeline region (referred to here as the "Hinge Zone" and commonly called the "Wasatch Line" in Utah) is a diffuse boundary that has, for the last 600 Ma, separated the multiply deformed edge of the North American craton from the more stable, less deformed interior of the continent. The hingeline (Fig. 1) is defined as the axis west of which late Precambrian and Paleozoic miogeoclinal strata thicken markedly [Kay, 1951; Stokes, 1976]. This axis was first formed during latest Precambrian rifting of the western margin of North America and the development of a passive continental margin during the early Paleozoic [Stewart, 1972]. The age of the initial rifting event has long been uncertain because the

---
[1] Present address: Pecten International, Box 205, Houston, TX 77001
[2] Present address: Union Oil Company of California, Box 6176, Ventura, CA 93006
[3] Present address: Sohio Petroleum, 9401 Southwest Freeway, Houston, TX 77074

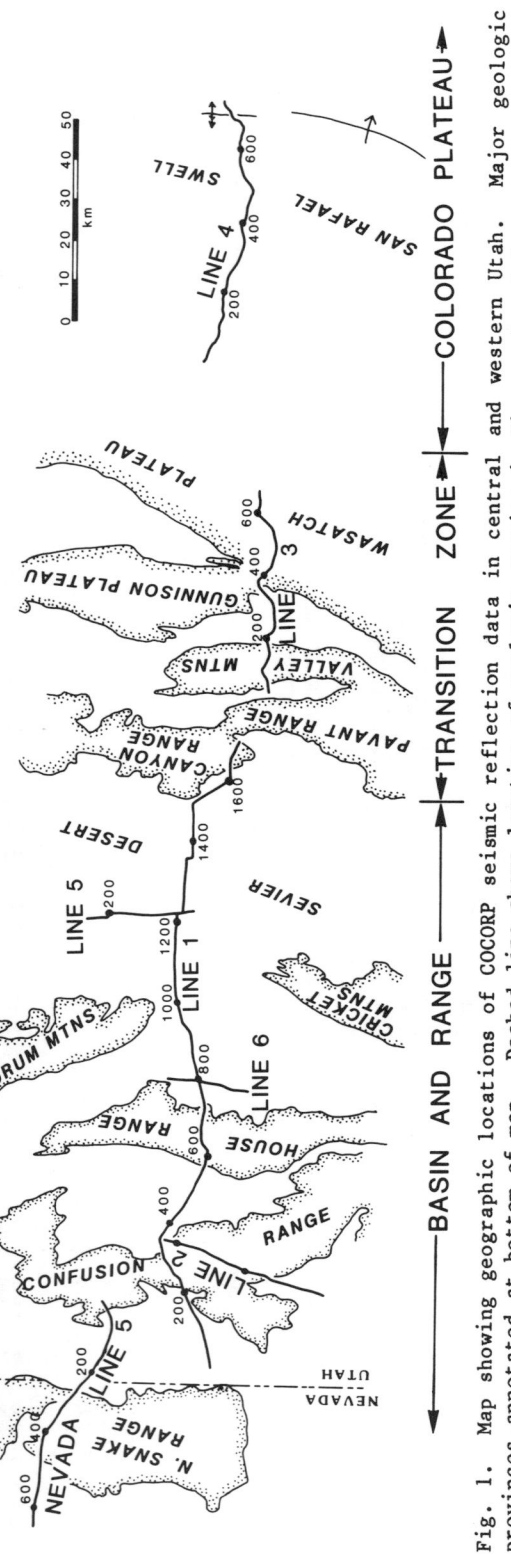

Fig. 1. Map showing geographic locations of COCORP seismic reflection data in central and western Utah. Major geologic provinces annotated at bottom of map. Dashed line shows location of geologic section in Figures 5 and 6.

syn- and early post-rift deposits are poorly dated. However, recent application of back-stripping techniques suggest that rifting occurred in the latest Proterozoic, less than 650 Ma ago [Armin and Mayer, 1983; Bond and Komniz, 1984]. The passive margin persisted, virtually uninterrupted, until the end of the Devonian and the onset of the Antler orogeny [Dickinson, 1977; Speed, 1982]. Thus, the passive margin of western North America existed for about 300 Ma, almost twice as long as its modern counterpart on the east coast of North America. The present width of this ancient passive margin, measured from the hingeline on the east to the westernmost miogeoclinal deposits and the initial $^{87}Sr/^{86}Sr = .706$ ratio that define the westernmost North American basement [Speed, 1982], is about 400 km (250 mi) (Fig. 1). Due to superposed younger deformations, the original width is unknown. In comparison, the equivalent part of the modern passive margin on the East Coast is only about 100 km wide [Grow, 1981].

Although middle and late Paleozoic orogenies (probably representing the accretion of allochthonous terranes) occurred in central and western Nevada, the Hinge Zone in central Utah remained undeformed until the Sevier orogeny in the Cretaceous and Early Tertiary [Armstrong, 1968]. At least three major thrust faults of the Sevier belt occur in central Utah (Figs. 1, 2): 1) the structurally highest Canyon Range thrust with a thick sequence of Upper Proterozoic and Paleozoic strata in its hanging wall, 2) the middle Pavant Range thrust that, where exposed, contains Lower Cambrian and younger clastic strata in its upper plate, and 3) the lowest thrust (informally referred to here as the "sub-Pavant thrust") which is nowhere exposed at the surface, but is known from industry drilling and seismic reflection data [Standlee, 1982; Sharp, 1984]. The thrusts are not well dated, but probably range in age from the Cretaceous (Albion) to the Early Tertiary [Speiker, 1946; Lawton, 1983; Villien and Kligfield, in press]. The Mesozoic-Early Tertiary thrust belt extends at least as far east as the west side of the Wasatch Plateau and may extend farther east as a blind thrust fault in the Jurassic section [Standlee, 1982; Villien and Kligfield, in press]. No major thrust faults structurally higher than the Canyon Range thrust have been recognized between the Canyon Range and the Utah-Nevada border, although some small thrust faults occur in the Confusion Range [Armstrong, 1968; Hintze, 1974].

Most recently, the Hinge Zone has been deformed by extensional tectonism that produced the Basin and Range Province. Zoback and others [1981] proposed that this extension took place in two discrete pulses: an earlier phase, beginning in the Oligocene, of low-angle normal faulting and accompanying calc-alkaline volcanism that occurred during the waning phases of subduction beneath the continent; and a later, Miocene and younger phase of higher angle normal faulting and resulted in

UTAH LINE 4

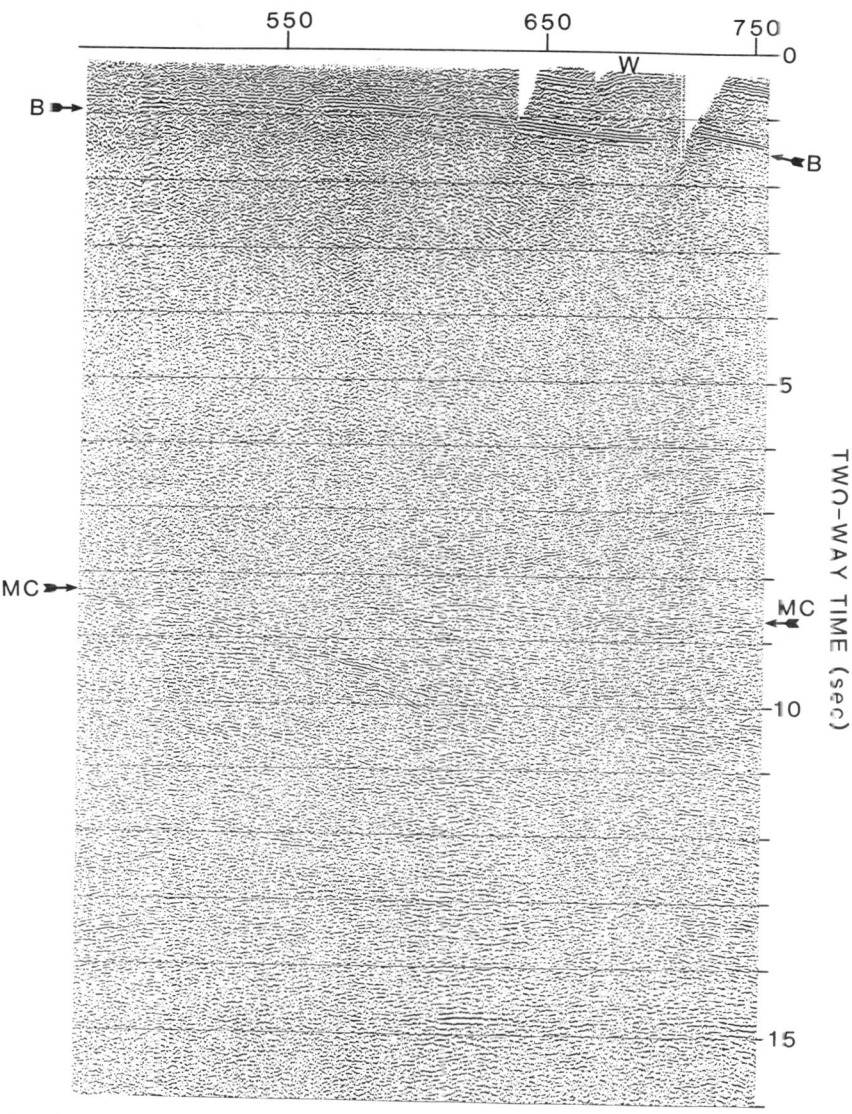

Fig. 2. Seismic reflection data from the eastern part of Utah Line 4. The section shows the entire crust beneath the eastern part of the San Rafael Swell. "W" shows the Woodside anticline. "B" indicates top of basement. "MC" shows the mid crustal horizon discussed in the text. Note dipping and crossing features in the middle and lower crust. Section processed as shown in Table 2. Not migrated. Processed by H. Farmer.

the formation of the present basin-range morphology of the province. Data presented by Von Tish et al., [1985] and reviewed here indicate episodic deformation but only one style of extension, dominated by low-angle normal faulting that probably began in the Late Oligocene and is at least as young as 1 m.y.b.p. Low-angle normal faults in eastern Nevada and west-central Utah have been described by Armstrong [1972], McDonald [1976], Allmendinger et al., [1983], and Smith and Bruhn [1984].

COCORP Data

The processing and acquisition parameters for the COCORP Utah lines, upon which this review is

TABLE 1: Data Acquisition Parameters, COCORP Utah Lines

	Line 1	Line 3	Line 4
Energy Source	Vibroseis	*	*
No. of Vibrators	5 (4 minimum)		
Vibrator Spacing	12.6 m	30.5 m	30.5 m
Vibrator Move-up	12.6 m		
No. of Sweeps/VP	8		
Sweep Frequencies	8-32 Hz		
Sweep Length	32 sec		
Geophones/Channel	24		
Geophone Frequency	7.5 Hz		
Channel Spacing	100.6 m		
Source Spacing	100.6 m		
No. of Channels	96		
Recording Filter	31.25 Hz		
Sample Rate	.008 sec		
Field Record Length	52 sec		
Correlated Rec. Length	20 sec		

*All parameters the same as Line 1, except as shown.

based, are summarized in Tables 1 and 2. A detailed discussion is beyond the scope of this review. Here we touch briefly on a few of the main points of the data, particularly as they relate to the Phanerozoic tectonics of the region. Excerpts of the original profiles and line drawings of all of the data are shown in Figures 2, 3, and 4.

Colorado Plateau

COCORP Utah Line 4 on the Colorado Plateau (Figs. 2, 4; Tables 1, 2) imaged cratonal-type continental crust that was relatively little deformed during the Phanerozoic. Because the line is directly adjacent to the orogenically modified edge of North America, it provides an excellent basis for comparison of "cratonic" or "pericratonic" vs. "orogenic" crust. Four characteristic features of the continental crust of the northwest Colorado Plateau shown on Line 4 are: 1) The shallow geometry of the Laramide San Rafael Swell, 2) a prominent lower-middle crustal horizon, 3) deep events interpreted to lie at the base of the reflective lithosphere, and 4) the general seismic fabric of the crust.

The San Rafael Swell appears on Line 4 as a broad, nearly symmetrical arch of reflectors associated with Paleozoic and Mesozoic strata that overly the Precambrian basement (Fig. 2). This arch has a half wavelength of about 50 km and a vertical structural relief of about 1.5 km. No events in the basement show a direct relation to the swell structure; the nature of Laramide basement deformation is thus enigmatic. Unless the sedimentary rocks responsible for the arched reflections are completely detached from the basement, the upper part of the crust must also be arched. There is little evidence for a west-dipping thrust or reverse fault in the basement beneath the Swell (Fig. 2). This could be due to 1) lack of a basement fault, 2) a fault too steep to be imaged, or 3) insufficient impedance contrast, perhaps due to the fact that displacements are probably dying out at the north end of the Swell where it was crossed by the COCORP line. A small, antithetic east-dipping reverse fault in the basement on the east side of the Swell produced the Woodside anticline (Fig. 2).

At ~8.5 s (27-29 km), a prominent horizontal zone of reflections occurs somewhat discontinuously across most of the section (Figs. 2,4). This band of reflections is at about the same time as a velocity discontinuity or gradient in the lower-middle crust of the Plateau identified by Roller [1965], and Prodehl [1979]. Unlike the Swell above it, the reflections are horizontal, not arched, suggesting that corresponding reflectors may be either younger than the Early Tertiary Swell or that the upper crust is detached at or above it. As noted below, the 8.5 s event is at about the same depth in the crust as the deepest reflections in the eastern Basin and Range Province to the west.

The deepest reflections on Line 4 are at about 14.8 s (~47 km); the crooked line geometry on the eastern side of the profile where these events occur suggests that they are directly beneath the line and not a result of sideswipe. These reflections are considerably deeper than what Roller [1965] and Prodehl [1979] interpreted to be Moho from refraction data, the mid-point of which was located 200 km south of the COCORP line. The velocity structure at this depth in the crust is not known in detail; however, the reflections are probably more than 15 km deeper than any reflections observed on the COCORP transect in the Basin and Range.

TABLE 2: Processing Sequence, COCORP Utah Lines

	Line 1	Line 3 43-366	Line 3 378-639	Line 4
Demultiplex	1	1	1	1
Vibroseis Correlation	2	2	2	2
Trace Editing		3 (a)		
F-K Filter	3			
CDP Gather	5	6	5	4
Datum	1900 m	2450 m	2450 m	1896 m
Datum Velocity	4500 m/s	3500 m/s	3500 m/s	4000 m/s
Datum Statics	4	5	4	5
Velocity Analysis (b)	7	7	6	6
Normal Moveout	8	8	8	8
Mute	9	9	9	10
Trace Equalization	6	10	10	9
Automatic Residual Statics			7	7
Stack, 4800%	10	11	11	11
Filter		12 (c)	12 (c)	12 (d)
Predictive Deconvolution		4	3	3
Automatic Gain Control	11 (e)	(f)	(f)	(f)

Notes
a) extreme highway-cultural noise
b) velocity spectra and constant velocity stacks; picked every 10 to 50 CDPs
c) 0 to 4 s -- 12-32 Hz; 6 to 14 s -- 8-32 Hz interpolated in between
d) 0 to 5 s -- 12-32 Hz; 7 s -- 8-32 Hz; 12 s -- 8-28 Hz; 14 s -- 8-22 Hz
e) 1 s AGC window
f) trace equalization used 1 s window; AGC not applied

Aside from the three features discussed above, the general character of the Colorado Plateau crust is one of numerous dipping, curving, and crossing events. Many of these features are reasonably interpreted as diffractions, although some may be dipping reflections or reflections from out of the plane of the section. These events occur both above and below the 8.5 s reflection (Fig. 2).

Wasatch Plateau and Transition Zone

COCORP Utah Line 3 (Fig. 1) crosses the Wasatch Plateau, Sevier Valley, and the Valley Mountains. Good shallow data were obtained from beneath the Wasatch Plateau (Fig. 3), but farther west and deeper than about 4 s the data provide little insight into crustal structure. The data show that the Mesozoic-early Cenozoic thrust faults and accompanying structural duplication occur at least as far east as the west side of the Wasatch Plateau (Fig. 3); blind thrusts in the Jurassic section could extend farther east [Standlee, 1982; Villien and Kligfield, in press]. The COCORP seismic data also suggest that the normal faults on the Plateau do not cut into basement, but may instead sole into and reactivate a blind thrust in the Jurassic section (Fig. 3). This conclusion is somewhat tenuous given the resolution of the seismic data and the small displacements on most of the normal faults.

A normal fault does offset the basement by 1.4-1.8 km beneath the prominent syncline on the west side of the Wasatch Plateau (Fig. 3). This normal fault does not cut above the Jurassic section; the syncline of Cretaceous rocks above it is unbroken. Thus, two interpretations of the fault are possible. First, it may be a Jurassic normal fault that controlled evaporite deposition in the Arapien basin directly to the west (i.e., the "Ancient Ephraim fault" of Moulton, 1975). Second, the fault could be Cenozoic in age, representing the location where the shallow normal faults of the Wasatch Plateau depart from their detachment horizon in the Jurassic section and cut down-section into the crust. Directly west of this normal fault the basement surface occurs at about 3.5 s (~9 km) depth (Fig. 3); farther west the top of the basement is only poorly imaged, if at all, on Line 3. Drilling results from the Placid WXC No. 1 well on the west side of the Valley Mountains, however, indicate that basement may be at approximately 9 km there as well, assuming no major thrusts and a normal stratigraphic section below total depth of the borehole.

Eastern Basin and Range

COCORP Utah Line 1 (Fig. 4) from the eastern Basin and Range showed remarkable overlapping lateral continuity of gently dipping structures across more than 120 km perpendicular to strike.

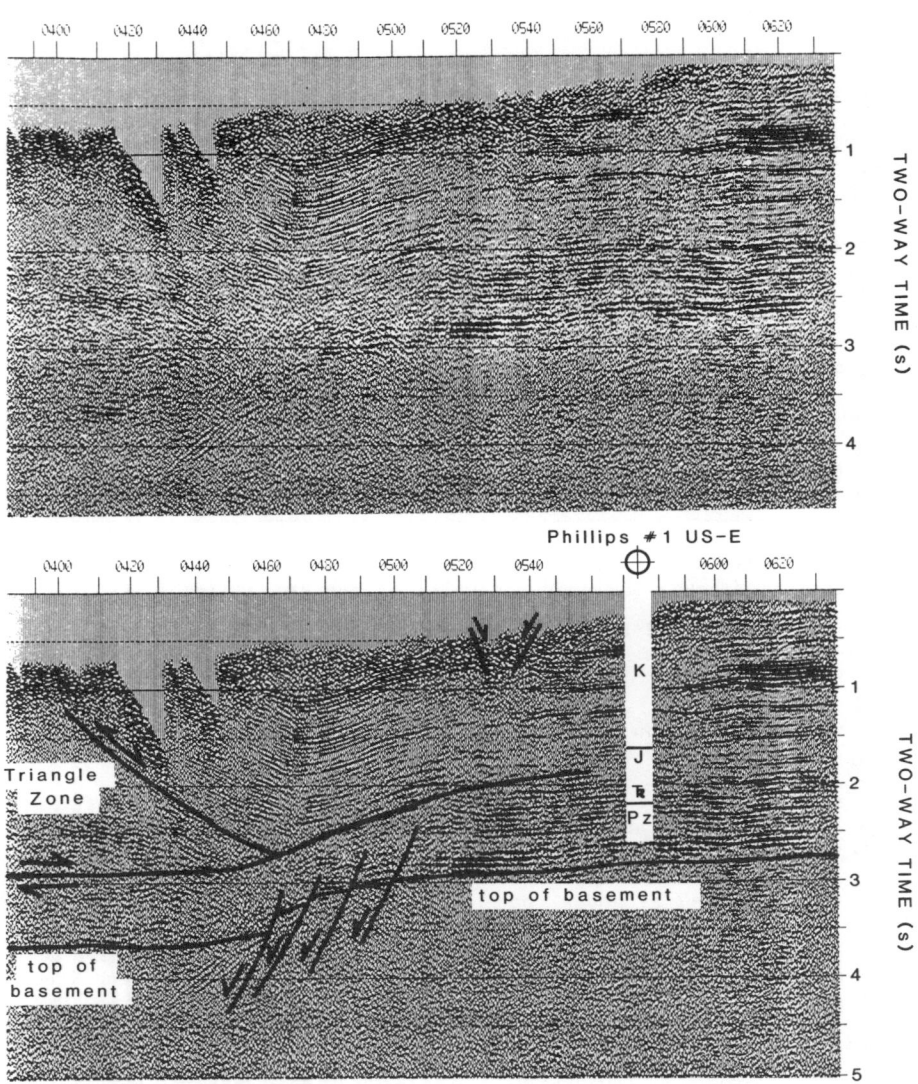

Fig. 3. Seismic reflection data from the eastern part of Utah Line 3; both interpreted (bottom) and uninterpreted (top) sections shown. Tops of units shown in Phillips #1 US-E well converted to time using the well sonic logs. Small graben interpreted in the bottom section is the Water Hollow fault zone. The Wasatch monocline is within the mute zone at the top of the section. Section processed as shown in Table 2. Not migrated. Processed by H. Farmer.

The Sevier Desert detachment is a low-angle normal fault that underlies the Cenozoic Sevier Desert basin. Normal faults within the basin that cut volcanic and sedimentary rocks at least as young as 4 m.y., but do not significantly offset (by more than about 300 m [Peddy and Brown, this volume] the detachment. The basinal geometries and ages in combination with palinspastically restored sections (discussed below) show that the long term rate of extensional displacement on the detachment is between 0.7 and 1.9 mm/yr [Von Tish et al., 1985]. These rates are the same order of magnitude as those determined for foreland fold-thrust belts (~1.5 mm/yr for the Idaho-Wyoming thrust belt [Allmendinger and Jordan, 1981]) and both have rates that are, on the average, one to two orders of magnitude slower than plate convergence rates. Hiatuses are recorded in the basin sediments [Von Tish et al., 1985; Lindsey et al., 1981], but there is no resolvable change in the rates of extension.

One of the major remaining problems posed by

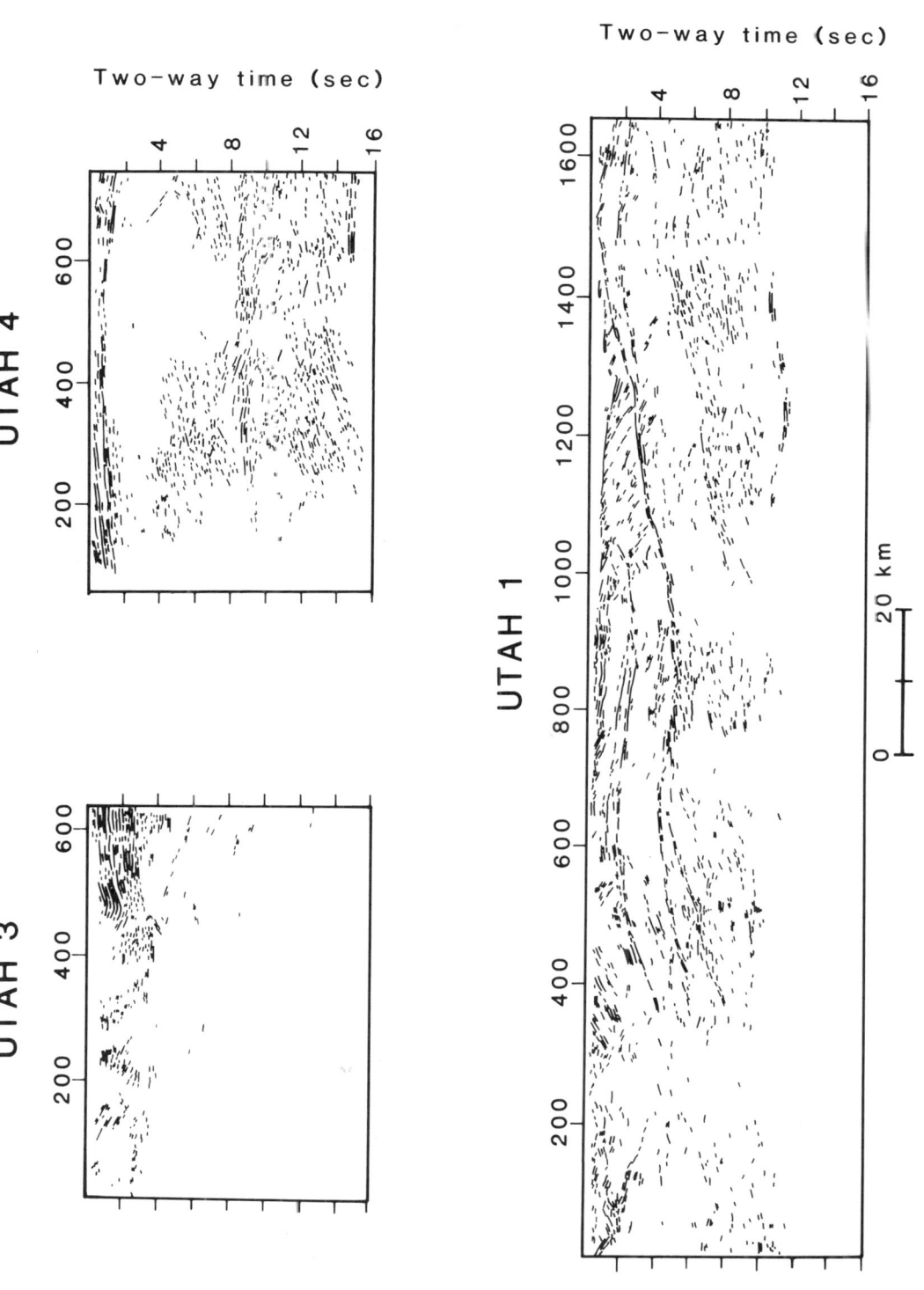

Fig. 4. Composite line drawings of Utah Lines 1,3, and 4. Line drawing of Line 1 from Allmendinger et al. [1983]. Approximately 1:1 at a velocity of 5 km/s.

BALANCED & RESTORED SECTIONS -- WEST-CENTRAL UTAH
SEVIER DESERT DETACHMENT AS A NEW CENOZOIC NORMAL FAULT

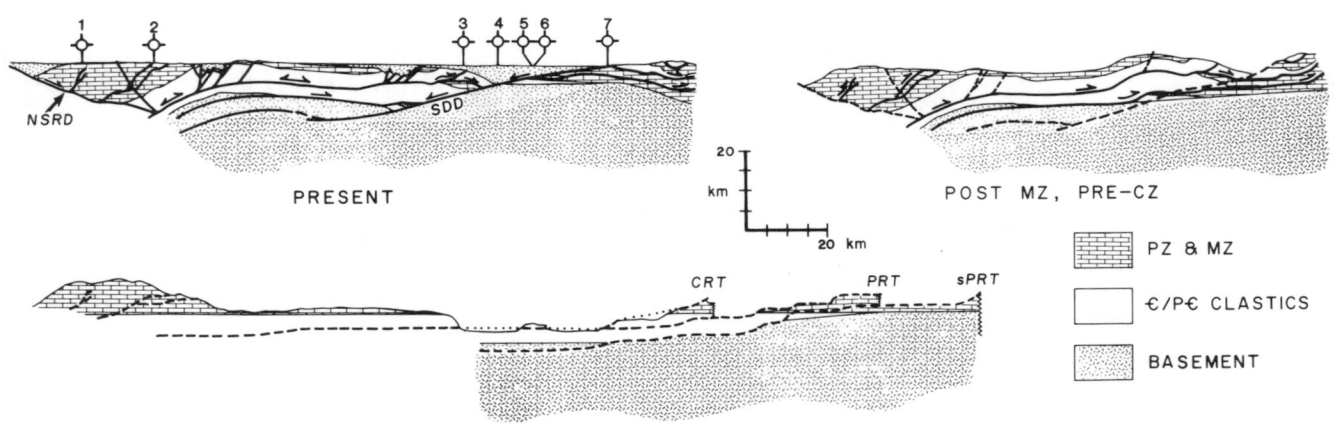

Fig. 5. Geologic section and palinspastic restoration constructed assuming the Sevier Desert detachment is a new Cenozoic normal fault. Section simplified from Sharp [1984]. The restoration indicates about 28 km extension on the Sevier Desert detachment and 35 km overall. About 110 km of Mesozoic-early Cenozoic thrust shortening are interpreted. Wells shown are: 1) Tiger Oil #1 USA Bishop Springs; 2) Cities Services #1 State-AB; 3) Cominco-American #2 Bear River; 4) Gulf-Gronning #1; 5) Argonaut Energy #1 Federal; 6) Arco Oil and Gas Pavant Butte #1; 7) Placid Oil Henley #1.

COCORP Line 1 and also by recent field studies is whether or not low-angle normal faults are reactivated thrust faults or new Cenozoic faults. To test these alternatives, balanced cross-sections were constructed [Sharp, 1984; R. Allmendinger, unpublished data, 1984] following standard techniques outlined by Dahlstrom [1970]. These cross-sections used the constant bed length technique and were constrained by the COCORP data, 10 industry boreholes in the region, and surface geology. Of six alternative sections constructed, two end members are presented here (Figs. 5, 6). The sections proved ambiguous regarding the question of reactivation, but both show that the geometry of the miogeocline may have exerted a fundamental control of the subsequent geometry of both compressional and extensional structures.

In the fully restored state, both end-member sections show a sharp westward increase in thickness of upper Proterozoic and Lower Cambrian clastic strata (Figs. 5,6). These rocks, the initial syn- and early post-rift deposits in the Cordilleran miogeocline, thicken from less than 1 km to more than 5 km in a lateral distance of only 20-30 km. Such gradients generally occur in fault-bounded sedimentary basins, and given the tectonic setting of these deposits, those faults were undoubtedly normal faults. The hingeline occurs at this gradient of rapid thickness change and corresponds to the point west of which the North American craton was significantly thinned during the late Proterozoic rifting that formed the Cordilleran passive margin and miogeocline. The autochthonous position of the hingeline lies directly beneath the Sevier Desert basin.

Despite the differences in initial assumptions, the two cross-sections both show that trajectories of the basal Mesozoic-early Cenozoic thrust fault track along the contact between basement and upper Proterozoic strata where the gradient of thickening was steepest; this is the region where Precambrian normal faults were most likely to be present (Figs. 5,6). Though it cannot be shown that the younger thrust faults reactivated the Precambrian normal faults, the structures and basin geometries of the Proterozoic rifted margin controlled the geometry of the compressional structures. More importantly, surface geology, drilling, and seismic data all indicate that both thrust faults and low-angle normal faults of Mesozoic and Cenozoic age cut deeply into the crust only west of the hingeline.

Summary--The Transition from Cratonic to Orogenic Province

COCORP Lines 1, 3, and 4 in Utah and Line 5 in Nevada, in combination with surface and subsurface data, enable us to construct a simplified structural cross-section of the crust from the northwestern Colorado Plateau and eastern Basin and Range (Fig. 7). This cross-section illustrates

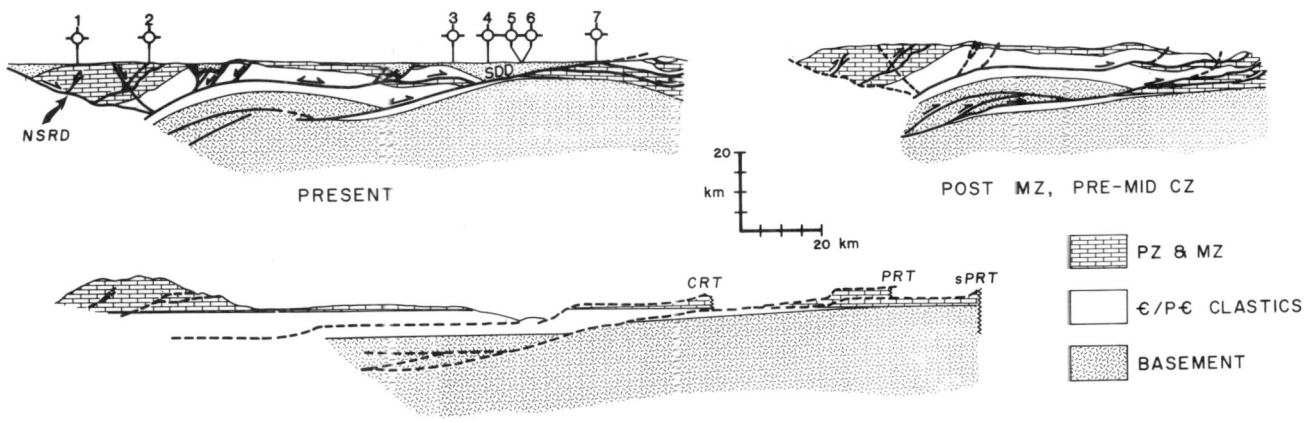

Fig. 6. Geologic section and palinspastic restoration constructed assuming the Cenozoic Sevier Desert detachment reactivates the Pavant Range thrust. The section, simplified from R. Allmendinger [unpublished data], indicates about 40 km of Cenozoic offset on the Sevier Desert detachment and 45 km of extension overall. Section has about 110 km of thrust shortening. Location and wells same as in Figure 5.

the changes in crustal structural geometry that occur across the transition from a province that has largely (although not completely) maintained its cratonal character to a province that has been deformed at times throughout the Phanerozoic by both compressional and extensional tectonism. A number of key features are high-lighted by the cross-section:

1. In the transition zone between the Basin and Range and Colorado Plateau, most of the faulting occurred above the sediment-basement interface. Basement in this zone is gently warped or arched, probably because of loading or unloading due to thrust faulting or low-angle normal faulting (respectively) that occurred in the overlying sedimentary column [see Jordan, 1981; Spencer, 1984]. A notable exception to this conclusion is the normal fault that offsets the basement surface beneath the western edge of the Wasatch Plateau. The origin of this normal fault remains ambiguous.

2. Significant faulting of the basement by thrusts of the Mesozoic-early Cenozoic Cordilleran fold-thrust belt and by younger low angle normal faults in the Basin and Range Province occurred mostly west of the autochthonous position of the late Proterozoic-early Paleozoic Hingeline. Thus, the late Proterozoic rifting event controlled the general character and location of subsequent low-angle thrusting and extension. The Laramide basement thrust faults that occur well east of the Hingeline in Wyoming, Colorado, and eastern Utah are tectonically very different than the Cordilleran fold-thrust belt.

3. The seismic data identify a reflective horizon at 27-30 km depth in both the Basin and Range and the Colorado Plateau, although it cannot be traced continuously between the two provinces. In both provinces, this horizon could be interpreted either as a structural decoupling zone for Mesozoic-Cenozoic structures, or as a younger feature of unknown origin, superimposed on or truncating those structures. In the Basin and Range, the 30 km horizon marks the boundary between reflective and nonreflective lithosphere. In the Colorado Plateau, however, the horizon appears to occur within a zone of otherwise similar seismic fabric above and below it, and the base of the reflective lithosphere is 15-20 km deeper.

4. The general character of the crust in the Colorado Plateau is one that appears dominated by numerous diffractions and reflections that dip in both directions and may be partly out of the plane. In contrast, the eastern Basin and Range crust has a west-dipping or horizontal fabric, although there are also some east-dipping features that are probably related to Cenozoic structures. We interpret this difference in character to represent a dominantly "cratonic" fabric in the Plateau versus a dominantly orogenic fabric in the Basin and Range. The orogenic fabric in the Basin and Range crust is due to Cenozoic, Mesozoic, and perhaps even late Proterozoic extension and compression. The cratonic fabric of the Plateau is related to structural, lithologic or perhaps metamorphic features dominantly of middle Proterozoic or older age. The 27-30 km horizon in the Colorado Plateau appears to be superimposed on and younger than this cratonic fabric. In the Basin

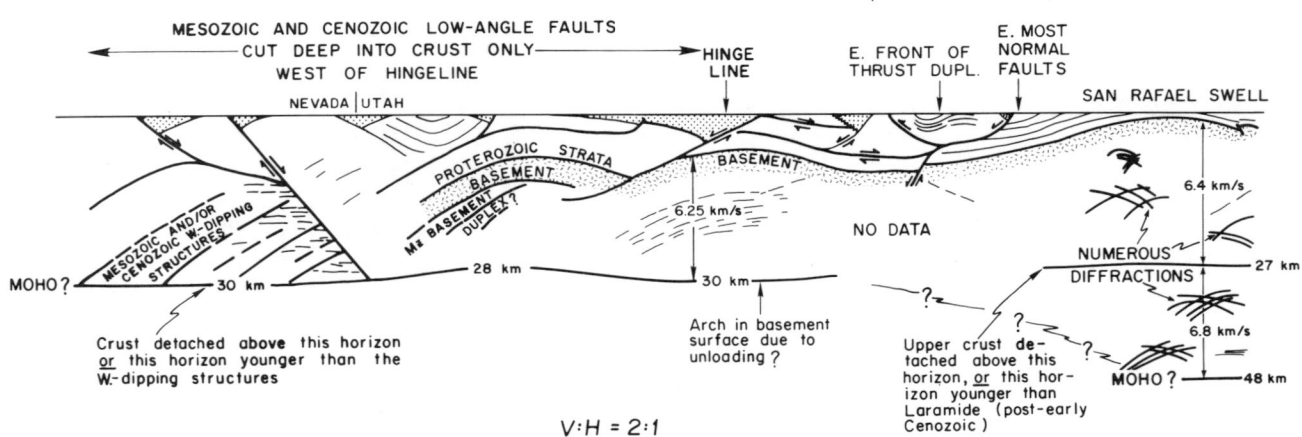

Fig. 7. Simplified and idealized crustal scale structural cross-section from the northwest Colorado Plateau to easternmost Nevada. Velocities beneath San Rafael Swell from Roller [1965] and Prodehl [1979]. Velocity beneath hingeline from Liu et al., [this volume]. Easternmost Nevada from Hauser et al., [unpublished data]. Section has 2:1 vertical exaggeration.

and Range, however, the 27-30 km horizon appears to be spatially and perhaps genetically related to the orogenic fabric; it is probably related to the Cenozoic tectonics of the region and may also be related to the Mesozoic tectonics.

5. The reflective lithosphere is 15-20 km thicker in the Colorado Plateau than in the Basin and Range. When this difference in thickness developed and by what processes is unknown. This is, perhaps, one of the great remaining mysteries of the region.

Acknowledgements. The COCORP Project is sponsored by the National Academy of Sciences and funded by the National Science Foundation (Grants EAR 80-18363 and EAR 82-12445). The field data were collected by Crew 6834 of Petty-Ray, Division of Geosource, Inc., and were processed on the MEGASEIS System at Cornell University. We thank the numerous people connected with the COCORP Project. Discussions with G. Thompson, M.L. Zoback, R.B. Smith, W. Arabaz, B. Wernicke, F. Royse, L. Brown, T. Hauge, and L. Serpa have been particularly helpful. Cornell contribution no. 812.

References

Allmendinger, R. W. and T. E. Jordan, Mesozoic evolution, hinterland of Sevier orogenic belt, Geology, 9, 308-313, 1981.

Allmendinger, R. W., J. W. Sharp, D. Von Tish, L. Serpa, L. Brown, S. Kaufman, J. Oliver, and R. B. Smith, Cenozoic and Mesozoic structure of the eastern Basin and Range Province, Utah, from COCORP seismic reflection data, Geology, 11, 532-536, 1983.

Armin, R. and L. Mayer, Subsidence analysis of the Cordilleran miogeocline: Implications for timing of late Proterozoic rifting and amount of extension, Geology, 11, 702-705, 1983.

Armstrong, R. L., Sevier orogenic belt in Nevada and Utah, Geol. Soc. Am. Bull., 79, 429-458, 1968.

Armstrong, R. L., Low-angle (denudation) faults, hinterland of the Sevier orogenic belt, eastern Nevada and western Utah, Geol. Soc. Am. Bull., 83, 1729-1754, 1972.

Bond, G. and M. Komniz, Construction of tectonic subsidence curves for the early Paleozoic miogeocline, southern Canadian Rocky Mountains: Implications for subsidence mechanisms, age of breakup, and crustal thinning, Geol. Soc. Am. Bull., 95, 155-173, 1984.

Dahlstrom, C. A. D., Structural geology in the eastern margin of the Canadian Rocky Mountains, Bull. Can. Petrol. Geol., 18, 332-406, 1970.

Dickinson, W. R., Paleozoic plate tectonics and the evolution of the Cordilleran continental margin, Pacific Coast Paleogeography Symposium 1, Paleozoic Paleogeography of the Western U.S., Soc. Econ. Paleont. Min., 137-157, 1977.

Grow, J. A., Structure of the Atlantic margin of the United States, in Geology of Passive Continental Margins, Am. Assoc. Petrol. Geol. Course Notes, 19, 3-1 - 3-41, 1981.

Hintze, L. F., Preliminary geologic map of the Conger Mountain Quadrangle, Millard County, Utah, scale 1:48,000, U.S. Geol. Surv. Misc. Field Studies Map, MF-643, 1974.

Hose, R. K., Structural geology of the Confusion Range, west-central Utah, U.S. Geol. Surv. Prof. Paper, 971, 9, 1977.

Jordan, T. E., Thrust loads and foreland basin evolution, Cretaceous, western United States, Am. Assoc. Petrol. Geol. Bull., 65, 2506-2520, 1981.

Kay, M., North American geosynclines, Geol. Soc. Am. Mem. 48, 143 p., 1951.

Lawton, T., Lithofacies correlations within the Upper Cretaceous Indianola Group, central Utah, Utah Geol. Assoc. Pub. 10, 199-213, 1982.

Lindsey, D. A., R. K. Glanzman, C. W. Naeser, and D. J. Nichols, Upper Oliogocene evaporites in basin fill of Sevier Desert region, western Utah, Am. Assoc. Petrol. Geol. Bull., 62, 251-260, 1981.

McDonald, R. E., Tertiary tectonics and sedimentary rocks along the transition, Basin and Range Province to plateau and thrust belt province, Utah, Rocky Mtn. Assoc. Geol. Symp., 281-317, 1976.

Miller, E. L., P. B. Gans, and J. Garing, The Snake Range décollement, an exhumed mid-Tertiary Ductile-Brittle transition, Tectonics, 2, 239-263, 1983.

Misch, P., Regional structural reconnaissance in central-northeastern Nevada and some adjacent areas, Observations and interpretations, Intermount. Assoc. Petrol. Geol., 11th Ann. Field Conf., Guidebook, 17-42, 1960.

Mitchell, G. C., Stratigraphy and regional implications of the Argonaut Energy No. 1 Federal, Millard County, Utah, Rocky Mtn. Assoc. Geol. Symp., 503-514, 1979.

Moulton, F., Lower Mesozoic and upper Paleozoic petroleum potential of the Hingeline area, central Utah, Rocky Mtn. Assoc. Geol. Symp., 87-97, 1975.

Peddy, C., L. Brown, and S. Klemperer, Interpreting the deep structure of rifts with synthetic seismic sections, this volume.

Prodehl, C., Crustal structure of the western United States, U.S. Geol. Surv. Prof. Paper 1037, 74p., 1979.

Ritzma, H., Six Utah "Hingeline" wells, Utah Geol. Assoc. Pub. 2, 75-80, 1972.

Roller, J. C., Crustal structure in the eastern Colorado Plateaus province from seismic-refraction measurements, Bull. Seis. Soc. Am., 55, 107-119, 1965.

Royse, F., M. A. Warner, and D. L. Reese, Thrust belt structural geometry and related stratigraphic problems, Wyoming-Idaho-northern Utah, Rocky Mtn. Assoc. Geol. 1975 Symp., 41-54, 1975.

Sharp, J., West-central Utah: Palinspastically restored sections constrained by COCORP seismic reflection data, M.S. thesis, 60 pp., Cornell Univ., Ithaca, NY, January 1984.

Smith, R. B. and R. L. Bruhn, Intraplate extensional tectonics of the eastern Basin-Range: Inferences on structural style from seismic reflection data, regional tectonics and thermal mechanical models of brittle-ductile deformation, Jour. Geophys. Res., 89, 5735-5762, 1984.

Speed, R., Evolution of the sialic margin in the central western United States, in Studies in continental margin geology, edited by J. S. Watkins and C. L. Drake, Am. Assoc. Petrol. Geol. Mem. 34, 457-468, 1982.

Spencer, J. E., Role of tectonic denudation in warping and uplift of low-angle normal faults, Geology, 12, 95-98, 1984.

Spieker, E. M., Late Mesozoic and Early Cenozoic history of central Utah, U.S. Geol. Surv. Prof. Pap. 205-D, 117-160, 1946.

Sprinkel, D., Twin Creek Limestone-Arapien Shale relations in central Utah, Utah Geol. Assoc. Pub. 10, 169-179, 1982.

Standlee, L., Structure and stratigraphy of Jurassic rocks in central Utah: Their influence on tectonic development of the Cordilleran foreland thrust belt, Rocky Mtn. Assoc. Geol. Symp., 357-382, 1982.

Stewart, J. H., Initial deposits in the Cordilleran geosyncline, evidence of a Late Precambrian (<850 m.y.) continental separation, Geol. Soc. Am. Bull., 83, 1345-1360, 1972.

Stokes, W., What is the Wasatch Line?, Rocky Mtn. Assoc. Geol. Symp., 11-25, 1976.

Villien, A. and R. Kligfield, Structural overview of thrusting and sedimentation in central Utah, in Paleotectonics and sedimentation in the Rocky Mountain Region, United States, edited by J. A. Peterson, and D. L. Smith, Am. Assoc. Petrol. Geol. Mem., in press.

Von Tish, D., R. W. Allmendinger, and J. Sharp, History of Cenozoic extension in the central Sevier Desert, west-central Utah, from COCORP seismic reflection data, Am. Assoc. Petrol. Geol. Bull., 1985 in press.

Zoback, M. L., R. E. Anderson, and G. A. Thompson, Cainozoic evolution of the state of stress and style of tectonism of the Basin and Range province of the western United States, Philo. Trans. Roy. Astron. Soc., A300, 407-434, 1981.

SEISMIC PROFILING OF THE LOWER CRUST: DIXIE VALLEY, NEVADA

David A. Okaya[1]

Department of Geophysics, Stanford University, Stanford, California 94305

Abstract. Shallow industry reflection profiles may be used to image the lower crust provided (1) a Vibroseis (registered trademark of CONOCO, Inc.) source was used, (2) the source signal was an upsweep, composed from low to high frequencies, and (3) the original uncorrelated field gathers are available for extended correlation. Extended correlation increases Vibroseis gather lengths by using a correlation sweep shorter in time than the original field sweep. The correlation sweep may either contain a fixed frequency bandwidth or be a "self-truncating" sweep whose frequency content diminishes with time. Recorrelation of three seismic lines in Dixie Valley using a "self-truncating sweep" converts profile travel-time from 4 seconds to 12 seconds. Conventional CDP stacking of the recorrelated data reveals basin reflections and many short, sub-horizontal reflections present in the intermediate to deep crust. These reflections are present in adjacent recorrelated field gathers, suggesting they are not seismic artifacts. Reflections in the intermediate crust may be due to Mesozoic basinal shelf sediments or their metamorphic equivalent. Lower crustal reflections may be due to some combination of internally layered or extensionally elongated magmatic intrusions, banded or laminated schists or gneisses, or compositionally varied granulites derived from recrystallization of lower crustal mafic rocks. A zone of reflections at the base of the lower crust may be related to the Moho transition zone. Possibilities to account for such a thick or laminated zone of reflections include crystallized layering from multiple melt, layering of partial melt, mantle-derived intrusion, delamination of upper mantle material, cumulate layering, and metasedimentary layering. A sharp drop in the density of these reflections occurs below 10 seconds.

Introduction

The increased use of seismic reflection profiling during the past decade illustrates the difficulty in imaging both the shallow and deep crust. Typically, seismic surveys achieve either deep penetration (time/depth) or detailed resolution at the expense of the other. Recording parameters associated with industry-obtained data yield profiles of high resolution but shallow depths. Recording parameters associated with academic surveys such as those collected by COCORP yield reflections at large travel-times (depths) but at the expense of finer

[1] Now at CALCRUST, Earth Sciences, LBL, Berkeley, CA 94720

structure usually located in the shallow crust. Few crustal surveys have resulted in both deep penetration and high resolution.

Often, one wishes that shallow industry profiles could extend to deep crustal times. In certain cases, shallow seismic reflection data may be processed to greater travel times. Vibroseis (registered trademark of CONOCO, Inc.) field gathers recorded with upsweep sources can be recorrelated with shortened upsweeps which contain lower frequency content. The resulting correlated gathers will extend to deeper two-way travel time. These extended field gathers can then be used to create crustal seismic profiles using conventional CDP stacking methods.

Deep seismic profiles across the Dixie Valley region have been recently recorded [Tom Hauge, personal communication, 1983]. We shall investigate, here, a set of industry profiles which have been reprocessed to greater travel times. This set provides independent imaging of the continental crust below Dixie Valley.

Extraction of Deep Crustal Reflections

Deep crustal reflections may be extracted from shallow industry data provided 1) a Vibroseis source was used, 2) the source signal (sweep) was composed from low to high frequencies, and 3) the original uncorrelated field gathers are available for reprocessing.

Vibroseis Correlation

The energy wavelet from an explosive source is causal, minimum phase, and contains a broad range of frequencies. (Figure 1). The post-correlation wavelet from a Vibroseis source is non-causal, zero-phase and contains a fixed bandwidth of frequencies. A causal, minimum-phase representation of the Vibroseis wavelet is the Vibroseis sweep. This sweep is used in the field, and can be composed from low to high frequencies (upsweep) or from high to low frequencies (downsweep).

Given a Vibroseis sweep of t_{sweep} duration, a reflected event which arrives at t_{event} will end at $t_{event} + t_{sweep}$. Upon correlation, the sweep signal is compressed to a simple wavelet (Figure 2A) which is centered at t_{event}. For field gathers recorded to t_{record}, the last possible reflected event with a fully recorded sweep is determined by

$$t_{event} = t_{record} - t_{sweep}$$

This may be restated as the length of the resulting

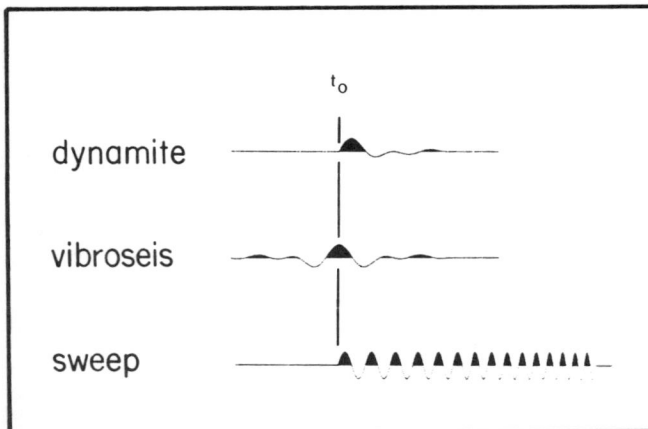

Fig. 1. Energy source wavelets. Dynamite wavelet is causal and has minimum phase. Vibroseis wavelet is non-causal and has zero-phase. The Vibroseis sweep is equivalent in frequency composition to the Vibroseis wavelet but is causal and minimum phase.

correlated seismic traces and hence stacked seismic profile is

$$t_{profile} = t_{record} - t_{sweep} \qquad (1)$$

This simple relation is true for nearly all Vibroseis data; conventional correlation creates data $t_{profile}$ seconds long. For example, Vibroseis data recorded for 16 seconds and shot using a 12 second sweep yields correlated data of 4 seconds duration.

A correlation artifact emerges from sweeps which contain harmonic frequencies. Correlation of frequencies with their harmonic frequenices produce secondary "reflections" which mimic the actual reflections. For downsweeps, these "correlation ghosts" appear in positive time, that is, in the recorded data. For upsweeps, they appear in negative time, before the data start time.

Vibroseis Extended Correlation

For a reflected event which occurs after $t_{profile}$, the tail-end of the reflected sweep arrives after the recording has stopped ($t_{event} + t_{sweep} > t_{record}$). For such an event, only a partial sweep is recorded. Using the previous example, an event present at 6 seconds will be missing the last two seconds of the reflected sweep (6 second arrival time + 12 seconds sweep duration = 18 seconds; this is greater than t_{record} = 16 seconds). Under normal processing conditions, an event at 4 seconds will be correlated, the event at 6 seconds will not.

Extended correlation of data using a shorter correlation sweep may extract the deeper signal. If the tail-end of a correlation sweep is truncated by Δt seconds, the resulting new profile is lengthened by this amount of time. Using relation (1):

$$t_{new-profile} = t_{record} - (t_{sweep} - \Delta t)$$
$$t_{new-profile} = (t_{record} - t_{sweep}) + \Delta t$$
$$t_{new-profile} = t_{profile} + \Delta t \qquad (2)$$

Shortening the correlation sweep by Δt has the effect of dropping the frequencies present in the interval Δt. For an upsweep source signal, the higher frequencies will be dropped from the shortened sweep. For deep reflections, a correlation sweep that is missing the higher frequencies is permissible as the higher frequencies do not necessarily return from the lower crust.

Two types of shortened sweeps are available for extended correlation. A "fixed" sweep has a time duration and frequency bandwidth lower than the original sweep (Figure 2B). This sweep is correlated with the recorded data so that the entire profile has a frequency bandwidth reduced from that of the original profile (compare the correlated traces of Figure 2A and 2B). This method was used by Oliver and Kaufman [1977] to add an extra 2 seconds to 16 second profiles recorded in Hardeman, Texas, and by Finckh et al. [this volume] to produce 11 second profiles in northern Switzerland.

A "self-truncating" sweep allows for maximum use of the original sweep. From $t = 0$ to $t_{profile}$, correlated reflections contain the original frequency bandwidth. Below $t_{profile}$, the reflected sweeps are not fully recorded, thus the sweeps needed for correlation are shorter in length than the original sweep. The net effect in the recorrelated data is that overall frequency content diminishes with later arrival time (Figure 2C). This method is applied to create 12 second industry profiles located in Dixie Valley; Wentworth [personal communication, 1985] describes profiles from the San Joaquin Valley, California, which were similarly extended to 12 seconds.

Using a "fixed" sweep for recorrelation rigidly defines the maximum length of the new seismic data. In theory, a "self-truncating" sweep could identify reflected events arriving as late as t_{record}. However, recorrelated events near t_{record} would be extremely band-limited in frequency content. Recorrelation should be performed to a travel-time as long as sufficient frequencies (at least one harmonic) are preserved to properly define seismic wavelets.

Seismic data shot using a downsweep can be recorrelated. However, as the length of the recorrelated data increases, so does the potential for "correlation ghosts" to appear.

Dixie Valley

Dixie Valley, in west-central Nevada (Figure 3), has become one of the best studied basins in the Basin & Range Province. Detailed geologic mapping of Dixie Valley and the surrounding Stillwater and Clan Alpine Ranges has been performed by Page [1965], Riehle et al. [1972], Speed [1976], and Stewart and Carlson [1976]. Geothermal studies combining hydrology, geochemistry, heat flow, magnetotellurics, resistivity, and well logging were performed by Bell et al. [1980] and by Parchman and Knox [1981]. Geophysical methods used in the Dixie Valley region include seismic refraction [Thompson et

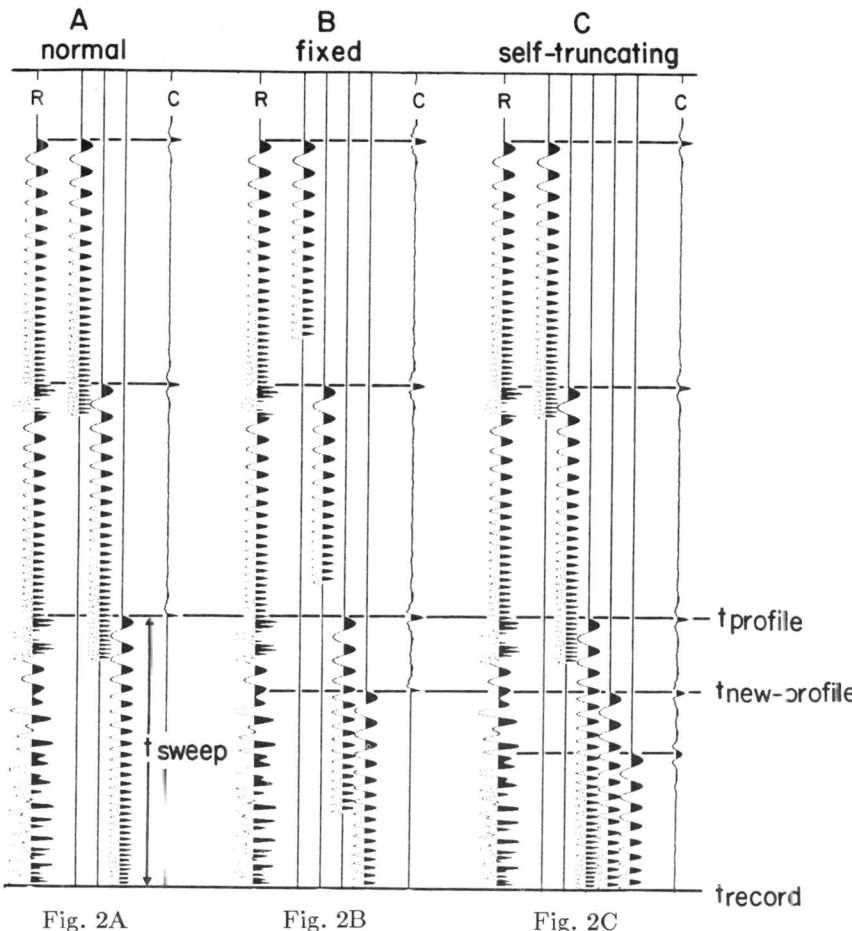

Fig. 2. Correlation of a recorded signal. R represents field recording. C represents result from correlation of field recording with a Vibroseis sweep. (A) Correlation with a normal (full) upsweep. Last possible correlation time using all of the sweep is at $t_{profile}$. Frequencies contained in the sweep are present in the correlated wavelet. Wavelet centered at $t_{profile}$ is missing lower half because correlation stopped at $t_{profile}$. (B) Correlation with a fixed (truncated) sweep which is shorter than the normal sweep. More correlation with the field recording is possible, but at the expense of frequency content of the correlated trace. The correlated wavelets here are broader than those in the normal correlation trace. Wavelet centered at $t_{new-profile}$ is missing lower half because correlation stopped at $t_{new-profile}$. (C) Correlation with a "self-truncating" sweep. Normal correlation occurs until $t_{profile}$. Subsequent correlation is performed with the sweep truncated at t_{record}. Frequency content drops with increasing time (as seen in the deeper correlated wavelets).

al., 1967; Meister, 1967], seismic reflection [Anderson et al., 1983], earthquake studies [Romney, 1957; Zoback et al., 1981], microearthquakes [Stauder and Ryall, 1967; Westphal and Lange, 1967; Ryall and Malone, 1971], aeromagnetics [Smith, 1968], geodetic measurements [Meister et al., 1968], and an integrated study combining seismic reflection, gravity, and synthetic earthquake seismograms [Okaya and Thompson, 1985]. In addition, aerophotographs have been analyzed by Burke [1967] and Whitney [1980].

Seismic Reflection Profiles

During the course of geothermal exploration in northern Dixie Valley, the Southland Royalty Co. obtained twenty-eight kilometers of high-resolution seismic reflection data (Figure 3). The four seismic lines, SRC-1N, SRC-1S, SRC-2 and SRC-3 represent a detailed view of the northwestern side of northern Dixie Valley.

Field and recording parameters include a 12-second, 12-60 Hz linear Vibroseis upsweep, 16-seconds recording

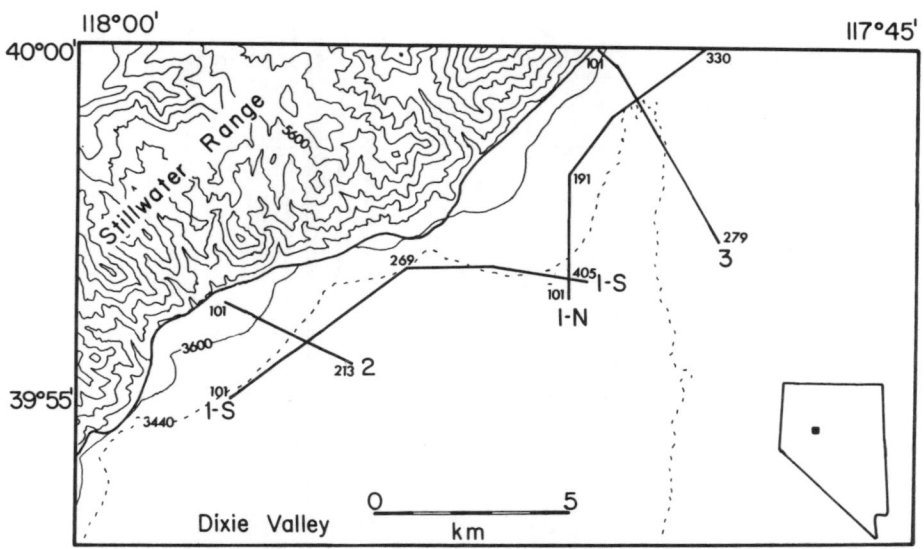

Fig. 3. Location of Southland Royalty Co. seismic lines in Dixie Valley. Small numbers at ends and bends of lines are VP stations. 50 stations approximately 1.7 km. Thick solid line represents location of range-front. Elevation taken from USGS 15° topographic sheet and is in feet; contour interval is 400 feet (121 m). Dashed contour (3440 feet or 1048 m) represents approximate limit of current playa.

time, 96-channel split-spread array with geophone spacing of 33 meters (far offset channel at 1700 meters), shot spacing of 66 meters, and 24-fold maximum CDP coverage.

For three of the four seismic lines, the original 4-second profiles were recorrelated to 12-seconds using a "self-truncating" sweep before conventional CDP-stacking. Frequency content ranges from 12-60 Hz between 0 and 4 seconds, tapering to 12-28 Hz at 12 seconds. Recorrelation and stack of line SRC-1S were performed by Western Geophysical/Denver. Recorrelation and stack of lines SRC-2 and SRC-3 were performed using DIGICON, Inc.'s DISCO software package at the Center for Computational Seismology, Lawrence Berkeley Lab. Processing steps applied to the seismic data include trace editing and mutes, spherical divergence correction, gather balancing, elevation statics, CDP-sort, velocity analysis, normal move-out, remute, stack, and bandpass filtering.

SRC-3

Shallow reflections describing the structural geometry of Dixie Valley have been identified by Okaya and Thompson [1985] using the original, non-recorrelated version of line SRC-3. These reflections, along with deeper events are seen in the recorrelated data.

An ungained, recorrelated field gather from SRC-3 is shown in Figure 4A. Channels (receivers) 1-6 were not recorded. Receivers for channels 7-40 were situated on the alluvial fan in front of the Stillwater Range as indicated by the fast direct arrival (A). Channels 41-96 were situated in the valley flat. Many reflections off basin sediments (B) are recorded in channels 48-96 between 0 - 1 1/2 seconds. Low-frequency surface waves (C) dominate the near-offset amplitudes in the upper section.

A faint arrival occurs at 9 1/2 seconds between channels 15-35 (D). Applying a correction for spherical divergence and then power-balancing the field gather reveals the relative strength of this arrival (Figure 4B). Similar but less-prominent reflections (E) may be found as shallow as 5 1/2 seconds for the same channels. These arrivals are not present after approximately 10 seconds. Dip on these events is due, in part, to velocity effects caused by near-surface inhomogeneity. Lack of lateral continuity of these events across the field gather may be due to changes in either near surface geology or surface coupling of the field instruments. Events D and E are present in over a dozen consecutive field gathers.

The steeper events (F) at 10 1/2 - 12 seconds in the middle channels (near-offsets) mimic the first arrivals and are artifacts of the correlation process. These are removed by CDP stacking.

A conventional stack of the recorrelated field gathers verifies the presence of the events found in the field gather (Figure 5). Within the basin, Okaya and Thompson [1985] have identified the alluvial fan (A in Figure 6), basin sediments (B), and a pre-extension volcanic horizon (G) which forms the basin floor. The top of the volcanic horizon was radiometrically dated in Carson Sink to be 8 (± 4) m.y. [Hastings, 1979]. The near surface position of the range-front fault is shown (H); Okaya and Thompson [1985] interpret the fault to extend to a depth of approximately 15 km. Projection of this fault to depth is difficult in the recorrelated stack.

Event D of the recorrelated field gather becomes enhanced with CDP stacking and appears to be several cycles in duration (wavelength of 230 meters using a

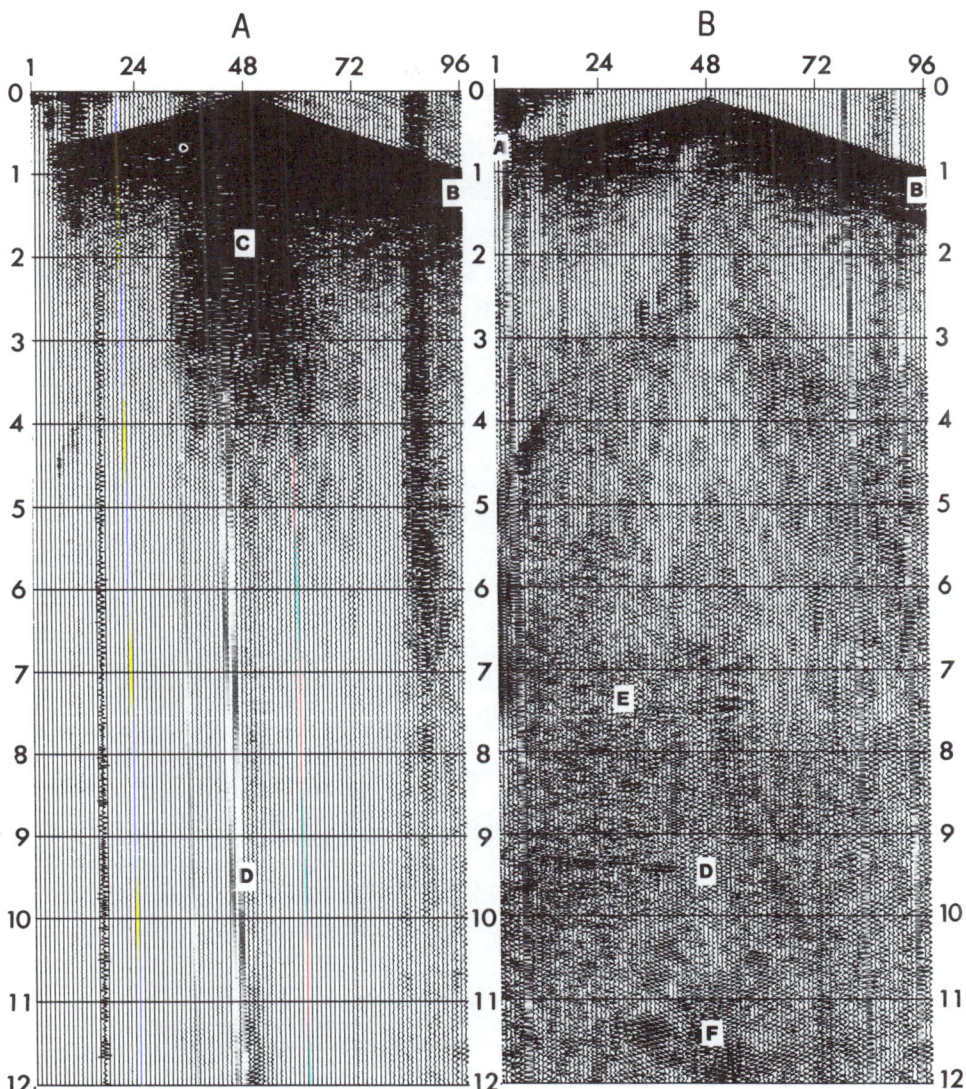

Fig. 4. 96-Channel shot gather from SRC-3 located at VP 154. Near-offset spacing not displayed. Channels 1-6 were not recorded. (A) Ungained but recorrelated to 12 seconds. Reflection symbols: A=alluvial fan; B=basin sediments; C=surface waves; D=deep event. (B) Spherical divergence correction plus trace power balancing. Additional symbols: E=intermediate to deep crustal reflections; F=correlation artifacts.

velocity of 7 km/s). Event D loses continuity across the profile, but similar events can be identified laterally. The magnitude and lateral variation in dip of this horizon mimics the shallow basin, suggesting distortion of a relatively sub-horizontal horizon due to lateral variation in near-surface velocities. As seen in the field gather, events (E) similar to event D are found between 5 1/2 - 9 seconds in the stacked profile. Lack of lateral continuity of these reflections is also found under the basin.

Low frequency events dominate the profile between 2 - 6 seconds. This is possibly due to either noise or the presence of low frequency surface waves (see Figure 4) which may not have been eliminated in CDP stacking.

A few scattered sub-horizontal reflections appear through the lower frequency energy.

A sharp contrast in the density of reflections occurs below event D. While many reflections appear above 10 seconds, few events, if any, arrive after event D.

SRC-2

Line SRC-2 is oblique to the Stillwater Range fault (see Figure 3). The recorrelated stack is shown in Figure 7. Traces with offsets less than 460 m from the source were selectively muted, removing strong, low frequency surface waves between 0 - 5 seconds from the

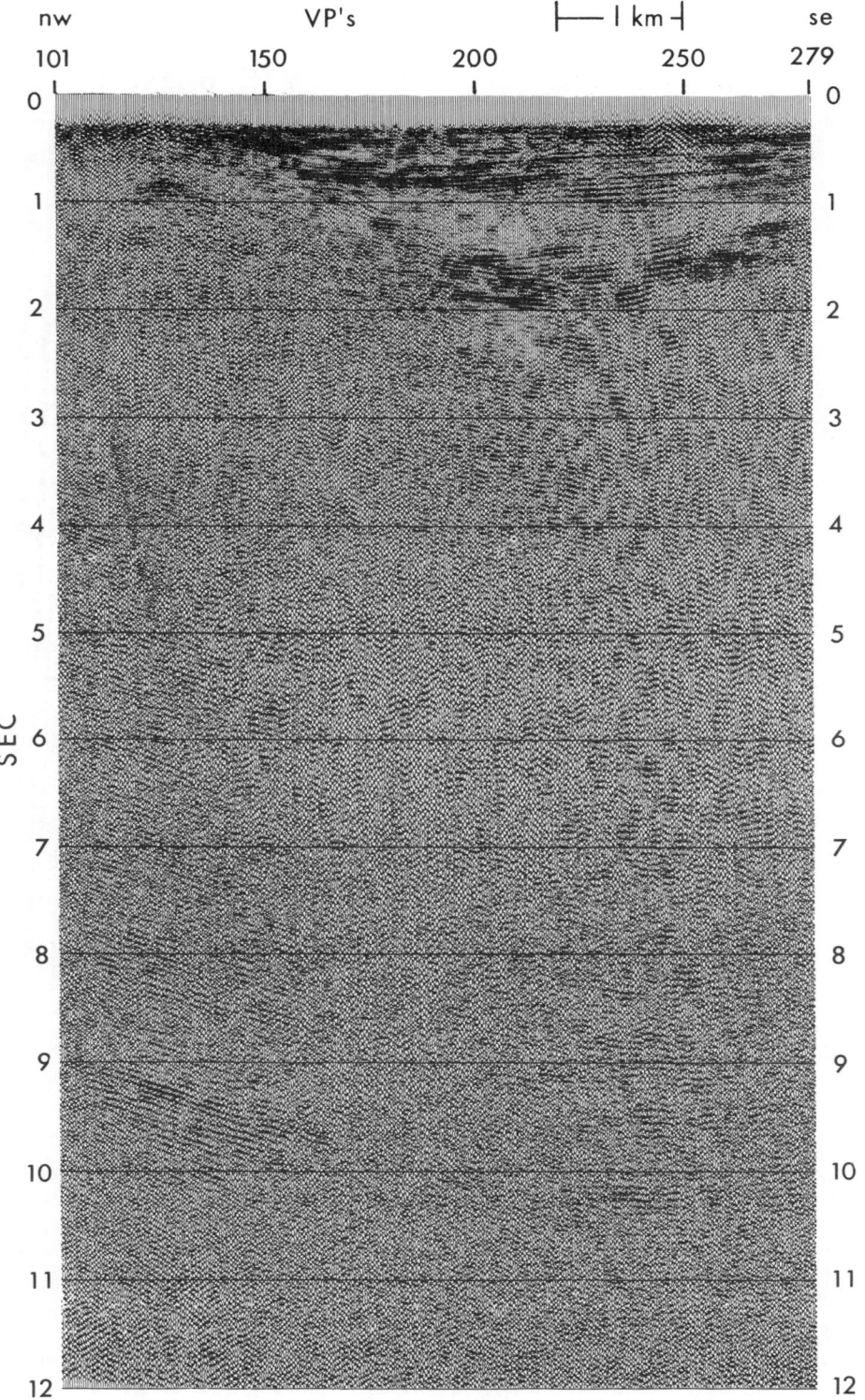

Fig. 5. Recorrelated stack of SRC-3.

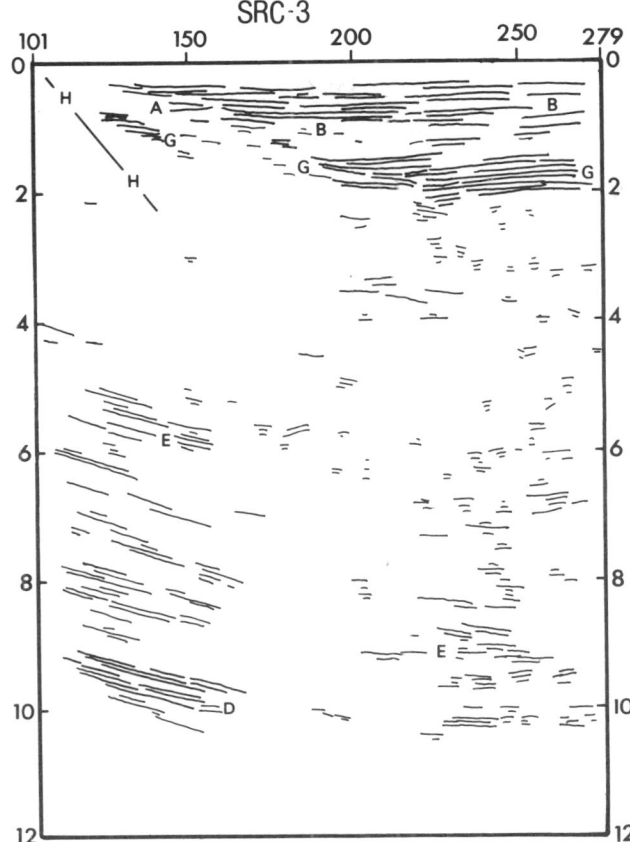

Fig. 6. Line drawing of SRC-3. 50 VP stations approximately 1.7 km. Symbols: A=alluvial fan; B=basin sediments; D=deep event; E=intermediate to deep crustal reflections; G=pre-extension volcanic horizon; H=range-front fault.

stacked profile. Similar to SRC-3, reflections from the alluvial fan, basin sediments, pre-extension volcanic horizon, and the range-front fault may be found in the shallow section [Okaya, 1985]. The stacked profile shows a predominance of sub-horizontal reflections between 3 1/2 to 9 1/2 seconds. Distinct layers of strong amplitude events exist at 5 - 6 and 8 seconds. The slight dip of the reflections mimic the basin floor (velocity pullup/sag). Similar to SRC-3, the sub-horizontal reflections do not occur after a certain travel-time - here at 9 1/2 seconds.

SRC-1S

Unlike SRC-2 and SRC-3, SRC-1S is essentially parallel to the Stillwater Range-front (Figure 3). The seismic line changes orientation twice, covering an alluvial fan in mid-line. Present in the shallow portion of the stacked section (Figure 8) are alluvial fan material, basin sediments, the pre-extension volcanic horizon, and fault reflections [Okaya, 1985; Figure 9].

Intermediate-to-deep reflections are present throughout the section between 3 - 9 1/2 seconds. A

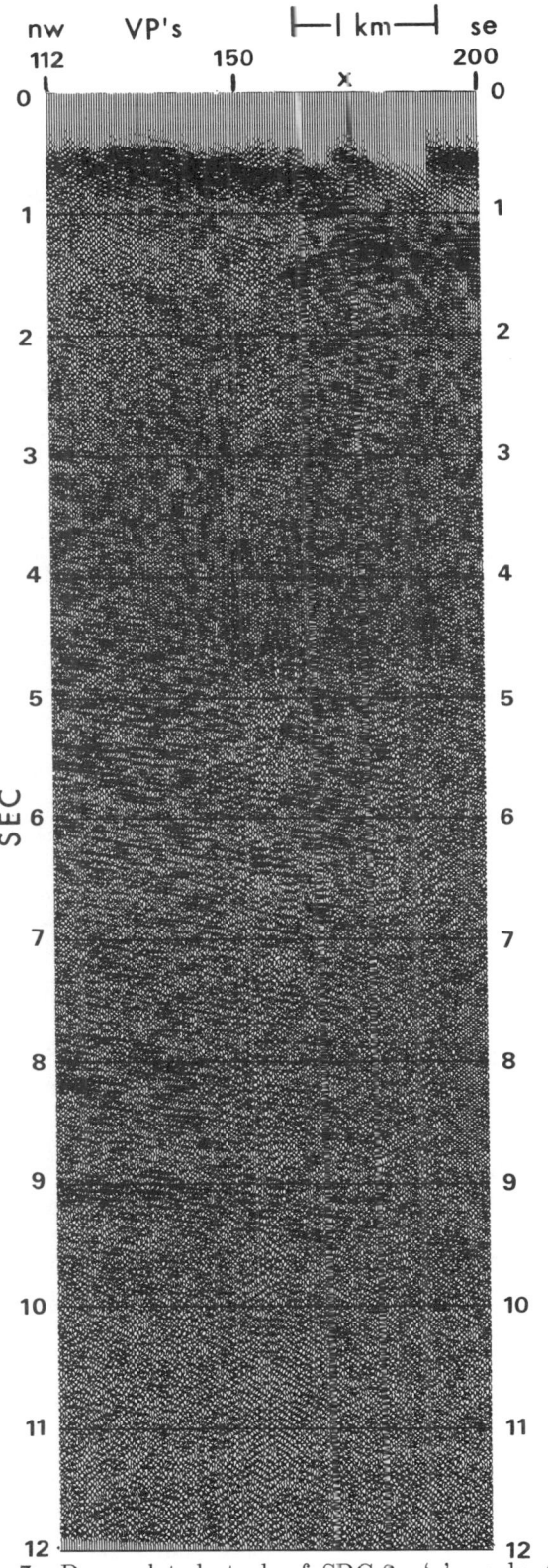

Fig 7. Recorrelated stack of SRC-2. 'x' marks the intersection with SRC-1S.

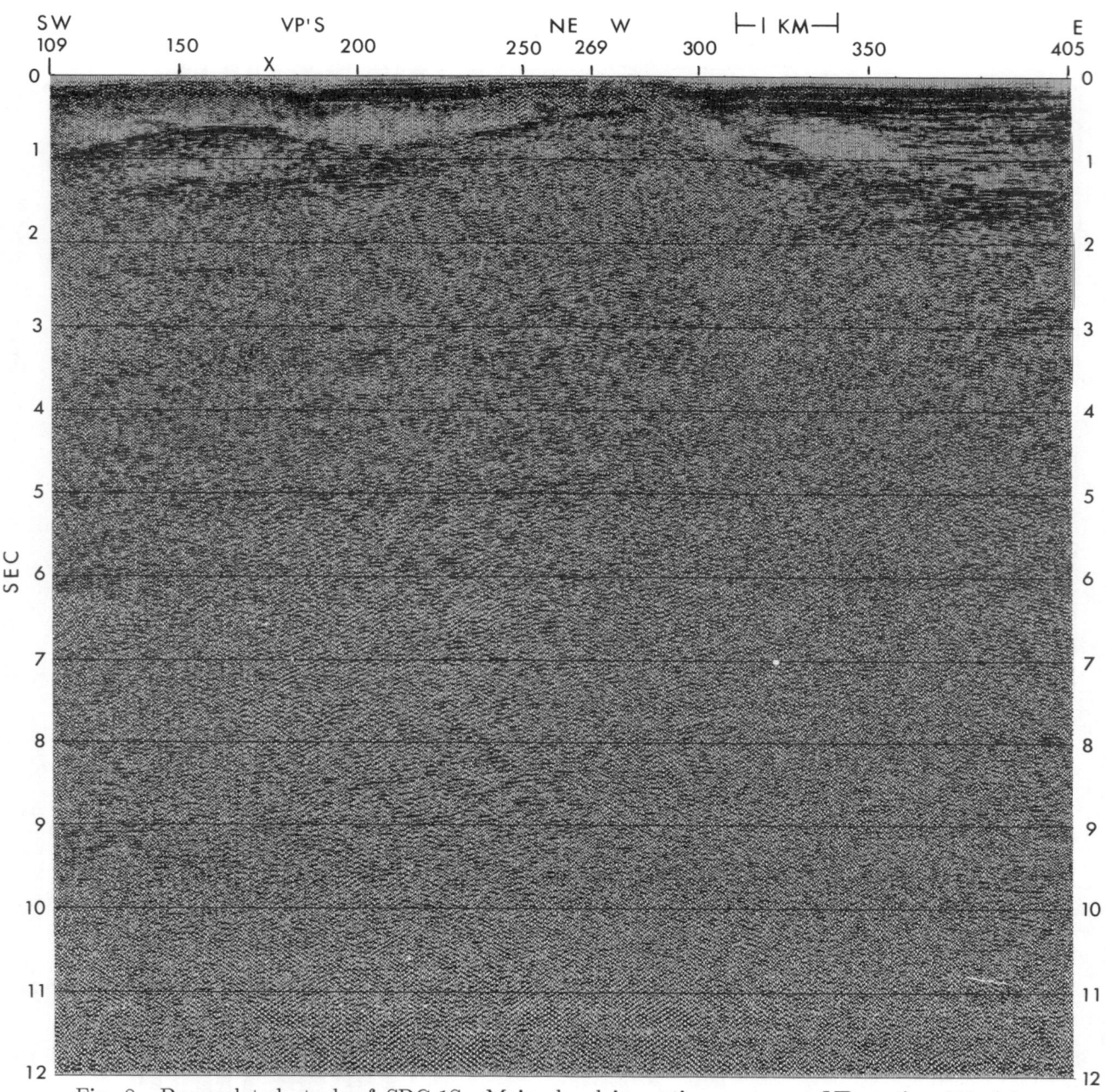

Fig. 8. Recorrelated stack of SRC-1S. Major bend in section occurs at VP station 269. 'x' marks the intersection with SRC-2.

more dense zone of reflections is found at 8 1/2 - 9 1/2 seconds. The reflections are short in lateral extent but are quite numerous. Like SRC-3 and SRC-2, a sudden drop in reflection density occurs below this zone.

A composite line drawing of SRC-2 and SRC-1S is shown in Figure 9. The identification of reflections in SRC-1S was based in part on the application of a tau-p coherency filter to the line (as independently outlined by Kong, et al. [1983] and Harlan, et al. [1984]). The line drawing highlights the spatial distribution of the reflections in the two lines.

Discussion

Extended correlation of the industry seismic lines in Dixie Valley reveal intermediate and deep crustal reflections. Exact geologic interpretation of these reflections is not well-constrained due to limited geolo-

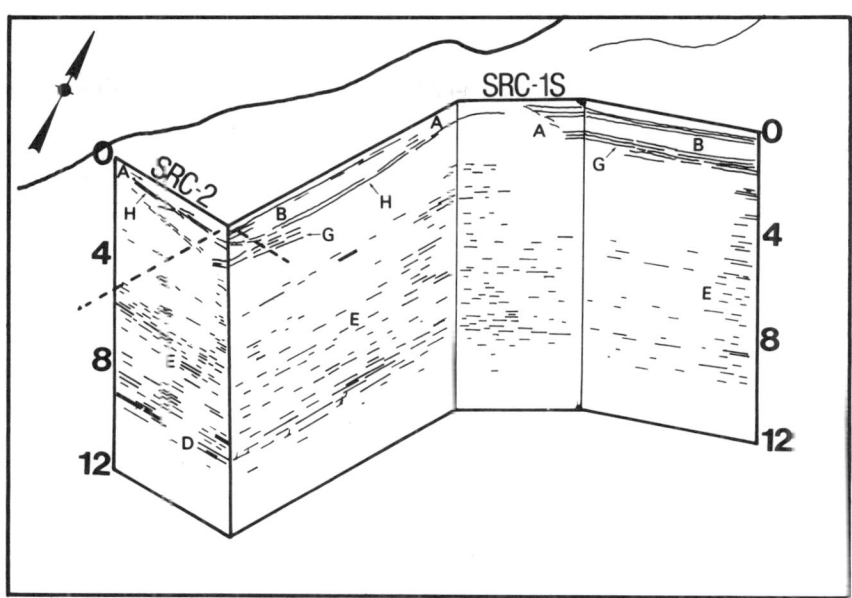

Fig. 9. Composite line drawing of SRC-2 and SRC-1S. Both profiles continue past their intersection point but are not drawn. Thick solid line represents approximate surface location of range-front. Symbols: A=alluvial fan; B=basin sediments; D=deep event; E=intermediate to deep crustal reflections; G=pre-extension volcanic horizon; H=range-front fault. Note abundance of intermediate-to-deep reflections, followed by a marked drop in reflection density below 9-10 seconds.

gic and geophysical information at these depths. Accurate normal move-out velocities are not available due to the relatively short receiver offsets. Other geophysical methods such as gravity or magnetics are not able to resolve individual reflectors. These methods can identify crustal layers; however, the seismic lines are not long enough to develop constrained models. Using information that is known, a generalized discussion regarding the intermediate and lower crust of Dixie Valley is possible.

Shallow reflections in the original three seismic lines were identified by Okaya and Thompson [1985] and Okaya [1985] to define the near-surface basin-range sedimentary and structural geometries. Reflections from alluvial fans, basin sediments, and the basin floor show an asymmetric, sagged valley (Figure 10). Reflections off the range-front fault appear on SRC-1S and SRC-2, which are oblique to the strike of the fault. The position of the range-front fault in SRC-3 is based not on actual fault reflections but is inferred based on changes in seismic character between the basement and the adjacent alluvial fan. The recorrelated seismic lines do not extend far enough across the basin to image the fault at intermediate or lower crustal depths.

Below the basin floor, the lines show a zone of low frequency energy or noise down to 3 - 5 seconds. Structurally below the basin floor is a Jurassic gabbroic intrusion that is several hundred meters thick [Speed, 1976]. Reflections from the intrusion may be obscured by the low frequency energy or the intrusion may be seismically transparent over the lateral extent of the lines. The intermediate crust contains earlier Mesozoic basinal shelf sediments that may be several kilometers thick [Speed, 1976]. Metamorphism of the deeper section of these sediments may have produced schists or gneisses. Lamination or foliation surfaces of the metamorphic rocks could produce some of the reflections (events E) seen at 3 - 6 seconds in the profiles. Alternatively, reflective sub-horizontal segments of highly folded rocks could account for the lateral shortness of many of the reflections.

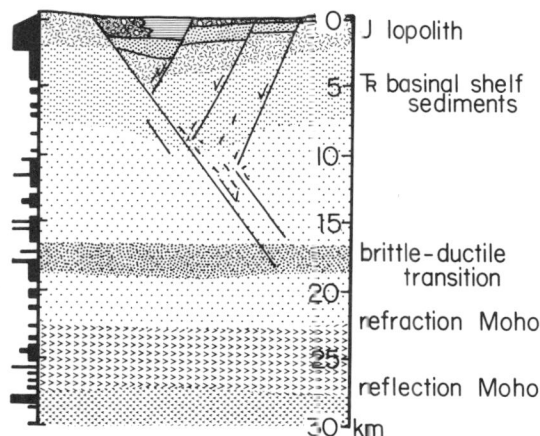

Fig. 10. Schematic cross-section of Dixie Valley. Tick marks on left depth scale represent density and relative amplitude of reflections seen in the seismic profiles.

The brittle-ductile transition zone under Dixie Valley occurs at a depth of about 15 km [Okaya and Thompson, 1985]. However, there is no distinct zone of reflections or a change in seismic character at 5-6 seconds to denote this transition.

Extension in the lower crust may be accomodated by magmatic intrusion [Thompson, 1959; Thompson and Burke, 1973]. These potentially reflective surfaces may intrude horizontally or come to a sub-horizontal position by rotating during extension. Compositional layering, crystal-liquid fractionation, gravitational layering, or magma immiscibility could produce interfaces within the intrusions which may be reflective. Granulites can form from lower crustal mafic rocks due to temperature-pressure conditions [Ringwood and Green, 1966]. Christensen and Fountain [1975] compared the ranges in observed seismic refraction velocities below 10 km to their experimental seismic velocities for various granulite samples. From this, they suggest the composition of the lower crust is quite varied, ranging from hornblende to pryroxene granulites with possible amphibolite facies assemblages; this can be considered as a possibility for the Dixie Valley lower crust.

The increase and successive sharp drop in reflection density at 8 1/2 - 10 seconds in the Dixie Valley profiles may be the seismic response to the lower crust-upper mantle transition zone. Synthetic seismic modeling by Meissner [1967], Fuchs [1969], Clowes and Kanasewich [1970], Daydova [1975], and Hale and Thompson [1982], among others, have suggested a laminated interface for the Moho transition zone. Meissner [1973] suggested crystallized layers from a multiple melt, layering of partial melt, mantle intrusion, or delamination of upper mantle material as possible reasons for a laminated Moho. Hale and Thompson [1982] also suggest cumulate and/or metasedimentary layering. The nearest refraction experiment interpreted the Moho transition zone to be at a depth of 23 km [Stauber and Boore, 1977]. The reflection zone here occurs at approximately 28 km. This discrepancy has been observed in reflection and refraction data throughout the Basin and Range province, although Klemperer (oral communication, 1985) suggests the difference is artificial and is due to the lower resolution of previous refraction surveys.

Conclusions

Shallow industry reflection profiles may be used to image the lower crust provided 1) a Vibroseis source was used, 2) the source signal was composed from low to high frequencies, and 3) the original uncorrelated field records are available for reprocessing. Extended correlation of three seismic lines in Dixie Valley converts profile travel-time from 4 seconds to 12 seconds.

Intermediate to deep crustal sub-horizontal reflections are present between approximately 3 to 10 seconds in the field gathers and the stacked profiles. Mesozoic basinal shelf sediments and a gabbroic intrusion (Speed, 1976) can account for the intermediate crustal reflections. No distinct seismic change occurs at the brittle-ductile transition zone. A sharp drop in the density of reflections occurs below 10 seconds, suggestive of the crust-mantle transition zone.

Various structural or petrological processes exist which could create seismically reflective horizons in the lower crust. Selection of a particular process is not well-constrained due to limited geologic and geophysical information for these depths. Most likely, a combination of processes, some related to crustal extension, can explain the presence of the reflections.

Industry seismic profiles present the basin-range sedimentary and structural pattern of Dixie Valley. Recorrelation of these profiles to greater travel-time yield a glimpse into the intermediate and lower crust. With additional reflection profiling, the lateral extent and reflecting surface of the sub-horizontal reflections found under Dixie Valley can be known. This in turn will lead to our increased understanding about lower crustal structure and the extensional process which formed Dixie Valley.

Acknowledgements. Tsvi Meidav of Trans-Pacific Geothermal Co., provided the original Southland Royalty field tapes. Bill Brown smoothed the road for Western Geophysical/Denver to handle the initial recorrelation; Chris Wheeler processed line SRC-1S. Permission was given by Tom McEvilly to process lines SRC-2 and SRC-3 using the Vax-11/780 owned by the Center for Computational Seismology, Lawrence Berkeley Laboratory. Tom Wood of DIGICON, Inc., spent a week of his time introducing me to the labrynthial world of DISCO; Tom Henyey of U.S.C. provided me with the funds to cover the incurred expenses. Preliminary analysis of the field tapes was performed at Stanford using the facilities of the Stanford Exploration Project; I would like to thank Jon Claerbout and Dave Hale, whose processing programs I had adapted to handle the Dixie Valley data. Discussion with Stew Levin and George Thompson were both invaluable and encouraging. This work was supported by National Science Foundation Grant #EAR81-09294 and #EAR-83-06406.

References

Anderson, R. E., M. L. Zoback, and G. A. Thompson, Implications of selected subsurface data on the structural form and evolution of some basins in the northern Basin and Range province, Nevada and Utah, *Geol. Soc. Am. Bull., 94,* 1055-1072, 1983.

Bell, E. J., M. E. Campana, R. L. Jacobson, L. T. Larson, D. B. Slemmons, T. R. Bard, B. W. Bohm, N. L. Ingraham, R. W. Juncal, and R. A. Whitney, Geothermal reservoir assessment case study, Northern Basin and Range province, Northern Dixie Valley, Nevada, *Report, U.S. Dept. of Energy,* 223 p., Mackay Minerals Research Institute, Univ. Nevada, Reno, 1980.

Burke, D. B., Aerial photography survey of Dixie Valley, Nevada, in Geophysical study of Basin-Range structure, Dixie Valley region, Nevada, edited by G. A. Thompson, et al., 1-36, *U. S. Air Force Cambridge Research Labs. Spec. Rept. 66-848,* Part IV, 1967.

Christensen, N.I. and D.M. Fountain, Constitution of the lower continental crust based on experimental studies of seismic velocities in granulite, *Geol. Soc. Am. Bull., 86,* 227-236, 1975.

Clowes, R.M. and E.R. Kanasewich, Seismic attenuation and the nature of reflecting horizons within the crust, *J. Geophys. Res. 86,* 2545-2555, 1981.

Davydova, N.I., Possibilities of the DSS technique in studying properties of deep-seated seismic interfaces, in Seismic Properties of the Mohorovicic Discontinuity, N.I. Davydova, ed., *National Technical Information Service, U.S. Dept. of Commerce,* Springfield, Va., 4-22, 1975.

Fuchs, K., On the properties of deep crustal reflectors, *Z. Geophys., 35,* 133-149, 1969.

Finckh, P., W. Frei, B. Fuller, R. Johnson, S. Mueller, S. Smithson, C. Sprecher, Detailed crustal strucutres from a seismic reflection survey in Northern Switzerland, this volume.

Hale, L.D. and G.A. Thompson, The seismic reflection character of the continental Mohorovicic discontinuity, *J. Geophys. Res., 87,* 4625-4635, 1982.

Harlan, W. S., J. F. Claerbout, F. Rocca, Signal/noise separation and velocity estimation, *Geophysics, 49,* 1869-1880, 1984.

Hastings, D. D., Results of exploratory drilling, northern Fallon Basin, Western Nevada, in *Basin and Range Symposium,* edited by G. W. Newman and H. D. Good, pp 515-522, Rocky Mt. Assoc. Geol. - Utah Geol. Assoc., Denver, 1979.

Kong, S., R. A. Phinney, and K. Roy-Chowdhury, A non-linear signal detector for noisy sections [abstr], *EOS Trans. AGU, 64,* 769, 1983.

Meissner, R., Exploring deep interfaces by seismic wide angle measurements, *Geophys. Prosp., 15,* 598-617, 1967.

Meissner, R., The "Moho" as a transition zone, *Geophys. Surveys, 1,* 195-216, 1973.

Meister, L. J., Seismic refraction study of Dixie Valley, Nevada, in Geophysical study of Basin-Range structure, Dixie Valley region, Nevada, edited by G. A. Thompson, et al., 1-72, *U. S. Air Force Cambridge Research Labs. Spec. Rept. 66-848,* Part I, 1967.

Meister, L. J., R. O. Burford, G. A. Thompson, and R. L. Kovach, Surface strain changes and strain energy release in the Dixie Valley - Fairview Peak area, Nevada, *Jour. Geoph. Res., 73,* 5981-5994, 1968.

Okaya, D. A., Seismic reflections off high-angle faults in the Basin and Range province, in Seismic reflection studies in the Basin and Range Province and Panama, Ph.D. thesis, 79-102, Stanford Univ., Stanford, 1985.

Okaya, D. A. and G. A. Thompson, Geometry of Cenozoic extensional faulting: Dixie Valley, Nevada, *Tectonics, 4,* 107-125, 1985.

Oliver, J. and S. Kaufman, Complexities of the deep basement from seismic reflection profiling, in The Earth's Crust, J. G. Heacock, ed., *Am Geophys. Union, Geophys. Monograph 20,* 243-253, 1977.

Page, B. M., Preliminary geologic map of a part of the Stillwater Range, Churchill County, Nevada, *Map 28,* Nev. Bur. Mines and Geol., 1965.

Parchman, W. L., Jr., and J. W. Knox, Exploration for geothermal resources in Dixie Valley, Nevada: a case history, *Geotherm Resour. Counc. Bull., 10,* 3-8, June 1981.

Riehle, J. R., E. H. McKee, and R. C. Speed, Tertiary volcanic center, west-central Nevada, *Geol. Soc. Am. Bull., 83,* 1383-1396, 1972.

Ringwood, A. E., and D. H. Green, Petrological nature of the stable continental crust, in The Earth Beneath the Continents, J. Steinhart and T. Smith, eds., *Geophys. Monograph Ser. 10,* 611-619, AGU, Washington, D.C., 1966.

Romney, C., Seismic waves from the Dixie Valley - Fairview Peak earthquakes, *Bull. Seismol. Soc. Am., 47,* 301-319, 1957.

Ryall, A., and S. D. Malone, Earthquake distribution and mechanism of faulting in the Rainbow Mountain - Dixie Valley - Fairview Peak area, Central Nevada, *Jour. Geoph. Res., 76,* 7421-7428, 1971.

Smith, T. E., Aeromagnetic measurements in Dixie Valley, Nevada; implications on Basin-Range structure, *Jour. Geoph. Res., 73,* 1321-1331, 1968.

Speed, R., Geology of Humboldt Lopolith & vicinity, *Geol. Soc. Am. MC-14,* 1976.

Stauber, D. and D. M. Boore, Crustal structure in the Battle Mountain heat flow high from seismic refraction experiments [abstr], *EOS Trans. AGU, 58,* 1238, 1977.

Stauder, W. and A. Ryall, Spatial distribution and source mechanism of microearthquakes in central Nevada, *Seismol. Soc. Am. Bull. 57,* 1317-1345, 1967.

Stewart, J. H. and J. E. Carlson, Geologic map of North-Central Nevada, *Map 50,* Nevada Bureau of Mines and Geology, 1976.

Thompson, G. A., Gravity measurements between Hazen and Austin, Nevada: a study of basin-range structure, *Jour. Geophys. Res., 64,* 217-229, 1959.

Thompson, G. A., and D. B. Burke, Rate and direction of spreading in Dixie Valley, Basin and Range province, Nevada, *Geol. Soc. Am. Bull., 84,* 627-632, 1973.

Thompson, G. A., L. J. Meister, A. T. Herring, T. E. Smith, D. B. Burke, R. L. Kovach, R. O. Burford, I. A. Salehi, and M. D. Wood, Geophysical study of Basin-Range structure, Dixie Valley region, Nevada, *U. S. Air Force Cambridge Research Labs. Spec. Rept. 66-848.* 244 p., 1967.

Westphal, W. H. and A. L. Lange, Local seismic monitoring- Fairview Peak area, Nevada, *Seismol. Soc. America Bull., 57,* 1279-1298, 1967.

Whitney, R. A., Structural-tectonic analysis of Northern Dixie Valley, Nevada, M.S. thesis, 65 pp., Univ. Nevada, Reno, 1980.

Zoback, M.L., R. E. Anderson, and G. A. Thompson, Cainozoic evolution of the state of stress and style of tectonism of the Basin and Range province of the western United States, *Phil. Trans. R. Soc. London A 300,* 407-434, 1981.

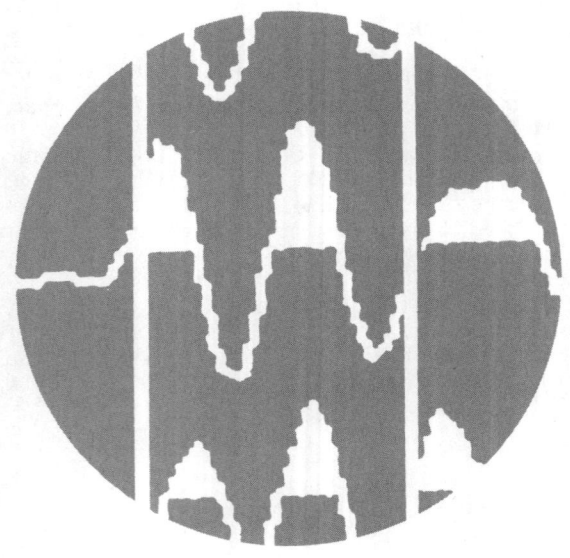

REFLECTION PROFILES FROM THE SNAKE RANGE METAMORPHIC CORE COMPLEX: A WINDOW INTO THE MID-CRUST

Jill McCarthy

Geology Department, Stanford University, Stanford, California 94305

Abstract. The northern Snake Range (NSR) metamorphic core complex in eastern Nevada is characterized by a detached and distended cover of Paleozoic strata overlying ductilely-strained and metamorphosed Precambrian sediments and Mesozoic and Tertiary(?) plutons. A gently-dipping to subhorizontal zone of detachment (the northern Snake Range decollement (NSRD)) separates these rheologically contrasting units. This decollement is best developed on the eastern flank of the range where lower plate ductile strain is greatest. Seismic profiles from the Consortium for Continental Reflection Profiling (COCORP) and Sohio Petroleum Company have traced this shallowly-dipping (5-10°) reflecting horizon over 10 km to the east beneath the Confusion Range, where it dies out at a two-way travel time of 3.0 seconds. Along the western flank of the range, however, a 128-fold sign bit seismic line shot across Spring Valley between the Snake and Schell Creek Ranges was unable to image the westward continuation of the NSRD for any appreciable distance. The absence of ductile deformation exposed in the Schell Creek Range and the correlation of the NSRD with a purely brittle fault in this region suggest that the westward-disappearance of the NSRD as a major reflecting horizon is due to a decrease in ductile strain to the west, away from the Snake Range. Although a westward transition of the rheological character of the NSRD from completely ductile deformation on the east flank of the NSR to completely brittle deformation on the west flank is compatible with a low-angle zone of simple shear rooted to the east into the lower crust or upper mantle, a major shear zone with displacements on the order of 60-100 km should be laterally more extensive and resolvable at greater crustal depths than those imaged on the seismic reflection profiles.

A highly-reflective middle and lower crust has also been imaged on COCORP and Sohio Petroleum seismic reflection profiles beneath the NSR. These strong, laminated reflections which extend from 4.0 seconds two-way travel time to the Moho are believed to represent a compositionally-layered and structurally-deformed fabric imposed on the middle and lower crust during Tertiary extension. If true, this implies that extension in the NSR has not been localized along the NSRD, but has been distributed throughout the entire crust down to the Moho.

Introduction

Metamorphic core complexes within the Basin and Range Province (BRP) of the North American Cordillera have been recognized as uplifted exposures of a regional belt of deep-seated metamorphism and crustal deformation (Fig. 1; Coney, 1980). As such, these core complexes serve as a window into the mid-crust, exposing rocks uplifted from depths of as much as 5-15 km. Seismic reflection studies across these uplifted areas reveal an abundance of layered, high-amplitude reflections extending throughout the crust down to the crust-mantle boundary. This reflectivity is typically confined to the lower crust in other less highly-extended regions of the BRP, but rises beneath metamorphic core complexes and can be correlated to rocks exposed in the field or sampled in deep drill holes. For this reason, studies of metamorphic core complexes provide important constraints on the mechanisms of continental extension in the deep crust. The seismic reflection data from the northern Snake Range metamorphic core complex are reviewed with this in mind, and are discussed in light of recent geologic studies completed by Miller et al. (1983), Gans and Miller (1983), and Gans et al. (1985) in eastern-most Nevada.

Background

The northern Snake Range of eastern Nevada (NSR), like most core complexes of the western Cordilleran (see Coney, 1980; Crittenden, 1980; and Armstrong, 1982 for a complete summary) is a domal uplift of detached and distended Paleozoic strata overlying ductilely-strained and metamorphosed Cambrian and Precambrian sediments and Mesozoic and Tertiary(?) plutons. A gently-dipping to subhorizontal zone of detachment (the northern Snake Range decollement (NSRD)) separates these rheologically-contrasting units and is stratigraphically-controlled across the southern portion of the northern Snake Range. Imbricate normal faults within the originally 7-km-thick upper plate have extended the Paleozoic section by 400-500% (Gans and Miller, 1983; Miller et al., 1983). These faults terminate in, but do not cut, the NSRD. Rocks In the lower plate are ductilely thinned to one-sixth to one-tenth their original thickness, while older (Mesozoic?) fabrics are cut or tran-

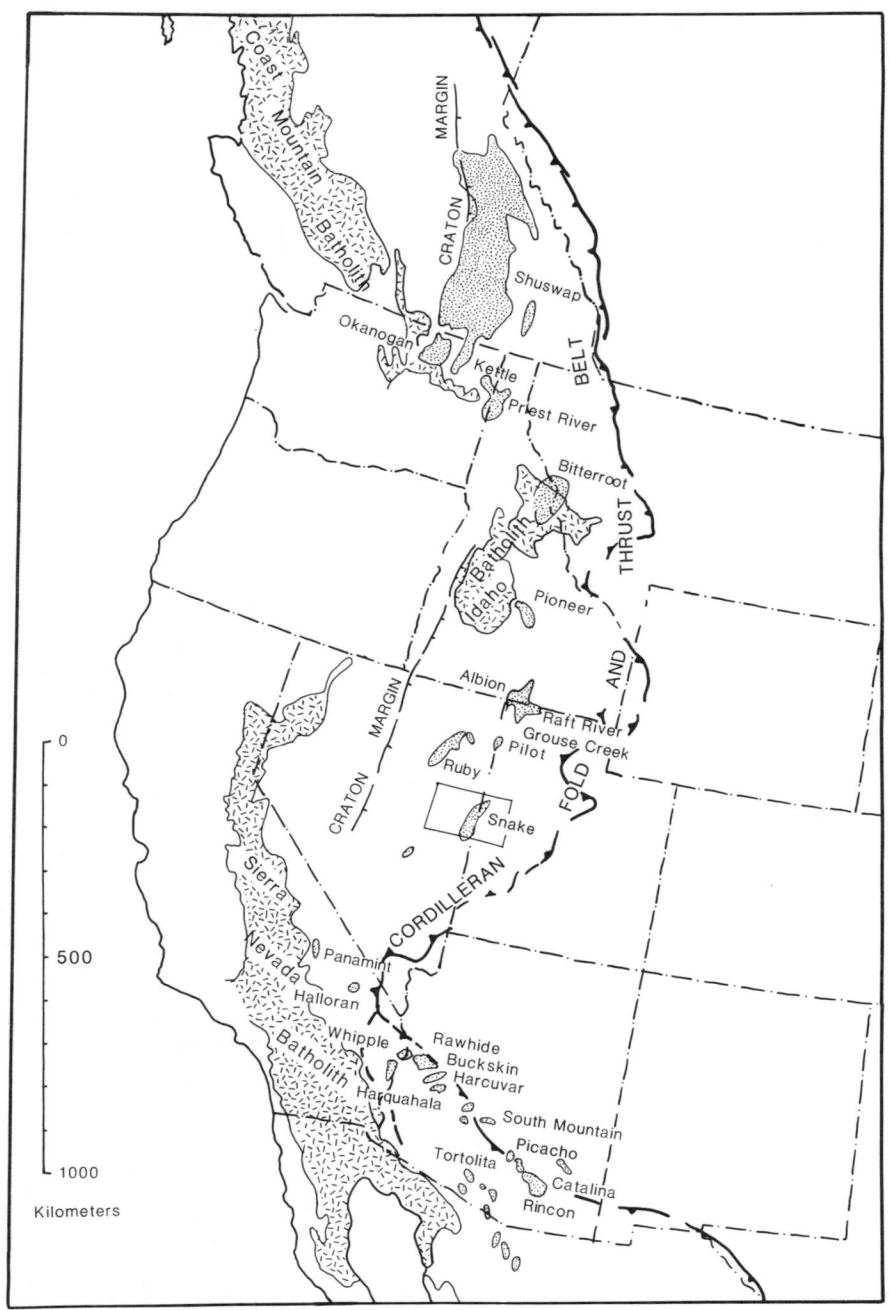

Fig. 1. Distribution of metamorphic core complexes (dotted) relative to Mesozoic batholiths (turkey track) and thrusts of the Cordilleran foreland fold and thrust belt. The box encompassing the Snake Range area has been enlarged in Figure 2. (Adapted from Armstrong, 1982).

sposed by a younger, subhorizontal fabric whose orientation is parallel to the extension direction (N55W-S55E) in the upper plate (Gans and Miller, 1983; Miller et al., 1983).

Snake Range Seismic Reflection Profiles

Until recently our knowledge of the NSR metamorphic core complex and its relation to present-

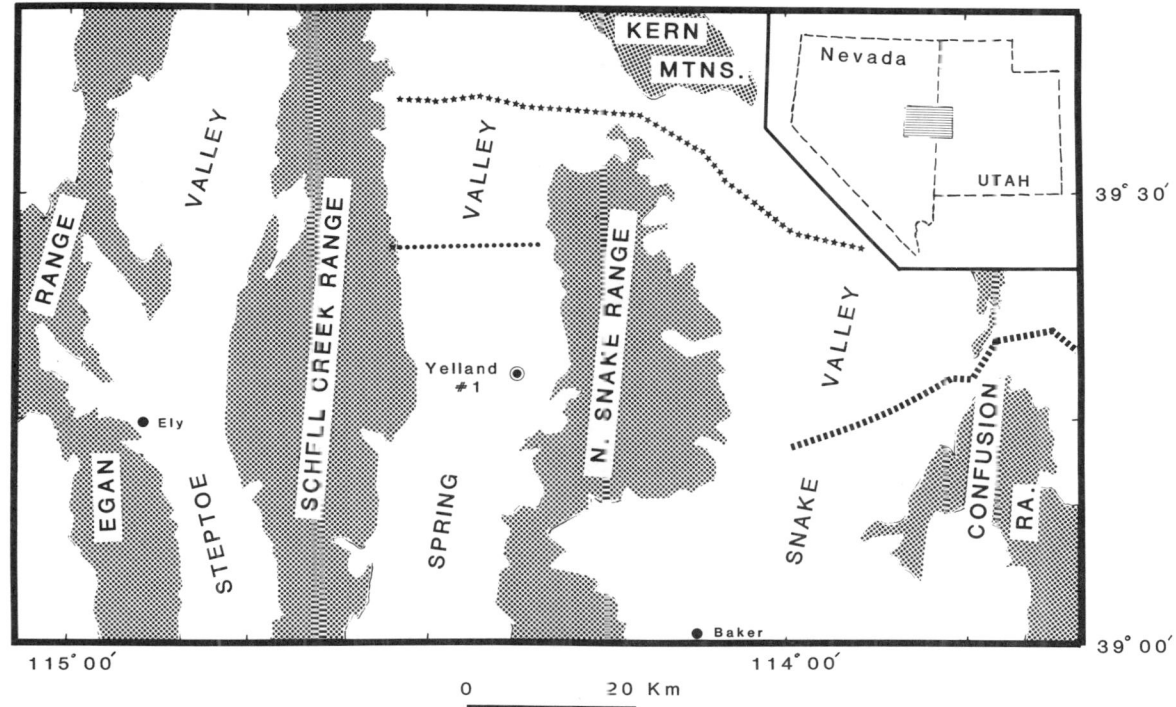

Fig. 2. Location map of seismic reflection profiles from the Snake Range region in eastern Nevada. The COCORP Utah line 1 is dashed, the COCORP Nevada line 5 is starred, and the Sohio Petroleum Spring Valley sign bit line is dotted. The location of the second Sohio profile, recorded on the eastern flank of the Snake Range (Fig. 3), is proprietary and cannot be shown. The Yelland Well #1 is also marked.

day properties of the deep-crust had been restricted to the limited exposures within the range. Recently acquired geophysical information, however, permits a deeper look into the crustal fabrics and geologic structures that underlie the range and their relationship to crustal extension. Four seismic reflection profiles -- two from COCORP and two collected by Sohio Petroleum Company -- provide both detailed, shallow subsurface information and important deep-crustal information from the Snake Range and vicinity (see Fig. 2 for location). The results of these seismic reflection studies are discussed here, first with regard to the behavior of the NSRD away from the Snake Range, and then with regard to the deeper reflections beneath the range.

The East-West Extent of the NSRD

The NSRD is best developed on the eastern flank of the NSR, where ductile strain in the lower plate is at a maximum and where Cambrian and Precambrian strata have been thinned to one-sixth to one-tenth their original thickness. Here the decollement is a gently east-dipping surface projecting beneath Snake Valley at 5-10°. A 60-fold dynamite reflection profile, collected by Sohio Petroleum Company in 1981, traces this reflecting horizon laterally from its eastern-most surface exposure, 10 km down-dip beneath Snake Valley (Fig. 3; note that this survey is proprietary and its location cannot be show on Fig. 2). The decollement reflection imaged on this profile is a continuous feature that ultimately dies out beneath the deepest portion of Snake Valley. The decrease in reflectivity of the NSRD to the east is coincident with a change in the seismic source used in the field (see Fig. 3 for an explanation). A second seismic profile (Utah line 1, Fig. 4, recorded in 1982 by COCORP, begins further to the south in Snake Valley (Fig. 2) and extends east 120 km. Like the Sohio profile, the COCORP line also imaged a prominent east-dipping reflector that has been correlated with the NSRD (Fig. 4; Allmendinger et al., 1983). The COCORP survey, however, was successful in imaging this feature as much as 10 km further to the east beneath the Confusion Range where it ultimately dies out at approximately 3.0 seconds two-way travel time. The ultimate fate of the NSRD beneath the Confusion Range is unclear. The reflectivity of the decollement dies out gradually to the east on the COCORP profile without any apparent structural break. Poor signal returns and/or bends in the survey line may account for these decreasing amplitudes, but other reflecting horizons are imaged beneath the Confusion Range (Allmendinger et al., 1983) suggesting that the NSRD

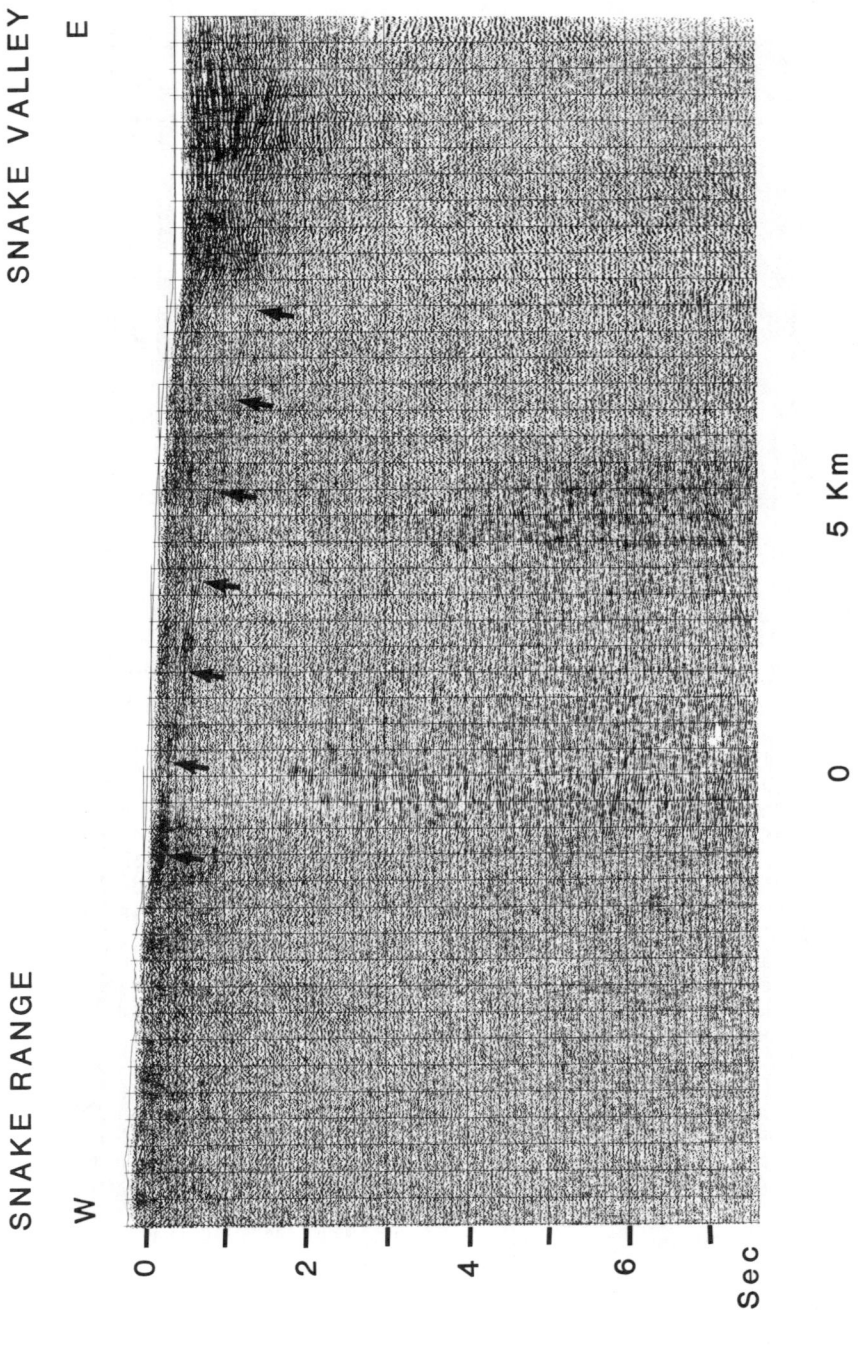

Fig. 3. Sohio Petroleum unmigrated profile recorded off the eastern flank of the northern Snake Range. The data are 60-fold and the vertical exaggeration is 1.2:1 at 6.0 km/sec. The dip on the NSRD is approximately 4–8°. The line was recorded with a variable source. The eastern and western thirds of the line were shot with a surface array while the more-reflective middle third was recorded with a subsurface dynamite source. The dynamite source was far more effective in transmitting energy into the ground, thus accounting for the apparent increase in the reflectivity in that region.

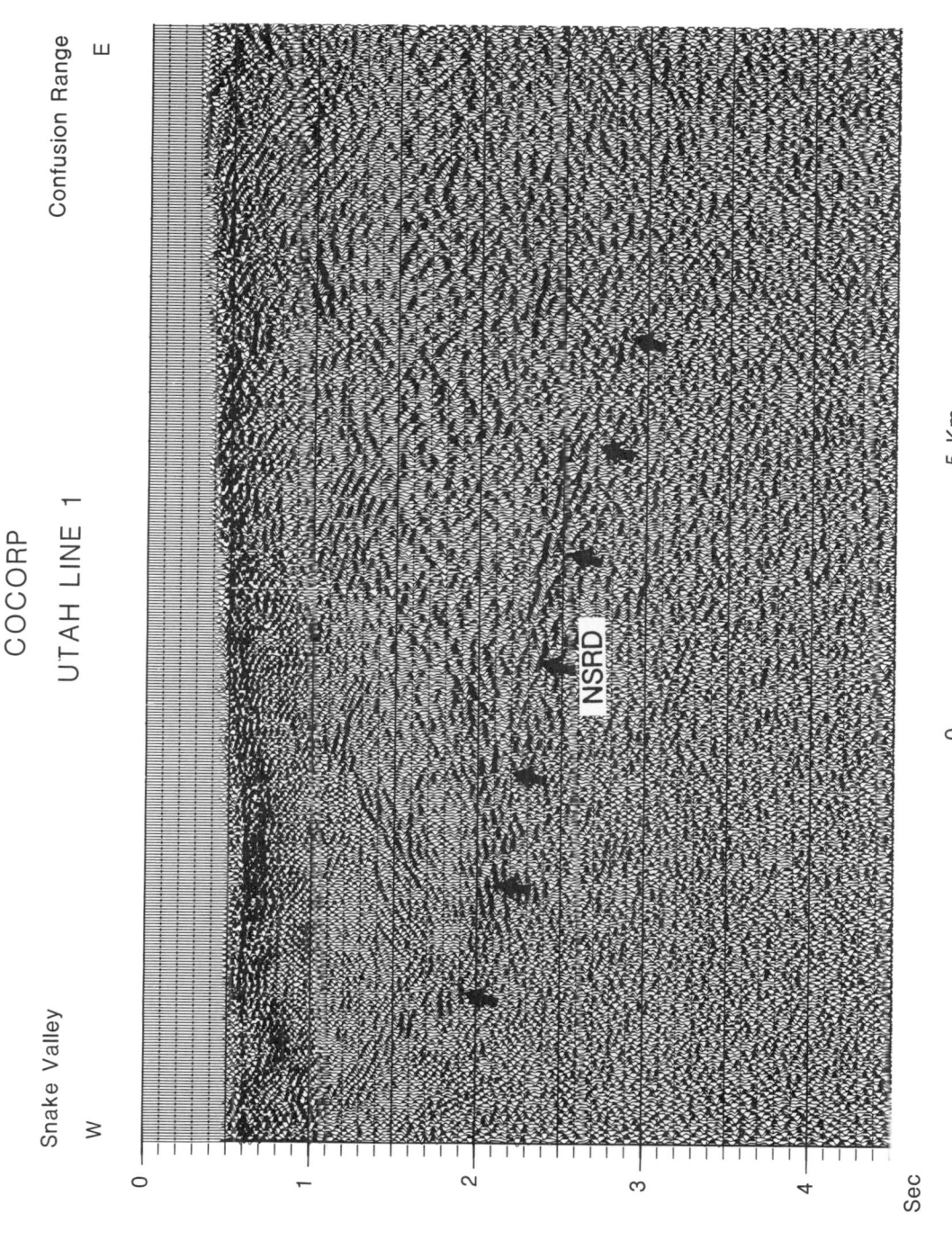

Fig. 4. COCORP unmigrated Vibroseis (registered trademark of Conoco Inc.) profile recorded over Snake Valley and the Confusion Range. The data are 48-fold and the vertical exaggeration is approximately 1:1 at 6.0 km/sec. The dip on the NSRD is 6 to 9°.

THE SNAKE RANGE METAMORPHIC CORE COMPLEX 285

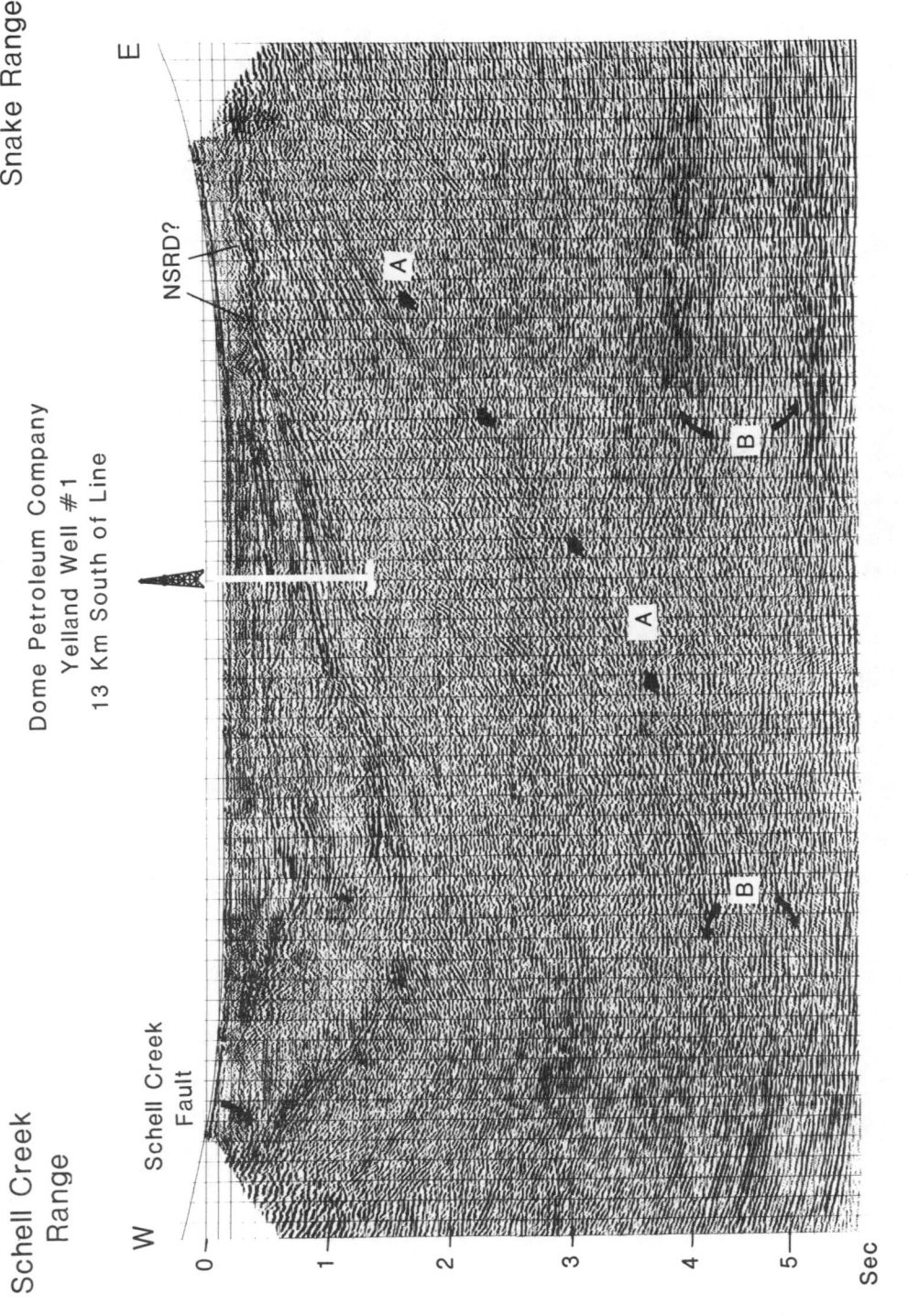

Fig. 5. Frequency-spatial (F-K) wave number migration of Sohio Petroleum's sign bit profile recorded across Spring Valley. The data are 128-fold and the vertical exaggeration at 6.0 km/sec is approximately 1.6:1. The location of Dome Petroleum's Yelland Well #1 is marked. Reflections A and B are discussed in the text. Note that the deep, laminated reflections labeled B are strongest beneath both the west and the east ends of the line and decrease in amplitude towards the center, where attenuation due to the thick basin fill is greatest. In addition, the reflective zone is gently dipping on the west end of the section due to a velocity pull-down associated with the eastward-thickening wedge of overlying basin sediments.

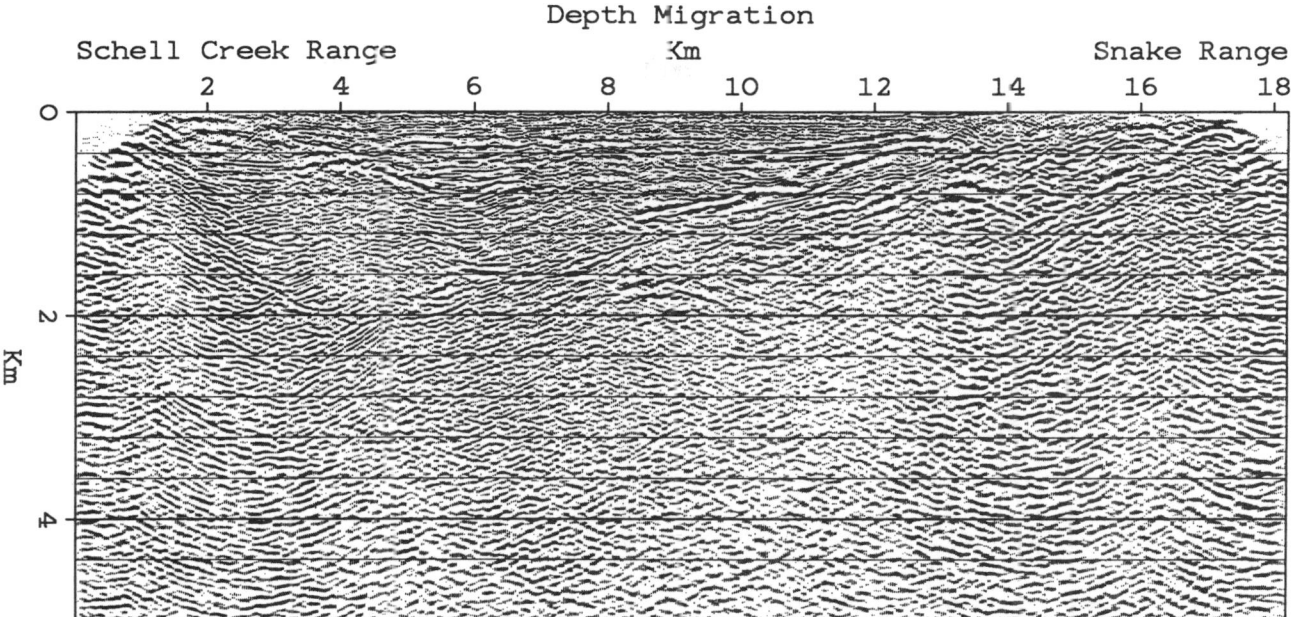

Fig. 6. Finite difference depth migration of the Sohio sign bit profile. The vertical exaggeration is 0.5:1. The reflections associated with event B of Figure 5 begin just beneath the bottom of this figure at 6 km depth and remain subhorizontal beneath Spring Valley.

does not continue beyond this point.

On the west flank of the NSR the decollement is cut by several down-to-the-east and down-to-the-west normal faults (Gans et al., 1985) and increases dip to 20-30°, cutting down-section to the west. Even further to the west, in the Schell Creek Range, normal faulting has resulted in significant extension, but no ductile deformation has been observed (Gans et al., 1985). Lower plate Cambrian and Precambrian units that are ductilely stretched and thinned in the lower plate of the NSR are brittlely faulted in the Schell Creek Range. Based on these observations, Gans et al. (1985) have suggested that the NSRD becomes a purely brittle feature to the west beneath Spring Valley.

In 1983 Sohio Petroleum recorded a 18-km-long seismic reflection profile from the west flank of the NSR across Spring Valley (Gans et al., 1985; Fig. 2). This profile, collected with a 1024-channel, 128-fold sign bit recording system (only reflection polarity recorded), has a CDP-trace spacing of 10 feet, and as such, represents one of the highest resolution profiles yet recorded in the BRP. The data shown in Figure 5 were stacked at Sohio and the results were later depth migrated at the facilities of the Stanford Exploration Project (Fig. 6).

In contrast to the sharp decollement reflection visible on the eastern flank of the northern Snake Range, the western down-dip projection of the NSRD is broken by down-to-the-east and down-to-the-west normal faults and cannot be convincingly traced more than 2.5 km west of its last surface outcrop (Gans et al., 1985). The location of the NSRD beneath Spring Valley is thus unconstrained by the reflection data and must be inferred from drilling results from the Yelland Well #1 (Fig. 2).

The Yelland Well, located on the east side of Spring Valley (Fig. 2), penetrated five major lithologic units, defined by well reports and geophysical logs (Fig. 7). The top unit consisted of 1082 meters of conglomerate interbedded with lacustrine limestone and sandstone. This unit was then followed by a second, 394-meter-thick unit of volcanics, tuffaceous sandstones, and conglomerates. The volcanics of unit two are correlated in part with the Kalamazoo volcanics of early Tertiary age (34 my). Beneath unit two is a 48-meter-thick crystalline limestone (unit three) believed to be Paleozoic(?) in age. The well then continued through 209 meters of unconsolidated sandstone and conglomerates (unit four) before crossing a fault(?) at a depth of 1733 meters. Beneath the fault the well penetrated 267 meters of Lower Cambrian Pioche shale and Prospect Mountain quartzite (unit five) and ultimately bottomed in Prospect Mountain quartzite at a depth of 2000 meters.

A synthetic seismogram (Fig. 8) was computed from the acoustic sonic log (Fig. 7) recorded in the Yelland Well and is here used to correlate these five basic lithologies to the reflections imaged on the Spring Valley seismic profile. The results, projected onto the seismic line by following the -230 mgal contour 13 km to the north, indicate that reflections from within the valley fill (units one and two) extend down to .98 seconds two-way travel time and are followed by reflections from the Paleozoic(?) limestone (unit three) and the Lower Cambrian Prospect Mountain quartzite (unit five) at .98 and 1.15 seconds, respectively. Thus all of the major reflections in the vicinity of the Yelland Well correspond to lithologic boundaries either within the valley fill or from the underlying Lower Cambrian stratigraphy. This has important implications for the west-

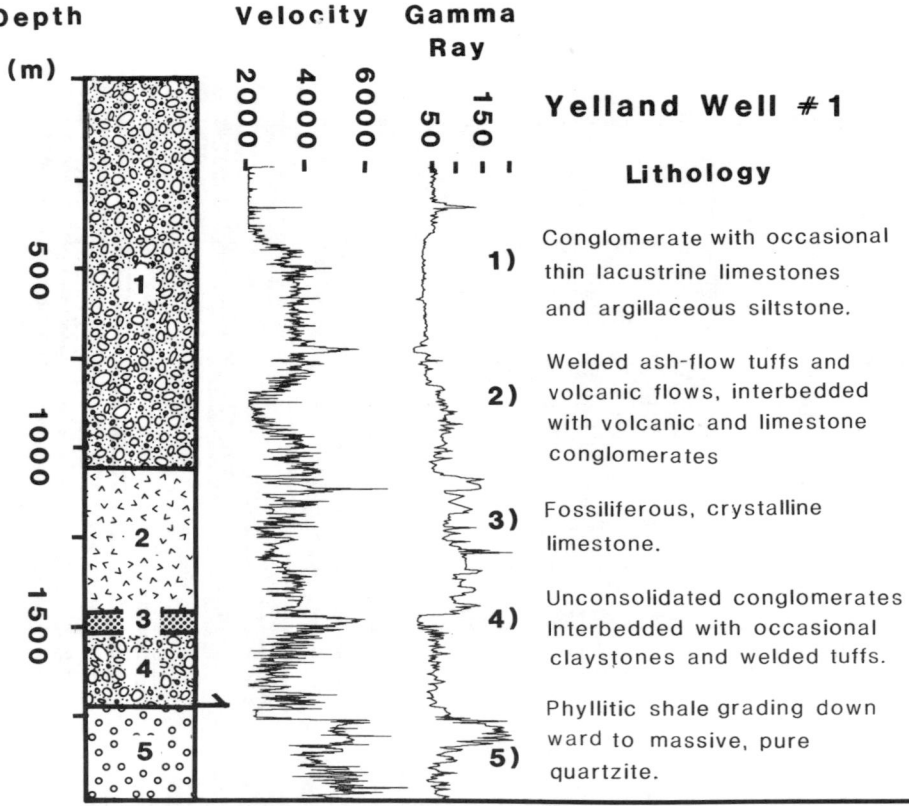

Fig. 7. Summary of the five major lithologies penetrated by Dome Petroleum's Yelland Well #1. The sonic velocity (meters/sec) and gamma ray (API units) logs are also shown as a function of depth in the well. A fault has been inferred at 1733 meters depth based on drilling rate, log response, and lithology contrasts.

ward continuation of the NSRD. Either the fault(?) encountered at 1733 meters depth in the Yelland Well is the NSRD (Gans et al., 1985), or the decollement has cut down-section to the west and is not penetrated in the Yelland Well at all. In either case, it is apparent from the Spring Valley seismic reflection data that the northern Snake Range Decollement is not a major reflecting horizon to the west, away from the NSR. This, together with the decreasing strain observed on the western flank of the NSR and the absence of a ductile lower plate in the adjacent Schell Creek Range, suggest that the NSRD has changed character to the west, becoming a purely brittle feature that no longer separates brittlely-deformed rocks above from ductilely-deformed rocks below.

Deep Reflections Beneath the Northern Snake Range

Horizon A. Deeper reflections are also prominent on the Sohio reflection profile. Beneath the eastern portion of Spring Valley a strong reflection event (A, Fig. 5; Fig. 6) extends at a constant dip of approximately 40° to 3.5 seconds or possibly even deeper before being replaced by a more prominent zone of subhorizontal reflectors (event B). Upon closer inspection this dipping event A can be characterized by two or more strong, subparallel reflections that broaden on the eastern-most 5 km of the profile to include the upper 2 seconds of reflections. The seismic line does not extend far enough into the range to trace event A any further.

Gans et al. (1985) interpreted event A to be a stratigraphic boundary within the alternating quartzite and schist sequence of the late Precambrian McCoy Creek Group. This interpretation is based on the westward-increasing (10-20°) dips of late Precambrian and Cambrian strata exposed on the western flank of the NSR. Because the amount of ductile strain in the lower plate can only be inferred, the deeper lithologies and the total thickness of these units beneath the western flank of the Snake Range remain unknown. In addition, the nature and composition of the basement rocks underlying the McCoy Creek Group can only be estimated, as its base is nowhere exposed in Nevada. Thus it is difficult to test their intra- or inter-formational reflection interpretation based on either travel time-depth estimates or synthetic seismogram models. Nevertheless, their interpretation is supported by the subparallel (0-2 sec) band of reflectors that can be correlated laterally with the lower plate McCoy Creek and Prospect Mountain formations exposed in the range.

Another explanation that cannot be discounted is that event A represents a lower detachment surface beneath the NSRD. This model of "stacked detachments" is not unlike that proposed by Rehrig (1982) to explain shingled mylonite zones in the Tortolita and Picacho core complexes of south-central Arizona. Multiple detachments have also been identified in the Chemehuevi (John, 1982) and Whipple (G. A. Davis, 1980) metamorphic core complexes of eastern-most California, while a 1-km-thick zone of detachments has been identified in the Ruby Mountains of eastern Nevada (Snoke and Lush, 1984). The presence of a deeper ductile shear zone beneath the NSRD in Spring Valley would help account for the transition of the NSRD into a more-brittle feature to the west, even though large amounts of upper plate extension still continue as far west as the Egan Range.

A third possible explanation for event A is that it represents a a fault or fault zone of Mesozoic(?) age. Miller et al. (1983) and Gans et al. (1985) argue against the existence of Mesozoic thrust faults in the Snake Range region based on the disconformable contact between Paleozoic and Tertiary strata, the absence of any thrust faults exposed in the range, and the lack of stratigraphic omission across the decollement. Nevertheless, Bartley and Wernicke (1985) call upon an east-verging Mesozoic thrust beneath Spring Valley to account for the stratigraphic relief observed between the Confusion and House Ranges. They also interpret this thrust to be bedding-parallel, located either within, or at the base, of the Precambrian McCoy Creek Group. Their model is difficult to test, not only because of the uncertainties, mentioned above, in the lower Precambrian stratigraphy, but also because of the ambiguity in distinguishing between intra- or inter-formational reflections and bedding-parallel fault reflections. Bartley and Wernicke's model of a Mesozoic thrust beneath Spring Valley, however, would require the preservation of a thick (7 km) thrust plate between the base of the valley fill and the top of the Cambrian Pioche shale (see Fig. 2 of Bartley and Wernicke, 1985). As Gans and Miller point out (in press), the existence of such a thrust plate is not substantiated by the Spring Valley seismic profile (Fig. 5 and 6) which shows only a thin, 1.5 km pre-Tertiary section preserved above the top of the Cambrian Pioche shale. Although an intra-formational or stacked detachment interpretation for event A is preferred over a Mesozoic thrust interpretation, the ultimate identification of event A must await further study.

Horizon B. Probably the most striking feature on the seismic profiles from the Snake and Schell Creek Ranges (Fig. 3, 5, and 6) is a zone of high-amplitude, laminated reflections (event B, Fig. 5) that begin at shallow depths (~10 km) and extend down to the base of the crust. This subhorizontal, reflective zone is most prominent on the Sohio sign bit profile (Fig. 5) from 4.0 seconds two-way travel time down to the bottom of the profile at 6.0 seconds. Proprietary seismic reflection profiles from the same area confirm the presence of this reflective zone and trace it to even greater travel times of 8.0 seconds. Beneath the eastern flank of the NSR

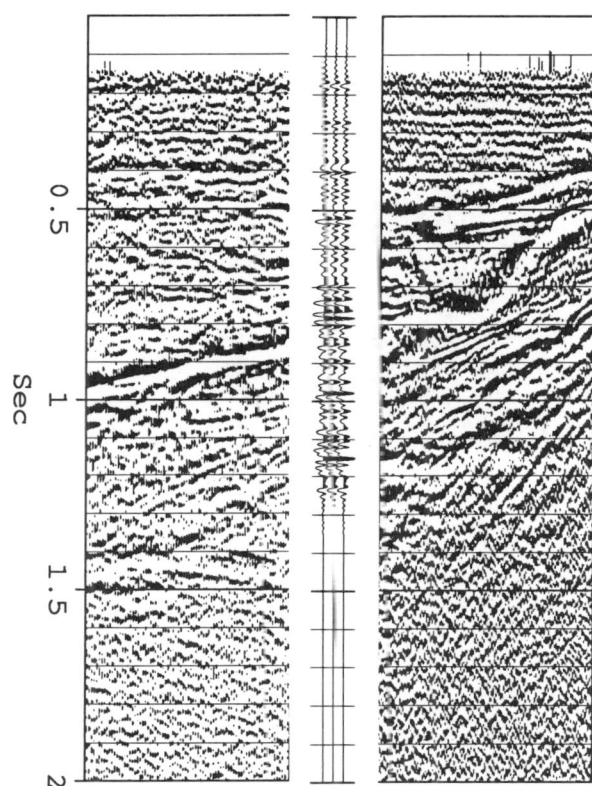

Fig. 8. Comparison of the Spring Valley seismic reflection data (Fig. 5) to the synthetic seismogram computed from sonic velocities measured in the Yelland Well #1 (Fig. 7). The synthetic seismogram has been triplicated for clarity.

the laminated reflections are not as strong (due largely to the surface seismic source used during acquisition) but are, nevertheless, easily identifiable from 4 seconds to the bottom of the profile at 8 seconds two-way travel time. COCORP results from Nevada line 5 (Klemperer et al., in press) show that this reflective fabric is dome-shaped, is centered beneath the west flank of the NSR, and can be extended beneath the entire width of the core complex. In addition, COCORP Nevada line 5 imaged these strong, laminated reflections from 4.0 seconds two-way travel time down to the Moho (10 seconds), where they end abruptly above a seismically-transparent upper mantle (Klemperer et al., in press).

The deep reflective fabric imaged on the NSR seismic profiles can be compared to results from seismic reflection profiles collected near the Picacho Mountain core complex in south-central Arizona (Fig 9.; see Fig. 1

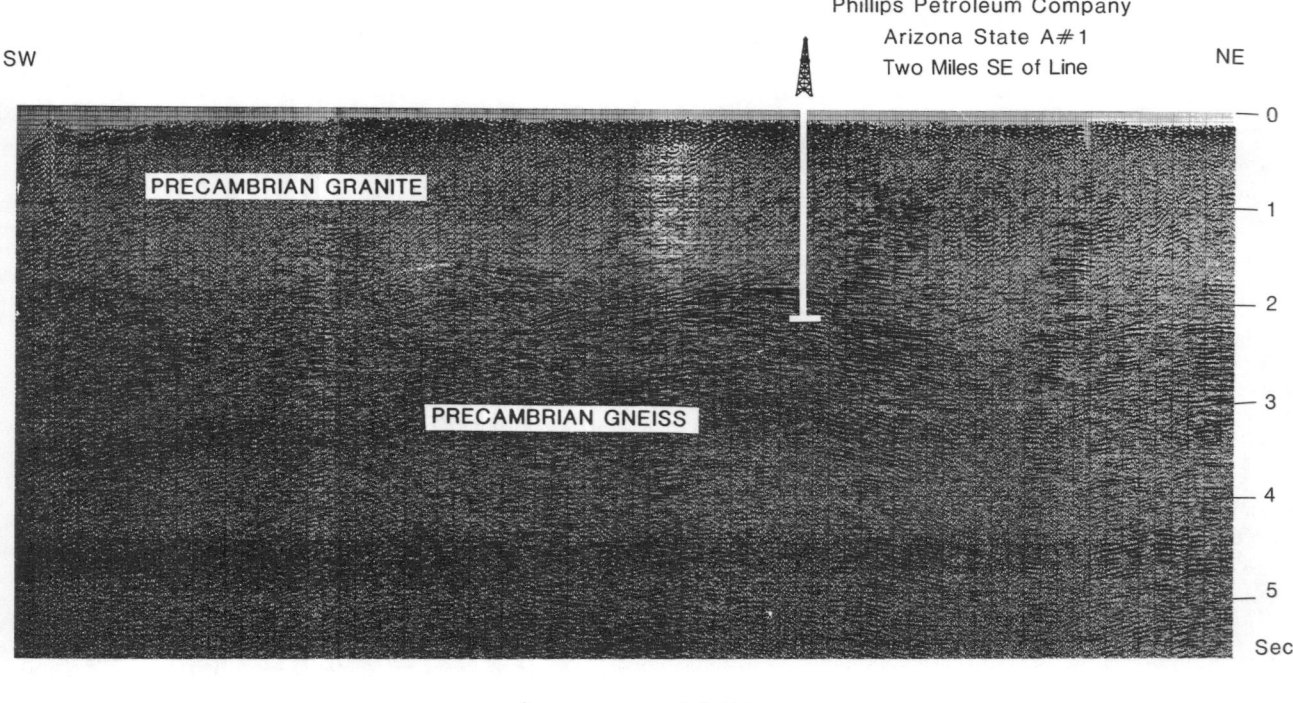

Fig. 9. 48-fold reflection profile recorded by Phillips Petroleum northwest of the Picacho Mountain metamorphic core complex. Precambrian granite exposed in the shallow section is replaced by Precambrian compositionally-layered and ductilely-deformed granodioritic gneiss at depth. The transition between these two lithologies is marked by an intervening Tertiary intrusion.

for location). These profiles were shot by Phillips Petroleum in 1980 to test the southern extension of the Mesozoic foreland fold and thrust belt into Arizona (Reiff and Robinson, 1981; Robinson, 1982). The results from the seismic survey reveal a transparent zone from 0-2 seconds two-way travel time, underlain by a more laminated fabric of high-amplitude reflections. Although originally interpreted as allochthonous Precambrian granite thrust over Paleozoic and Mesozoic sediments, a subsequent 5492 meter drill hole penetrated three crystalline units consisting of an upper Precambrian granite, a lower metamorphosed Precambrian gneiss, and an intervening Tertiary intrusion localized along the fault contact. Thus, the strong laminated fabric below 2.0 seconds was due not to Paleozoic and Mesozoic sediments, but to the textural and compositional fabric within the Precambrian gneiss. This pervasive, laminated fabric seen from 2-4 seconds is identical to the 4-10 second laminated fabric noted beneath the Snake Range and is subparallel to the topography of the range and the trend of the decollement.

Discussion

The Snake Range reflection profiles described above provide a rare opportunity to study the geologic structures and reflection character across a metamorphic core complex. The differences among these profiles are striking. Reflection data from the east flank of the range imaged a strong, continuous reflector extending approximately 20 km to the east. On the western flank, however, the Sohio sign bit profile was unable to image the decollement for any appreciable distance at all. The preferential development of a decollement reflection on the eastern flank of the NSR is just one example of asymmetry observed within the range. Other examples include: 1) an increase in ductile strain and a thinning of units on the east flank of the range, 2) a progressive decrease in K-Ar ages eastward within the lower plate, and 3) a dominance of down-to-the-east normal faults in the upper plate (Miller et al., 1983).

The asymmetry observed within the northern Snake Range has been observed in other core complexes as well, including the Harcuvar, Harquahala, and Buckskin metamorphic core complexes in western Arizona (Rehrig and Reynolds, 1980), the Albion and Raft River complexes in northwestern Utah and southern Idaho (Miller, 1980; Compton et al., 1977), the Bitterroot complex of eastern Idaho (Hyndman, 1980), and the Okanogan and Kettle complexes of Washington (Cheney, 1980). G. H. Davis (1983) argues that this asymmetry is the result of movement along a low-angle zone of simple shear located between the upper and lower plates of the core complex. Although the eastward transition from brittle deformation to purely ductile deformation along the decollement surface in the NSR is compatible with

Davis' model of a low-angle normal fault rooted to the east, other assumptions implicit in the Wernicke (1981) and Davis' (1983) shear zone model are not satisfied by the seismic reflection data discussed above.

First, the reflection profiles along the eastern flank of the NSR do not image the detachment surface below 3 seconds two-way travel time (approximately 9 km), nor do they trace the decollement reflection more than 20 km east. If the NSRD is a low-angle normal fault that has experienced 60 km of displacement (Bartley and Wernicke, 1985), then it should continue as a major reflecting horizon beneath the Confusion and House Ranges down to a depth of at least 15 km (see Fig. 2, Bartley and Wernicke, 1985). The fact that the NSRD does not continue this far suggests that, if their model is correct, either: 1) the data quality is not good enough to trace the decollement further east or 2) the rheology of the fault surface has changed and is no longer reflective. In the case of the former, the quality of the COCORP Utah line 1 reflection profile does degrade to the east in the vicinity of the Confusion Range (due to numerous bends in the recording line, see Fig. 2), but other deep reflections, described by Allmendinger et al. (1983), have been imaged in this area to depths of 10-15 km. Thus the data quality is not believed to be a controlling factor in the eastward disappearance of the NSRD reflection. As to a rheological transition, evidence from other reflection surveys across major dislocation surfaces such as the COCORP profiles across the Wind River thrust (Lynn et al., 1983) and Sevier Desert Detachment, (Allmendinger et al., 1983) and the British Institutions Reflection Profiling Syndicate (BIRPS) profile across the Outer Isles thrust (Peddy, 1985) suggests that deep crustal shear zones are, in fact, highly reflective in both the middle and lower crust, well below the brittle-ductile transition. If true, then the NSRD must, instead, be a local feature that does not cut down-section through the crust and upper mantle, but rather, is confined to the middle and upper crust in a narrow (20-km-wide) region centered about the NSR.

Second, Wernicke's model of a low-angle normal fault rooted into the lower crust or upper mantle (1981) requires that extension in the upper plate be physically removed from extension in the lower plate. This cannot be the case in the NSR, however, if the strong mid- and lower-crustal laminated reflections described above are a by-product of Cenozoic extension (like their lower-crustal counterparts imaged on COCORP profiles across Nevada (Klemperer et al., in press)), as they are situated directly beneath the range and are not offset from the upper plate imbricate normal faults. The predominance of these reflections throughout the crust, and their location directly beneath the NSR, implies that extension in the northern Snake Range has been distributed throughout the entire crustal column and has not been accommodated solely by movement along the NSRD.

Conclusion

The seismic profiles from the Snake Range and south central Arizona suggest that the middle and lower crust in regions of the BRP that have undergone large amounts of Tertiary extension can be characterized by laminated reflections often comparable or greater in amplitude than anything observed in the near surface (Fig. 5, 9). Correlation of these reflections with those imaged in the lower crust by COCORP across all of Nevada suggests that this reflectivity is post-Mesozoic in age (Klemperer et al., in press) and must therefore be a by-product of Tertiary extension.

The prominent reflectivity beneath the metamorphic core complexes is important in that it underscores the relative importance of lower-crustal ductile stretching and contemporaneous magmatic intrusion in the overall process of crustal extension. A model of extension based solely on simple shear along a low-angle normal fault (Wernicke, 1981) cannot adequately account for this strong mid- and lower-crustal reflectivity. Although the existence of low-angle normal faults is not being disputed, this study suggests that they may not be through-going features that bisect the entire crust, nor are they the dominant mechanism of crustal extension. Instead, the lower-crustal reflectivity is better explained by a combination of ductile stretching and contemporaneous magmatic intrusion that has resulted in a compositionally- and texturally-layered lower crust underlying a brittlely-faulted and extended upper crust. Anastomosing shear zones, layered intrusives, stretching fabrics, and transposed compositional layering are presumably all by-products of this extensional process and are thus all inferred to contribute to the overall reflectivity of these highly extended terranes.

Acknowledgments. The author gratefully acknowledges Sohio Petroleum Company for providing much of the seismic reflection data presented in this study. The above ideas benefited from many inspiring discussions with George Thompson, Phil Gans, Elizabeth Miller, and Erik Goodwin. Phil Gans, Elizabeth Miller, and Greg Davis also provided constructive reviews of the manuscript. The work was supported under National Science Foundation grants EAR 81-09294 and EAR 83-06406.

Reference

Allmendinger, R. W., J. W. Sharp, D. Von Tish, L. Serpa, L. Brown, S. Kaufman, J. Oliver, and R. B. Smith, Cenozoic and Mesozoic structure of the eastern Basin and Range from COCORP seismic reflection data, *Geology, 11,* 532-536, 1983.

Armstrong, R. L., Cordilleran metamorphic core complexes - From Arizona to southern Canada, *Annual Review of Earth and Planetary Science, 10,* 129-154, 1982.

Bartley, J. M., and B. P. Wernicke, The Snake Range Decollement interpreted as a major extensional shear zone, *Tectonics, 3,* 647-658, 1985.

Cheney, E. S., Kettle dome and related structures of northeastern Washington, in *Cordilleran Metamorphic Core Complexes,* edited by M. D. Crittenden, Jr., P. J. Coney and G. H. Davis, pp. 463-484, Geological Society of America Memoir 153, Boulder, Colorado, 1980.

Compton, R. R., V. R. Todd, R. E Zartman, and C. W. Naeser, Oligocene and Miocene metamorphism, folding and low-angle faulting in northwestern Utah, *Geol. Soc. Am. Bull., 88,* 1237-1250, 1977.

Coney, P. J., Cordilleran metamorphic core complexes: An overview, in *Cordilleran Metamorphic Core Complexes,* edited by M. D. Crittenden, Jr., P. J. Coney and G. H. Davis, pp. 7-34, Geological Society of America Memoir 153, Boulder, Colorado, 1980.

Crittenden, M. D., Jr., Metamorphic core complexes of the North American Cordillera: summary, in *Cordilleran Metamorphic Core Complexes,* edited by M. D. Crittenden, P. J. Coney and G. H. Davis, pp. 485-490, Geological Society of America Memoir 153, 1980.

Davis, G. H., Shear-zone model for the origin of metamorphic core complexes, *Geology, 11,* 342-347, 1983.

Davis, G. A., J. L. Anderson, E. G. Frost, and T. J. Shackelford, Mylonitization and detachment faulting in the Whipple-Buckskin-Rawhide Mountains terrane, southeastern California and western Arizona, in *Cordilleran Metamorphic Core Complexes,* edited by M. D. Crittenden, P. J. Coney and G. H. Davis, pp. 79-129, Geological Society of America Memoir 153, 1980.

Gans, P. B., and E. L. Miller, Style of mid-Tertiary extension in east-central Nevada, *Utah Geological and Mineral Survey Special Studies 59, Guidebook Part 1,,* pp. 107-160, GSA Rocky Mountain and Cordilleran Sections Meeting, Salt Lake City, Utah, 1983.

Gans, P. G., E. L. Miller, J. McCarthy, and M. L. Ouldcott, Tectonic evolution of the northern Snake Range and vicinity: New insights from seismic data, *Geology, 13,* 189-193, 1985.

Gans, P. G., and E. L. Miller, The Snake Range decollement interpreted as "a major extensional shear zone", *Tectonics,* in press, 1985.

Hyndman, D. W., Bitterroot dome-Sapphire tectonic block, an example of a plutonic-core gneiss-dome complex with its detached suprastructure, in *Cordilleran Metamorphic Core Complexes,* edited by M. D. Crittenden, Jr., P. J. Coney and G. H. Davis, pp. 427-444, Geological Society of America Memoir 153, Boulder, Colorado, 1980.

John, B. E., Geologic framework of the Chemehuevi Mountains, southeastern California, in *Mesozoic-Cenozoic Tectonic Evolution of the Colorado River Region, California, Arizona, and Nevada,* edited by E. G. Frost and D. L. Martin, pp. 317-325, Cordilleran Publishers, San Diego, California, 1982.

Klemperer, S. L., T. A. Hauge, E. C. Hauser, J. E. Oliver, and C. J. Potter, The Moho in the northern Basin and Range Province, Nevada, along the COCORP 40N seismic reflection transect, *Geol. Soc. Am. Bull.,* in press, 1985.

Lynn, H. B., S. Quam, and G. A. Thompson, Depth migration and interpretation of the COCORP Wind River, Wyoming, seismic reflection data, *Geology, 11,* 462-469, 1983.

Miller, D. M., Structural Geology of the northern Albion Mountains, south-central Idaho, in *Cordilleran Metamorphic Core Complexes,* edited by M. D. Crittenden, Jr., P. J. Coney and G. H. Davis, pp. 399-423, Geological Society of America Memoir 153, Boulder, Colorado, 1980.

Miller, E. L., P. B. Gans, and J. D. Garing, The Snake Range Decollement: an exhumed mid-Tertiary ductile-brittle transition, *Tectonics, 2,* 239-263, 1983.

Peddy, C. P., Displacement of the Moho by the Outer Isles thrust show by seismic modeling, *Nature, 312,* 628-630, 1984.

Rehrig, W. A., Metamorphic core complexes of the southwest United States - an updated analysis, in *Mesozoic-Cenozoic Tectonic Evolution of the Colorado River Region, California, Arizona, and Nevada,* edited by E. G. Frost and D. L. Martin, pp. 551-560, Cordilleran Publishers, San Diego, California, 1982.

Rehrig, W. A., and S. J. Reynolds, Geologic and geochronologic reconnaissance of a northwest-trending zone of metamorphic core complexes in southern and western Arizona, in *Cordilleran Metamorphic Core Complexes,* edited by M. D. Crittenden, Jr., P. J. Coney and G. H. Davis, pp. 131-158, Geological Society of America Memoir 153, Boulder, Colorado, 1980.

Reiff, D. M., and J. P. Robinson, Geophysical, geochemical, and petrographic data and regional correlation from the Arizona state A-1 well, Pinal County, Arizona, *Arizona Geological Society Digest, 13,* 99-109, 1981.

Robinson, J. P., Petroleum exploration in southeastern Arizona: anatomy of an overthrust play, *Rocky Mountain Association of Geologists,* 665-674, 1982.

Snoke, A. W., and A. P. Lush, Polyphase Mesozoic-Cenozoic deformational history of the northern Ruby Mountains-East Humboldt Range, Nevada, in *Western Geological Excursions, GSA Annual Meeting, 4,* edited by J. Lintz, Jr., pp. 232-260, Geological Society of America and Mackay School of Mines, University of Nevada, Reno, Nevada, 1984.

Wernicke, B., Low-angle normal faults in the Basin and Range province: nappe tectonics in an extending orogen, *Nature, 291,* 645-648, 1981.

SHALLOW STRUCTURE OF THE SOUTHERN ALBUQUERQUE BASIN (RIO GRANDE RIFT), NEW MEXICO, FROM COCORP SEISMIC REFLECTION DATA

Zhengwen Wu

Institute for the Study of the Continents, Cornell University, Ithaca, New York 14853
and
Beijing Graduate School, Wuhan College of Geology, Beijing 100083, China

Abstract. Detailed examination of COCORP Line 1A across the Rio Grande rift suggests that in contrast to listric faults and close-spaced steep normal faults proposed by others, a folded detachment surface and a set of fan-shaped reverse faults exist beneath the Albuquerque Basin. The folded detachment is thought to be the subsurface projection of the curved, low-angle Jeter fault exposed on the northeastern corner of the Ladron Mountains. Evidence indicates that both the folding of the detachment and thrusting along the reverse fault took place about 10 m.y. ago. This interpretation infers that at that time the area occupied by the present Albuquerque Basin was undergoing compression rather than extension. This compression might have resulted from the clockwise rotation of the Colorado Plateau relative to the Great Plains. Subsequently, compression apparently gave way to extension. This may have been caused by crustal relaxation as rotation of the plateau gradually ceased.

Introduction

The Rio Grande rift is a conspicuous tectonic element in the western part of North America. Despite detailed mapping and publication of many papers since the 1930's, its origin still evokes much controversy. Seismic reflection surveys carried out by the Consortium for Continental Reflection Profiling (COCORP) in the rift near Socorro, New Mexico have successfully obtained deep structural data along six profiles. The detailed structure of the upper crust has been revealed much more clearly than is possible by other geophysical methods. We are now in a better position to consider the geological development of the rift, based on this new three-dimensional data.

Two interpretations have been suggested for the COCORP Rio Grande rift data. Brown et al., [1979, 1980] inferred from the shallow portion of Line 1A and Line 1 that a large, buried "intergraben horst" exists beneath the Albuquerque Basin north of the Sierra Ladron, and that high-angle normal faults characterize the shallow crust, especially on the flanks of the "horst". They believe that the key structural features represented by the reflection data are consistent with crustal extension. Cape et al., [1983], on the other hand, suggested that extension in this region has been accommodated along listric faults. In addition, Jurdy and Brocher [1980] presented a velocity model for the portion of the same seismic sections between 0 and 1.4 s derived from calculation of refracted wave velocities. In their model, the distribution of time-stratigraphic units confirms the existence of extreme structural relief and also indicates that the Paleozoic cover is thin or even missing on the "horst".

Both Brown et al., [1979, 1980] and Cape et al., [1983] associated the structural features on the sections across the Albuquerque Basin with extensional tectonics, although their geometric interpretations are quite different. After reviewing the seismic reflection data on Line 1A and considering the regional geological setting, it seems to the writer that the Cenozoic structural evolution of the rift is much more complicated than either interpretation suggested earlier, and that a simple extensional model cannot explain all the structural features shown on the seismic reflection section or indicated by regional investigation of surface geology. For this reason, a new geometric and mechanical model is presented here. The writer hopes that the presentation of this model will contribute a new perspective to the ongoing discussion about the origin of the Rio Grande rift.

Interpretation and Analysis

Line 1A is located within the southern part of the Albuquerque Basin, the largest basin within the Rio Grande rift system. The line begins in the east on Tertiary-Quaternary deposits, then runs west and northwest to the Sierra Lucero, on the edge of the Colorado Plateau (Figure 1). Line 1A is perhaps the most important of the profiles in this area, not only because it is the longest (61 km), but also because it reveals the

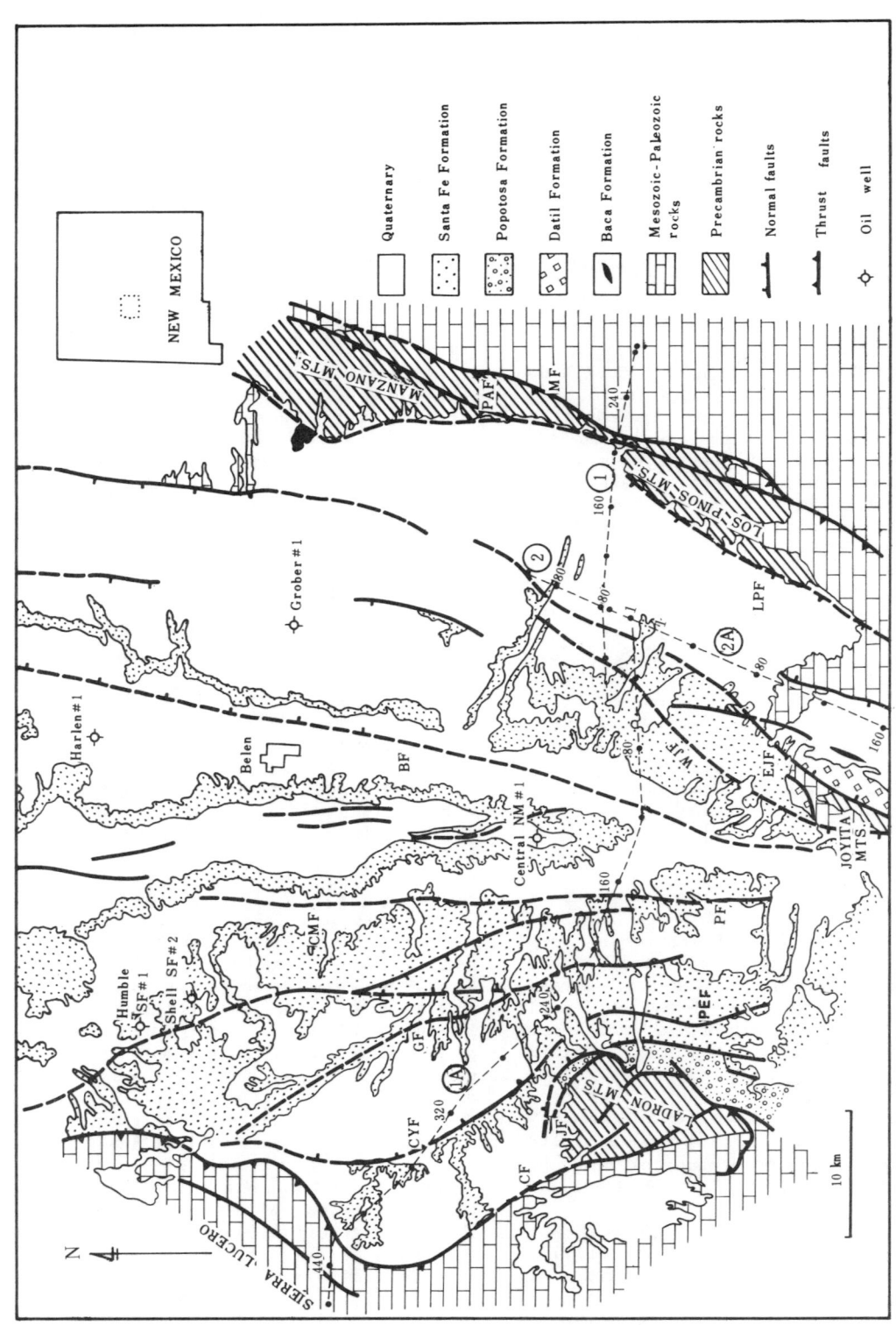

Fig. 1. Location of COCORP seismic reflection lines and geology of the southern Albuquerque Basin. Line numbers are circled, and vibration points are labeled. Geology is simplified from Kelley [1977]. CF = Comanche fault, JF = Jeter fault, CYF = Coyote fault, PEF = Pelado fault, GF = Gabaldon fault, CMF = Cat Mesa fault, PF = Puerco fault, BF = Belen fault, WJF = West Joyita fault, EJF = East Joyita fault, LPF = Los Pinos fault, PAF = Paloma fault, MF = Montosa fault.

294 WU

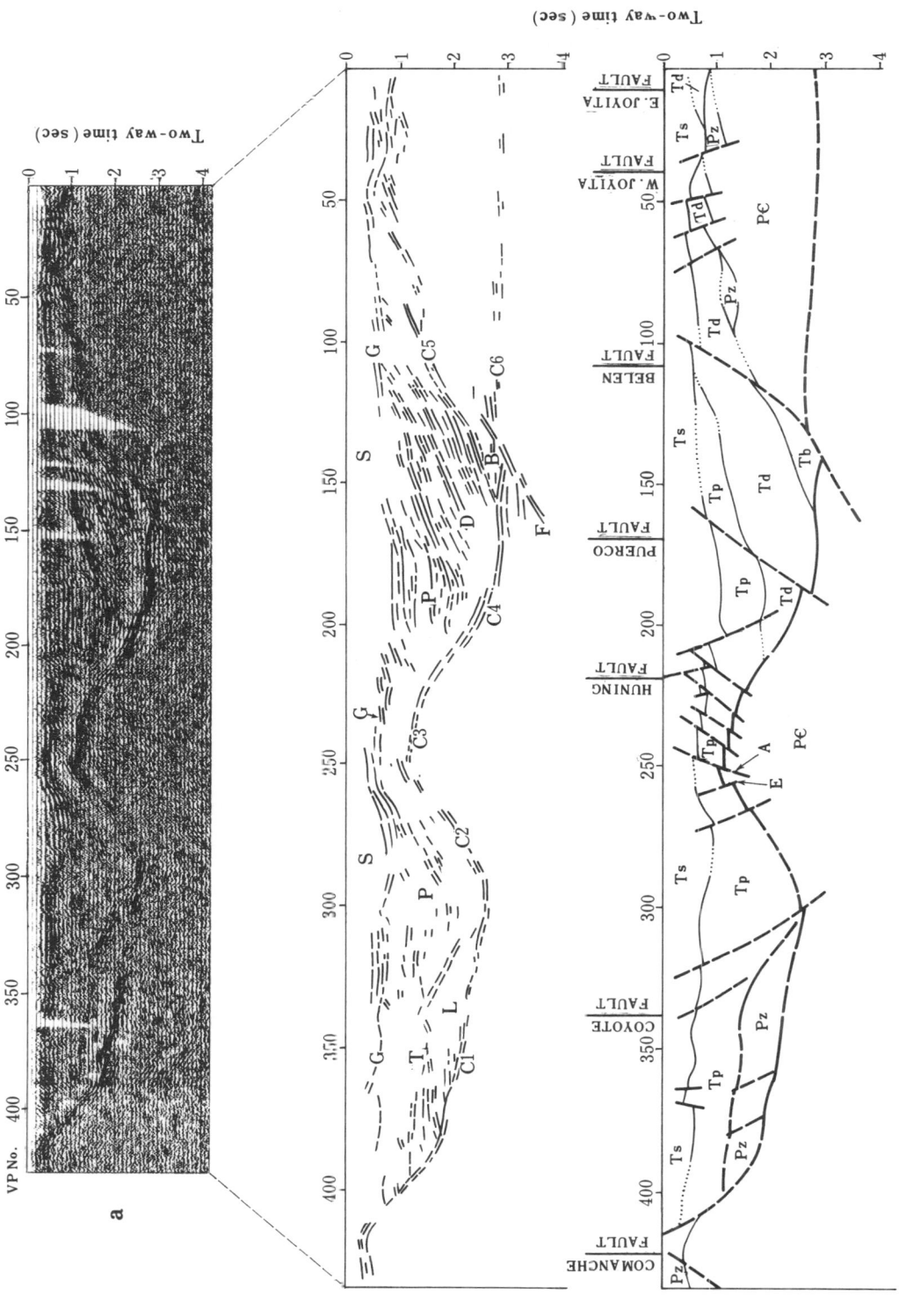

Fig. 2. Seismic section from Line 1A and its interpretation. (a) Unmigrated seismic reflection section. (b) Line drawing emphasizing prominent features on the same seismic section and showing labeled reflections discussed in text. (c) Interpreted time section. VP spacing is 134 m, vertical scale in two-way travel time. PЄ = Precambrian rocks, Pz = Paleozoic rocks, Tb = Baca Formation, Td = Datil volcanics, TP = Popotosa Formation, Ts = upper Santa Fe group, A and E = typical reverse faults.

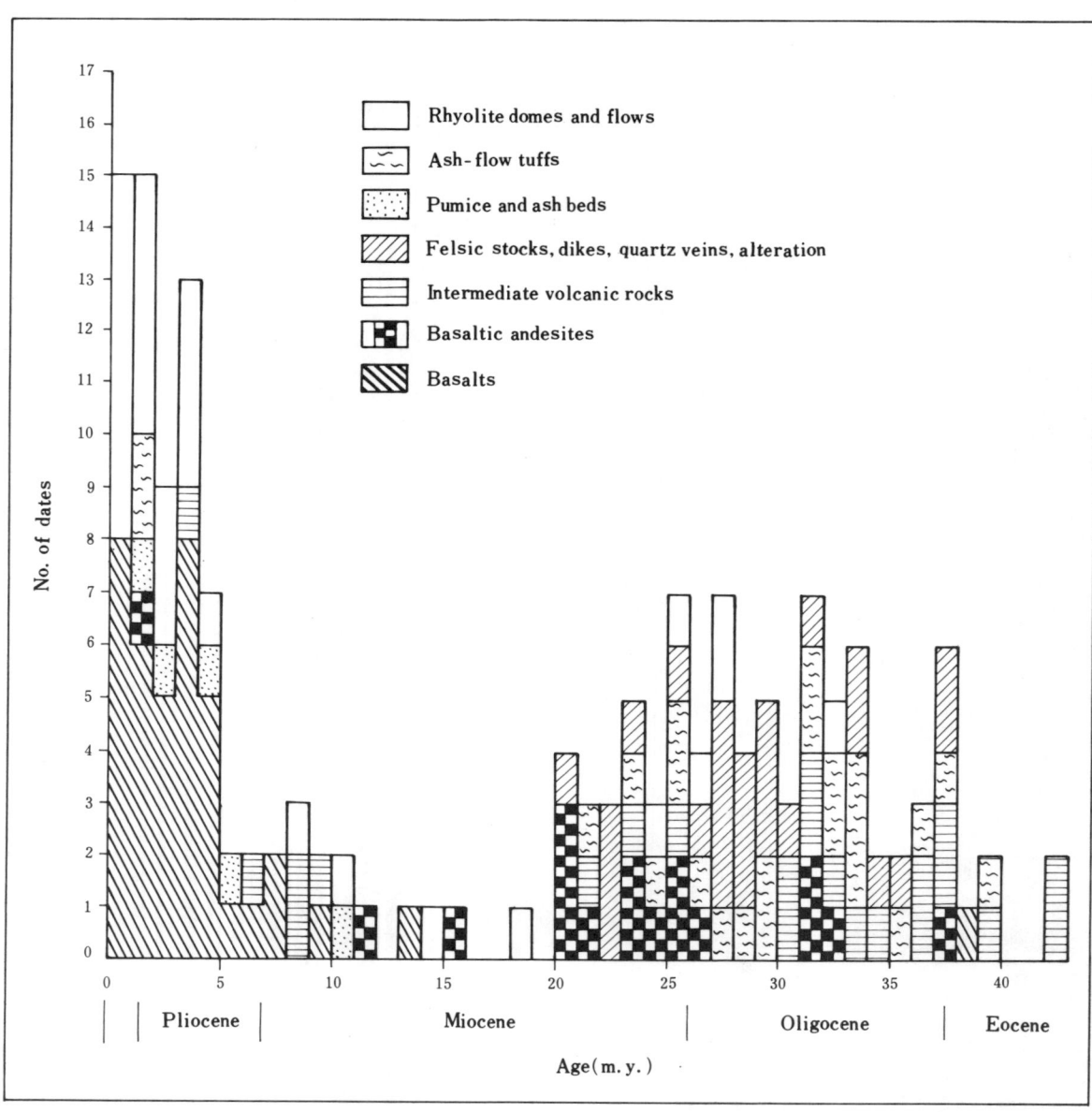

Fig. 3. Histogram of 161 K/Ar and fission-track datas on igneous rocks of late Eocene to Holocene age in New Mexico [Chapin and Seager, 1975].

spectacular structure of the central basin. Both migrated and unmigrated seismic reflection sections of Line 1A are the basis of the interpretation presented here. Because the reflections shown on the unmigrated section are much clearer than those on the migrated one, and in the interest of saving space, the latter is not displayed in this paper.

Reflections S

In the uppermost portion of the seismic section (Figure 2), a set of prominent reflections labeled S can be traced. Though they are offset by a series of high-angle faults, these reflections appear to have little structural relief and low apparent dips across the entire section. The strata represented by S rest on the underlying rocks unconformably (VP 10-130, 270-340) and near conformably (VP 130-210). These reflections are interpreted here to represent the upper Santa Fe Group (Sierra Ladron Formation of Machette, 1978). At the surface the upper Santa Fe Group (late Miocene-Pliocene) is widely exposed along Line 1A and exhibits gentle dips. Even at the eastern edge of the Ladron Mountains, where a

Quaternary	Alluvial sand and gravels.	
	~~	5th disturbance
Pliocene	Upper Santa Fe Group Sandstone, mudstone, arkose conglomerate and fanglomerate. 600—2000 m.	
	— — — — —	4th disturbance
Miocene	Popotosa Formation Fanglomerate, mudstone, sandstone and local andesitic flows. 900—1520 m.	
Oligocene	Datil volcanics Volcanic fanglomerate and tuff. >2000m.	
Eocene	Baca Formation Conglomerate, sandstone and mudstone >100m.	
	~~	3th disturbance (Laramide orogeny)
Cretaceous	Sandstone, shale and coal (marine and nonmarine facies). 1100—1250 m.	
	~~	2nd disturbance
Jurassic	Sandstone, mudstone, gypsum and limestone (continental facies). 100—340 m.	
Triassic	Mudstone, sandstone and conglomerate (continental facies). 0—340 m.	
Permian	Sandstone, limestone, and mudstone. 1140—1600 m.	
	~~	1st disturbance
Pennsylvanian	Limestone, shale and sandstone. 440—640 m.	
Precambrian	Gneiss, schist, greenstone, quartzite and plutonic granitic rocks.	

TABLE 1. Stratigraphic column, and interpreted tectonic events for the southern Albuquerque Basin and its outer region. Compilation based mainly on the information given by Kelley [1977].

series of faults are developed, the dips in the upper Santa Fe Group rarely exceed 35 degrees. The exposed contact relationship between the upper Santa Fe Group and older sequences is commonly an angular unconformity except in the La Sencia Basin, southwest of the Ladron uplift, and perhaps in some portion of the Albuquerque Basin [Chamberlin et al., 1983; Kelley, 1977; Chapin and Seager, 1975]. These characteristics are in accord with those of reflections observed on the seismic section.

Reflection G is interpreted to represent the basal boundary of the upper Santa Fe Group. It probably represents an erosional unconformity between the upper Santa Fe Group and older rock sequences (see the fourth disturbance in Table 1). Since the underlying uppermost Popotosa Formation yields a K-Ar age of 10.7 m.y. [cited in Chapin and Seager, 1975], it is believed that uplift and erosion likely took place about 10 m.y. ago. The fact that reflection G can be clearly distinguished on the seismic section is quite significant. First, its occurrence implies that the upper Santa Fe Group in the vicinity of the Ladron uplift can be separated in the subsurface as a unit distinct from the underlying Popotosa Formation. Second, since reflection G represents a structural disturbance, it is logical to expect that a change of tectonic environment may have occurred after deposition of the Popotosa Formation.

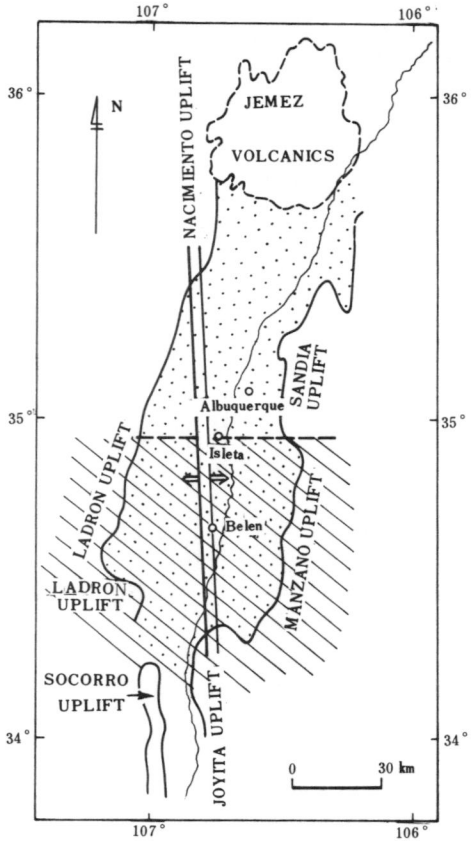

Fig. 4. Reconstruction of paleotectonics within the Albuquerque Basin during late Jurassic – early Cretaceous. Compilation based on Kelley's opinion [1977, 1979]. Fine dots represent the Albuquerque Basin; an anticline pattern with N-S trending – paleo-uplift in late Paleozoic; the oblique line – upwarped area in late Jurassic – early Cretaceous.

The variation in thickness of the upper Santa Fe Group can be estimated from depth conversion of the two-way travel time to reflection G. Above the "horst" the thickness of the upper Santa Fe Group is only about 600 m. Stratigraphic thickness of 1200 to 1400 m occurs on both sides of the "horst" at about VP 280 and VP 200, respectively. Elsewhere, the upper Santa Fe Group tends to be thinner. It is difficult to compare the thickness expressed above with those measured in drill holes, because it is unknown whether the basin sequences sampled in the holes include the Popotosa Formation. It is clear, however, that the upper Santa Fe Group is at least 1400 m thick at VP 200 and may be as much as 2000 m thick east of the Pelado fault [Kelley, 1977]. The upper Santa Fe Group is the true fill of the rift, as its distribution is generally limited to the present basin [Chapin and Seager, 1975; Woodward, 1977]. The basal time boundary of the upper Santa Fe Group is believed to be about 10 m.y. This correlates approximately with the time of transition from andesitic volcanism to basaltic volcanism in New Mexico (Figure 3). Also, as Kelley [1977] pointed out, gravelly facies, regardless of composition, occur mostly in the upper part of the upper Santa Fe Group. Thus the stratigraphic record reveals that the intensity of block-faulting within the basin increased during the later period of sedimentation of the upper Santa Fe Group. The coincidence of block-faulting and change to basaltic volcanism suggests some genetic relation between them.

Reflections P, D and B

Reflections P, D and B are sandwiched between events G and C (Figure 2). Brown et al., [1980] interpreted these reflections as part of the Cenozoic graben-fill, and Cape et al., [1983] believe that they consist of Mesozoic-Paleozoic and Tertiary sequences. Jurdy and Brocher [1980] stated that the Paleozoic rocks (4.7 km/s) are limited to areas near VP 340 and east of VP 160, whereas the Mesozoic rocks (3.7 km/s) occur only at the top of the "horst". In general, Jurdy and Brocher's opinion is supported by local field relationships. Kelley [1977] mentioned that variations in thickness of the pre-basin formations are mainly the result of several erosional episodes. He emphasized three important erosion surfaces, designated events 1, 2, and 3 in Table 1. He suggested that the first hiatus lasted from late Pennsylvanian to Permian, during which time the orientation of the crest line of the uplifted area was approximately coincident with the longitudinal axis of the present basin. The northern end of the crest line is located at the Nacimiento uplift, and the southern end at the Joyita uplift (Figure 4). I infer that the Pennsylvanian rocks in the subsurface along the crest line are thin or even absent. The second disturbance, as Kelley [1977] stated, which may have influenced the existing thicknesses of pre-rift rocks, took place in late Jurassic to early Cretaceous. In this interval, an east-west hinge line was upwarped broadly, north of which the crust relatively subsided. As a result, most of the Triassic and some Permian beds in the southern part of the Albuquerque Basin were eroded. Late Cretaceous to middle Eocene Laramide deformation was the third disturbance that reduced the thickness of the pre-rift beds. Apparently, the limited distribution of the Cretaceous rocks at the surface is closely related to the Laramide orogeny. Based on the analyses discussed above, it is unlikely that Mesozoic and Paleozoic rocks are widespread beneath the Albuquerque Basin at the latitude of the COCORP survey. Therefore, in this paper the reflections P and D are interpreted to represent middle and late Tertiary rocks rather than older rocks.

The middle and late Tertiary sequences of this

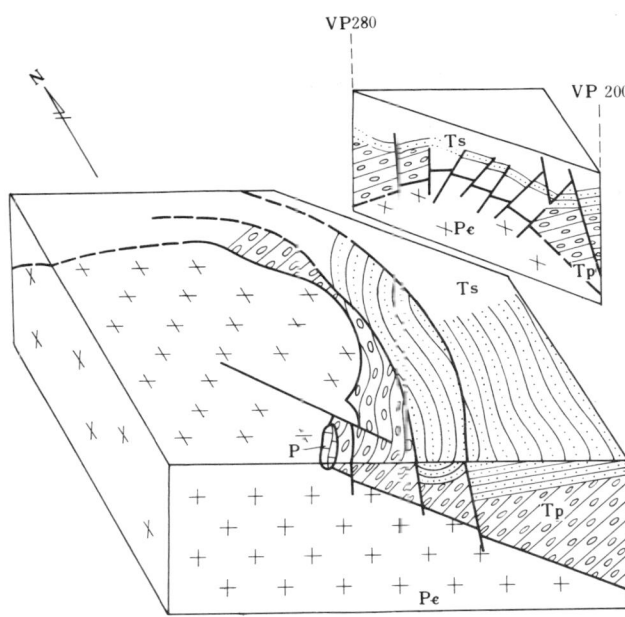

Fig. 5. Three-dimensional diagram showing the plunging feature of the Jeter fault and the relationship between it and the event C2-C4 on the seismic section of Line 1A. VP - projected vibrator point, P€ - Precambrian rocks, P - Pennsylvanian rocks, Tp - Popotosa Formation, Ts - upper Santa Fe group. Compilation based mainly on the information provided by Kelley [1977].

area comprise the Datil volcanics and the overlying Popotosa Formation (Table 1). The former is mainly exposed within the Joyita uplift [Kelley, 1977]; and the latter along the east and west flanks of the Ladron uplift. On the seismic section, the strong layered reflections D probably represent the Datil volcanics. This interpretation is based on the following considerations. First, the Datil volcanics contain much interbedded lavas and tuffs, which would give the large impedance contrasts necessary to produce the bright reflection events that appear on the seismic section. Second, the strike of the Datil volcanics exposed in the Joyita uplift is NNE through the eastern segment of Line 1A. Third, the La Jara Peak basaltic andesite, which is the major unit in the upper portion of the Datil volcanics [Osburn and Chapin, 1983], is exposed at Black Butte, only 2 km south of the eastern end of Line 1A. These observations indicate that the Datil volcanics are present on the eastern section of Line 1A.

According to the local stratigraphic column, reflections P may originate from within the Popotosa Formation. Their weak reflection character probably result from monotonic lithology of these clastic rocks. The relatively steep dips of these reflections and the apparently unconformable relationship between them and the overlying upper Santa Fe Group match the characteristics of the Popotosa Formation observed at the surface.

If the interpretation given above is correct, both the Datil volcanics and the Popotosa Formation have considerable thicknesses within the basin. On the basis of rough estimates, the thickness of the Datil volcanics beneath the surface reaches about 2600 m. According to Osburn and Chapin's data [1983], in the southern margin of the Albuquerque Basin the Datil volcanics above the Baca Formation and below the Popotosa Formation are as thick as 4000 m or so. Hence, it is not surprising that they occupy as much as 1.3 s two-way time on the seismic section. No precise thickness for the Popotosa Formation can be proposed, since its interior structural features are not yet well understood [Kelley, 1979]. Osburn and Chapin [1983] estimated a 900 m thickness for the surface exposures of the Popotosa Formation in the Socorro area. Chapin and Seager [1975] stated that the thick Popotosa Formation was deposited in an enclosed basin, and referred to this as the "Popotosa Basin" in order to distinguish it from the present rift basin. They inferred that the Popotosa Basin had a width in Miocene time of about 65 km, and extended from the Los Pinos Mountains on the east to the Gallinas Mountains on the west. This large basin is generally thought to have undergone segmentation into several narrow basins and uplifts during late Miocene-Pliocene time. This segmentation corresponds to the fourth tectonic disturbance shown in Table 1. It is reasonable to assume that the segmentation resulted from a significant change in the regional tectonic framework.

At about VP 150, a weak reflection area (B) is sandwiched between C4, C5 and F. This area probably represents either conglomerate or coarse sandstone rocks, belonging to the Baca Formation.

Event C

The most spectacular shallow features of Line 1A are reflectors C1-C4, which outline the "midgraben horst" (Figure 2). Brown et al., [1980] presented three possible interpretations of C1-C4, including: 1) a possible correlation with the La Jara Peak basaltic andesite of the Datil volcanics, 2) an upper Cretaceous sedimentary layer, and 3) the top of the Precambrian basement or a basal horizon within the Paleozoic sequence. Cape et al., [1983] suggested that C1 and C4 represent major listric faults, while C2 and C5 are depositional contacts between the Precambrian basement and the overlying Paleozoic and Mesozoic sediments. What is suggested here is that C1-C4 represents a detachment plane that separates the underlying Precambrian rocks, having no apparent reflection events, from the overlying younger sediments. At VP 140 this detachment is apparently truncated by a west-dipping fault (F); thereafter, it (C6) appears to be discontinuous and obscured eastward.

As pointed out by Cape et al. [1983], it is

clear that C1 and C4 are faults, because the overlying reflections L, P and D are truncated by C1 and C4. To determine whether the two faults are independent of one another or different segments of the same fault, it is necessary to review the surface geology in the Ladron Mountains. The mountains lie along strike with the midgraben "horst" a few kilometers to the southwest.

Kelley [1979, 1977] described the curved, low-angle Jeter fault which extends approximately 10 km around the northeastern corner of the Ladron Mountains (Figures 1 and 5). He reported that on the east side of the mountains the fault dips 20°-30° toward S 80° E. Northward the strike of the fault surface curves around toward the west so that at the north end of the mountains the fault dips 20°-30° toward N 10° E. This geometry implies that the hinge of the convex Jeter fault plunges approximately N 55° E. This plunge direction is consistent with the orientation of the Ladron uplift implied by its associated Bouguer gravity anomaly [Cordell et al., 1978]. The hanging wall of the Jeter fault, where exposed, is primarily composed of Popotosa Formation and faulted blocks of Paleozoic, Mesozoic, and Cenozoic rocks, including various volcanic units [Chamberlin et al., 1983]. The footwall, where exposed, is composed of Precambrian crystalline rocks. These relationships are consistent with the geometry of strata above and below reflections C2, C3 and C4 on the seismic section. Accordingly, it is likely that C2-C4 is the projection of the Jeter fault on the seismic section. In other words, C3 can be correlated with the hinge position of the convex Jeter fault surface, and C2 (VP 260-300) and C4 (VP 150-230) correspond to its E-W trending segment on the north side of the mountains and its N-S trending segment on the east side of the mountains, respectively. Each of these segments are therefore thought to represent parts of a single, continuous fault surface referred to here as the Jeter detachment.

Reflection C1 extends continuously from VP 300 to VP 410. At about VP 380 C1 begins to shallow steeply toward the west. Though C1 is actually separated from C2 by an eastward dipping fault, there is a strong suggestion that the two were originally continuous. Cape et al., [1983] described the same geometry, but on their generalized cross-section of the southern Albuquerque Basin they show a listric fault that would cut C2 and extend eastward beneath the "horst". Their interpretation seems to be less probable since it does not accord with the outcrop features just described.

Reflection T is also interpreted as a fault here, based on the observation that reflections above and below are clearly truncated by it. Toward the east and west, T is intersected by C1 at VP 310 and VP 400 (Figure 2). T may be a branch of the detachment, and therefore the rock mass (L) bounded by T and C1 could be a structural lens formed during the period of displacement of the Jeter detachment. The writer presumes reflection L to represent Paleozoic rocks. Jurdy and Brocher [1980] came to a similar inference, based on their finding a layer with a refraction velocity of 4.7 km/s at this position. Two faults have been recognized within the structural lens. They may indicate eastward displacement of the upper plate along the detachment surface relative to the lower one if they are normal faults.

In the above discussion I have taken C1-C4 as a continuous detachment fault. However, this fault appears to be truncated by a high-angle fault (F). East of F a short reflection (C6) with strong amplitude occurs at 2.7 s. Farther east the reflection becomes obscured. This may not be surprising since east of VP 140 the detachment occurs within Precambrian rocks. The similarity in lithology above and below the detachment along that portion results in little impedance contrast across the fault.

In the Ladron Mountains the youngest rocks cut by the Jeter detachment belong to the Popotosa Formation (early-middle Miocene). The upper Santa Fe Group was not subjected to significant deformation. Therefore, movement on the Jeter detachment must have largely predated deposition of the upper Santa Fe Group.

The seismic section indicates large structural relief on the Jeter detachment. At the top of the "horst", the detachment surface occurs at 1 s, whereas on both sides of it, the lowest points occur at 2.6-2.8 s. This means that the structural relief on the detachment surface is about 3 to 4 km within a horizontal distance of 6 km. What mechanism produced such large structural relief is a very important question that should be answered if one wants to reconstruct the evolutionary history of the basin. Brown et al., [1980] postulated that the "horst" was cut by a series of steeply dipping, step-like normal faults, implying that its formation was related to block faulting. Cape et al., [1983] attributed the origin of the "horst" to listric faulting. Neither explanation appears likely to the author since the existence of these kinds of faults cannot be distinguished unequivocally on the seismic section. Rather, it appears that the large structural relief of the detachment surface was mainly caused by folding, and subsequent high-angle reverse faulting.

High-angle faults

The seismic section of Line 1A reveals a number of high-angle faults, as shown on Figure 2. Their high apparent dips suggest that they are nearly perpendicular to the section. They can be classified into two different types according to the depth to which they extend. One type of fault offsets and extends beneath the Jeter detachment surface; the other type appear to die out at relatively shallow depth within the sedi-

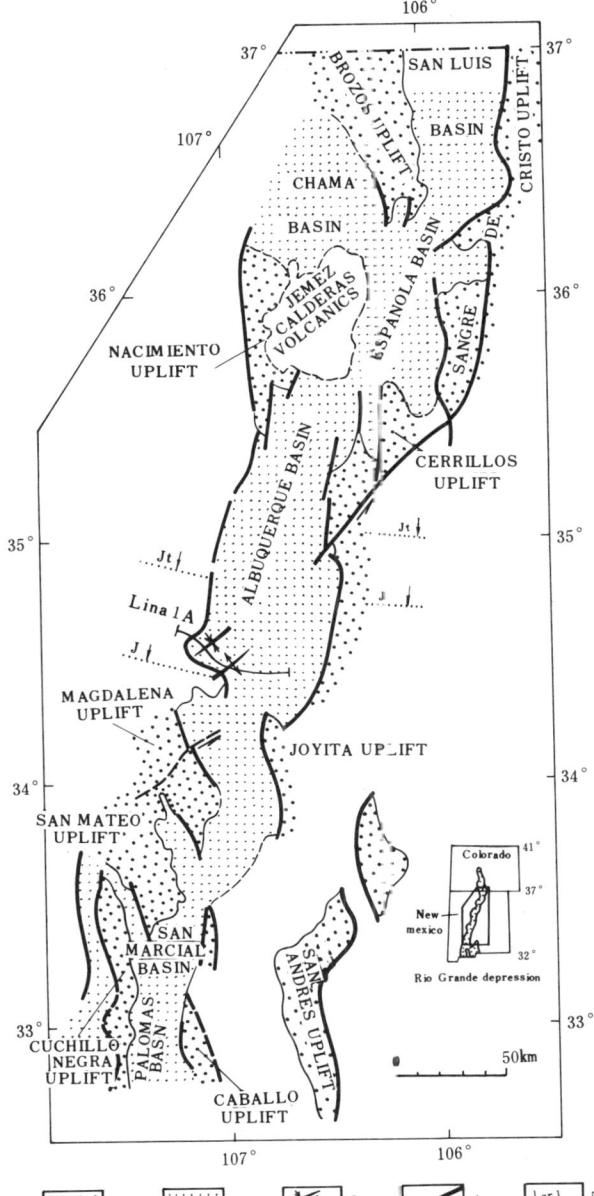

Fig. 6. Schematic tectonic map of the Rio Grande rift in New Mexico. 1 - uplift area, 2 - basin area, 3 - antiform and synform, 4 - fault, 5 - wedge edge of Jurassic formation (Jt - Todilto Formation, J - Morrision Formation). Modified and simplified from Kelley [1979].

mentary section. The latter are characteristically normal faults with small offsets, such as the faults east of VP 80 and west of VP 300. Undoubtedly, they were formed under extensional stress, though the amounts of their horizontal extensions appear limited.

The first kind of fault is of greater interest, for the characteristics of these faults indicate that they had a complicated history. They occur mostly between VP 150 and VP 330, and are especially centered on the "midgraben horst". A critical observation is that they dip toward the west on the eastern flank of the antiform and toward the east on the western flank of the antiform. Thus, their spatial distribution is a fan-shaped pattern. Another interesting characteristic is that at the top of the "horst" the hanging walls of these faults appear to be displaced upward relative to the footwalls along the folded detachment surface. The basal boundary of the upper Santa Fe Group also appears to be displaced upwards. On Figure 2, faults E and A are representatives of this type of structure. All of these appear to be reverse faults. Since the point of change of dipping direction is just at VP 250, the crest of the antiform, it is reasonable to infer that the formation of these faults is related to folding of the detachment.

Further away from the antiform, however, the displacement situation on steep faults looks more complicated (Figure 2). The direction of displacement of hanging walls relative to footwalls along the basal boundary of the upper Santa Fe Group is in the opposite sense from that along the detachment surface. The coexistence of these contradictory offset indicators implies that these faults have been reactivated under different stress conditions.

This implies at least two stress periods. During the first period (compressional), fan-shaped reverse faults were formed subsequent to the detachment generation. This compressional episode probably continued through the beginning of upper Santa Fe Group deposition. After that, the stress field became extensional and a number of small-scale normal faults were formed. That the offsets of all the normal faults appear small on the seismic section, and that most of these faults observed at the surface cannot be traced deeply into the subsurface (Figure 2), are indications that the high-angle normal faulting was not as significant as is generally assumed.

Initiation of Rifting

As yet, the problem of when onset of Rio Grande rift extension occurred has not been solved. This is because the distribution and structural features of the early-middle Miocene rocks beneath the upper Santa Fe Group, and correlative sequences are unknown. Chapin and Seager [1975] and Chapin [1979] suggested that the inception of the rift in the vicinity of Socorro occurred 29 m.y. ago. Their inference has been generally accepted and cited by many geologists [Brown et al., 1980; Cape et al., 1983; Woodward, 1977; Chamberlin et al., 1983]. Their conclusion is based mainly on the following facts: numerous normal faults with an average trend of N 10° W are distributed across a belt as much as 17 km wide in the Magdalena district

(Figure 6), about 40 km SSW of the Ladron Mountains. Along these faults a number of stocks and dikes are intruded that range in age from 30 to 28 m.y. The youngest sequences penetrated by the intrusive rocks, called the Potato Canyon tuff, is as old as about 31 m.y. In addition, some faults and intrusive rocks are unconformably overlain by the oldest Popotosa basin deposits, fanglomerates and interbedded lava flows of the Arroy Montosa unit (25 m.y.).

There are two problems with Chapin and Seager's conclusion. First, the orientation of these normal faults (N 10° W) is different from the general NNE trend of the rift in New Mexico. This difference implies that these faults probably represent second or even third order structures derived from, and controlled by, some first order structure. Their orientation probably does not accurately reflect the regional extension direction since the local stress field usually differs from the general one. The local extension represented by a set of N-10°-W-trending normal faults, therefore, should not be used to indicate the nature of the regional deformation.

Second, they argued that the existence of an enclosed Popotosa Basin in the Socorro area is an important criterion for setting an initial time of rifting at 29-26 m.y. At present this argument is suspect, since the true aerial distribution and detailed deformation pattern of the Popotosa Formation is unknown.

As mentioned earlier, folding of the Jeter detachment beneath the southern Albuquerque Basin probably took place about 10 m.y. ago, and reverse faulting probably occurred a little later than the folding, since these faults displace the basal layer of the upper Santa Fe group. Both the folding and the faulting indicate the existence of a compressional stress field, possibly NWW-trending, in the vicinity of the Ladron Mountains area about 10 m.y. ago. The detachment itself was also possibly formed during the same stress phase. These relationships, along with the fact that the true rift fill, the upper Santa Fe group, is strictly controlled by the present rift boundaries, suggest that extension across the rift started only about 10 m.y. ago. This time for initial rifting coincides with the previously mentioned transition from andesitic to basaltic volcanism.

Mechanism of Rifting

As with the Rio Grande rift as a whole, the mechanism of formation of the Albuquerque Basin has long been in debate. The key to the problem may lie in our understanding of the Jeter fault's behavior. Possible origins [cited from Chamberlin et al., 1983] of the Jeter fault include: (1) younger-over-older westward thrusting, (2) gravity sliding associated with domal uplift of the Ladron block, and (3) low-angle normal faulting related to crustal extension. I propose yet another potential model for it. That is, the Jeter fault may be the exposed portion of a major compressional detachment fault that underlies much of the Socorro area. De Voogd et al., [Cornell University, unpublished data, 1985] have re-interpreted Abo Pass Line 1, which overlaps the eastern segment of Line 1A and runs east through Los Pinos Mountains onto the Paleozoic rocks. Their reprocessed seismic section clearly shows a west-dipping fault. It seems to the writer that this fault may project eastward to the Paloma and Montosa faults, commonly interpreted as Laramide thrust, and to the west it might connect with reflector C6 at 2.7 s on the seismic section of Line 1A. If this inference is correct, the Jeter fault and the Paloma and Montosa faults could be different segments of the same detachment. Therefore, the compressional detachment fault may underlie all of the southern Albuquerque Basin. Since the 1930's, various mechanisms have been suggested for the Rio Grande rift. This is not surprising, as geologists emphasize the different geological phenomena they have observed and have viewpoints with different tendencies. The principal differences among the theories are primarily whether the rift formed by extension across its length, or by transcurrent faulting along its length, though either class of explanation includes a number of variations. The opinions of Chapin and Seager [1975], Brown et al. [1980], and Cape et al. [1983] belong to the former, and Kelley's hypothesis [1977, 1979] that the Rio Grande rift formed by left-lateral transcurrent movement, which took place during the time from the Laramide orogeny to the early Cenozoic, belongs to the latter.

Kelley's arguments in support of a transcurrent faulting model are as follows: first, along a number of border faults of the Albuquerque Basin sinistral offsets can be observed in the field. For example, the wedge edge of Jurassic formations on the western side of the rift are displaced southward relative to those of the eastern side (Figure 6). Kelley [1977] estimated the displacement value of the left offset to be as much as 35-40 km; second, a series of NW-trending, en-echelon uplifts and grabens occur on both sides of the basin (Figure 6). These uplifts and grabens are oriented approximately N 5°-20° W, and are bounded by high-angle faults, indicating local extension normal to uplifts. Since these faults are arranged as a series of right-stepping, en-echelon faults and have an average angle of 35° between them and the axial trend of the rift, Kelley inferred that both the faults and uplifts resulted from left-lateral, transcurrent movement parallel to the axis of the Albuquerque Basin. On this basis, the faults with N 5°-30° W trend, including those in the Magdalena district reported by both Chapin and Seager [1975] and Laughlin et al. [1983], can be viewed as second-order structures relative to the longitudinal transcurrent faults, as shown on Figure 6.

Kelley [1977, 1979] attributed all these

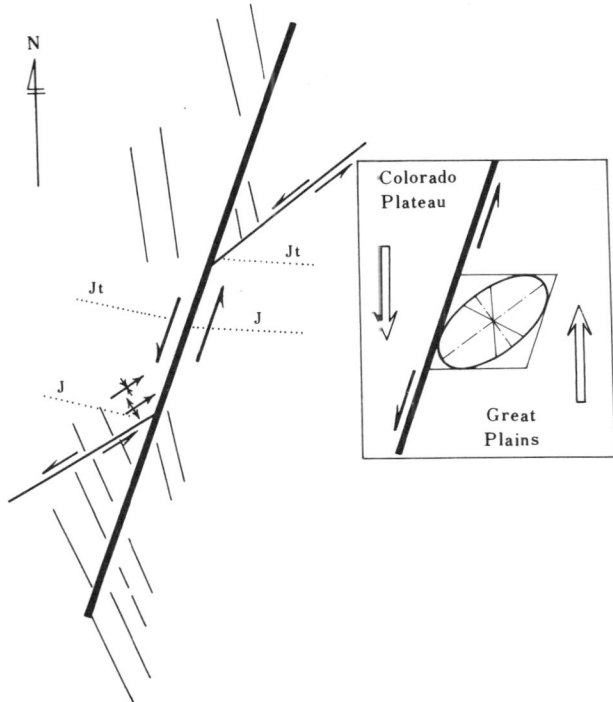

Fig. 7. Schematic tectonic interpretation of the pre-late Miocene structures in the Rio Grande rift area, northern and central New Mexico, based on structures shown in Figure 6. Heavy line - the major transcurrent fault; fine NNW-trending line - the second order boundary fault of uplift; fine NE-trending line - the second order transpressional fault; dotted line - wedge edge of Jurassic formation (Jt = Todilto Formation, J - Morrision Formation); antiform and synform pattern - the structural relief on the Jeter detachment surface. The right mounted figure showing that the various kinds of structures in this area may have resulted from a left-lateral shear couple.

structural phenomena to Laramide-age movement and suggested that later rifting initiated in Miocene time. He did not postulate what relationship existed between the Laramide structural setting and later rifting geometry, or what happened during the time interval from the Laramide movement to late Miocene. The information derived from COCORP suggests an answer. That is, the left-lateral, transcurrent displacement might have lasted until about 10 m.y. ago. The best indications are the existence of the folded detachment plane and the fan-shaped reverse faults.

This left-lateral, transcurrent movement could have resulted from clockwise rotation of the Colorado Plateau relative to the Great Plains (Figure 7). The clockwise rotation of the Colorado Plateau possibly initiated during the Laramide orogeny and continued intermittently until late Miocene. Since 10 m.y. ago, the rotation may have gradually ceased, and the compression-shear stress field in this area may have been replaced by a relaxed stress field. Under this condition, preexisting faults were reactivated in normal sense, and some new faults were formed. The NNE trend of the Albuquerque Basin indicates that the basin is probably controlled by the first-order structure, namely, the left-lateral, transcurrent faults. The second, or subordinate, faults also show their effect by complicating the outline of the basin. Associated with relaxation, predominantly basaltic lava was erupted in the area. The culmination of the relaxation might have occurred in Pliocene time, for both the strongest basaltic eruption and most rapid accumulation of the upper Santa Fe group occurred during that period.

Summary

1. Neither a simple listric fault pattern, nor a close-spaced, step-like fault pattern appears adequate to explain the complicated structures revealed by COCORP Line 1A, especially after a folded detachment surface, associated with a set of fan-shaped reverse faults, is recognized.
2. That the detachment has only involved rocks older than late Miocene and that reverse faults have mainly affected the same sequence indicate that there was a compressional stress field instead of an extensional field in the basin area before about 10 m.y. ago.
3. The compressional condition prior to late Miocene was probably caused by a couple that resulted from the clockwise rotation of the Colorado Plateau relative to the Great Plains. The rotation possibly began during the Laramide orogeny, though subsequently it might have been intermittent.
4. Since 10 m.y. ago the compressional stress field may have gradually been replaced by a normal one, a phenomenon that was set off by the cessation of clockwise rotation of the Colorado Plateau. This change of the stress field from compression to relaxation brought about the apparent variation of the style of tectonic deformation. During the latter period the tectonics has been characterized by normal faulting and eruption of basaltic flows.

Acknowledgements. This paper is a result of reinterpretation of COCORP seismic reflection data on Line 1A in the vicinity of Socorro, New Mexico, carried out while the author was a visiting scientist at the Institute for the Study of the Continents and Department of Geological Sciences at Cornell University. I wish to thank J. Oliver and S. Kaufman for their reviews of the manuscript and their encouragement. I also acknowledge D. Nelson, T. Hauge, L. Brown, C. Merey, B. DeVoogd and T. Byrne for valuable discussions and editorial efforts. The views expressed in this paper, however, are wholly my own. Institute for the Study of the Continents Contribution No. 18.

References

Brown, L. D., P. A. Krumhansl, C. E. Chapin, A. R. Sanford, F. A. Cook, S. Kaufman, J. E. OLiver, and F. S. Schilt, COCORP seismic reflection studies of the Rio Grande rift, in Rio Grande Rift: Tectonics and Magmatism, edited by R. E. Riecker, pp. 169-184, AGU, Washington, D.C., 1979.

Brown, L. D., C. E. Chapin, A. R. Sanford, S. Kaufman, and J. Oliver, Deep structure of the Rio Grande rift from seismic reflection profiling, J. Geophys. Res., 85, 4773-4800, 1980.

Cape, C. D., S. McGeary, and G. A. Thompson, Cenozoic normal faulting and the shallow structure of the Rio Grande rift near Socorro, New Mexico, Geol. Soc. Amer. Bull., 94, 3-14, 1983.

Chamberlin, R. M., G. R. Osburn, C. E. Chapin, M. N. Machette, J. M. Barker, J. W. Hawley, S. M. Cather, J. C. Osburn, and O. J. Anderson, N. Mex. Geol. Soc. 34th Guidebook, 29-59, 1983.

Chapin, C. E., and W. R. Seager, Evolution of the Rio Grande rift in the Socorro and Las Cruces areas, N. Mex. Geol. Soc. Field Conf. Guideb. 26, 297-321, 1975.

Chapin, C. E., Evolution of the Rio Grande rift, in Rio Grande Rift: Tectonics and Magmatism, edited by R.E. Riecker, pp. 1-5, AGU, Washington, D.C., 1979.

Cordell, L., G. R. Keller, and T. G. Hildebrand, Complete Bouguer Gravity Map of the Rio Grande rift, Open File Report 79-958, U.S. Geol. Surv., Denver, Colorado, 1978.

Jurdy, D. M., and T. M. Brocher, Shallow velocity structure of the Rio Grande rift near Socorro, New Mexico, Geology, 8, 185-189, 1980.

Kelley, V. C., Geology of Albuquerque Basin, New Mexico, N. Mex. Bur. Mines Miner. Resour. Mem. 33, 60 pp., 1977.

Kelley, V. C., Tectonics, middle Rio Grande rift, New Mexico, in Rio Grande Rift: Tectonics and Magmatism, edited by R. E. Riecker, pp. 57-70, AGU, Washington, D. C., 1979.

Laughlin, A. W., M. J. Aldrich, and D. T. Vaniman, Tectonic implications of mid-Tertiary dikes in west-central New Mexico: Geology, 8, 45-48, 1983.

Machette, M. N., compiler, Preliminary geologic map of the Socorro 1°x2° quadrangle, central New Mexico, U.S. Geological Survey Open-file Report 78-607, 1978.

Osburn, G. R., and C. E. Chapin, Nomenclature for Cenozoic rocks of northeast Mogollon-Datil volcanic field, New Mexico, N. Mex. Bur. Mines Miner. Resour., Stratigraphic Chart 1, 1983.

Woodward, L. A., Rate of crustal extension across the Rio Grande rift near Albuquerque, New Mexico: Geology, 5, 269-272, 1977.

GEOMETRIES OF DEEP CRUSTAL FAULTS: EVIDENCE FROM THE COCORP MOJAVE SURVEY

M. J. Cheadle[1], B. L. Czuchra[2], C. J. Ando[3], T. Byrne[4], L. D. Brown, J. E. Oliver and S. Kaufman

Institute for the Study of the Continents and Department of Geological Sciences
Cornell University, Ithaca, NY 14853

Abstract. Several reflecting horizons imaged during deep seismic reflection profiling in the western and northern Mojave Desert are interpreted as fault zones which penetrate the deep crust of that region. The most prominent is a complex, though laterally correlatable midcrustal horizon (9-20 km) which extends over the northern area of the Mojave Survey into the Basin and Range Province and is interpreted to be a major southwesterly dipping crustal fault zone. Its shape resembles ramp and flat geometry, which suggests that deep "faults" in crystalline terranes can have geometries similar to thrusts mapped in foreland thrust belts.

The crust-mantle transition appears to be represented by a continuous series of reflections which occur at about 10 s (33 km) in the north of the survey, and at about 8-9 s (26-29 km) in the south. The change in two-way travel time to this horizon, the base of which is interpreted to be the Moho, provides evidence for a fault which offsets the Moho.

The COCORP survey also traversed the two major strike-slip faults that bound the Mojave block. The San Andreas fault zone, though poorly constrained by the seismic data, appears to be a major vertical feature separating Mojave basement, with numerous discontinuous reflections down to 30 km depth, from basement to the south, which is devoid of such reflections. Conversely the Garlock fault appears to be a relatively shallow feature, extending to less than 9 km depth, because it does not offset an underlying reflecting horizon.

---

[1]Present address: Dept. of Earth Sciences, Cambridge University, Cambridge, CB3 0EZ

[2]Present address: Tenneco Oil Company, PO Box 51345, Lafayette, LA 70505

[3]Present address: Shell Development Company, PO Box 481, Houston, TX 77001

[4]Present address: Dept. of Geological Sciences, Brown University, Providence, RI 02912

Introduction

Deep seismic reflection profiling over the continents, in conjunction with earthquake [Sibson, 1982; Chen and Molnar, 1983] and crustal rheological studies [Sibson, 1983; Kusznir and Park, 1984], has begun to provide important information about the geometry and extent of major faults or shear zones at depth. However, many questions still remain: How deep do discrete faults penetrate? What do strike-slip faults look like at depth? Are the "rules of faulting" derived from surface geological mapping in foreland thrust belts [Dahlstrom, 1970; Boyer and Elliot, 1982] applicable at depth?

This paper describes results from the COCORP Mojave survey that pertain to the above questions by discussing briefly the detailed geometry of possible low-angle, intracrustal fault zones in the Mojave block, the possiblity of faults at Moho depths and the depth and geometry of two major strike-slip faults, the San Andreas and Garlock faults. In this paper the term "fault zone" refers to major, low-angle structures imaged by the seismic survey. Whether these structures are brittle faults or discrete ductile shear zones is unclear and cannot be determined from seismic data.

Burchfiel and Davis [1981] reviewed the geology of the Mojave block, which appears to be a predominantly crystalline province dominated by Mesozoic granites which intrude Precambrian crust and overlying Paleozoic cover rocks. The presence within the region of the Pelona-Orocopia-Rand Schist, whose age, origin, and mode of emplacement remain enigmatic [Mukasa et al., 1984; Burchfiel and Davis, 1981], gives an indication of the geological complexity of the Mojave block. Several relatively shallow (1-2 km) Tertiary basins have been superimposed on this older crystalline basement [Dibblee, 1967].

In 1982 the Consortium for Continental Reflection Profiling (COCORP) collected six deep seismic reflection profiles totalling 300 km in length within the Mojave Desert region (Figure 1). The results and processing of this survey are described in detail by Cheadle et al.

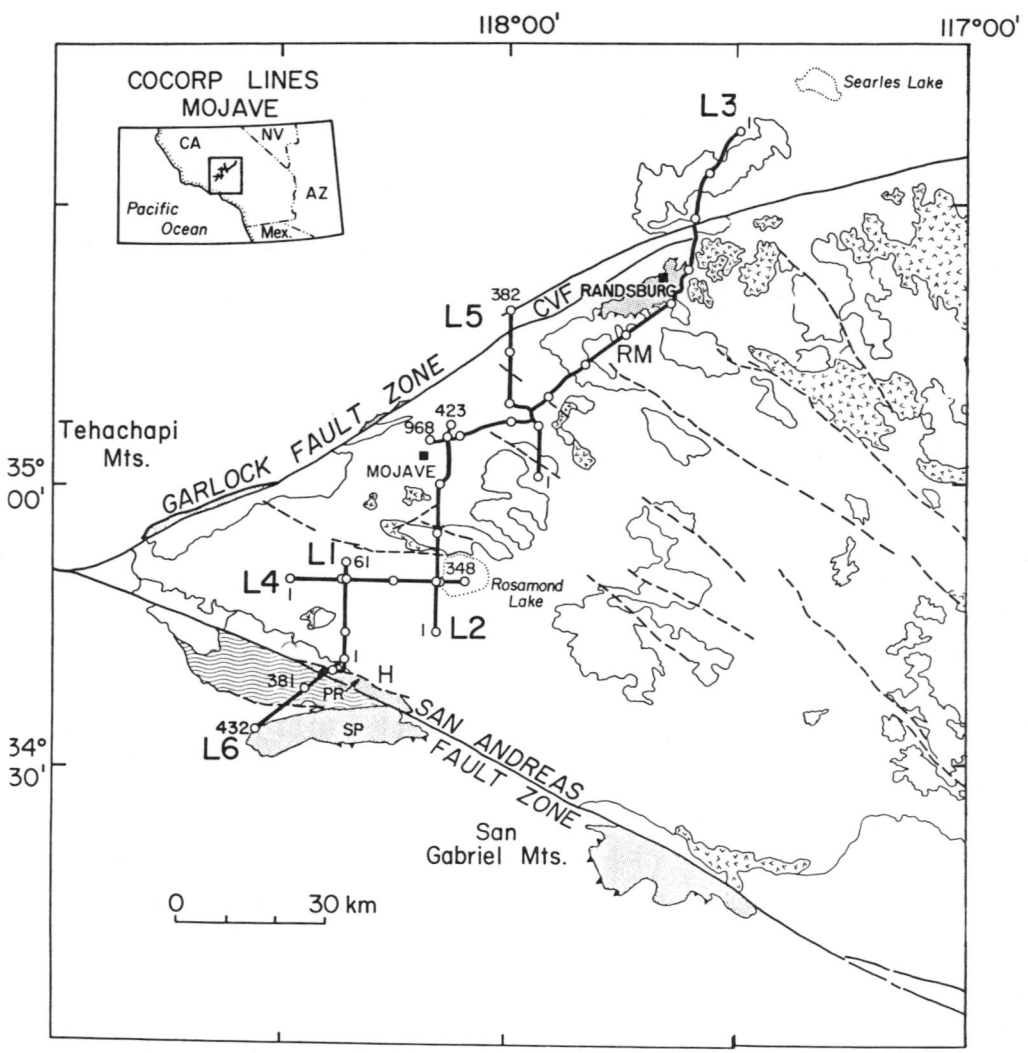

Fig. 1. Generalized geologic map of the Mojave Desert showing the positions of the COCORP seismic reflection lines. Inset shows position of map in southern California. Dotted pattern = Rand-Pelona-Orocopia Schist; V pattern = Tertiary sediments and volcanics; wavy pattern = Precambrian outcrops; white = Mesozoic granites; unshaded = alluvium; faults shown as bold lines. Abbreviations for faults are: CVF, Cantil Valley fault; H, Hitchbrook fault; abbreviations for mountains are: PR, Portal Ridge; RM, Rand Mountains, SP, Sierra Pelona.

[1985]. The survey revealed several major southwesterly-dipping reflecting horizons and a prominent Moho-depth horizon which appears to be offset beneath the northwestern Mojave Desert (Figure 2). The uppermost of these horizons can be traced to the surface and has been identified as the Rand thrust; however, the deep horizons are not traceable to the surface. Cheadle et al. interpreted these deeper horizons to be major fault zones and proposed three models for their origin. The horizons may be 1) the westward deepening crustal continuation of the system of Mesozoic thrusts which crops out in southern Nevada and southeastern California; 2) late Cretaceous to early Cenozoic northeast-vergent thrusts related to the emplacement of the Pelona-Orocopia-Rand Schist; or 3) low-angle normal faults related to early Miocene northeast-southwest directed crustal extension.

Results

The Geometry of the Mid- to Deep-Crustal Reflecting Horizon

Figure 2 summarizes the COCORP Mojave data, and delineates the major reflecting horizons

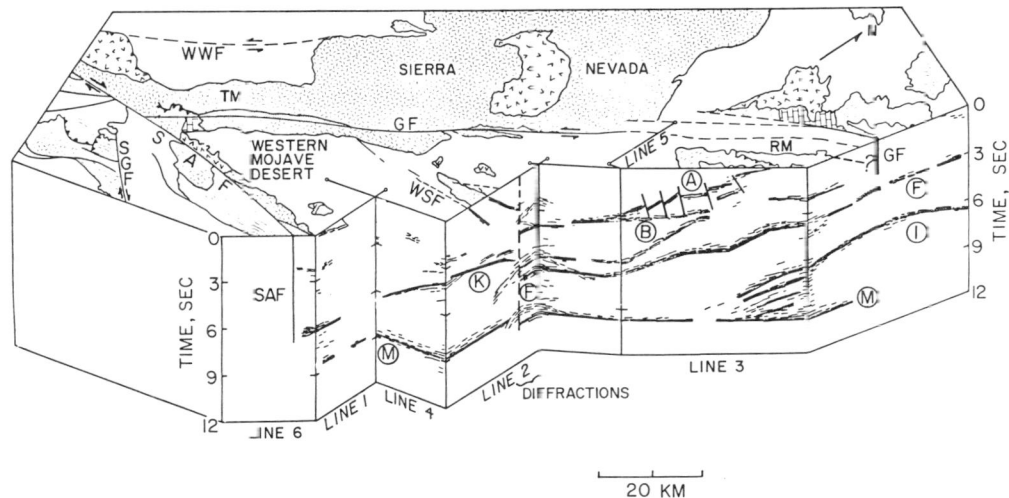

Fig. 2. Summary block diagram showing major interpreted features of the Mojave survey. Abbreviations: GF, Garlock fault; SAF, San Andreas fault; SGF, San Gabriel fault; WSF, Willow Springs fault; WWF, White Wolf fault; R = Rand Mountains; THM, Tehachapi Mountains; A and B = Rand thrust and related horizon; M = Moho-depth horizon, F, I, and K = Mid- and Deep-crustal horizons. Dotted pattern = Rand; Tehachapi Schist, wavy pattern = Precambrian rocks; brick pattern = Paleozoic sediments; V pattern = Tertiary volcanic and sedimentary rocks; dashed pattern = Mesozoic granites.

identified by Cheadle et al. [1985]. The reflecting horizons are discontinuous, consisting of discrete, but laterally correlative, events which are often multicyclic with a duration of up to 1 s two-way travel time (TWTT). Most of these horizons cannot easily be correlated with the known surface geology because they are not traceable to the surface over the length of the survey. In this respect the data are typical of many other deep seismic reflection profiles [Gibbs et al., 1984; Klemperer et al., 1985]. When interpreting such data, care must first be taken to distinguish diffractions, sideswipe and reflected refractions from "real" reflections. Then, often only a combination of the characteristic geometry and extent of the reflections, and

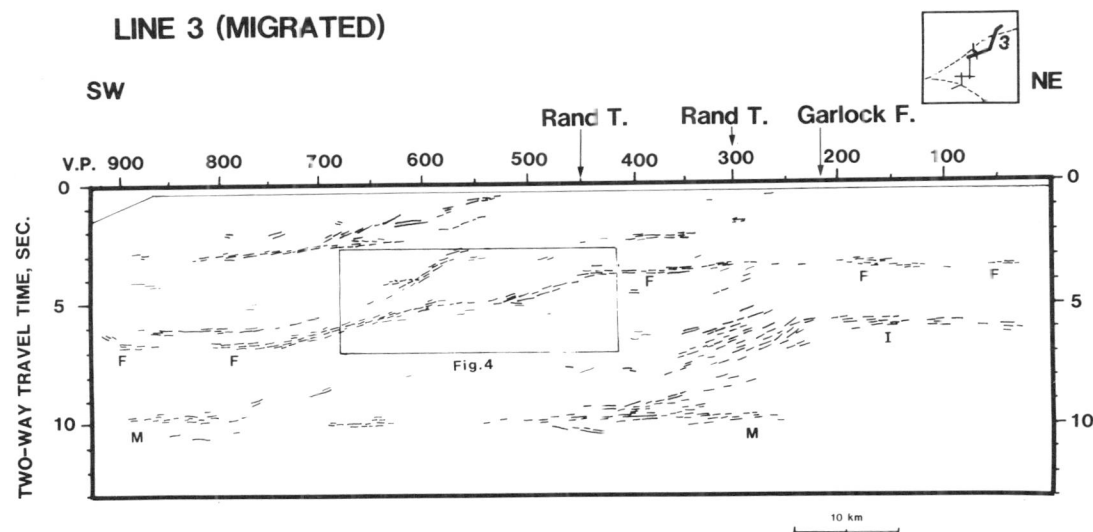

Fig. 3. Line 3: line drawing of the complete migrated line. Lettered reflections are discussed in the text. Numbers along top of section are VP numbers which key to figure 1. Scale is 1:1 for 4.5 km/s. To convert vertical scale to approximate depth scale multiply TWTT by 3 km/s.

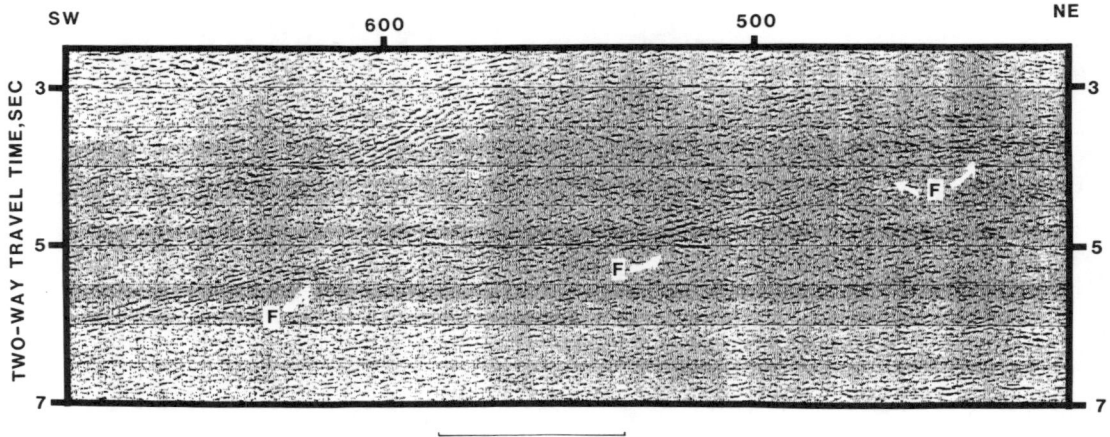

Fig. 4. Detail of COCORP Mojave Line 3, VPs 420-680, illustrating reflections discussed in text (2.8 to 7 s TWTT only). Scale is 1:1 for 4.5 km/s. To convert vertical scale to approximate depth scale multiply TWTT by 3 km/s.

the geological history of the region can provide an indication of their origin. Such an approach suffers from the limitations of current geological knowledge. For example the form of the bottom of a batholith, the form of a continent-continent suture, and the geometries of faults at depth are largely subjects for speculation. In fact, it is questionable whether reflective features with complicated shapes are even recognizable in deep seismic data [e.g., Wong et al., 1982].

Several of the reflecting horizons visible on the Mojave data (Figure 2) are gently-dipping and extend over relatively large distances (> 50 km) and so are plausibly interpreted to be major low-angle faults. Horizon F (Figure 2) is the best example, and the other gently-dipping horizons are described in Cheadle et al. [1985]. On Line 3 (Figure 3) horizon F is interpreted to extend from VP 50 at 3.2 s TWTT to VP 900 at 6.8 s TWTT, and therefore it is traceable over approximately 100 km of the survey to depths of at least 20 km.

Three-dimensional control on the geometry of horizon F is provided by the network of seismic lines that cross F. Two-dimensional wave equation migration of the data employing interval velocities calculated from the data for the upper crust, and refraction velocities [Roller and Healey, 1963; Prodehl, 1979] for the lower crust, indicate that horizon F on Line 3 (Figures 3 and 4) has a ramp and flat geometry typical of thrusts in foreland thrust belts and possibly characteristic of low-angle normal faults [Dahlstrom, 1970]. The ramps dip at approximately 30° south-southwest, an angle characteristic of the ramps of thrust faults exposed at the surface [Boyer and Elliot, 1982]. The dip directions of the ramps imply that the fault zone is either a north-northeast-vergent thrust fault or a low-angle normal fault along which the direction of movement was to the south-southwest. Further seismic reflection profiling is required to trace this structure to the surface and thus distinguish between these two possibilities.

The identification of horizon F is based on its extent and form and on the geology of the region. If this interpretation is correct, it suggests that the geometry of major, deep thrust or normal faults can be similar to the geometry of thrusts mapped in foreland thrust belts and that they can exhibit this geometry in crystalline terranes. The existence of major thrusts with ramp and flat geometry in crystalline terranes has been suggested by Bartley [1981] and Hodges et al. [1982] for the Scandinavian Caledonides.

Moho Depth Reflections

A conspicuous feature of much of the Mojave data is the presence of a series of horizontal reflections between 8.5 and 10 s TWTT (28-33 km). These events occur as both individual reflections and as packets of reflections of up to 1 s duration, and they define a discontinuous horizon M (Figure 2) which is mainly horizontal [Cheadle et al., 1985]. Horizon M is distinguished from the other predominantly southwesterly-dipping horizons seen in the Mojave data by its depth and its lack of an apparent dip.

Cheadle et al. [1985] compared the two-way travel times to horizon M with two-way travel times to the Moho calculated from the crustal models derived from refraction data across the Mojave Desert. They concluded that horizon M occurs at depths appropriate for the crust-mantle transition and that the Moho itself may corre-

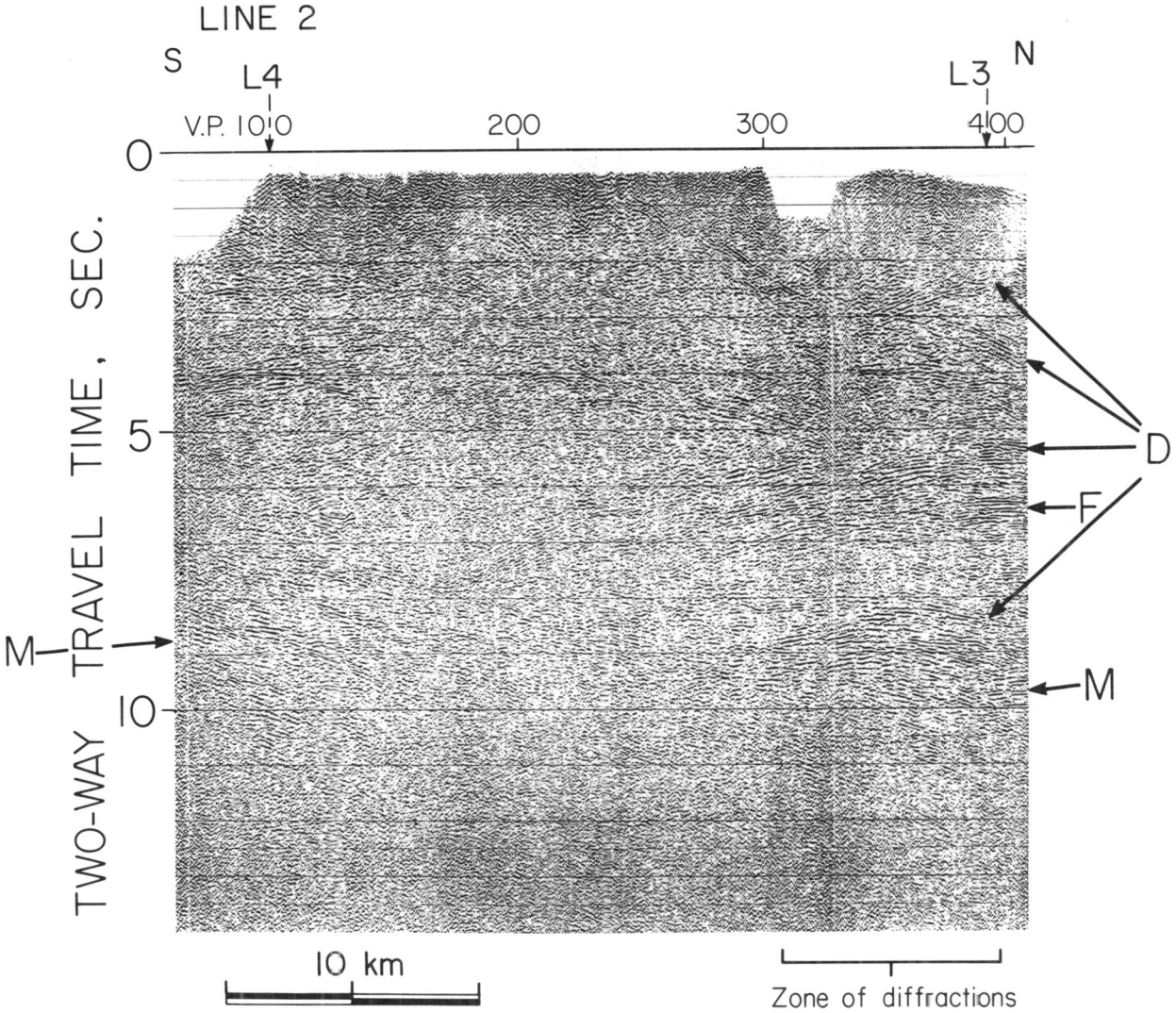

Fig. 5. Line 2: unmigrated time section. Numbers along top of section are VP numbers which key to figure 1. Scale is 1:1 for 4.5 km/s. To convert vertical scale to approximate depth scale multiply TWTT by 3 km/s. Abbreviations: D, Diffractions; F, Horizon F; M, Horizon M.

spond to the base of horizon M, as suggested by other seismic refraction surveys [Barton et al., 1984].

There is an abrupt change in TWTT to the base of horizon M from 9.5-10 s on Line 3 (Figure 3) and the northern part of Line 2 (Figure 5) to 8.5-9.0 s on the southern part of Line 2 (Figures 2 and 5). Realistic crustal velocity variations cannot explain this change in TWTT, therefore it probably reflects a real change in depth to the Moho. This Moho offset occurs below a complicated vertical zone of numerous diffractions and reflections on Line 2, through which none of the intracrustal horizons (B, F and K, Figure 2) can be traced (Figure 5), suggesting that a near-vertical, crustal penetrating fault exists near the northern end of Line 2 in the vicinity of the town of Mojave. A strike-slip fault, possibly the northern continuation of the Mojave-Sonoran Megashear [Silver and Anderson, 1974] or a proto San Andreas fault [Graham, 1976], is a possible identity. Faults which offset the Moho have been reported from active collisional orogens [the Himalayas and the Apennines - Hirn et al., 1984; Morelli, this volume] and stable cratons [India - Roy Chowdhury and Hargraves, 1981].

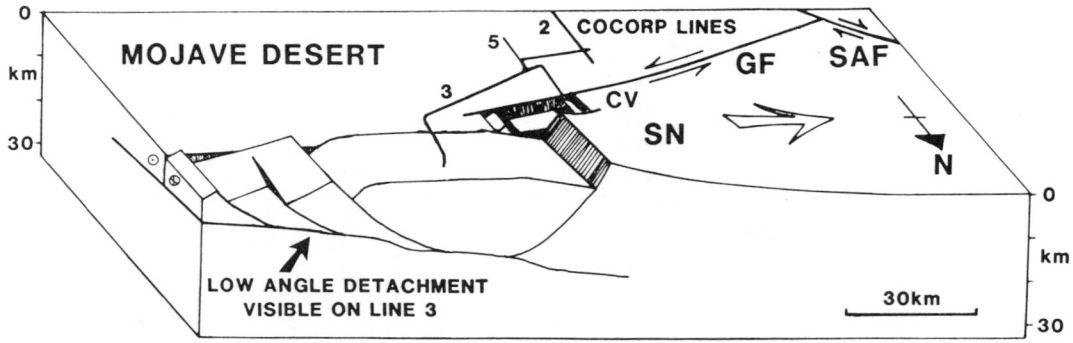

Fig. 6. Schematic model, based on Wernicke (1981) and Stewart (1983), illustrating the possible relationship of the Garlock fault to the low-angle detachment (horizon F) discussed in the text. Abbreviations: GF, Garlock fault; SAF, San Andreas fault; CV, Cantil Valley; SN, Sierra Nevada.

Strike-Slip Faults

The geometry of strike-slip faults at upper crustal levels is relatively well-known. However, the deep crustal geometry of strike-slip faults is poorly understood. Sibson [1983] reviewed the possible configuration of strike-slip faults at depth and concluded that they either merge into steeply-dipping shear zones which may remain constant in width, or alternatively may widen with depth or, less likely, merge into subhorizontal decoupling horizons.

The imaging of vertical or very steeply dipping faults, particularly strike-slip faults, is a major problem in deep seismic reflection profiling. Without fault plane reflections, which are virtually impossible geometrically, diffractions and reflected refractions can sometimes be used to suggest the presence of very steep faults. But perhaps the best constraints on the shape and depth of strike-slip faults are provided by marker horizons (for example, sedimentary or structural reflections) which may be truncated by the fault or pass under the surface trace of the fault. Unfortunately, such marker horizons are often rare on deep seismic profiles.

The COCORP Mojave survey crossed two major strike-slip faults, the San Andreas and Garlock faults, and both appear to show different geometries. The Garlock fault appears to be confined to the upper crust by two reflection horizons, I and F, which pass beneath its surface trace without any apparent offset (Figure 2), though the reflections which comprise the horizons vary in amplitude and continuity [Cheadle et al., 1985]. The geometry of the Garlock fault above horizon F is unknown because marker events, fault plane reflections, diffractions or reflected refractions, which might delineate its geometry are absent. It is possible that the fault extends vertically down to 9 km and terminates at horizon F. Horizon F, or at least the part of it to the north of the Garlock fault, may therefore be a low-angle detachment associated with extension within the Basin and Range Province (Figure 6). Such a low-angle detachment may be an older thrust which has been reactivated to the north of the Garlock fault during Basin and Range extension.

A shallow Garlock fault is consistent with field mapping of the eastern end of the fault by Burchfiel et al. [1983], who suggest that the Garlock fault is a shallow tear fault bounding the Kingston Range Detachment to the north. It is also compatible with models of detachment faulting in the Death Valley region [for example, Stewart, 1983].

The San Andreas fault is poorly constrained by the seismic data, but the presence of reflections down to 30 km depth to the north of the fault and the complete absence of reflections to the south of the fault within the San Gabriel Mountains (Figure 2) is consistent with the San Andreas fault extending as a vertical feature down to at least 30 km.

Two other deep seismic profiles cross the San Andreas fault and are equally disappointing: the COCORP Parkfield Line [Long, 1981] and the USGS San Benito County Line [McEvilly, 1981]. Both lines provide little information about the detailed geometry of the San Andreas fault, but they reveal gently-dipping midcrustal reflections and a Moho-depth reflection within the Salinian block to the west of the fault and few reflections from the Franciscan terrane to the east. Removal of the late Tertiary displacement of the San Andreas fault suggests that the Salinian block may once have been contiguous with the Mojave block [Graham, 1978] and therefore that the reflectors imaged by the COCORP Parkfield and USGS lines may represent the continuation of the major low-angle structures imaged by the COCORP Mojave survey. The similar depth Moho reflections support this speculative interpretation.

Conclusions

COCORP deep seismic reflection profiling in the Mojave Desert of Southern California has revealed deep crustal structures which provide constraints on the relatively poorly understood subject of the geometry and extent of deep crustal faults. The data reveal:

1) Several major low-angle faults within the Mojave block, one of which may exhibit ramp and flat geometry to depths of 20 km. This suggests that faults may develop ramp and flat geometries at depth within crystalline terranes.
2) That the Moho may be offset, possibly by a major strike-slip fault, in the Mojave region.
3) That some major strike-slip faults, such as the Garlock fault, are relatively shallow crustal features. The San Andreas fault is poorly imaged, but the data is consistent with it extending to at least 30 km depth.

Acknowledgments. We thank R. Renner and P.E. Malin for their valuable advice and assistance. The COCORP project is supported by National Science Foundation Grant EAR-8212445. Data collection was carried out by crew 6834 of Petty-Ray Geophysical, a division of Geosource Inc. The data were processed on the MEGASEIS (trademark, Seiscom Delta, Inc.) computing system of Cornell University. Cornell Institute for the Study of the Continents Contribution No. 24.

References

Bartley, J. M., Basement thrust ramps and redeformation of the Caledonian nappe stack, North Norway, Geol. Soc. Am., Abstr. with Programs, 13, 1981.

Barton, P., D. Matthews, J. Hall, and M. Warner, Moho beneath the North Sea compared on normal incidence and wide-angle seismic records, Nature, 308, 55-56, 1984.

Boyer, S. E., and D. Elliot, Thrust systems, AAPG Bull. 66, 1196-1230, 1982.

Burchfiel, B. C., and G. A. Davis, Mojave desert and environs, in The Geotectonic Development of California, (Rubey; vol. 1), edited by W. G. Ernst, pp. 218-252, Prentice-Hall, Englewood Cliffs, N.J., 1981.

Burchfiel, B. C., J. D. Walker, G. A. Davis, and B. Wernicke, Kingston Range and related detachment faults - a major "breakaway" zone in the southern Great Basin, Geol. Soc. Am., Abstr. with Programs, 15, 536, 1983.

Cheadle, M. J., B. L. Czuchra, T. Byrne, C. J. Ando, J. E. Oliver, L. D. Brown, S. Kaufman, P. E. Malin, and R. A. Phinney, The deep crustal structure of the Mojave Desert, California, from COCORP seismic reflection data, Tectonics, in press, 1985.

Chen, W. P., and P. Molnar, Focal depths of intracontinental and intraplate earthquakes and their implications for the thermal and mechanical properties of the lithosphere, J. Geophys. Res., 88, 4183-4214, 1983.

Dahlstrom, C. D. A., Structural geology in the eastern margin of the Canadian Rocky Mountains, Bull. Can. Pet. Geol., 84, 1407-1422, 1970.

Dibblee, T. W., Jr., Areal geology of the western Mojave Desert, Calif., U.S. Geol. Survey Prof. Paper 522, 153 pp., 1967.

Gibbs, A. K., B. Payne, T. Setzer, L. D. Brown, J. E. Oliver, and S. Kaufman, Seismic-reflection study of the Precambrian crust of central Minnesota, Geol. Soc. Am. Bull. 95, 280-294, 1984.

Graham, S. A., Role of the Salinian block in the evolution of the San Andreas fault system, AAPG Bull., 62, 2214-2231, 1978.

Hirn, A., J.-C. Lepine, G. Jobert, M. Sapin, G. Wittlinger, X. Z. Xin, G. E. Yuan, W. X. Jing, T. J. Wen, X. S. Bai, M. R. Pandey, and J. M. Tater, Crustal structure and variability of the Himalayas border of Tibet, Nature, 307, 23-25, 1984.

Hodges, K. V., J. M. Bartley, and B. C. Burchfiel, Structural evolution of an A-type subduction zone, Lofoten-Rombak area, Northern Scandinavian Caledonides, Tectonics, 441-462, 1982.

Klemperer, S. L., L. D. Brown, J. E. Oliver, C. J. Ando, B. L. Czuchra, and S. Kaufman, Some results of COCORP seismic reflection profiling in the Grenville-age Adirondack Mountains, New York State, Can. Jour. Earth Sci., 22, 141-153, 1985.

Kusznir, N. J., and R. G. Park, Intraplate lithosphere deformation and the strength of the lithosphere, Geophys. J.R. Astr. Soc., 78, 1984.

Long, G. H., A COCORP deep seismic reflection profile across the San Andreas fault, Parkfield, California, M.A. Thesis, Cornell University, Ithaca, New York, 1981.

McEvilly, T. V., Extended reflection survey of the San Andreas fault zone, USGS Open File Report 81-388, 38 pp., 1981.

Morelli, C., Deep crustal knowledge in Italy, this volume.

Mukasa, S. B., J. T. Dillon, and R. M. Tosdal, A Late Jurassic minimum age for the Pelona-Orocopia Schist protolith, southern California, Geol. Soc. Am., Abstr. with Programs, 16, 323, 1984.

Prodehl, C., Crustal structure of the western United States, U.S. Geol. Sur. Prof. Paper 1034, 74 pp., 1979.

Roller, J. C., and J. H. Healy, Seismic refraction measurements of crustal structure between Santa Monica Bay and Lake Mead, J. Geophys. Res., 68, 5837-5849, 1963.

Roy Chowdhury, K., and R. B. Hargraves, Deep seismic sounding in India and the origin of continental crust, Nature, 291, 648-650, 1981.

Sibson, R. H., Fault zone models, heat flow and the depth distribution of earthquakes in the continental crust of the United States, Bull. Seis. Soc. Am., 72 (1), 151-163, 1982.

Sibson, R. H., Continental fault structure and the shallow earthquake source, J. Geol. Soc. Lon., 140, 741-767, 1983.

Silver, L. T. and T. H. Anderson, Possible left-lateral early to middle Mesozoic disruption of the southwestern North America craton margin, Geol. Soc. Am., Abstr. with Programs, 6, 955-956, 1974.

Stewart, J. H., Extensional tectonics in the Death Valley area, California, transport of the Panamint Range structural block 80 km northwestward, Geology, 11, 153-157, 1983.

Wernicke, B., Low-angle normal faults in the Basin and Range Province: Nappe tectonics in an extending orogen, Nature 291, 645-648, 1981.

Wong, Y. K., S. B. Smithson, S.B., and R. L Zawislak, The role of seismic modelling in deep crustal reflection interpretation. Part 1: Contributions to Geology, University of Wyoming, 20, 91-109, 1982.

STRUCTURE OF THE LITHOSPHERE IN A YOUNG SUBDUCTION ZONE:
RESULTS FROM REFLECTION AND REFRACTION STUDIES

Ron M. Clowes, George D. Spence[1], Robert M. Ellis and David A. Waldron[2]

Department of Geophysics and Astronomy, University of British Columbia, Vancouver

Abstract. Vancouver Island, on the west coast of Canada, is believed to have formed through the process of accretionary tectonics and represents a dispersed block of the Wrangellia terrane. Presently, the oceanic Juan de Fuca plate is subducting beneath the continental plate. A series of refraction and reflection experiments has been carried out to provide better delineation of lithospheric structure in this complex zone of convergence. An offshore-onshore refraction profile has enabled development of a lithospheric model from the deep ocean to near the volcanic arc. Seaward of the continental slope, the plate which includes a 9 km thick crust dips ~1° landward; below the slope the angle of dip increases to ~3°; and below the continental shelf it increases substantially to ~15°. At depths of 30 to 40 km below the shelf, a landward-dipping upper mantle boundary has been inferred from wide-angle reflection phases, and is interpreted as the base of the subducting oceanic lithosphere. In the overlying continental crust a large block of high velocity (~7.7 km/s) material is embedded and may represent a detached remnant of subducted slab. A test multichannel explosion profile shows reflections to two-way traveltimes of 11 s; four reflectors correspond closely to boundaries in the refraction model. A 1984 follow-up Vibroseis program has produced high quality data, which, combined with the refraction data, illustrate the complementary nature of the two data sets.

Introduction

An active zone of plate convergence exists off the west coast of Canada. The oceanic Juan de Fuca and Explorer plates are being subducted beneath the continental America plate (Figure 1) at convergence rates of about 40 and less than 20 mm/yr, respectively (Riddihough, 1977, 1984). Geophysical and other evidence (Riddihough and Hyndman, 1976; Keen and Hyndman, 1979) support the argument that subduction has occurred, and is presently occurring, along the western margin of Canada and the northwestern United States. Seismicity studies confirm present subduction of the Juan de Fuca plate. Analyses by Crosson (1981) and Taber (1983) show a classic Benioff zone dipping at 11° under the Olympic Peninsula and Puget Sound. Similarly, Rogers (1983) determines the Benioff zone to be dipping northeast at 12° below southern Vancouver Island and Georgia Strait.

Above the Benioff zone lies the westernmost portion of the Canadian Cordillera, formed dominantly by accretionary tectonics. Wrangellia (Jones et al., 1977) is the principal terrane forming Vancouver Island, although other blocks are recognized (Monger et al., 1985). According to Yorath and Chase (1981), the Wrangellia terrane collided and amalgamated with the Alexander terrane, a diverse assemblage of Paleozoic rocks, about 140 Ma ago as they both were drifting northward. The two joined terranes collided and amalgamated with the North American continent from about 90 to 40 Ma ago. Little information exists concerning the structures or tectonics associated with the amalgamation.

The 1980 Vancouver Island Seismic Project (VISP) was undertaken to provide a seismic structural model from the deep ocean of the Juan de Fuca plate to the inland volcanic arc of the America plate, and in particular to provide better delineation of the subduction zone and other structures in the region of Vancouver Island. A series of onshore-offshore refraction and reflection experiments was carried out; details are included in Ellis et al. (1983). Figure 1 shows the location of the profiles for which interpretations have been completed. McMechan and Spence (1983) interpreted profile NAF along Vancouver Island, while Waldron (1982) interpreted the marine profile P-P' recorded on ocean bottom seismographs (OBSs). These interpretations provided constraints for the analysis of the onshore-offshore profile PJ and subsequent interpretation of the subduction zone architecture (Spence, 1984; Spence et al., 1985). The 1980 reflection program (RL on Figure 1) was a feasibility study which included a 10 km, 1200% common reflection point explosion survey and a test using offshore airgun

---

[1]Now at Bullard Laboratories, Department of Earth Sciences, University of Cambridge, England.
[2]Now with Helix Limited, London, England.

313

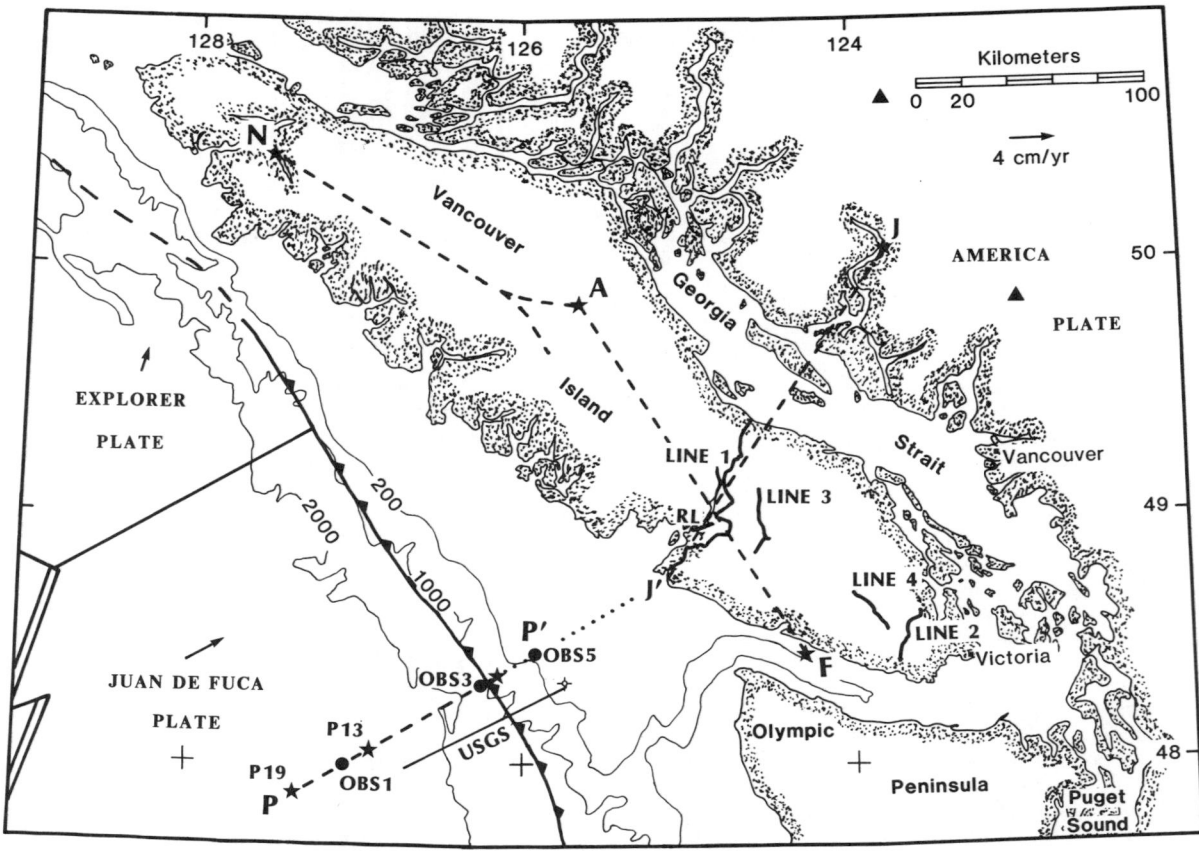

Fig. 1. Tectonic and profile location map. Arrows show directions and magnitudes of plate motions relative to North America. Solid triangles identify the location of two inland volcanos. Heavy, short dashed lines show refraction profiles. Stars on Vancouver Island show individual shot locations; offshore, stars near P and P' identify the ends of a line of explosive charges. Solid circles are ocean bottom seismograph locations. P13 and P19 identify two particular shots mentioned in the paper. USGS identifies the location of a 2400% off shore multichannel reflection profile; the well symbol on the shelf is the location of the Shell-Anglo Cygnet drill hole. On Vancouver Island, RL is a 1200% explosion test reflection profile; Lines 1 to 4 are 3000% Vibroseis reflection profiles. Bathymetric contours are in meters.

shots recorded on a 4.8 km spread on land (Clowes et al., 1983).

The success of the VISP program in terms of the feasibility reflection experiments and the existence of a seismic structural model from the refraction experiments led to the selection of southern Vancouver Island as the locale of a major seismic reflection survey for Phase 1 Lithoprobe. Lithoprobe (CANDEL, 1981; Clowes, 1984) is a new Canadian research initiative to study the interrelationship between the structure of the lithosphere and surface geology. Seismic reflection technology will spearhead the program; refraction experiments, other geophysical, geological and geochemical investigations will provide supporting data to complement the reflection results and enable integrated interpretations. In the 1984 program, 206 km of 30-fold Vibroseis reflection data were acquired (Figure 1). These new data, combined with the refraction data, illustrate the complementary nature of the two data sets and enable a more complete interpretation of both.

Offshore Crustal Model

The 100 km marine portion of line PJ consisted of 37 shots recorded on three OBSs deployed in the deep ocean, on the continental slope, and outer continental shelf (Figure 1). To provide more details of upper crustal structure, shots from a 32 litre airgun were fired into the OBSs at intervals of 250 m for distances from 15 to 20 km. A continuous seismic profile (CSP) using a 5 litre airgun provided constraints on sediment structure from OBS1 to the base of the slope. The interpreted seismic models were also constrained by two

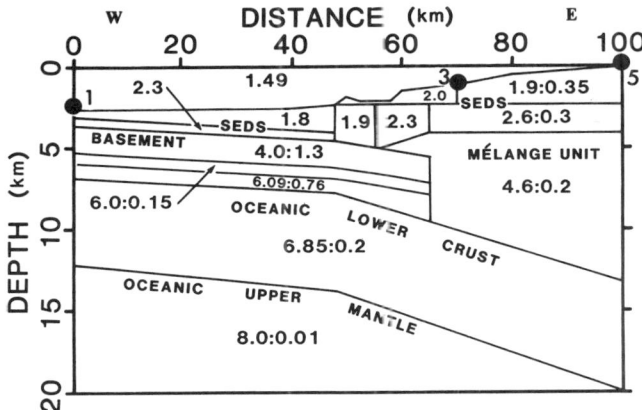

Fig. 2. Seismic interpretation for the continental margin from OBS1 to OBS5, interpreted from data on OBSs 1, 3 and 5 (solid circles with numbers). Velocities (km/s) are given for the top of each region, followed after the colon by the velocity gradient (km/s/km) if one was used.

nearby multichannel reflection sections, log data from a well on the outer shelf, and gravity data. One- and two-dimensional seismic interpretation procedures, with an emphasis on ray tracing (Whittall and Clowes, 1979) and 2-d synthetic seismograms (McMechan and Mooney, 1980) were used to evolve a final model which was consistent with all of the information contained in the various data sets. Waldron (1982) provides details of the interpretation.

Figure 2 shows the final structural model between OBSs 1 and 5. Each of the blocks has a fixed P-wave velocity at the upper boundary and a constant velocity gradient normal to it. Vertical boundaries in the model are an artifact of the modelling procedure and are intended to approximate lateral changes in velocity. In the sediments, velocities are constrained by traveltimes from CSP and multichannel reflection sections, stacking velocity determinations from the multichannel data, sonic logs in a nearby well and the interpretation of the airgun-CBS data. All other velocities arise from the refraction modelling of

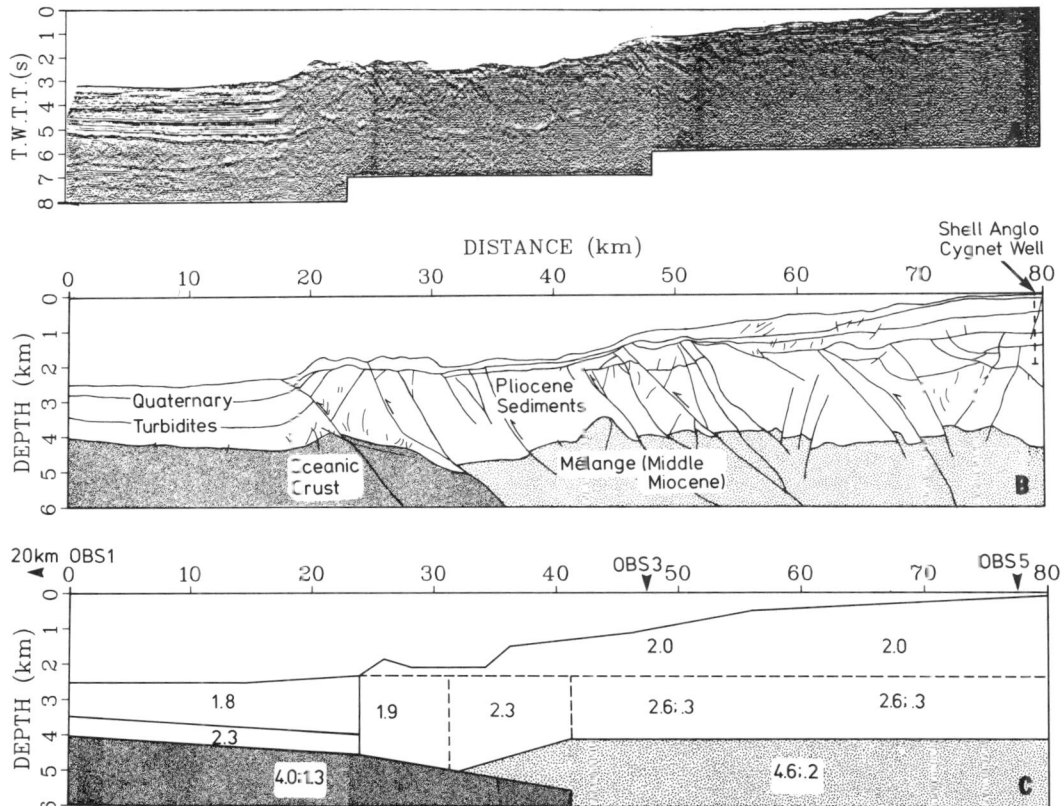

Fig. 3. (a) Unmigrated 2400% reflection profile along line USGS of Figure 1 (from Snavely and Wagner, 1981). (b) Geological interpretation of (a). The interpretation of the distance-time section by Snavely and Wagner (1981) was converted to a distance-depth section using the refraction velocities of Figure 2. (c) Upper crustal velocity structure from Figure 2 to be compared with (b). Velocities and gradients as in Figure 2. Note the good correspondence between the two interpretations.

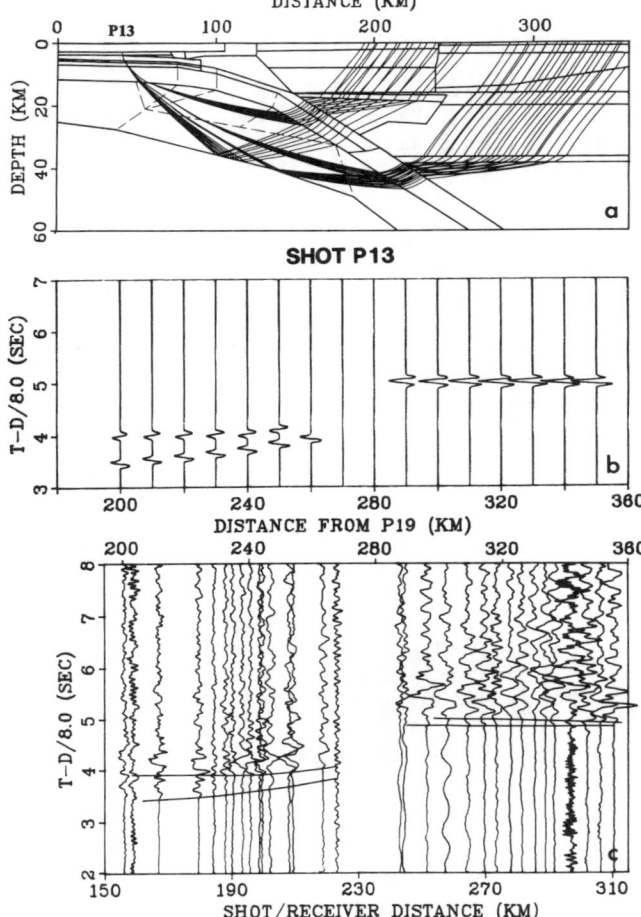

Fig. 4. (a) Structural configuration for the interpretation of line PJ. For shot P13 turning rays and rays reflected from a mantle discontinuity (the base of the lithosphere) are traced to locations on Vancouver Island and on the mainland. (b) Synthetic seismograms for shot P13. (c) Observed record section for shot P13. Model traveltimes from the synthetic seismograms are superimposed. Primary arrivals are turning rays through the mantle. Secondary arrivals correspond to rays reflected from an upper mantle boundary, with the velocity below the boundary smaller than the velocity above (see Figure 5a).

the explosion data. The sediment-basement interface dips eastward at 1.4°; sub-basement layers are assumed to dip at the same angle. These layers increase in dip to ~3° under the continental rise. Beneath the outer edge of the continental shelf there is a structural change from the relatively high velocity layers of the upper oceanic crust (4.0 to 6.85 km/s) to a block in which the reduced gradient of 0.2 km/s/km results in a lowered average velocity; the block extends down more than 5 km from a depth of 4 km, at which the velocity is 4.6 km/s. No evidence for a sharp transition from one region to the next exists.

The position of this block agrees well with that of a middle Miocene melange unit as interpreted from a multichannel reflection profile (Figure 1) by Snaveley and Wagner (1981). They interpreted the seismic section (Figure 3a) in terms of two-way traveltime. Using the velocities from our refraction model we have converted their interpretation to a depth section (Figure 3b). Figure 3c shows the equivalent part of the offshore seismic structural model for comparison. Clearly there is good correspondence between Figures 3b and 3c. This represents an example of the importance of combining refraction data with reflection data to provide a more complete interpretation of both.

Onshore-Offshore Subduction Zone Model

The 350 km offshore-onshore line PJ consisted of 32 land-based seismographs deployed across Vancouver Island and on the mainland. These recorded two shots (J series) at the eastern end of the line and 17 shots (P series) along the westernmost 100 km of the profile (Figure 1). Spence et al. (1985) and Spence (1984) present a complete interpretation of this extensive data set.

An important control for the interpretation of line PJ was the two-dimensional interpretation of profile NAF along Vancouver Island by McMechan and Spence (1983). This provided a crustal velocity model for the region near the intersection of the two profiles, which occurs at a distance of 235 km on Figures 4a and 5a. To a depth of 20 km, the model was well constrained by the data, but ambiguity in lower crustal structure persisted. McMechan and Spence's (1983) preferred interpretation contained a low velocity zone throughout the lower crust and an upper mantle velocity of 7.5 km/s at 37 km depth.

Interpretation was based on traveltime modeling and calculation of synthetic seismograms for 2-d models. For the former, the method described by Spence (1984) is an iterative inversion technique for explosion traveltime data in which shots at several locations are recorded on the same set of receivers. Synthetic seismograms were calculated using the asymptotic ray theory approach of Spence et al. (1984).

Three shots of the P series, with particularly good arrivals at the 32 stations to enable accurate picking of the traveltimes of first arrival and secondary phases, were selected for use in the traveltime inversion procedure. Multiple record sections compiled from the data for one shot recorded at all receivers and for one receiver recording all shots (somewhat equivalent to reversed profiles) were compared with similar theoretical sections. As an example, Figure 4a shows the final structural model derived from these interpretive methods. In the inversion

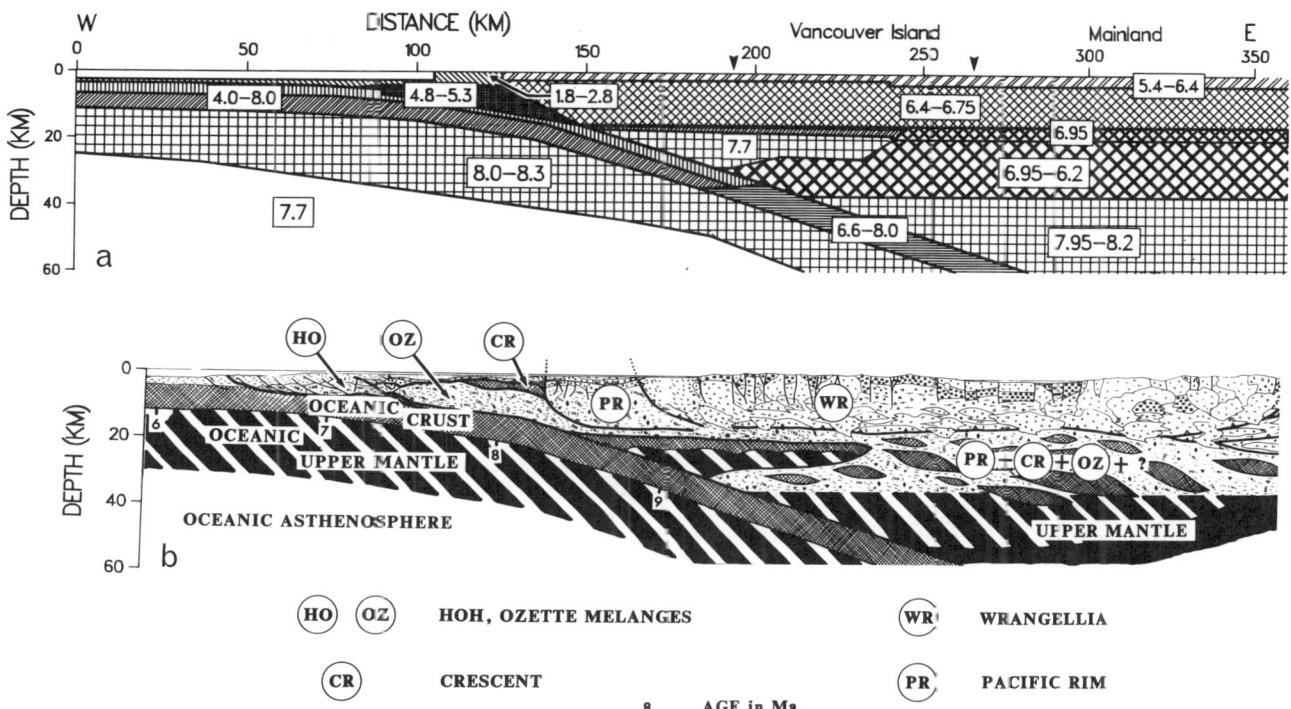

Fig. 5. (a) Generalized seismic structural model for line PJ (Spence et al., 1985). Numbers are velocities (km/s) at the top and bottom of a block distinguished by its shading. Vancouver Island lies between the arrowheads. (b) Stylized tectonic cross section adapted from Monger et al. (1985). No vertical exaggeration in either section.

procedure, the rms residual for the 3 shots at the 32 stations was 0.079 s, compared with an estimated measurement error of 0.075 s. Rays from shot P13 are traced through the model of Figure 4a. The synthetic seismograms corresponding to these ray paths are shown in Figure 4b, which can be compared with the observed record section of Figure 4c. Stations with greater shot-receiver distances (240-310 km on Figure 4c) recorded clearer arrivals than at shorter distances, and the arrival times were delayed by about 1 s. On the Vancouver Island stations, two phases can be distinguished, the first due to turning rays in the subducting lithosphere, the second due to reflected rays from the base of the lithosphere. On the mainland stations, the two phases have nearly identical traveltimes so that they overlap and only one event is observed. The 1 s traveltime delay to the mainland stations arises from travel paths through a thick low-velocity continental crust while paths to the Vancouver Island stations traverse a high velocity wedge.

Figure 5a shows the final interpreted seismic model for line PJ across the subduction zone. Figure 5b is a stylized tectonic model (Monger et al., 1985) based on surface geology, other geophysics and the geometry represented in Figure 5a. Included in the tectonic model is the speculative concept that assemblages corresponding to older terranes are vertically stacked and are underlain by the descending Juan de Fuca plate.

The interpretation of profile PJ has extended the crustal models of Waldron (1982) and McMechan and Spence (1983) to the upper mantle region, and has added other features. Data recorded at the onshore stations from shots on the outer continental shelf provide additional constraint for a dip of 3° or less at the base of the melange unit. Assuming other oceanic layers dip with similar angles, the bend in the subducting slab thus must occur more than 35 km landward of the base of the continental slope. At the bend the dip of the oceanic plate must increase to at least 15° to enable the plate to subduct beneath the overriding continental plate, which has a minimum thickness of 37 km below Vancouver Island (McMechan and Spence, 1983). The upper mantle reflector (8.3 to 7.7 km/s) in the velocity model may correspond to the base of the subducting oceanic lithosphere (Figure 5). An anomalous, but necessary, feature of the model is a block of high velocity material (7.7 km/s) above the downgoing crust in the depth range 20 to 25 km. Along profile NAF (Figure 1), McMechan and Spence (1983) independently inter-

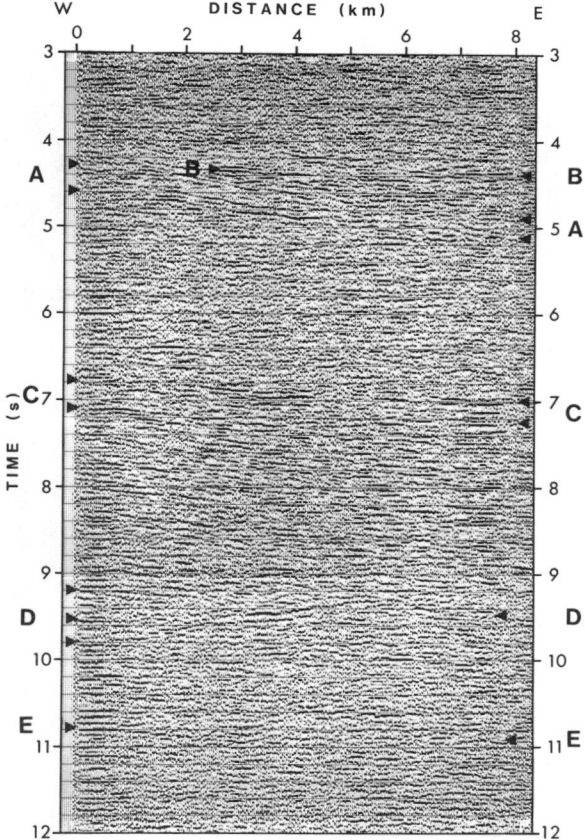

Fig. 6. Processed 1200% reflection profile along RL of Figure 1. Reflections discussed in the text are identified with symbols and labelled A to E. D and E are more speculative than the shallower events.

preted a similar region of high-velocity material at 20 km depth. Spence et al. (1985) speculate that such features may represent remnants of a subducted slab, perhaps detached when the subduction zone jumped westward to its present position.

Reflection Studies

One component of VISP was a feasibility reflection experiment on the west coast of Vancouver Island (RL on Figure 1). Figure 6 shows the corrected and stacked (1200%) explosion section to which automatic gain control, crooked line processing, 8-30 Hz bandpass filtering and a coherency filter have been applied. The first prominent reflections, A and B, occur at two-way traveltimes of ~4.4 s. A good reflection, C, at ~7.0 s is followed by a part of the section which contains a plethora of short, coherent reflection segments. Existence of reflectors A and C has been corroborated by an experiment in which these reflections were observed from an airgun source (Clowes et al., 1983). Using the refraction velo-

city model of Figure 5a at the location of RL, the reflections can be converted to approximate depths. Figure 7 shows the model below Vancouver Island; the depths of the reflectors from Figure 6 have been super-imposed at their recording location. The significant results from the short line are (1) reflected energy is obtained from depths to 40 km in this geotectonic environment and (2) the reflections correspond closely with boundaries interpreted from a refraction survey.

The 1984 Lithoprobe Vibroseis profiles have confirmed our expectations based on the test line; data quality is excellent. Two data sections from Line 1 (Figure 1) are shown in Figure 8. Clowes et al. (1984) provide details of the data acquisition and basic processing. A number of reflections are identified on the two sections. Using the refraction velocity model of Figure 7, we have converted these reflections to depth and superimposed them on the model. A line drawing interpretation of Line 1, on which the high velocity slab and top of the subducting oceanic plate from Figure 7 have been outlined, is shown in Figure 9. Figures 7 and 9 illustrate both the similarities and the discrepancies between the reflection data and refraction interpretation. On Figure 7,

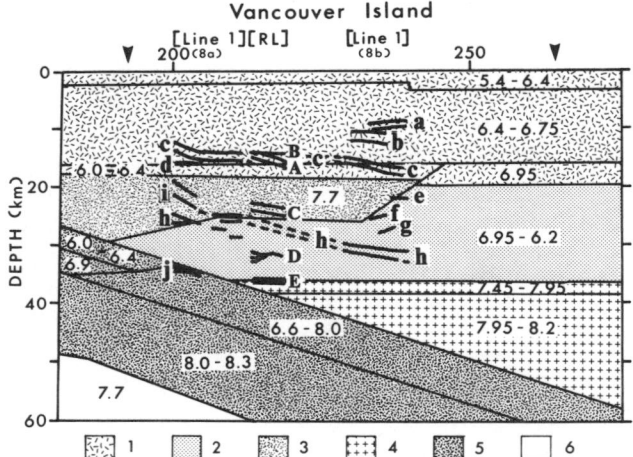

Fig. 7. Expanded version of the Vancouver Island segment of Figure 5a (no exaggerations). Reflectors identified with upper case letters in Figure 6 have been converted to depth and placed on the model at their approximate location relative to line PJ. Reflectors from two parts of Line 1 (Figures 8a and 8b) also have been converted to depth and placed at their appropriate location on the model. The lower case letters correspond to the reflectors identified in Figure 8. Reflectors c and h are continuous between the two panels (Figure 9) and are so indicated by dashed lines. Shading delineates major components of the model. 1 - upper continental crust; 2 - lower continental crust; 3 - remnant of subducted slab; 4 - continental upper mantle; 5 - subducting oceanic plate; 6 - oceanic asthenosphere.

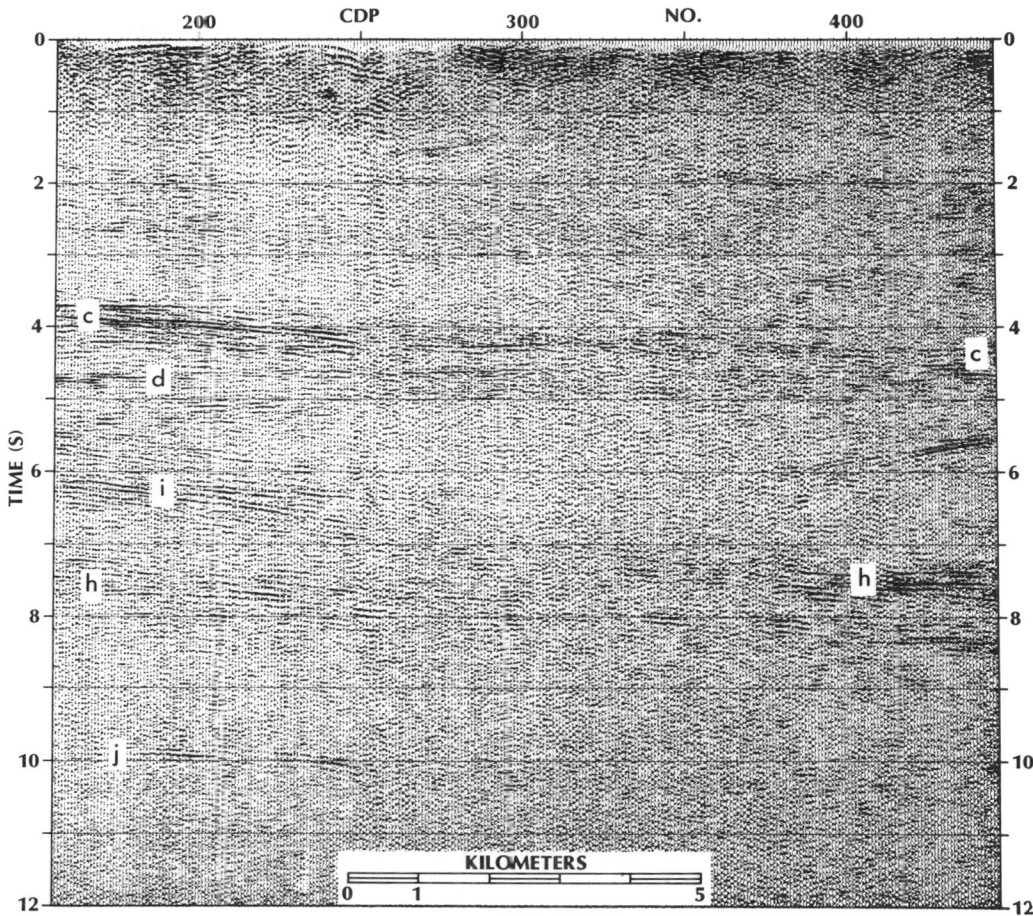

Fig. 8a. **Segment of the 1984 Lithoprobe 3000% Vibroseis Line 1 (Figure 1) located near the southwestern end of the line.**

reflections c and A, B of Figure 6 correspond closely with the refraction model interface at 16 km depth. The other continuous reflection, h, corresponds with the base of the 7.7 km/s block on the western side of the island but dips eastward to significantly greater depths. Of the other reflections, some correspond to refraction model boundaries; others do not.

Yorath et al. (1985), Clowes et al. (1984) and Green et al. (1985) provide additional data examples and preliminary interpretations. More complete interpretations which integrate Lithoprobe geological studies have been prepared (C.J. Yorath et al., submitted; A.G. Green et al., submitted; R.M. Clowes et al., submitted). With reference to Figure 9, reflections above c are associated with the Mesozoic and Paleozoic rocks of Wrangellia. As shown, eastward dipping faults are inferred. These verge into the reflective zone c which is interpreted as a decollement zone at or beneath the base of Wrangellia. The layered sequence may represent intercalated materials of sedimentary and volcanic origin that overlay and were emplaced with an underplated zone of oceanic crust/mantle material. The latter is associated with the thick interval (5 to 8 s) having few reflections; it also corresponds approximately to the high velocity block of the refraction model which is inferred to be an older oceanic slab now accreted to the continental crust. Below the slab is a relatively thick zone of closely spaced high amplitude reflections (Figure 8). The unpublished interpretations noted above suggest that the reflective sequence represents the upper part of a zone of dynamic subduction accretion in which oceanic materials of the downgoing plate are presently being underplated to the overriding continental plate. That is, the reflective zone delimits a region of decoupling between the overriding and subducting plates. This would place the subducting slab at least 5 km above that interpreted from the refraction data (Figure 9).

Clearly a reconciliation of the reflection and refraction data sets is necessary. The new information from the Lithoprobe program is being incorporated into revised refraction interpretations

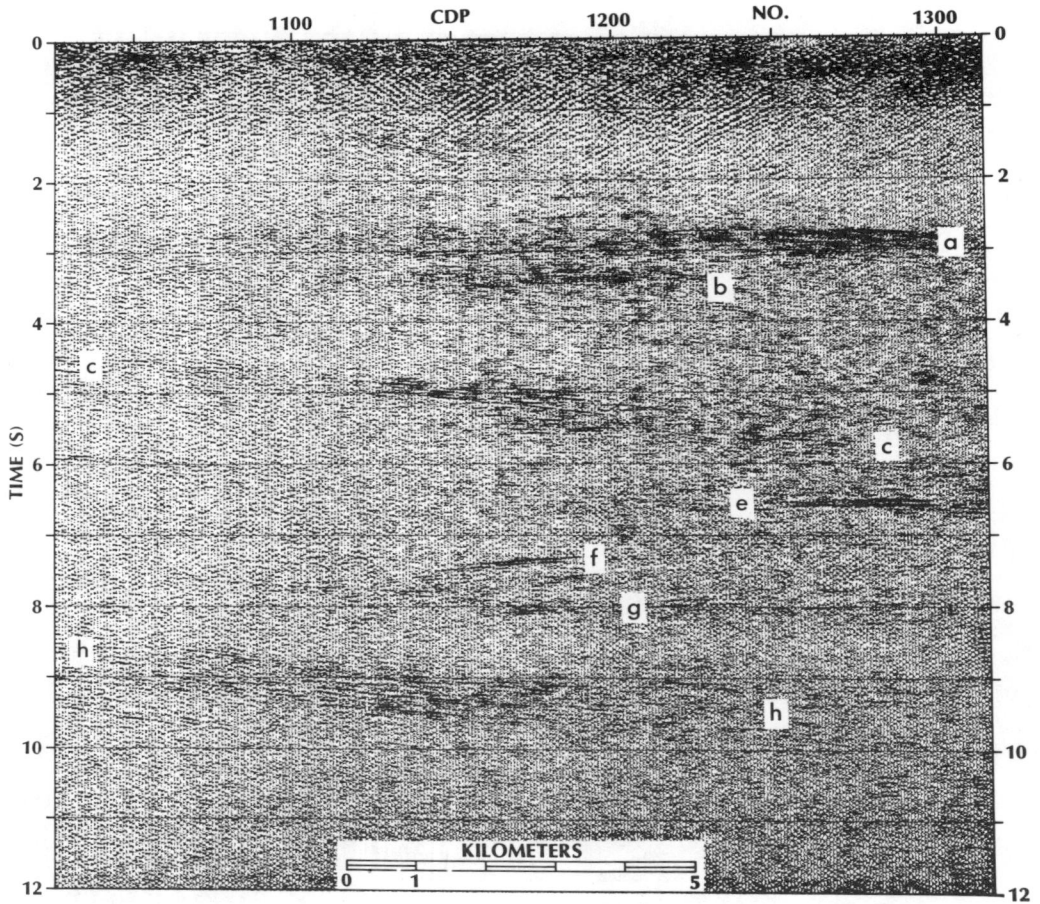

Fig. 8b. Segment from Line 1 starting 45 km from the southwest end of the line.

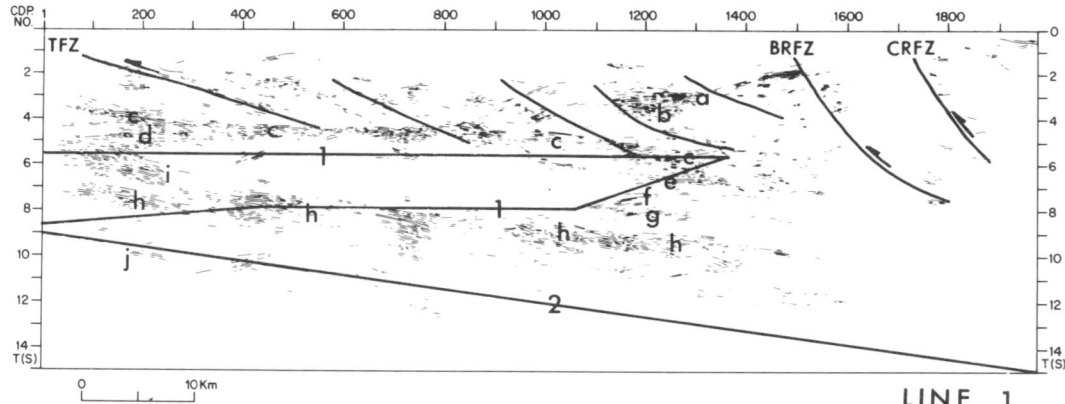

Fig. 9. Line drawing diagram for Lithoprobe Line 1. Locations of data sections in Figure 8 are marked. 1 - outline of high velocity block; 2 - top of subducting plate (from Figures 5a and 7). Other lines are faults or boundaries interpreted from the data and surface geology: TFZ - Tofino fault zone; BRFZ - Beaufort Range fault zone; CRFZ - Cameron River Fault zone. (Adapted from Yorath et al., 1985; and A.G. Green et al., submitted manuscript.)

along PJ and NAF. But the refraction models have provided valuable velocity information and enable a realistic comparison of the reflections with possible tectonic or other boundaries in the lithosphere. The two data sets applied in conjunction will enable a more complete interpretation of both.

Acknowledgements. This research was supported by Natural Sciences and Engineering Research Council of Canada (NSERC) grants A2617, A7707 and G0738 and by Energy, Mines and Resources Canada (EMR) Research Agreements 214/3/81, 41/3/82 and 42/3/82. Phase 1 Lithoprobe is funded by an NSERC Collaborative Special Project grant and by the Earth Sciences Sector, EMR. Personal financial support for G.D.S. was provided by an NSERC Postgraduate Scholarship and an H.R. MacMillan Family Fellowship and for D.A.W. by a Canadian Commonwealth Scholarship.
Lithoprobe Publication No. 3.

References

CANDEL, Lithoprobe: Geoscience studies of the third dimension - a coordinated national geoscience project for the 1980s, Geosci. Can., 8, 117-125, 1981.

Clowes, R.M., Phase 1 Lithoprobe - a coordinated national geoscience project, Geosci. Can., 11, 122-126, 1984.

Clowes, R.M., R.M. Ellis, Z. Hajnal, and I.F. Jones, Seismic reflections from subducting lithosphere?, Nature, 303, 668-670, 1983.

Clowes, R.M., A.G. Green, C.J. Yorath, E.R. Kanasewich, G.F. West and G.D. Garland, Lithoprobe - a national program for studying the third dimension of geology, J. Can. Soc. Expl. Geophys., 20, in press, 1984.

Crosson, R.S., Review of seismicity in the Puget Sound region from 1970 through 1978: a brief summary, in Earthquake Hazards of the Puget Sound Region, Washington State, edited by J.C. Yount, U.S. Geol. Surv. Open File Rep. 6-18, 1981.

Ellis, R.M., G.D. Spence, R.M. Clowes, D.A. Waldron, I.F. Jones, A.G. Green, D.A. Forsyth, J.A. Mair, M.J. Berry, R.F. Mereu, E.R. Kanasewich, G.L. Cumming, Z. Hajnal, R.D. Hyndman, G.A. McMechan, and B.D. Loncarevic, The Vancouver Island Seismic Project: A CO-CRUST onshore-offshore study of a convergent margin, Can. J. Earth Sci., 20, 719-741, 1983.

Green, A.G., M.J. Berry, C.P. Spencer, E.R. Kanasewich, S. Chiu, R.M. Clowes, C.J. Yorath, D.B. Stewart, J.D. Unger and W.H. Poole, Recent seismic reflection studies in Canada, this volume, 1985.

Jones, D.L., N.J. Silberling, and J. Hillhouse, Wrangellia - a displaced terrane in northwestern North America, Can. J. Earth Sci., 14, 2565-2577, 1977.

Keen, C.E., and R.D. Hyndman, Geophysical review of the continental margins of eastern and western Canada, Can. J. Earth Sci., 16, 712-747, 1979.

McMechan, G.A., and W.D. Mooney, Asymptotic ray theory and synthetic seismograms for laterally varying structures: theory and application to the Imperial Valley, California, Bull Seism. Soc. Am., 70, 2021-2035, 1980.

McMechan, G.A., and G.D. Spence, P-wave velocity structure of the Earth's crust beneath Vancouver Island, Can. J. Earth Sci., 20, 742-752, 1983.

Monger, J.W.H., R.M. Clowes, R.A. Price, P.S. Simony, R.P. Riddihough, and G.J. Woodsworth, Continent-ocean Transect B2: Juan de Fuca plate to Alberta plains, International Geodynamics Transect Program, Geol. Soc. Am. Spec. Publ., 1985.

Riddihough, R.P., A model for recent plate interactions off Canada's west coast, Can. J. Earth Sci., 14, 384-396, 1977.

Riddihough, R.P., Recent movements of the Juan de Fuca plate, J. Geophys. Res., 89, 6980-6994, 1984.

Riddihough, R.P., and R.D. Hyndman, Canada's active western margin - the case for subduction, Geosci. Can., 3, 269-278, 1976.

Rogers, G.C., Seismotectonics of British Columbia, Ph.D. thesis, Univ. of British Columbia, 247 pp., January 1983.

Snaveley, P.D. Jr., and H.C. Wagner, Geological cross section across the continental margin off Cape Flattery, British Columbia, U.S. Geol. Surv. Open File Rep. 81-978, 5 pp., 1981.

Spence, G.D., Seismic structure across the active subduction zone of western Canada, Ph.D. thesis, 191 pp., Univ. of British Columbia, Vancouver, July 1984.

Spence, G.D., R.M. Clowes, and R.M. Ellis, Seismic structure across the active subduction zone of western Canada, J. Geophys. Res., 90, in press, 1985.

Spence, G.D., K.P. Whittall, and R.M. Clowes, Practical synthetic seismograms for laterally varying media calculated by asymptotic ray theory, Bull. Seism. Soc. Am., 74, 1209-1223, 1984.

Taber, J.J. Jr., Crustal structure and seismology of the Washington continental margin, Ph.D. thesis, 159 pp., Univ. of Washington, Seattle, August 1983.

Waldron, D.A., Structural characteristics of a subducting oceanic plate, M. Sc. thesis, 121 pp., Univ. of British Columbia, Vancouver, December 1982.

Whittall, K.P., and R.M. Clowes, A simple, efficient method for the calculation of traveltimes and raypaths in laterally inhomogeneous media, J. Can. Soc. Expl. Geophys., 5, 21-29, 1979.

Yorath, C.J., and R.L. Chase, Tectonic history of the Queen Charlotte Islands and adjacent areas - A model, Can. J. Earth Sci., 18, 1717-1739, 1981.

Yorath, C.J., R.M. Clowes, A.G. Green, A. Sutherland Brown, M.T. Brandon, N.W.D. Massey, C. Spencer, E.R. Kanasewich, and R.D. Hyndman, LITHOPROBE - Phase 1: southern Vancouver Island: preliminary analyses of reflection seismic profiles and surface geological studies, Geol. Surv. Can. Paper 85-1A, 543-554, 1985.

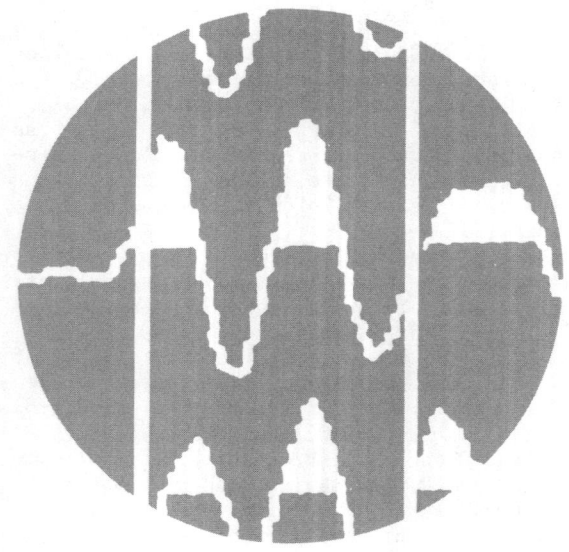

THE VICTORIA LAND BASIN: PART OF AN EXTENDED CRUSTAL COMPLEX BETWEEN EAST AND WEST ANTARCTICA

Yeadong Kim, L. D. McGinnis, and R. H. Bowen

Department of Geology, Louisiana State Univeristy, Baton Rouge, LA 70803

Abstract. Seismic reflection soundings to 12 seconds two-way time in the southern Victoria Land Basin of the western Ross Sea indicate the presence of a deep sedimentary basin overlying a thinned crust. In addition to reflection data, seismic refraction and gravity studies provide control on the configuration of crystalline basement and depth to the Mohorovicic Discontinuity. Dipping reflectors suggest a basin depth of 13 km and a 200 km long reversed refraction profile provides a MOHO depth of 21 km below sea level. The basin contains undeformed sediments dipping seaward and is similar to continental margins which were formed by rifting. Since the only apparent periods of rifting occurred during the emplacement of the Ferrar dolerites and the McMurdo Volcanics, it is believed that the deep, layered strata are synrift sediments of Jurassic age and younger. Two areas of normal faulting bound the rifted basin on the east and west. Flat-lying glacial marine sediments with few internal reflections cover the basin. Present day high heat flow and active volcanism suggest that the basin beneath McMurdo Sound is undergoing a second phase of rifting. The Victoria Land Basin and the Wilkes subglacial basin lying west of the Transantarctic Mountains form part of an extensional complex along the boundary of East and West Antarctica.

Introduction

A 24-channel, 12-fold seismic reflection profile in the Southern Victoria Land Basin (Figure 1) has been shot from sea ice using 5 kg of dynamite suspended in water at a depth of 5 meters. Twelve second records were collected with a 24-channel Texas Instruments DFS-III at a two millisecond sampling rate. Initial processing was done by Mobil E & P Services, Inc. with final processing completed on a Western Geophysics Pre/ Seis system at Louisiana State University. The location of transects in McMurdo Sound is shown in Figure 2.

McMurdo Sound, a channel lying between Ross Island and the Transantarctic Mountains, lies over a deep sedimentary basin in the western Ross Sea. The basin, one of three lying in the Ross Embayment, is referred to as the Victoria Land Basin by Davey (1983) and Davey et al. (1982). The Ross Embayment is part of an extensional complex (Steed, 1983) which also includes the Transantarctic Mountains and the Wilkes subglacial basin lying to the west of the mountains. The Ross Embayment has been explored intensively with marine reflection profiling the last several field seasons (Sato et al., 1984; Eittreim et al., 1984; and Hinz, in press).

The Wilkes basin was first discovered by the Victoria Land Traverse of 1958 and was reported by Crary (1963) who referred to the Wilkes Basin as a geosyncline; however, reflection quality at the 30 stations in Victoria Land where the ice-rock interface was measured was poor and no reflections below the interface were observed at that time. The work of Crary (1963) and Bentley (1964) has been refined through radio-echo sounding studies by Drewry (1983). Depth to magnetic basement (i.e. base of sedimentary layer) over the Wilkes Basin-Transantarctic Mountains-Victoria Land Basin complex was published by Behrendt (1983). These studies provide a continental scale perspective of the lands flanking the Transantarctic Mountains and illustrate in a general way the basin-range-basin relationship of the region lying in the Ross Embayment and to the west. Sedimentary thicknesses up to eight kilometers are suggested in the Wilkes Basin.

Single channel seismic reflection studies in McMurdo Sound have been reported by Wong and Christoffel (1981). Refraction data (McGinnis et al., 1985) indicate ocean floor velocities ranging from 1.7 to 3.0 km/sec increasing to 5 km/sec at depths of 4 km in eastern McMurdo Sound. A 6.5 km/sec layer at greater depths was interpreted to represent an intracrustal boundary consisting of granulites. Granulite-type rocks have been identified in inclusions of volcanic rocks from Ross Island (Berg, personal communication). The interpretation presented here is based on a deep

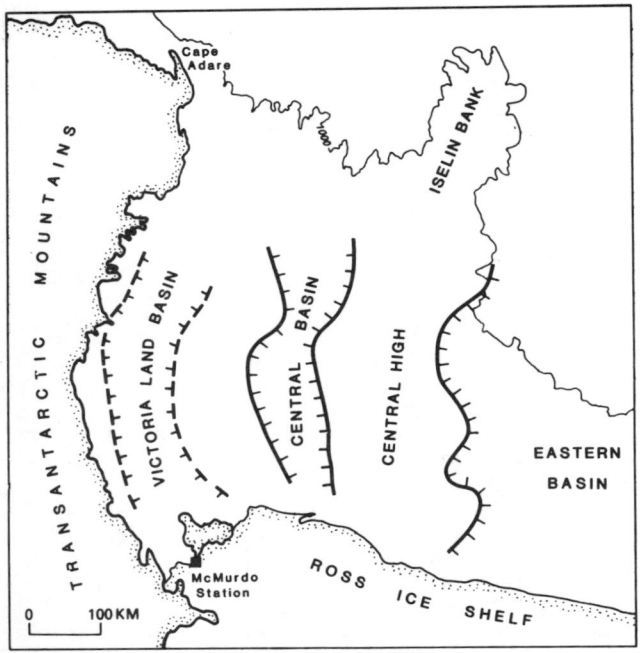

Fig. 1. Map of the Ross Sea taken from Eittreim (et al, 1984) showing the location of the three large sedimentary basins in the Ross Embayment.

reflection profile crossing McMurdo Sound and is constrained by gravity, magnetic, and reversed seismic refraction data.

Seismic Processing

Severe noise problems were generated from a variety of causes including the bubble, sideswipe from Ross Island and the mainland coast, multiples from the ocean floor, and sideswipe from the McMurdo Ice Shelf. Additional noise below seven seconds is attributed to flexural waves travelling through the sea ice. Field data were processed in the standard manner and a spatially variant deconvolution filter was applied to attenuate water reverberation and multiples. Data were then filtered with a 15 to 50 Hz band pass filter with 18 and 36 db/octave low-cut and high-cut roll-off slope to remove high frequency noise generated by the deconvolution. The low frequency cut-off was used to eliminate the low frequency ice flexural waves at the expense of signal. Velocities determined in previous refraction work (McGinnis et al., 1983) were used to constrain the velocities below six seconds two-way reflection time.

Interpretation

An unmigrated section of the 1982-83 data extending from Hut Point Peninsula on Ross Island to the Strand Moraine on the mainland coast is shown in Figure 3. Prominent east-dipping reflectors extend from the coast to km 21, measured from Ross Island, where they flatten out and end abruptly near a series of chaotic reflectors believed to be caused by normal faulting dipping back toward the coast. The deepest, continuous reflector begins at about one second on the west and extends to about six seconds beneath km 21. Discontinuous reflectors are apparent below the deep reflector between km 30 to about km 34. Enlargements of representative reflectors in the outlined squares in Figure 3 are shown in Figures 4, 5, 6, and 7.

Shallow reflectors at two-way travel times less than 3 seconds across the entire profile are generally noncontinuous and relatively weak, in contrast with deep reflectors from this study and

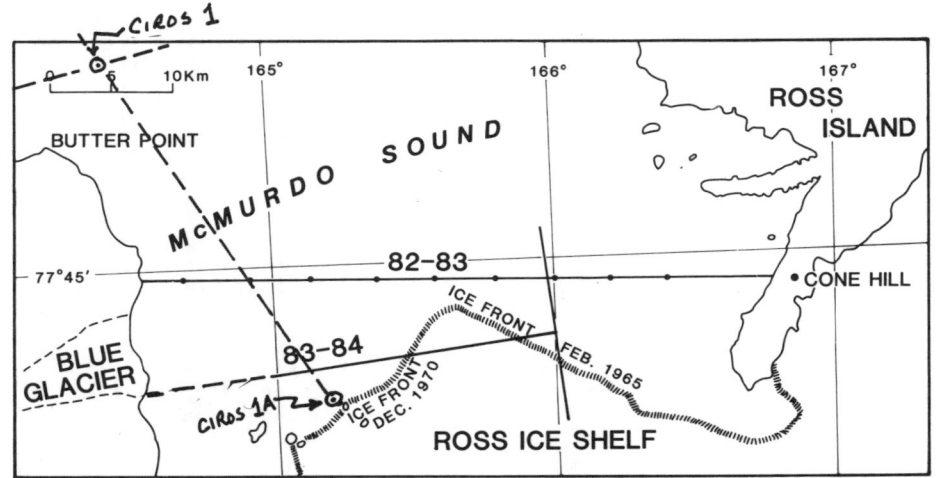

Fig 2. McMurdo Sound geophysical transects. Solid lines give the locations of reflection transects. Dashed lines are proposed transects. CIROS 1 and 1A are proposed drill sites. Dots on the 82-83 reflection transect give the locations of recording sites for refraction profiles reported by McGinnis et al (1985).

from the high frequency, single channel study of Wong and Christoffel (1981). The profiles obtained by Wong and Christoffel clearly show the complexity of shallow layering and unconformities of glacial deposits above the first ocean floor multiple. A good reflector lying between 2.1 and 2.4 seconds, bending down toward Ross Island is shown in Figure 4. The reflector is interpreted here as the planar crystalline basement surface, displaced by some minor faults, that is flexed downward by crustal loading of the volcanic pile composing Ross Island. It is possible that the sedimentary section on the east side of the profile extends deeper than the prominent reflector shown in Figure 4; however, the deeper reflectors are not as pronounced as those between 2.1 to 2.4 seconds and are therefore assumed to be sub-basement.

A region of chaotic reflectors is shown in Figure 5. Reflectors are discontinuous, dip in both east and west directions, and are assumed to represent displaced strata extending to the faulted basement which is downdropped to the west from two seconds to four seconds below km 17.

A deep, flat-topped reflector at seven seconds two-way time is shown in Figure 6. This reflector was originally thought to be caused by a mantle pluton intruding the crust below the deepest segment of the Victoria Land basin; however, it has a 1.4 km/s stacking velocity. The distance of the seismic line from the northernmost point of the McMurdo Ice Shelf suggests that this reflector, and similar ones less well pronounced, are caused by sideswipe from the McMurdo Ice Shelf.

An example of layered reflectors, some truncated or pinching out to the west, is shown in Figure 7. A disconformity at 2.4 seconds below km 35 is thought to represent the boundary between synrift sediments of Mesozoic age and the glacial marine sediments of Tertiary age and younger.

Discussion

Deep seismic reflection profiling shown here, along with magnetics (Pederson, et al, 1981), gravity, and seismic refraction profiling (McGinnis et al, 1983) are interpreted to show that the uplifted Transantarctic Mountains (Gleadow, 1983) occur on the west flank of a rifted and thinned crust (Figure 8). The rifted margin, lying beneath the Victoria Land sedimentary basin is floored by a crystalline basement lying at a maximum depth of 13 km. Because rifting is suggested, it was hoped that deep reflection profiling would show evidence of shallowly dipping reflectors suggestive of detachment-type faulting within the crust and perhaps older metasediments due to underthrusting of the Ross Embayment crust under the Transantarctic Mountains. We did not observe any such features, (see Figure 3), but instead found only subparallel reflectors dipping steeply offshore, with some high-angle normal faulting common in regions of extensional tectonics.

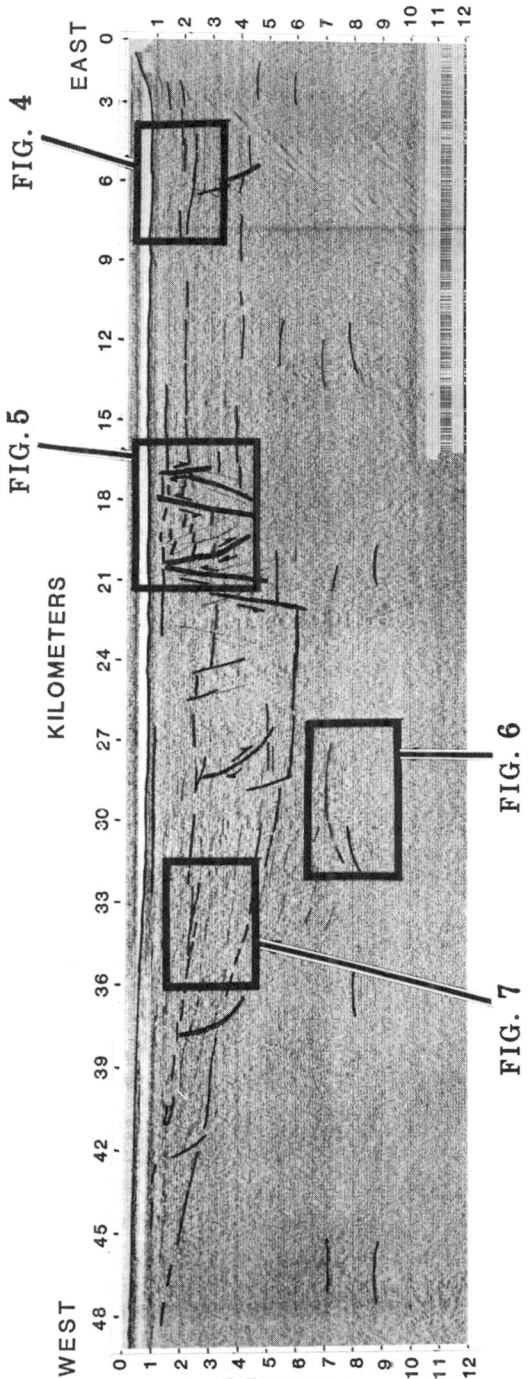

Fig. 3. Reflection profile of 1982-83 data along the same line as the simplified version shown in Figure 8. A deconvolution filter has been applied to the data; however, it is unmigrated. Crystalline basement is believed to extend to six seconds between km 21-30. Layered reflectors immediately west of this zone greater than six seconds may represent a down-dropped wedge of synrift sediments. Vertical exaggeration is about 0.6x.

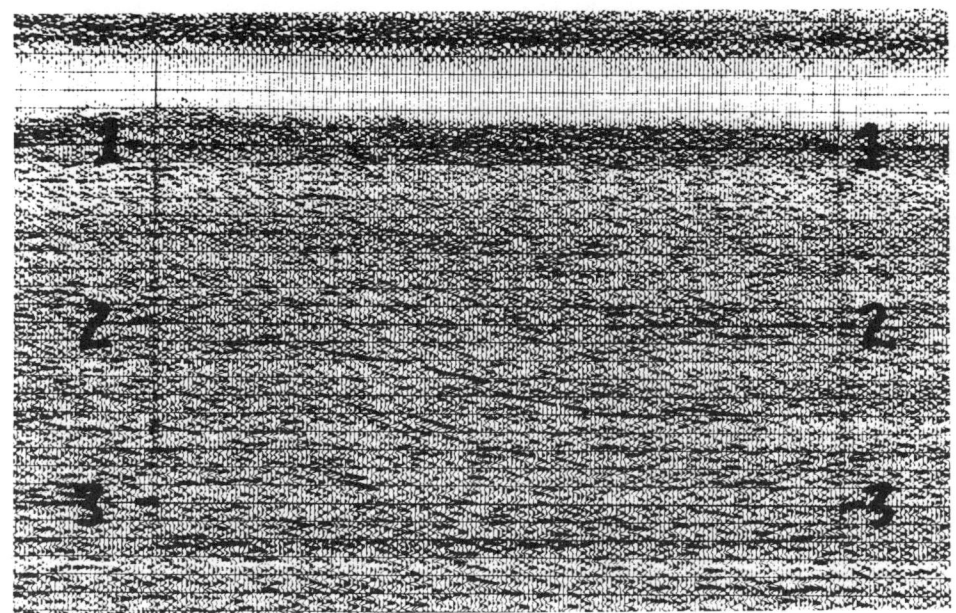

Fig. 4. Shallow reflectors bending downward toward Ross Island. See Figure 3 for location. The reflector below two seconds is "pre-volcanic" basement and may be caused by crustal subsidence beneath the island.

Subparallel reflectors are underlain by a 6.5 km/s refractor we interpret to be igneous and metamorphic rocks of the Ross Supergroup. The 6.5 km/s refractor rises from 13 km below sea level in the basin center to 2 km beneath the Strand Moraine. Large amplitude reflectors (Figure 6) at seven seconds (two-way) time beneath the rift were originally thought to be MOHO since they occur near depths where an 8.2 km/s refractor was found from a 200 km refraction profile (McGinnis, et al, 1985). As discussed earlier, it is now thought that this reflector is due to sideswipe from the McMurdo Ice Shelf. Alternative interpretations of these deep events may still be

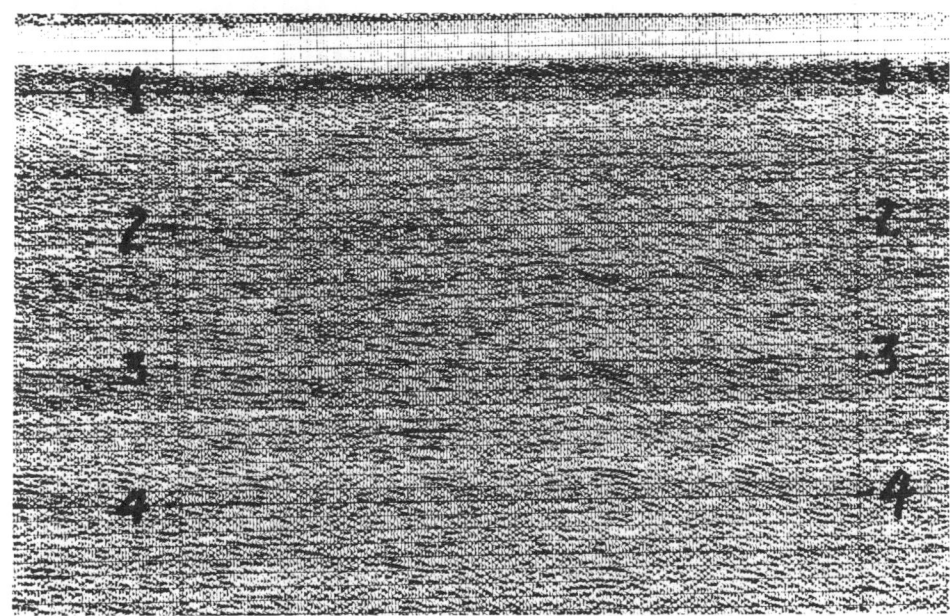

Fig. 5. Shallow reflectors in a "chaotic", faulted zone. See Figure 3 for location.

Fig. 6. Example of a deep diffractor near mantle depth (i.e. 7 seconds). The location is shown in Figure 3. The diffractor is located beneath the deepest part of the Victoria Land Basin but is most likely caused by sideswipe from the McMurdo Ice Shelf.

viable. For example, Schilt et al (1981) have observed deep diffraction patterns similar to those below seven seconds in McMurdo Sound. They attribute the diffractions to deep plutons. It is possible that they may be due to lithosphere "necking" due to extension beneath deep basins (see Fletcher and Hallett, 1983).

We have not observed low-angle detachment

Fig. 7. Dipping reflectors on the western side of the profile taken from Figure 3. An unconformity represented by reflectors converging to the west is located between two and three seconds in the west-central part of the illustration.

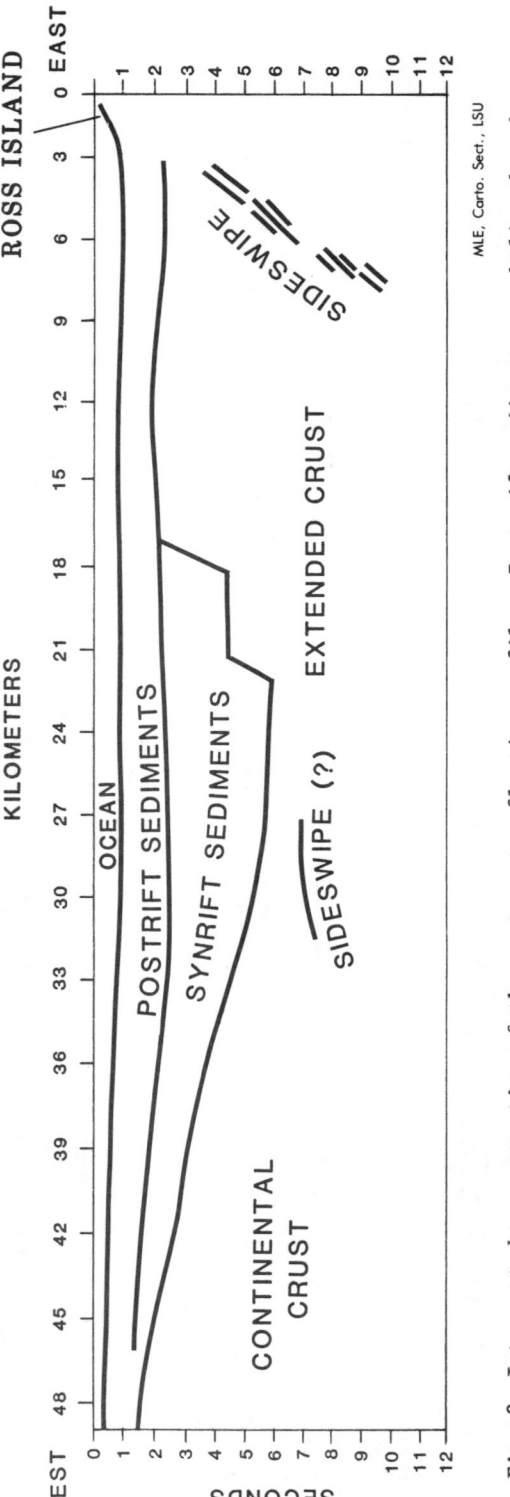

Fig. 8. Interpreted cross section of the east-west reflection profile. Post-rift sediments are believed to be of glacial marine origin of Cenozoic age. Synrift sediments are believed to be Mesozoic in age and were deposited following intrusion of the Ferrar dolerites in the Transantarctic Mountains.

faults which are increasingly being recognized as an important mode of extension of continental regions (e.g. Allmendinger et al, 1983) nor have we observed the 7.3 to 7.4 km/s layer beneath McMurdo Sound which is believed to represent evidence for a rift cushion frequently found beneath other features of intracontinental, rift origin (Ervin and McGinnis, 1976). Deep refraction data indicate only the normal crustal velocities overlying a normal mantle of 8.2 km/s velocity (McGinnis et al, 1983). Despite the fact that "rift cushion" velocities were not observed, it is quite possible that in areas of half-graben extension, the crust is not intruded by anomalous mantle and therefore the 7.4 km/s velocities associated with the lower crust in other areas of continental rifting, are not observed.

The presence of a deep basin of extensional origin running the length of the Transantarctic Mountains suggests that the regional interpretation of the tectonic framework of the East-West Antarctic boundary of Steed (1983) may be essentially correct.

Ivanov (1983) has made an attempt to classify sedimentary basins in Antarctica and has typed the Victoria Land and Wilkes Basins as "intracratonic". Because of their close association with the orogenic belts parallel to the Transantarctic Mountains this interpretation is questionable. Intracratonic basins, such as the Illinois and Michigan basins of North America, have their origins in incipient rifts which are discordant with older structures (McGinnis, 1970; Hinze and Braile, in press). These basins do not display parallelism to long orogenic belts similar to that of the Ross Orogen (Craddock, 1983). Steed (1983) interprets the Wilkes Basin as lying on the eastern edge of the Precambrian craton of East Antarctica and west of the early Paleozoic orogen. He suggests the Wilkes Basin might be analogous to the Mesozoic basin of the North Sea.

Conclusions

Deep reflection profiling conducted on the sea ice of McMurdo Sound has established further details on the tectonic history of the Transantarctic Mountains and other extensional features associated with the mountains. A large half-graben, bounded on the east by a normal fault complex, down-stepping to the west, is located between 17 and 48 km from Hut Point Peninsula, Ross Island. The graben probably formed during intrusion of the Ferrar dolerites and therefore would contain synrift sediments of Jurassic and younger age. Maximum sediment thickness of 13 km occurs between 23 and 30 km from Ross Island. A prominent diffraction at seven seconds below the deepest part of the basin may be due to a mantle diapiric plume piercing the crust; however, a more probable alternative to the plume hypothesis is simply that the diffraction is caused by sideswipe from the tip of the McMurdo Ice Shelf.

Results of the study suggest that the Wilkes

subglacial basin, Transantarctic Mountains, and Victoria Land basin are parallel lineaments extending the length of the mountains that formed as a result of crustal thinning and extension. This would support the observations of Steed (1983) who suggests that the true edge of the Precambrian craton of East Antarctica lies along the western boundary of the Wilkes basin. The Victoria Land basin in that case would be one of several linear basins paralleling the mountains and would include the Central and Eastern basins of the Ross Embayment.

Since the publication of maps illustrating sub-ice topography in Antarctica (Crary, 1963; Bentley, 1964, and Drewry, 1983) it has become clear that the East Antarctic craton is not a structureless slab of Precambrian shield rocks, but may contain deep sedimentary basins of varying age. Webb et al (1984) have suggested that glacial debris found in the Transantarctic Mountains, containing late Cenozoic marine microfauna was derived from the Wilkes-Pensacola basin, a lowland extending the length of the Transantarctic Mountains on the East Antarctic side of the range. A seismic transect across the center of this basin at some future time could determine if it has the geophysical characteristics of intracratonic basins found in other continental interiors (McGinnis, 1970) or if it bears some relation in age and tectonics to the extensional terranes of the Ross Embayment, in particular to the Victoria Land basin. It has become quite clear that many sedimentary basins have graben or half-graben origins and that they might bear quite different geophysical signatures. An array of geophysical measurements including gravity, magnetics, and CDP profiling can test the hypothesis of Webb et al (1984) and Behrendt (1983) to see if the Wilkes Basin is indeed a deep geological basin containing a thick sediment record or if it is a flat, deep, shelf carved onto a crystalline basement floor with the usual rift anomalies being absent. Further definition of the geophysics of the Wilkes Basin is important, not only to establish the presence of a sedimentary source, but it would also provide information on the crust below the basin.

From gravity and magnetic characteristics it is quite probable that the crust lying beneath McMurdo Sound is similar petrologically to that underlying the Transantarctic Mountains. Pederson et al (1982), on the basis of an aeromagnetic study, suggested that the Ferrar dolerite sills common in the mountains did not occur beneath McMurdo Sound because of the absence of short wave length magnetic anomalies. The lack of anomalies can now be explained by the fact that the magnetic crust lies at depths as great as 13 km or more below sea level. In addition, Berg et al (1984) have recently found an inclusion of Ferrar dolerite in McMurdo volcanic rocks in the Dailey Islands in western McMurdo Sound.

The eastern-most Dailey Island, approximately 20 km from the coast, is located near a significant change in gradient of the crystalline basement reflector (0.1 s/km to 0.2 s/km). Kyle and Cole (1974) point out the close association of volcanism and faulting in the McMurdo Sound region. A normal fault dipping toward Ross Island is shown near the gradient change approximately 37 km from Ross Island as shown on Figure 3. Stern (1984), from seismic refraction data, suggests the possible presence of normal faults underlying the Dailey Islands; however, his study explored only to depths of two km, or to the 5 km/s refractor prevalent in McMurdo Sound. In the study presented here, the 5 km/s layer is thought to be the top of synrift sediments of Mesozoic age and the steep gradients observed on the basement in the reflection study would not have been seen in Stern's refraction data.

A multichannel seismic transect should be run into New Harbor (see Figure 2) where the geophone array could be placed on shore several kilometers. This profile would be invaluable since the geology is well known from drilling (Webb, personal communication) and from abundant local outcrops along the flanks of the Ferrar and Taylor Valleys. A reflection line into New Harbor would extend subsurface information westward over 10 km from the line terminating at the Strand Moraine (Figure 2). This line would be oriented to cross CIROS drill hole(s) planned for the 1984-85 field season and also the 48 channel reflection profile of the USGS cruise on the Lee (Eittreim, et al., 1984). Recording time on the New Harbor profile should be increased to 15 seconds, which would correspond to a depth of exploration over 40 km.

Acknowledgements. Field crews were staffed by graduate students from Northern Illinois and Louisiana State Universities. Field crew members, excluding authors, include Bryant Hinton, Stuart Wolf, John Isbel, Doug Layman, and Marcia Honz. Jim Kreamer with Mobil Exploration conducted the initial processing at Mobil's office in New Orleans. This study was supported by the Division of Polar Programs, National Science Foundation grant number DPP 8019995.

References

Allmendinger, R.W., J.W. Sharp, D. Von Tish, L. Serpa, L. Brown, S. Kaufman, J. Oliver, and R.B. Smith, Cenozoic and Mesozoic structure of the eastern Basin and Range Province, Utah, from COCORP seismic - reflection data, Geology, 11, 532-536, 1983.

Behrendt, J. C., Are there petroleum resources in Antarctica?, U. S. Geol. Survey Circ. 909, 3-24, 1983.

Bentley, C.R., The Structure of Antarctica and its ice cover; in Research in Geophysics, Odishaw, H. ed., Cambridge, Mass., MIT Press, 335-389, 1964.

Berg, J. H., R. A. Hank, D. L. Herz, and J. A. Gamble, Crustal inclusions from the Erebus volcanic province, Antarctica, Geol. Soc.

Am., Abstracts with Programs, 16, 443, 1984.

Craddock, C., The East Antarctica-West Antarctica boundary between ice shelves: a review, Antarctic Earth Science: Fourth International Symposium; R. L. Oliver, P. R. James, and J. B. Jago, Cambridge University Press, 94-97, 1983.

Crary, A. P., Results of United States Traverses in East Antarctica, 1958-1961, Rept. No. 7, American Geographical Society, IGY World Data Center A, 144 pp., 1963.

Davey, F. J., Geophysical Studies in the Ross Sea region: Jour. Royal Soc. New Zealand, 11, 465-479, 1983.

Davey, F. J., D. J. Bennett, and R. E. Houtz, Sedimentary basins of the Ross Sea, Antarctica, N. Z. Journal of Geol. and Geophys., 25, 245-255, 1982.

Drewry, D. J., Isostatically adjusted bedrock surface of Antarctica, Plate 6.1; Antarctica: Glaciological and Geophysical Folio; Scott Polar Research Institute, Univ. of Cambridge, 1983.

Eittreim, S. L. and A. K. Cooper, Marine geological and geophysical investigations of the Antarctic continental margin, U.S. Geol. Survey Circ. 935, 12 pp., 1984.

Ervin, C. P. and L. D. McGinnis, Reelfoot rift: Reactivated precursor to the Mississippi Embayment, Geol. Soc. Am. Bull., 86, 1287-1295, 1975.

Fletcher, R. C. and B. Hallet, Unstable extension of the lithosphere: A mechanical model for Basin-and-Range structure; Jour. Geophys. Res., 88, B9, 7457-7466, 1983.

Gleadow, A. J. W., Fission track geochronology of granitoids and uplift history of the Transantarctic Mountains, Victoria Land, Antarctica, Antarctic Earth Science: Fourth International Symposium; R. L. Oliver, P. R. James, and J. B. Jago, Cambridge University Press, 563 p., 1983.

Hinz, K., Results of geophysical investigations in the Weddell Sea and Ross Sea, Antarctica: Proceedings of the 11th World Petroleum Congress, London, in press.

Hinze, W.J. and L.W. Braile, Geophysical aspects of the craton; in GSA The Midcontinent Province, L.L. Sloss, et al, eds., Decade of North American Geology, in press.

Ivanov, V. L., Sedimentary basins of Antarctica and their preliminary structural and morphological classification; Antartica Earth Science: Fourth International Symposium; R. L. Oliver, P. R. James, and J. B. Jago, Cambridge University Press, 539-544, 1983.

Kyle, P. R. and J. W. Cole, Structural control of volcanism in the McMurdo Volcanic Group, Antarctica; Bull. Volcan., 38-1, 16-55, 1974.

McGinnis, L. D., Tectonics and the gravity field in the continental interior, Jour. Geophys. Res., 75, 317-332, 1970.

McGinnis, L. D., R. H. Bowen, J.M. Erickson, B.J. Allred, and J.L. Kreamer, East-West Antarctic boundary in McMurdo Sound, Tectonophysics, 114, 341-356, 1985.

McGinnis, L. D., D. D. Wilson, W. J. Burdelik, and T. H. Larson, Crust and upper mantle study in McMurdo Sound; in Antarctic Earth Sciences; R. L. Oliver, P. R. James, and J. B. Jago, Cambridge University press, 204-208, 1983.

Pederson, D. R., G. E. Montgomery, L. D. McGinnis, C. P. Ervin, and H. K. Wong, Aeromagnetic survey of Ross Island, McMurdo Sound, and the Dry Valleys; in Dry Valley Drilling Project, Am. Geophys. Union, Antarctic Research Series, 33, 7-26, 1981.

Sato, S., N. Asakura, and T. Saki, Preliminary results of geological and geophysical surveys in the Ross Sea and in the Dumont D'Vrille Sea, off Antarctica; Japan Memoirs of National Institute of Polar Research Special Issue No. 33; Proceedings of the Fourth Symposium on Antarctic Geosciences 1983; Tokyo, 66-92, 1984.

Schilt, F. S., S. Kaufman, and G. H. Long, A three-dimensional study of seismic diffraction patterns from deep basement sources; Geophysics, 46, 1673-1683, 1981.

Steed, R.H.N., Structural interpretations of Wilkes Land, Antarctica, Antarctic Earth Science; Fourth International Symposium, R.L. Oliver, P.R. James, and J.B. Jago, eds., Cambridge University Press, 567,1983.

Stern, T. A., A seismic refraction survey near the Dailey Islands, southwestern McMurdo Sound, Antarctica; N. Z. Dept. of Scientific and Industrial Research; Geophysics Division; Rept. No. 198, 32 pp., 1984.

Webb, P. N., Harwood, D. M., McKelvey, B. C., Mercer, J. H., and Scott, L. D., Cenozoic marine sedimentation and ice-volume variation on the East Antarctic Craton; Geology, 12, no. 5, 287-291, 1984.

Wong, H. K., and D. A. Christoffel, A reconnaissance seismic survey of McMurdo Sound and Terra Nova Bay, Ross Sea; in Dry Valley Drilling Project, Antarctic Research Series, 33, 37-62, Am. Geophys. Union; 1981.

WHOLE-LITHOSPHERE NORMAL SIMPLE SHEAR: AN INTERPRETATION OF DEEP-REFLECTION PROFILES IN GREAT BRITAIN

Brian Wernicke

Department of Geological Sciences,
Harvard University

Abstract. Marine deep-seismic reflection profiles from the British Isles acquired by the British Institutions' Seismic Reflection Profiling Syndicate (BIRPS) provide some of the best images to date of the deep structure of a zone of intracontinental extension. Simple quantitative considerations of finite strain indicate that the reflection geometry on the eastern half of the MOIST profile is consistent with the concept that large, low-angle normal faults persist as single zones of displacement through the entire lithosphere. It is proposed that the lower crust absorbs displacement by the formation of a foliation parallel to the maximum elongation direction within finite-width normal shear zones. In contrast, both up-dip and down-dip in stronger or more brittle layers, the shear zones not only narrow, causing foliation to become aligned with their boundaries, but they also absorb strain via a foliation comprised of discrete, boundary-parallel slip surfaces. Such a geometry of through-going zones of displacement predicts non-alignment of lower crustal reflections with those in the mantle and upper crust within the same displacement zone. This model runs counter to current interpretations of deep reflection data that ascribe a more fundamental role to rheological stratification for the localization and development of zones of displacement in the lithosphere.

Introduction

One of the most interesting and controversial aspects of research in continental tectonics is the nature of faults and shear zones at deep levels in the crust and mantle. While much is known about the kinematics of lithospheric strain in the upper 10 km of crust, it is poorly understood how major zones of displacement in the upper crust interact with deeper levels. Do large faults continue into the lower crust and mantle lithosphere as shear zones, or do they sole into decoupling horizons at various levels, controlled perhaps by changing rheology with depth [e.g. Sibson, 1983]? Recent deep-reflection data from the British Isles [Smythe et al., 1982; Brewer and Smythe, 1984; Brewer et al., 1983; Matthews and McGeary, 1984] provide some of the best seismic images to date of the deep structure of a zone of intracontinental extension. Some of these data are analyzed here in terms of the above question, and it is concluded that the data are consistent with either hypothesis. As such, they do not provide compelling evidence for the existence of decoupling horizons (as defined below) in the lithosphere.

Decoupling and Displacement in the Lithosphere

A distinction is made here between the terms zone of displacement, a structure that accommodates motion between two fragments of lithosphere, and decoupling horizon, a zone of compensation that develops between tiers of lithosphere that accommodate approximately equal strains, but by contrasting mechanisms (Figure 1). Much of the new reflection data from rift zones, including COCORP data from the Nevada Basin and Range [Hauge et al., 1984; Hauser et al., 1984; Potter et al., 1984] as well as the BIRPS data show 1) a relatively non-reflective upper crust characterized by dipping reflections indicative of sediment-filled half-grabens; 2) a highly reflective lower crust with many short, discontinuous events; 3) a very flat, abrupt downward termination of these events at an appropriate depth to be the Moho; and 4) a relatively transparent upper mantle which locally displays long, continuous, discrete bands of dipping reflections. The fact that reflection geometry shows this gross stratification may be interpreted in conjunction with laboratory [e.g. Brace and Kohlstedt, 1980] and seismicity [e.g. Chen and Molnar, 1983] studies to infer that the upper crust deforms by localized shear in strong material, the lower crust by penetrative flow in weak material, and the upper mantle by localized shear in strong material that gradually weakens toward the base of the lithosphere [e.g. Smith and Bruhn, 1984]. It is thus very appealing to assume that for a given bulk strain of the

CONCEPT OF A DECOUPLING HORIZON

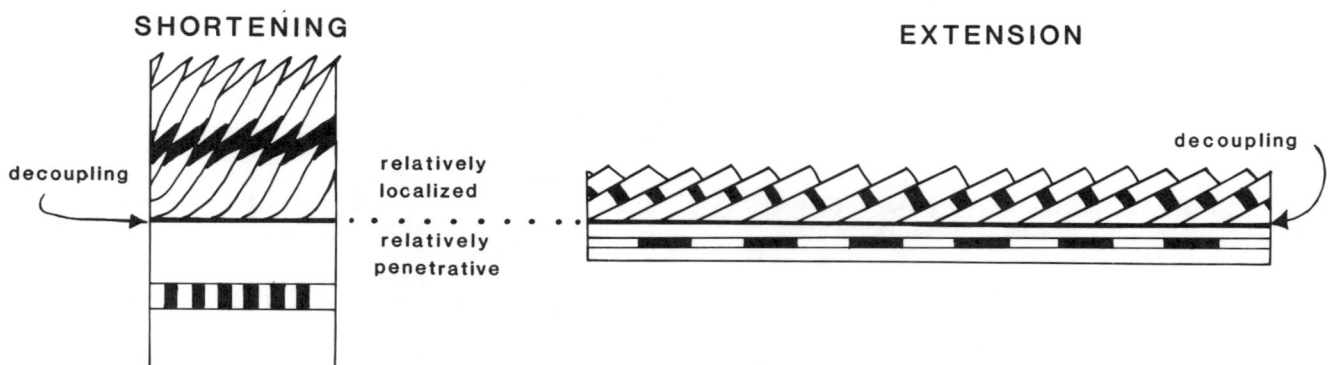

Fig. 1. Concept of a decoupling horizon.

lithosphere, these rheological tiers will respond differentially in some way, creating decoupling horizons between them.

As discussed elsewhere [Wernicke, 1985, Wernicke et al., 1985; Bartley and Wernicke, 1984], evidence from the Basin and Range is inadequate to demonstrate the existence of large-scale decoupling horizons in the upper and middle crust. It is now widely accepted that detachments (normal faults with large displacement that form at low-angle) in the Basin and Range form as single, relatively planar entities which pass from very high, brittle levels into upper greenschist and even lower amphibolite facies metamorphic conditions (ca. $500°$ C), regimes in which quartz and calcite experience dynamic recrystallization, and well below highly pressure-sensitive upper crustal rheologies [e.g. Davis et al., 1983; Bartley and Wernicke, 1984; Reynolds and Spencer, 1985].

Since detachments do transgress the upper crustal brittle-ductile transitions of quartzose rocks, it may be reasonable to suppose that they may do the same through other, deeper transitions [see Smith and Bruhn, 1984].

It is shown below that the reflection geometries on the MOIST profile and nearby profiles in Scotland are consistent with the concept that extension occurs via zones of displacement only, without the formation of decoupling horizons in the lithosphere. It is strongly emphasized that the demonstration of the consistency of MOIST with a displacement-zone hypothesis for the deep crust does not prove its validity, but it does show that the data do not require the existence of decoupling horizons. This point is important because crustal-scale displacement zones are proven entities, while the existence of decoupling horizons, at least as narrowly defined here, remains speculative.

Model and Application

I will focus the discussion on the MOIST profile, which are some of the highest quality data available. On this line (Figure 2), the characteristic half-grabens are found in the upper crust, which except for the graben-bounding faults and basin fill, are relatively transparent compared with the lower crust. The assumed Moho on MOIST is a continuous, discrete band of reflections, rather than a lower limit to numerous short, discontinuous reflections found on many other BIRPS profiles and on lines from the western Basin and Range. Noteworthy on the MOIST line are: 1) a graben-bounding zone of displacement which extends more or less unbroken into the lower crust, and projects into a zone of mylonites on-land [Outer Isles thrust of Sibson, 1977]; and 2) a discrete band of dipping reflections in the upper mantle which both widens and flattens upward as it enters the lower crust [Flannan "thrust" of Smythe et al., 1982, and Brewer and Smythe, 1984].

In a previous report [Wernicke et al., 1985], it was argued that the Outer Isles and Flannan structures are normal zones of displacement which penetrate the Moho and may not sole into any decoupling horizons or zones of horizontal laminar flow. Other workers [Brewer and Smythe,

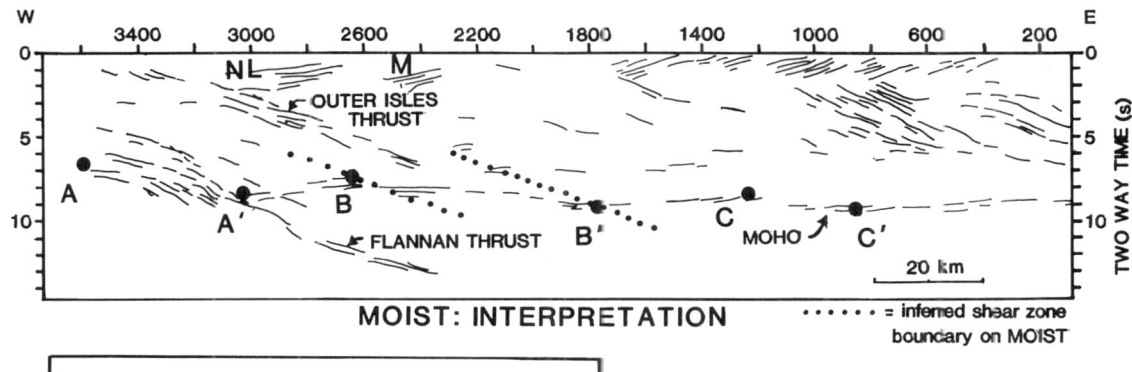

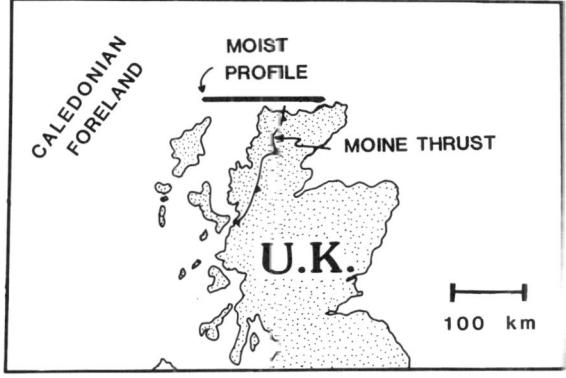

Fig. 2. Line drawing of the MOIST profile at 6.0 km/s [after Brewer and Smythe, 1984]. NL, North Lewis basin; M, Minch basin. Points B (7.5 s TWTT) and B' (9.0 s TWTT) define the proposed 1.5 s offset on the Moho caused by normal displacement on the Outer Isles "thrust". Points C-C' and D-D' bound other proposed normal offsets of the Moho, by the Flannan "thrust" and the fault bounding the Minch basin, respectively. Relief on the Moho is best seen by viewing the figure from one side with the eye about an inch above the page.

1984; Blundell et al., 1985; Smythe et al., 1982; Matthews and McGeary, 1984; Peddy, 1984] have commented on the fact that the total relief on the Moho (about 1.5 s two-way travel time) is only a fraction of the relief developed across the sedimentary half-grabens, and on this basis have rejected the interpretation that the Outer Isles structure displaces the Moho in a normal sense the same amount as the upper crustal half-graben fill.

It is clear from the reflection data presented in Brewer and Smythe [1984] that the Moho is comprised of long, east-dipping segments where the Flannan, Outer Isles, and next-higher graben-bounding structures project into it (segments AA', BB', and CC', respectively, on Figure 2). Peddy [1984] has analyzed MOIST and other lines nearby and concluded that the graben fill in the North Lewis and Minch basins (Figure 2) introduces pull-downs on the Moho which give rise to spurious offsets on time sections. However, her arguments do not change the geometry of the east-dipping segment between points B and B' on Figure 2, as one is at 7.5 s TWTT (two-way travel time) and the other is at 9.0 s TWTT, and neither point underlies thick graben fill. Thus, while the possible pull-down effects discussed by Peddy [1984] may slightly modify the shape of the east-dipping segment B-B', the total relief on the Moho (on the order of 5 km, assuming a lower crustal velocity of 6-7 km/s) could only be reduced by some sort of sideswipe effect, or marked lateral velocity changes in the crystalline rocks beneath the graben fill. The 5-10° west dip of the Moho to the west of point B on Figure 2 may flatten due to pull-down effects of the North Lewis basin, as suggested by Peddy [1984] based on analysis of the WINCH-1 line of Brewer et al. [1983]. Determining how much flatter this segment actually is depends upon a more detailed knowledge of the velocity structure of the North Lewis basin than is currently available. The flattening effect due to pull-down may also be partly reduced by migration. For example, in Blundell et al.'s [1985] model of MOIST, this segment of the Moho flattens from 8° on the time section to 4° on the depth section, rather than becoming absolutely horizontal as suggested by Peddy [1984].

It is shown below that the reflection geometry is quantitatively consistent with a model whereby the lithosphere extends along normal zones of displacement which have constant offset through the lithosphere. The model is contrasted with a decoupling-horizon model in Figure 3, showing how the rheology of the lithosphere could affect the

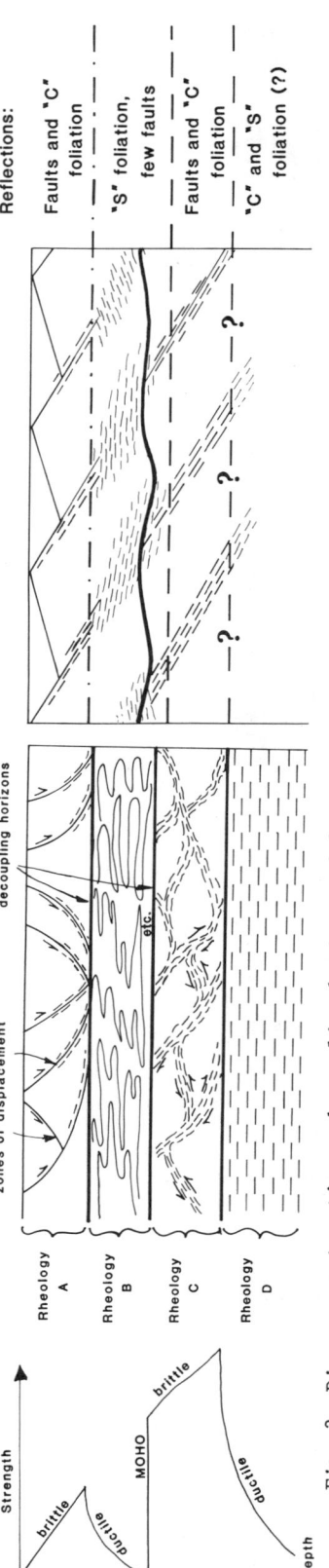

Fig. 3 Diagram contrasting a decoupling-horizon model with a displacement zone model for lithospheric extension.

nature of the shear zone with depth. Implicit in the model is the assumption that more ductile rheologies tend to widen the displacement zones, as well as decrease the tendency to form discrete offsets. In the parlance of "S-C mylonites" [Lister and Snoke, 1984], it is suggested that the displacement zones are comprised of discrete faults (C planes) in the upper crust, wide zones of S-C mylonite in the middle crust [cf. Sibson, 1983], but then in the lower crust the S foliation predominates. Brittle shear or the formation of S-C mylonites would then take over as the dominant mode in the upper mantle. The notion that wide, S-dominated shear zones occur in the lower crust is supported by the general preponderance of penetrative gneissosity and schistosity (i.e., S foliations) over mylonites with a strong C-plane fabric in very deeply eroded areas of continental crust.

Such a model resembles decoupling-horizon models only in the sense that penetrative subhorizontal foliation dominates over discrete shear in the deep crust, but contrasts with them in that 1) no decoupling horizons are present at any level; and 2) the strain path of the lower crustal rocks is characterized by progressively rotating non-coaxial laminar flow rather than simple coaxial flow. Thus, the transitions from dipping upper crustal and upper mantle reflections to "subhorizontal" lower crustal reflections may represent changing modes of accommodation within a continuous, dipping zone of displacement rather than a decoupling horizon. Key to the analysis below is the concept that foliation in ductile shear zones may develop at considerable angles to the shear zone boundary [Ramsay and Graham, 1970; Kligfield et al., 1984]. As such, a displacement zone that penetrates the lower crust would generally not be expressed on seismic profiles as bands of reflections parallel to those in the mantle and upper crust.

Theoretical Considerations

To test the hypothesis that the Flannan and Outer Isles structures penetrate the lower crust as wide zones of progressive simple shear, it is necessary to quantitatively model a lithospheric zone of normal displacement in order to calculate 1) how initially flat, planar markers such as the Moho would be affected; and 2) how the orientation of foliation in lower crustal rocks would evolve.

The model developed here depends upon the following geometrical assumptions (Figure 4): 1) the extensional allochthons deform as a series of "megadominos" with limited rotation (since the displacement zones initiate at shallow angle) whose upper corners define a surface parallel to the geoid; 2) the displacement zones in the lower crust are zones of homogeneous progressive simple shear and have the same total offset across them as their more discrete counterparts

GENERALIZED "MEGADOMINO" MODEL

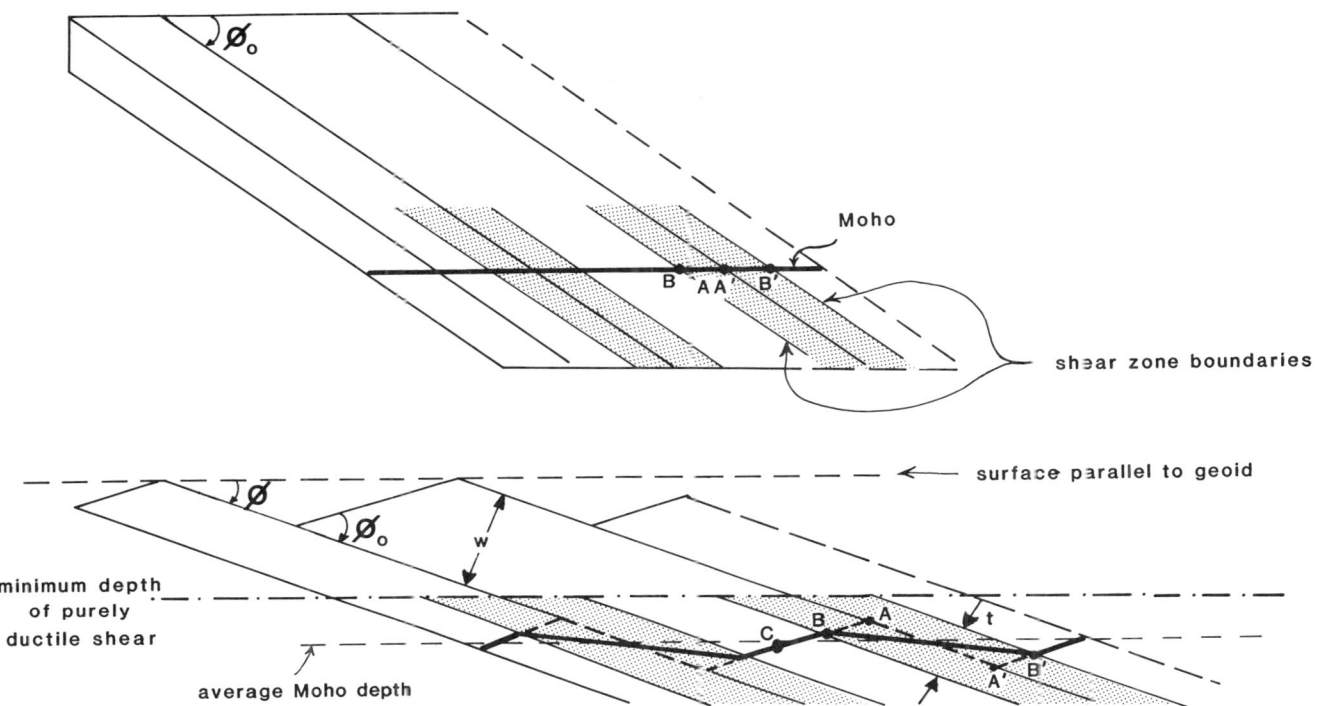

Fig. 4 Model of large-scale "dominos" whose bounding faults initiate at low angle. Points on this diagram are keyed to points on Figs. 5 and 6. ϕ_0, initial dip of shear-zone boundaries; ϕ, final dip of shear zone boundaries; w, width of block; t, width of lower crustal shear zone.

in the upper crust and upper mantle; 3) the foliation planes develop in the lower crust as a result of simple shear and contain the maximum elongation axis; and 4) all the blocks are of the same width. While these assumptions are not precisely duplicated by the half-graben geometry in Figure 2, they should give a reasonable first-order view of how the upper crustal geometry affects the lower crust and upper mantle if the model in Figure 4 is correct.

The geometrical problem is uniquely determined by specifying only two conditions: 1) the initial dip of the upper crustal fault and lower crustal shear zone boundaries, ϕ_0; and 2) the ratio of the width of the lower crustal shear zones, t, to the width of the extending blocks, w, measured in a direction perpendicular to the shear zone boundaries (Figure 4). By specifying only t/w, we find from a simple proof using similar triangles that, for any amount of extension or ϕ_0, the ratio of Moho relief to that in the upper crustal half-grabens and t/w sum to unity (Figure 5). Thus, for a perfectly discrete fault through the crust and upper mantle, the relief on the Moho is the same as that in the half grabens. If the shear zones are as wide as the blocks (t/w = 1), then there is no relief on the Moho. By specifying only t/w and ϕ, the final dip of the displacement zones, we obtain the dip of the maximum elongation direction θ (the S foliation) in the lower crust (Figure 6). The derivations of these relationships are similar to, but less complicated than, those derived in Kligfield et al. [1984].

Comparison of Theoretical Results with MOIST and WINCH

These theoretical results provide a simple explanation for why the Moho on the MOIST line has only a fraction of the relief of the upper crustal half-grabens. The deepest sediments in the North Lewis basin appear at about 3 s TWTT, or roughly 6 - 8 km depth in sediment [see velocity analysis of Blundell et al., 1985]. This much relief in the half-graben is slightly more than half that on the Moho (4.5 km assuming 6 km/s). These data predict that t/w should be slightly less than 1/2, according to the equation derived on Figure 5. This agrees well with Figure 2. The distance between the Outer Isles and Flannan structures is about 20 km on the 6

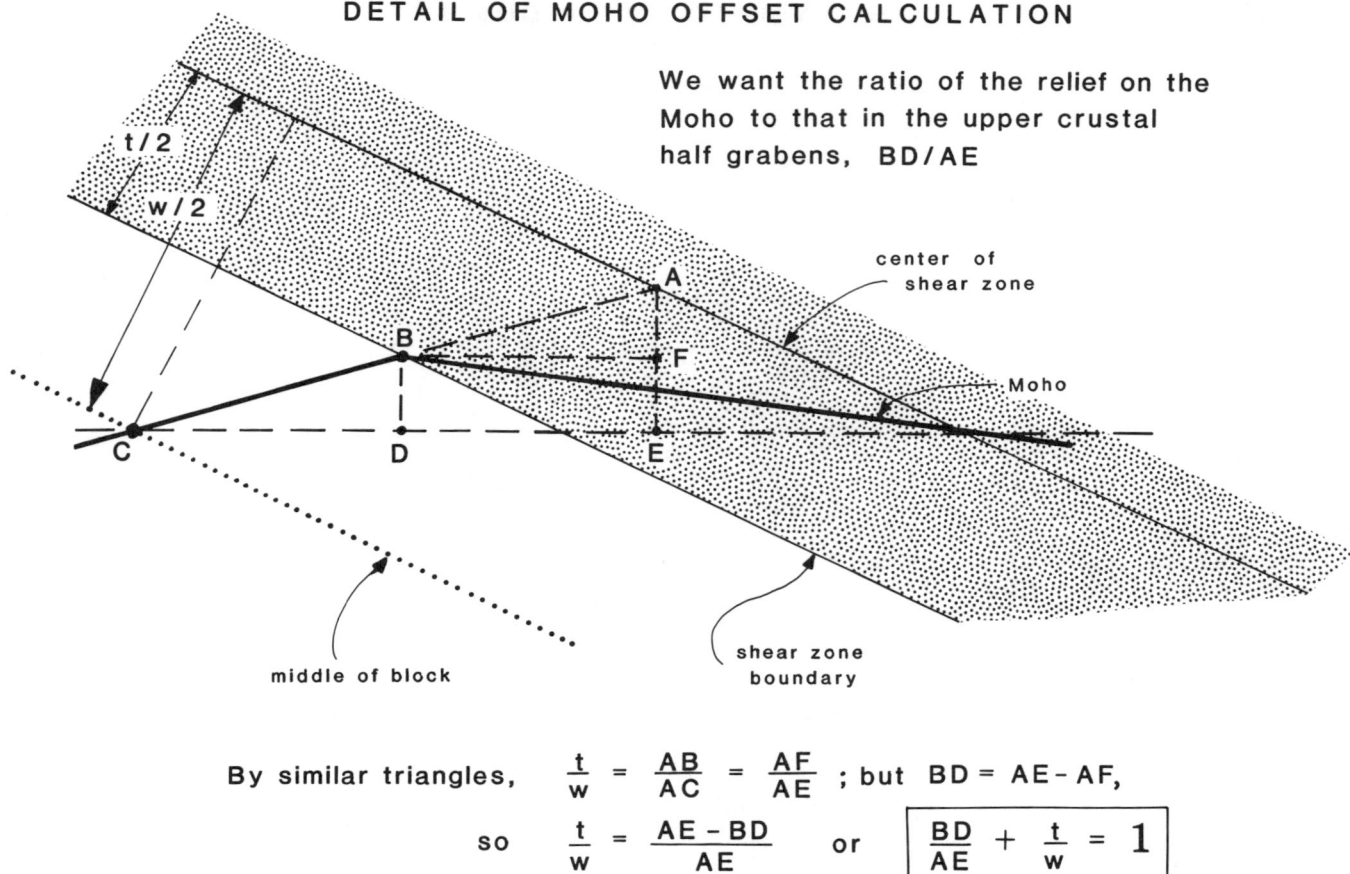

Fig. 5 Geometry showing the relief on a Moho displaced by a finite-width shear zone as compared to that of the upper crustal half-grabens. Points A, B and C are keyed to points on Figs. 4 and 6.

km/s time section, 21 km on Blundell et al.'s [1985] depth section, and 24 km on Peddy's [1984] depth section of WINCH-1. The width of the inferred shear zone on Figure 2 is about 10 km, and a similar width can be measured on the depth section of Blundell et al. [1985]; this parameter cannot be determined from Peddy's [1984] depth section, because it does not consider reflections as far east as those at point B on Figure 2]. It thus seems that within the uncertainties involved in depth conversion, the ratio t/w is about 1/2, in reasonable accord with the geometric model proposed here.

Other mechanisms, such as soling the normal faults into Moho-parallel simple or pure shear [e.g., Blundell et al., 1985], or "smoothing" the Moho through time by lateral flow (Brewer and Smythe, 1984) are also clearly plausible. One of the severest limitations in eliminating any of these mechanisms is that the velocity structure of the profile is only very poorly known. While the precise configuration of the reflectors may never be known, the analysis by Blundell et al. [1985] suggests there is little latitude in changing the basic geometry of the original time section in Brewer and Smythe [1984]. In particular, the upward flattening of the Flannan structure and the east-dipping segments of the Moho along deep projections of the Outer Isles and next-higher graben-bounding displacement zones are preserved. The dips of most reflections increase in Blundell et al.'s [1985] model, most notably the Flannan structure, which increases from $25°$ on the time section to $30-35°$ on their model. We wish to determine whether or not possible values of ϕ_0, ϕ, t/w, γ, and θ are internally consistent with the reflection data.

On the time section [Figure 2], ϕ for the Outer Isles structure is approximately $15-20°$ and the dip of the steepest sediments in the North Lewis basin is about $10-15°$, giving a ϕ_0 in the range of $25-35°$. However, the model of Blundell et al. [1985] suggests that ϕ is actually $23°$ and the bottom of the graben fill dips $17°$, suggesting a ϕ_0 of $40°$. The normal displacement on the Outer Isles structure is at least 20 km,

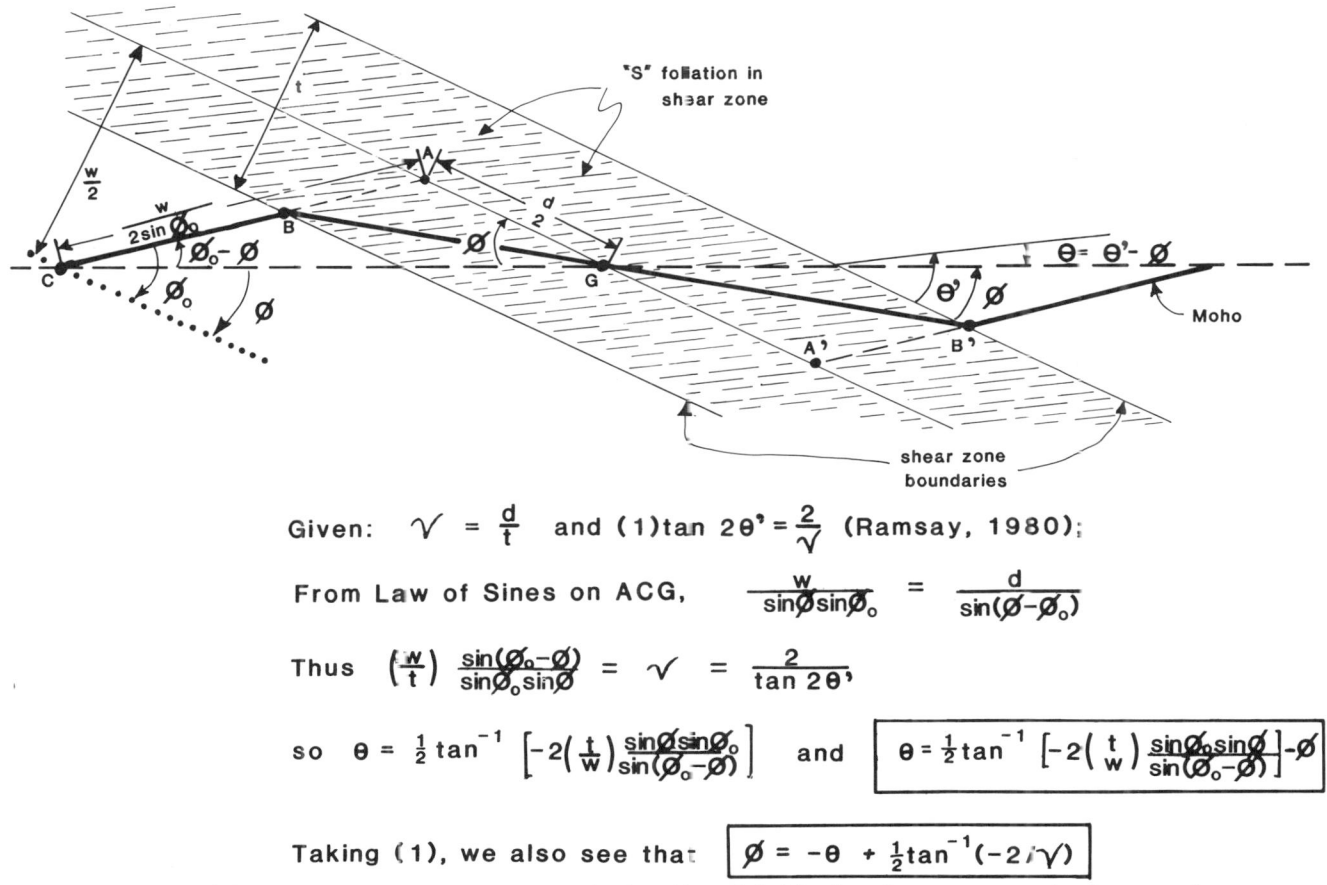

Fig. 6. (a) Geometry and calculation of the dip of foliation in a progressively rotating shear zone.

because this is the distance over which footwall basement is juxtaposed with the graben fill. The displacement could conceivably be on the order of 30 km or more.

The sigmoidal shape of the Moho where the Outer Isles structure projects to depth is consistent with a 20 km or greater offset. Assuming that t/w is on the order of 0.5, the approximate shear strain, γ, within the presumed lower crustal shear zone is 2. From Figure 6b, we note that a t/w of 1/2, $\gamma = 2$, $\phi_o = 40°$, and $\phi = 25°$ are internally consistent with one another, and suggest foliation dips in the lower crust within 5° of horizontal.

Some of the strongest reflections in the lower crust on MOIST occur where the Flannan structure projects into it. The reflections in the lower crust make an angle of about 15° with the Flannan on Brewer and Smythe's [1984] time section, but this angle becomes about 30° according to the model of Blundell et al. [1985]. Unfortunately, the total normal displacement on the Flannan structure is not well known, but the moderate east dip of the lower crustal reflections is certainly within the limits of the present model, because if we assume a migrated ϕ of 30° for the Flannan, then with even moderate rotation (implying $\phi_o = 40-45°$) it is very difficult to get the foliation to dip opposite the shear zone. However, in the case where ϕ_o is low and the shear zones wide, opposite dips in lower crustal foliation would be possible even at very high strains ($\gamma > 10$).

Conclusions

It is clear from this analysis that there are many degrees of freedom in interpreting the reflection data, and that the consistency of any model with the data does not constitute grounds

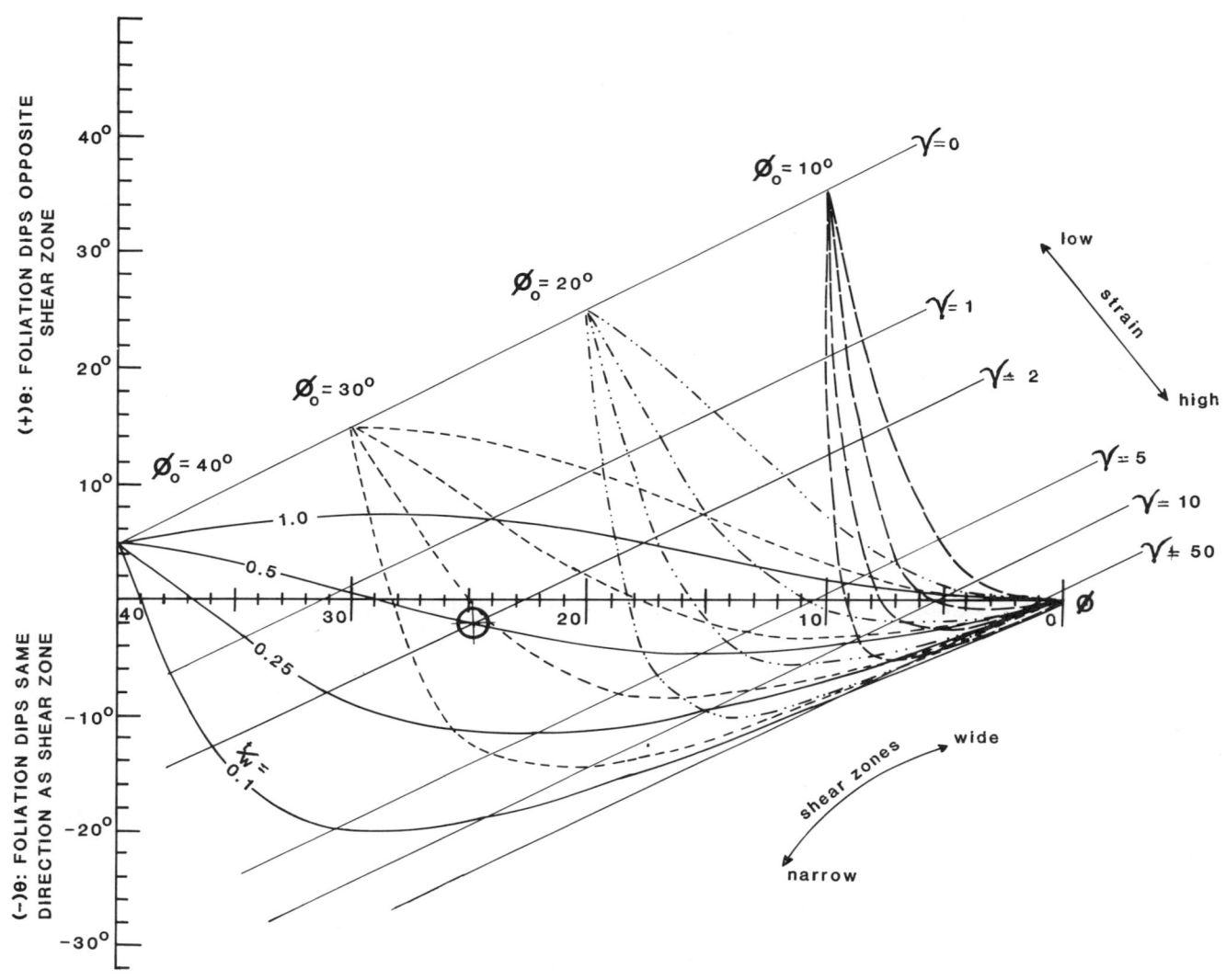

Fig. 6. (b) Graphical portrayal of the relationships between various parameters in (a). Circle with cross represents point discussed in text. Parameters not defined in Fig. 4. include: d, net displacement across shear zone; γ, shear strain within shear zone; θ', angle between foliation and shear zone boundary. Points A, A', B and B' are keyed to points on Fig. 4.

for ruling out other models. The purpose here is to introduce a more quantitative approach to thinking about lower crustal strain geometry that may eventually provide insight on how to test some of the many hypotheses about deep lithospheric extension.

One of the most attractive attributes of extending the lower crust on inclined shear zones is that it provides a mechanism for the development of shallow, variably dipping foliations. Although it is emphasized by many workers that the lower crust is dominated by many short, "subhorizontal" reflections, careful inspection of much of this data shows that the majority of these reflections actually dip, many of them as much as $20°$ or more [see, e.g., Brewer et al., 1983]. Upon migration, these dips will presumably increase. It is difficult to envision why a crust undergoing extension on the order of a factor of two or more by horizontal pure shear would preserve so many dipping reflections. The presence of pre-existing steep foliations presents major complications for both models, but the present one provides a means by which primary extensional foliation may have substantial dips, in either direction to the regional sense of shear (Figure 6b).

Deep reflection profiling by itself is clearly

a tantalizing but very limited method for trying to unravel the nature of the lower crust. Significant progress in understanding the deep crust will most likely come out of attempting to test some of the many interpretations of deep reflection data by direct surface observation of deeply eroded continental crust, about which precious little is currently known. Considering that the phenomenal strides recently made in understanding continental extension processes are principally the result of detailed field studies in the Basin and Range province (similar to the early understanding of compressional structure gained from field studies in the Alps), it seems that only a combination of direct, hands-on geology and seismic studies will eventually lead to our most detailed understanding of lower crustal processes.

Acknowledgements. I thank M. L. Zoback and D. K. Smythe for discussions which prompted the preparation of this report. Critical reviews of J. K. Snow, Tom Hauge, Rick Allmendinger, and an anonymous reviewer contributed significantly to the clarification of ideas presented herein. Research leading to this report was sponsored by NSF grant EAR-8319767 awarded to B. P. Wernicke, and a grant from the Clifford P. Hickock Junior Faculty Development Fund of Harvard University.

References

Bartley, J.M., and B.P. Wernicke, The Snake Range decollement interpreted as a major extensional shear zone, Tectonics, 3, 647-657, 1984.

Blundell, D.J., C.A. Hurich, and S.B. Smithson, A model for the MOIST seismic reflection profile, N. Scotland, J. Geol. Soc. London, 142, 245-258, 1985.

Brace, W.F. and D.L. Kohlstedt, Limits on lithospheric stress imposed by laboratory experiments, J. Geophys. Res., 85, 6248-6252, 1980.

Brewer, J.A., D.H. Matthews, M.R. Warner, J. Hall, D.K. Smythe, and R.J. Whittington, BIRPS deep seismic reflection studies of the British Caledonides: Nature, 305, 206-210, 1983.

Brewer, J.A., and D.K. Smythe, MOIST and the continuity of crustal reflector geometry along the Caledonian-Appalachian orogen, J. Geol. Soc. London, 141, 105-120, 1984.

Chen, W.P. and Peter Molnar, Focal depths of intracontinental and intraplate earthquakes and their implications for the thermal and mechanical properties of the lithosphere, J. Geophys. Res., 88, 4183-4214, 1983.

Davis, G.A., G.S. Lister, and S.J. Reynolds, Interpretation of Cordilleran core complexes as evolving crustal shear zones in an extending orogen, Geol. Soc. Am. Abstr. Programs, 15, 311, 1983.

Hauge, T., et al., The COCORP 40° N transect of the North American Cordillera: Part 1, Geol. Soc. America Astr. Prgms., 16, n. 6, 532, 1984.

Hauser, E. et al., The COCORP 40° N transect of the North American Cordillera: Part 2, Geol. Soc. America Abstr. Prgms., 16, n. 6, 532, 1984.

Kligfield, Roy, Jean Crespi, S. Naruk and G.H. Davis, Displacement and strain patterns of extensional orogens, Tectonics, 3, 577-609, 1984.

Lister, G. S., and A. W. Snoke, S-C mylonites: J. Structural Geol., 6, 617-638, 1984.

Matthews, D., and S. McGeary, Deep seismic reflections around Britain, EOS, 65, 1107-1108, 1984.

Peddy, C., Displacement of the Moho by the Outer Isles thrust shown by seismic modelling, Nature, 312, 628-630, 1984.

Potter, C., et al., The COCORP 40°N transect of the North American Cordillera: Part 3, Geol. Soc. America Abstr. Prgms., 16, n. 6, 626, 1984.

Ramsay, J.G., and R.H. Graham, Strain variations in shear belts, Can. J. Earth Sci., 7, 786-813, 1970.

Reynolds, S. J., and J. E. Spencer, Evidence for large-scale transport on the Bullard detachment fault, west-central Arizona, Geology, 13, 353-356, 1985.

Sibson, R.H., Fault rocks and fault mechanisms, J. Geol. Soc. Lond., 133, 191-213, 1977.

Sibson, R.H., Continental fault structure and the shallow earthquake source, J. Geol. Soc. Lond. 140, 741-761, 1983.

Smith, R. B., and R. L. Bruhn, Intraplate extensional tectonics of the eastern Basin-Range: Inferences on structural style from seismic reflection data, regional tectonics, and thermal-mechanical models of brittle-ductile deformation, J. Geophys. Res., 89, 5733-5762, 1984.

Smythe, D.K., A. Dobinson, R. McQuillan, J.A. Brewer, D.H. Matthews, D.J. Blundell, and B. Kelk, Deep structure of the Scottish Caledonides revealed by the MOIST reflection profile: Nature, 299, 338-340, 1982.

Wernicke, Brian, Uniform-sense normal simple shear of the continental lithosphere, Can. J. Earth Sci., 22, 108-125, 1985.

Wernicke, Brian, J.D. Walker, and M.S. Beaufait, Structural discordance between Neogene detachments and frontal Sevier thrusts, central Mormon Mountains, southern Nevada, Tectonics, 4, 213-246, 1985.

WITHDRAWN
FROM STCH LIBRARY